SUPRAMOLECULAR
SOFT MATTER

SUPRAMOLECULAR SOFT MATTER
Applications in Materials and Organic Electronics

EDITED BY

Takashi Nakanishi
National Institute for Materials Science (NIMS), Tsukuba, Japan

A JOHN WILEY & SONS, INC., PUBLICATION

Published by John Wiley & Sons, Inc., Hoboken, New Jersey.
Published simultaneously in Canada.

For general information on our other products and services or for technical support, please contact our
Customer Care Department within the United States at (800) 762-2974, outside the United States at
(317) 572-3993 or fax (317) 572-4002.

Wiley also publishes its books in a variety of electronic formats. Some content that appears in print
may not be available in electronic formats. For more information about Wiley products, visit our
web site at www.wiley.com.

Library of Congress Cataloging-in-Publication Data:

Supramolecular soft matter : applications in materials and organic electronics / edited
by Takashi Nakanishi.
    p. cm.
  ISBN 978-0-470-55974-1 (hardback)
  1. Supramolecular electrochemistry. 2. Molecular structure. 3. Molecular biology.
I. Nakanishi, Takashi.
  QD880.S87 2011
  541′.226–dc22

                                                          2011010584

oBook ISBN: 9781118095331
ePDF ISBN: 9781118095317
ePub ISBN: 9781118095324

10 9 8 7 6 5 4 3 2 1

CONTENTS

SECTION III

Dimension Controlled Organic Frameworks   **119**

PREFACE

Soft matter including organic assemblies and supramolecular objects are of high potential, but still seminal, because of their unique characteristics such as flexibility, self- and hierarchical-organization, self-repairing, self-healing, and stimuli-responsiveness. All their features are highly desired in practical applications in wet-processed electronic systems such as organic semiconducting devices, which are difficult to achieve by inorganic hard matter and organic crystals. Utilization of weak intermolecular forces for the construction of new supramolecular objects with controlled dimensionality and their implementation in practical applications is a major theme in contemporary chemistry, nanoscience, nanotechnology, and materials science. Moreover, fine-tuning of the interaction between π-conjugated molecules can enable the development of supramolecular materials with attractive electronic properties such as semiconductivity, redox activity, magnetism, and photoresponse. Better understanding and manipulation of these electronic properties together with the aforementioned characteristics of soft matter are necessary for the development of new generations of organic/polymeric soft materials toward organic electronic devices such as photovoltaics and field-effect transistors, representing some of the most desired materials for saving energy sources in the future. Under these conditions, the utilization of various performances achieved from supramolecular assemblies composed of *dyes, organic frameworks, organic radicals, gels, liquid crystals, conjugated polymers*, and *nanocarbon clusters (fullerenes* and *carbon nanotubes)* are promising approaches to open the door for ideal interdisciplinary treatment of organic or polymeric soft materials. Critical to the realization of the supramolecular materialization in organic electronic systems is the imaginative molecular design and synthesis as well as the development of smart molecular assemblies that are exemplified by researches highlighted in this book.

This book, *Supramolecular Soft Matter: Applications in Materials and Organic Electronics*, has a selection of subjects that do not aim to offer a plain compilation of current trends in supramolecular soft matter; it rather attempts to identify concepts that I believe hold promise for successful development of organic electronics with tremendous prospects. Hopefully, this book will stimulate all scientists working in broad research fields such as organic, polymer or supramolecular chemistry, nanomaterials, surface science, optics, photophysics, and materials science, among others, and will provide a useful guideline to direct research in supramolecular chemistry toward electronic soft materials. In addition, the chapters in this book will inspire the current and future generations of chemists to create ever more useful and diverse soft materials, because most chemists still have not understood what kind of organization of materials is suitable for bringing out the intrinsic features of organic materials.

TAKASHI NAKANISHI
National Institute for Materials Science, Tsukuba, Japan

CONTRIBUTORS

Takuzo Aida
Department of Chemistry and Biotechnology, School of Engineering,
The University of Tokyo, 7-3-1 Hongo, Bunkyo-ku, Tokyo 113-8656,
Japan

Ayyappanpillai Ajayaghosh
Photosciences and Photonics Group, Chemical Sciences and Technology
Division, National Institute for Interdisciplinary Science and Technology
(NIIST), CSIR, Trivandrum 695 019, India

Katsuhiko Ariga
World Premier International (WPI) Research Center for Materials
Nanoarchitectonics (MANA), National Institute for Materials Science,
1-1 Namiki, Tsukuba, Ibaraki 305-0044, Japan
Japan Science and Technology Agency (JST), Core Research
of Evolutional Science and Technology (CREST), Japan

Hidehiko Asanuma
Department of Interfaces, Max Planck Institute of Colloids and Interfaces,
Germany

Sukumaran Santhosh Babu
National Institute for Materials Science (NIMS), Japan

Dario M. Bassani
Institut des Sciences Moléculaires, Université Bordeaux 1, CNRS, 351,
Cours de la Libération, 33405 Talence, France

David Bilby
Department of Materials Science and Engineering, University of Michigan,
2098 H.H. Dow, 2300 Hayward St., Ann Arbor, MI 48109, USA

Michael J. Bojdys
Technische Universität Berlin, Englische Straße 20, 10587 Berlin, Germany

Suresh Das
Photosciences and Photonic Section, Chemical Sciences and Technology
Division, National Institute for Interdisciplinary Science and Technology
(CSIR), Thiruvananthapuram 695 019, Kerala, India

Juan Luis Delgado
Depatamento de Química Orgánica, Facultad de Ciencias Químicas,
Universidad Complutense de Madrid, Ciudad Universitaria s/n, 28040,
Madrid, Spain
IMDEA-Nanociencia, Facultad de Ciencias, Spain

Xuesong Ding
Department of Materials Molecular Science, Institute for Molecular Science,
National Institutes of Natural Sciences, 5-1 Higashiyama, Myodaiji, Okazaki
444-8787, Japan

Charl F. J. Faul
School of Chemistry, University of Bristol, Bristol BS8 1TS, UK
Bristol Centre for Nanoscience and Quantum Information, University
of Bristol, Bristol BS8 1FD, UK

Tsuyohiko Fujigaya
Department of Applied Chemistry, Graduate School of Engineering,
Kyushu University, 744 Motooka, Fukuoka 819-0395, Japan

John W. Goodby
Department of Chemistry, University of York, Heslington, York YO10 5DD,
UK

Dirk M. Guldi
Lehrstuhl für Physikalische Chemie I, Interdisciplinary Center for Molecular
Materials (ICMM), Friedrich-Alexander-Universität Erlangen-Nürnberg,
Egerlandstraße 3, 91058 Erlangen, Germany

Jia Guo
Department of Materials Molecular Science, Institute for Molecular Science,
National Institutes of Natural Sciences, Japan

Kenji Higashiguchi
Department of Synthetic Chemistry and Biological Chemistry,
Graduate School of Engineering, Kyoto University, Katsura, Nishikyo-ku,
Kyoto 615-8510, Japan

Jonathan P. Hill
World Premier International (WPI) Research Center for Materials
Nanoarchitectonics (MANA), National Institute for Materials Science
(NIMS), 1-1 Namiki, Tsukuba, Ibaraki 305-0044, Japan
Japan Science and Technology Agency (JST), Core Research of Evolutional
Science and Technology (CREST), Japan

Qingmin Ji
World Premier International (WPI) Research Center for Materials
Nanoarchitectonics (MANA), National Institute for Materials Science (NIMS),
1-1 Namiki, Tsukuba, Ibaraki 305-0044, Japan
Japan Science and Technology Agency (JST), Core Research of Evolutional
Science and Technology (CREST), Japan

Donglin Jiang
Department of Materials Molecular Science, Institute for Molecular Science,
National Institutes of Natural Sciences, 5-1 Higashiyama, Myodaiji, Okazaki
444-8787, Japan

Takashi Kato
Department of Chemistry and Biotechnology, School of Engineering,
The University of Tokyo, 7-3-1 Hongo, Bunkyo-ku, Tokyo 113-8656,
Japan

Jinsang Kim
Department of Materials Science and Engineering, University of Michigan,
2098 H.H. Dow, 2300 Hayward St., Ann Arbor, MI 48109, USA

Ju-Jin Kim
Department of Physics, Chonbul National University, Korea

Dirk G. Kurth
Chemische Technologie der Materialsynthese, Universität Würzburg,
Röntgenring 11, 97070 Würzburg, Germany

Jeong-O. Lee
NanoBio Fusion Research Center, Korea Research Institute of Chemical
Technology, Daejeon 305-343, Korea

Hiromitsu Maeda
College of Pharmaceutical Sciences, Institute of Science and Engineering,
Ritsumeikan University, Kusatsu 525-8577, Japan
Japan Science and Technology Agency (JST), Japan

Nazario Martín
Depatamento de Química Orgánica, Facultad de Ciencias Químicas,
Universidad Complutense de Madrid, Ciudad Universitaria s/n, 28040,
Madrid, Spain
IMDEA-Nanociencia, Facultad de Ciencias, Spain

Marta Mas-Torrent
Institut de Ciència de Materials de Barcelona (ICMAB-CSIC)
and Centro de Investigación Biomédica en Red en Bioingeniería,
Biomateriales y Nanomedicina (CIBER-BBN), Spain

Kenji Matsuda
Department of Synthetic Chemistry and Biological Chemistry,
Graduate School of Engineering, Kyoto University, Katsura, Nishikyo-ku,
Kyoto 615-8510, Japan

Veronica Mugnaini
Institut de Ciència de Materials de Barcelona (ICMAB-CSIC)
and Centro de Investigación Biomédica en Red en Bioingeniería,
Biomateriales y Nanomedicina (CIBER-BBN), Spain

Takashi Nakanishi
National Institute for Materials Science (NIMS), 1-2-1 Sengen, Tsukuba
305-0047, Japan

Naotoshi Nakashima
Department of Applied Chemistry, Graduate School of Engineering,
Kyushu University, 744 Motooka, Fukuoka 819-0395, Japan
Japan Science and Technology Agency (JST), CREST, Japan

Imma Ratera
Institut de Ciència de Materials de Barcelona (ICMAB-CSIC)
and Centro de Investigación Biomédica en Red en Bioingeniería,
Biomateriales y Nanomedicina (CIBER-BBN), Campus Universitari de la
Universtitat Autonoma de Barcelona, Bellaterra, Spain

Concepció Rovira
Institut de Ciència de Materials de Barcelona (ICMAB-CSIC)
and Centro de Investigación Biomédica en Red en Bioingeniería,
Biomateriales y Nanomedicina (CIBER-BBN), Campus Universitari de la
Universtitat Autonoma de Barcelona, Bellaterra, Spain

Isabel Saez
Department of Chemistry, University of York, Heslington, York YO10 5DD, Great Britain, UK

Yo Shimizu
National Institute of Advanced Industrial Science and Technology (AIST), Kansai Center, Midorigaoka, Ikeda, Osaka 563-8577, Japan

Sampath Srinivasan
Photosciences and Photonics Group, Chemical Sciences and Technology Division, National Institute for Interdisciplinary Science and Technology (NIIST), CSIR, Trivandrum 695 019, India

Yasuhiko Tanaka
Department of Applied Chemistry, Graduate School of Engineering, Kyushu University, 744 Motooka, Fukuoka 819-0395, Japan

Arne Thomas
Institute of Chemistry: Functional Materials, Technische Universität Berlin, Englische Straße 20, 10587 Berlin, Germany

Takashi Uemura
Department of Synthetic Chemistry and Biological Chemistry, Graduate School of Engineering, Kyoto University, Katsura, Nishikyo-ku, Kyoto 615-8510, Japan

Jaume Veciana
Institut de Ciència de Materials de Barcelona (ICMAB-CSIC) and Centro de Investigación Biomédica en Red en Bioingeniería, Biomateriales y Nanomedicina (CIBER-BBN), Campus Universitari de la Universtitat Autonoma de Barcelona, Bellaterra, Spain

Ratheesh K. Vijayaraghavan
Photosciences and Photonic Section, Chemical Sciences and Technology Division, National Institute for Interdisciplinary Science and Technology (CSIR), Thiruvananthapuram 695 019, Kerala, India

Jens Weber
Max Planck Institute of Colloids and Interfaces, Department of Colloid Chemistry, Science Park Golm, D-14424 Potsdam, Germany

Huaping Xu
Key Lab of Organic Optoelectronics & Molecular Engineering, Department of Chemistry, Tsinghua University, Beijing 100084, P. R. China

Shiki Yagai
Department of Applied Chemistry and Biotechnology, Graduate School of
Engineering, Chiba University, 1-33 Yayoi-cho, Inage-ku, Chiba 263-8522,
Japan

Takuma Yasuda
Department of Chemistry and Biotechnology, School of Engineering,
The University of Tokyo, 7-3-1 Hongo, Bunkyo-ku, Tokyo 113-8656,
Japan

Ryo Yoshida
Department of Materials Engineering, Graduate School of Engineering,
The University of Tokyo, 7-3-1 Hongo, Bunkyo-ku, Tokyo 113-8656,
Japan

Xi Zhang
Key Lab of Organic Optoelectronics & Molecular Engineering,
Department of Chemistry, Tsinghua University, Beijing 100084,
P. R. China

SUPRAMOLECULAR OBJECTS TOWARDS MULTI-TASK ORGANIC MATERIALS

CHAPTER 1

SUPRAMOLECULAR MATERIALIZATION OF FULLERENE ASSEMBLIES

Sukumaran S. Babu,[1] *Hidehiko Asanuma,*[2] *and Takashi Nakanishi*[1]

[1]National Institute for Materials Science, Sengen, Tsukuba, Japan
[2]Max Planck Institute of Colloids and Interfaces, Potsdam, Germany

1.1 INTRODUCTION

Self-organization of molecules using various supramolecular interactions has gained much attention in past decades because a large number of versatile assemblies have been created with excellent and tunable properties [1–3]. Hence, it is particularly important to study the various aspects of supramolecular chemistry, including soft functional assemblies [4]. The crucial and deciding factors while assembling molecules in a regular pattern are satisfied by the incorporation of suitable functional moieties [4]. It enables the molecules to recognize and self-organize in a programmed manner to gather amended properties, which is not attainable as monomers [5]. The reversibility of functional properties through monomer–aggregate transition is always an inspiration for scientists to design new functional assemblies. The challenging task is the control over the self-organization, whereby morphologies with defined size and shape are to be tuned. The incorporation of various hydrogen bonding units, ionic groups, chiral directing moieties, and solubility-controlling alkyl/glycol chains to various π-conjugated, commonly planar, chromophores such as oligo(p-phenylenevinylene)s (OPVs), perylene- or merocyanine-type dyes, porphyrins, hexabenzocoronene (HBC), and dendritic type mesogens enabled precise control over assembly formation [4].

Meijer *et al.* have shown that the incorporation of various self-recognition as well as hydrogen bonding units and various alkyl side chains resulted in functional OPV assemblies [6]. Interestingly, these assemblies acted as a good medium for energy or electron transfer studies [7]. In addition, this resulted in

Supramolecular Soft Matter: Applications in Materials and Organic Electronics, First Edition.
Edited by Takashi Nakanishi.
© 2011 John Wiley & Sons, Inc. Published 2011 by John Wiley & Sons, Inc.

the formation of assemblies with exciting supramorphologies [8]. Reports from the group of Ajayaghosh showed that the OPV scaffold is adequate for facile energy transfer mediated by tunable organogel medium with enhanced emission [9]. Organogelation of a series of OPVs with different end functional groups, which are donors (D) and acceptors (A), enabled tuning of the excited state properties, resulting in white-light-emitting organogels [10]. Würthner *et al.* have contributed much in the area of self-organization of perylene- and merocyanine-based dyes [11]. The extensive studies have shown that different self-assembled dyes with near-infrared (NIR) absorption features are potential candidates in creating supramolecular assemblies and applications in organic photovoltaic devices [12]. Aida and coworkers have studied the self-assemblies and functional properties of HBC [13]- and porphyrin [14]-based systems. Self-organization of amphiphilic HBC molecules has led to HBC nanotubes, which may find direct application in organic devices. Recently, the incorporation of various functional chromophores in the liquid crystalline assemblies have also been extensively studied [15].

Apart from the flat planar π-systems, self-organization of spherical π-systems such as fullerenes (C_{60} and C_{70}) have also been vastly studied in the past decades [16]. The abundant and unique optoelectronic properties of the curved π-surface of C_{60} has been utilized for the sensible design of inexpensive, lightweight, and durable organic photovoltaic devices [17]. The functionalization of fullerenes, reactive owing to its enhanced curvature, using different synthetic protocols resulted in a large number of derivatives with exceptionally good electron transfer and self-assembly properties [18]. The research area of fullerene self-assembly has formulated new dimensions through molecular design. Here, we focus on the new concept of hydrophobic amphiphilicity, which gained much attention recently [19].

1.2   HYDROPHOBIC–AMPHIPHILIC CONCEPT

The concept of molecular amphiphilicity is one of the widely exploited strategies in the area of interface science because of the versatile features of self-assembly [20]. Amphiphilic molecules consist of hydrophobic (hydrocarbon moiety) and hydrophilic units (charged anionic/cationic groups or uncharged polar groups), which can exhibit a large number of supramolecular architectures, such as micelles, vesicles, lamellae, tubular arrangements, as well as various cubic phases, depending on the relative balance between hydrophobic and hydrophilic interactions and on the solvophobic conditions [21–23]. The concept of amphiphilicity has been studied deeply in surfactants, detergents, and oils, and has great implications toward chemistry, biochemistry, biophysics, and colloid science [24]. The structural features of an amphiphile enable it to aggregate in an aqueous environment through aggregation of the hydrophobic apolar tails, which are protected from water by polar heads.

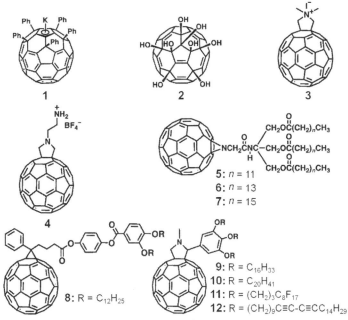

Scheme 1-1   Chemical structures of potassium salts of pentaphenyl fullerene (**1**), C_{60}-N,N-dimethylpyrrolidinium iodide (**2**), octahydroxy fullerene (**3**), fullerene containing ammonium cation and BF_4^- counter anion (**4**), C_{60} derivatives consisting of three alkyl chains with an amide and three ester groups (**5–7**), C_{60}-didodecyloxybenzene dyad (**8**), fulleropyrrolidine functionalized with 3,4,5-alkyloxyphenyl groups (**9, 10**), 3,4,5-semiperfluoroalkyl (**11**), and diacetylene (**12**) groups.

The research interest regarding self-assembly behaviors of amphiphilic fullerenes has been significantly increased. Reports from the group of Nakamura have shown that a potassium salt of pentaphenyl C_{60} (**1**) (Scheme 1-1) forms vesicles in aqueous conditions [25]. Zhang *et al.* reported that another amphiphile, octahydroxy C_{60} (**2**) (Scheme 1-1), forms spherical aggregates in water with a hydrodynamic radius R_h of about 100 nm [26]. Tour and coworkers investigated the self-assembly of C_{60}-N,N-dimethylpyrrolidinium iodide (**3**) (Scheme 1-1) leading to the formation of 1-D nanorods and vesicles under various experimental conditions [27]. Later, Shiga *et al.* studied the self-assembly of **3** in binary liquid mixtures of toluene and iodomethane, which resulted in the formation of nanosheets and precipitation as nanofibers in toluene/dimethyl sulfoxide (DMSO) mixture [28]. Interestingly, the matted nanosheets, several micrometers in length and about 100 nm in thickness, were formed from a large number of nanorods of 20-nm diameter. Another report from the group of Prato showed that C_{60} derivative with a short aliphatic chain containing an ammonium cation and a counter anion, BF_4^- (**4**), forms well-ordered nanorod-like aggregates in water [29]. All the above examples are less extended in shape and morphology.

In this context, a concept of "hydrophobic amphiphilicity" will be considered [19]. The structural modification of an amphiphile that relies on hydrophobic–hydrophobic balance has added a new dimension to amphiphilicity. The relative balance between two hydrophobic interactions, namely, $\pi-\pi$ (C_{60}) and van der Waals (alkyl chains), enabled delivery of diverse nano and microscopic architectures (supramolecular polymorphism) [30], which are never observed in the case of conventional amphiphilic fullerenes. A clear evidence of solvophobicity is exhibited by this C_{60}-based hydrophobic amphiphiles, leading to supramolecular assemblies with tunable properties and morphology.

1.3   SUPRAMOLECULAR ASSEMBLIES OF C_{60}-BEARING ALIPHATIC CHAINS

Nakashima and coworkers have reported C_{60} derivatives that bear three alkyl chains with an amide and three ester connectors (**5–7**, Scheme 1-1) [31–33]. These molecules lack the high hydrophilic nature of the conventional C_{60} amphiphiles and hence are soluble in various organic solvents. More interestingly, although the molecules are mainly composed of hydrophobic C_{60} and aliphatic chains, they were capable of forming Langmuir films on water, which behaves similar to lipid biomembranes [33]. Patnaik and coworkers reported the self-assembly of a partially ground-state charge-separated nonpolar–polar–nonpolar fullerene(C_{60})-didodecyloxybenzene dyad (**8**) (Scheme 1-1) to form micrometer-sized rod/sheet-like aggregates and in-plane bilayer vesicles with a head-to-head C_{60} packing conformation [34,35].

In addition, a series of alkylated C_{60} derivatives synthesized in our laboratory have exhibited interesting assembly phenomena when compared to the other alkylated C_{60} derivatives [36]. The relative balance of two hydrophobic interactions derived from C_{60} (less hydrophobic) and alkyl chains (more hydrophobic) [19] has resulted in dimensionally controlled supramolecular architectures with interesting assembly properties. Detailed assembly studies have revealed that this multi(alkyloxy)-phenyl substituted fulleropyrrolidines show supramolecular polymorphism in different organic solvents.

1.3.1   Hierarchical Supramorphology

The self-assembly of fulleropyrrolidine functionalized with a 3,4,5-tri(hexadecyloxy)phenyl group (**9**) exhibited supramolecular polymorphism: formation of different well-defined self-organized superstructures in various solvents and experimental conditions (Fig. 1-1) [30]. It is remarkable that a delicate balance between the hydrophobic interaction of chemically different C_{60} and saturated hydrocarbon chains drives the molecules to nanostructures with various interesting morphologies. The C_{60} composed of carbon atoms with sp^2 hybrid orbitals exhibits high affinity toward aromatic solvents (benzene, toluene, and xylenes), whereas the saturated hydrocarbon part consists of sp^3 hybridized carbons that have a higher affinity to alkanes than the aromatic compounds.

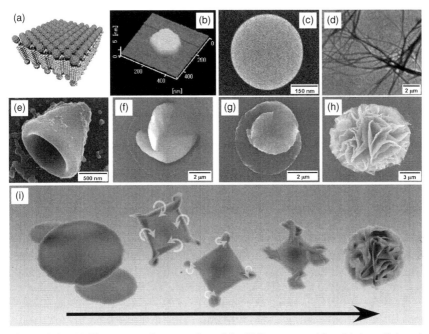

Figure 1-1   (a) The proposed structural model of bilayer assembly with interdigitated alkyl chains. (b) AFM image of disc-shaped assemblies of **9** formed in 1,4-dioxane as a precursor of the flower-shaped supramolecular assembly (h). SEM images of supramolecular assemblies of **9**, (c) spherical, (d) fibrous, (e) conical, (f) left-handed, and (g) right-handed spiral objects obtained from various solvent conditions. (i) Schematic representation of the formation mechanism of the flower-shaped supramolecular assembly. *Source:* Reprinted with permission [30,37].

This preferential affinity of the aromatic and aliphatic moieties toward different solvents is the basis of the unusual amphiphilicity observed in the molecular assemblies of C_{60} derivatives that bear aliphatic chains.

Self-assembled supramolecular objects were prepared by evaporation to dryness of a 1 mL chloroform solution of **9** ([**9**] = 1.0 mM) followed by the addition of 1 mL of solvents with different polarities. The field emission scanning electron microscopic (FE-SEM) images of the light brown mixtures from the respective solvents after heating at 60–70°C for 2 h indicated that **9** self-assembles into hierarchically ordered nano- and micro-superstructures [30]. It is assumed that the self-organization starts with a bilayer structure of self-assembled interdigitated bilayer of **9**, in which aromatic C_{60} layers in the upper and lower parts are separated by an aliphatic chain layer (Fig. 1-1a). Interestingly, when 1,4-dioxane is used for self-assembly, 2-D self-organized single bilayer discs (Fig. 1-1b) with diameters of 0.2–1.5 μm were obtained. In the case of 2-propanol/toluene system, **9** forms spherical aggregates with an average diameter of 250 nm (Fig. 1-1c). Fibers with partially twisted tapes appeared in 1-propanol (Fig. 1-1d), whereas conical objects with a diameter of 60 nm and perforated at the cone apex were developed from H_2O-tetrahydrofuran (THF) mixture (1:1) (Fig. 1-1e). In addition, left-handed spiral microstructures were obtained from 2-(R)-butanol (Fig. 1-1f), whereas 2-(S)-butanol provided right-handed spiral objects (Fig. 1-1g) of 3 to 6 μm diameter. The hierarchical organization of **9** on cooling down the 1,4-dioxane solution from 60 to 20°C, followed by cooling to 5°C, resulted in microscopic flowerlike superstructures (Fig. 1-1h) 3–10 μm in size, with crumpled-sheet-like or flakelike nanostructures of several tens of nanometers in thickness [37]. Figure 1-1i explains the transformation (shape shift) mechanism from assembled molecular bilayer discs to microscopic flower-shaped superstructures of the alkyl-conjugated C_{60}-derivative **9**. In order to understand the mechanism and intermediate assembly structures of the flowerlike objects in detail, a homogeneous 1,4-dioxane solution of **9** was cooled rapidly from 60 to 5°C. Interestingly, preformed disc objects were loosely rolled up at the edges (Fig. 1-1i). The rolling distortions at every quarter of the disc resulted in square-shaped objects having four corners with conical shapes developed by the encounter of the rolled or folded edges on the discs. As rolling up proceeds continuously, spatial congestions at the four corners lead to crumpling, bending, stretching, and fracture of the discs. Immediately after these transformations, the bilayer growth at the edges continues which fixes the spatial conformation of the crumpled sheets, and finally leads to the formation of flower-shaped superstructures (Fig. 1-1i) [37].

In order to characterize the self-organized objects, various optical, morphological, and analytical experimental techniques have been utilized. At first, the assemblies were observed through an optical microscope to confirm the formation and further the nano-micrometer-sized structures. A more magnified analysis was carried out by scanning electron microscopy (SEM). To get a clear approximation of the organization at the molecular level, X-ray diffraction (XRD) and transmission electron microscopy (TEM and cryo-TEM) experiments were performed. In our case, the bimolecular layer assembles through alkyl chain interdigitation. Analytical experiments such as differential scanning calorimetry (DSC) to understand the thermal phase transitions and stability of the assembly and optical measurements using UV-vis spectroscopy to monitor the $C_{60}-C_{60}$ interaction and FT-IR to record the alkyl chain conformation have been commonly used. The surface topography of the resulted assembly structures were observed through atomic force microscope (AFM) and occasionally by scanning tunneling microscope (STM) imaging [38–40].

1.3.2   Antiwetting Architectures

Antiwetting feature is a fundamentally important phenomenon. The tendency of water molecules to exclude or move away from nonpolar molecules leads to marginal segregation between water and nonpolar substances. The crucial deciding factor is the surface roughness of the nonpolar surface on which water is spread. A surface with micro- and nanostructured roughness generates superhydrophobicity with a water contact angle (CA) greater than 150°; the air is trapped between the surface and water droplets [41]. The development of functional assemblies of C_{60} with antiwetting properties is of considerable significance in current research interest because of the applications in durable organic devices. In this context, the well-defined three-dimensional (3-D) fractal architectures of alkyl-conjugated C_{60} find useful applications using the morphological features. Self-organization of a C_{60} derivative 10 (Scheme 1-1) with 3,4,5-tri(eicosyloxy)phenyl group in 1,4-dioxane solution led to spontaneous formation of micrometer-sized globular objects with wrinkled nanoflake structures at the outer surface (Fig. 1-2a) [42]. The dewetting ability of the globular objects was investigated by measuring the static water contact angle of the superhydrophobic surface obtained from **10**. Interestingly, in the case of globular objects, a water contact angle of 152° (inset of Fig. 1-2a) was observed, whereas a simple spin-coated film prepared from a homogeneous chloroform solution of **10** exhibited a static water contact angle of about 103°. The unique geometry of the "nano"-flakelike "micro" particles possesses tiny pockets at the surface that entrap air inside, exhibit two-tier roughness, and enhance the surface hydrophobicity. One of the advantages of these superhydrophobic systems is the reusability; the prepared thin films can readily be recovered by dissolving them in chloroform and reused.

It is a crucial question whether the C_{60} or the alkyl tails of **10** is exposed to the outer surface. In order to understand this, nano-flakelike microparticles of **10** have been prepared from the nonpolar solvent n-dodecane (Fig. 1-2b), which exhibited a static water contact angle of 164° (inset of Fig. 1-2b). This contact angle value is higher than that of the assembly prepared from the polar solvent 1,4-dioxane (152°, Fig. 1-2a), indicating that in the case of assemblies in a polar solvent, such as 1,4-dioxane, C_{60} moieties are exposed to the outer surface. The hydrocarbon tails are more hydrophobic than the moderately hydrophobic C_{60} moiety [19,33], and hence, owing to the mutual affinity, the outer surface components of the assembly of **10** from n-alkane solvents are presumably composed of the hydrocarbon tails [43]. This observation was further confirmed by the static water contact angle of the fractal-shaped microparticles ($\sim$148°) (Fig. 1-2c) of the fluoroalkyl conjugated C_{60} derivative (**11**, Scheme 1-1) assembled in a polar solvent, that is, diethoxyethane. The C_{60} moieties are exposed to the outer surface of the nanoflaked microparticles prepared from polar solvent conditions.

Even though the self-assembled objects exhibit an interesting surface morphology with water repellent properties, the robustness of the assemblies still remains a challenge. This may limit the potential use of these and most of the self-assembled organic objects for practical applications. In order to overcome this difficulty, C_{60} equipped with alkyl chains containing photo cross-linker (diacetylene) was synthesised (**12**, Scheme 1-1) [44]. UV light irradiation of the flakelike microparticles of **12** (Fig. 1-2d) resulted in the polymerization of both diacetylene and C_{60} moieties (Fig. 1-2e) and enabled getting chemically and mechanically robust assemblies (>29-fold compared to the nonpolymerized one) with antiwetting property (inset of Fig. 1-2d).

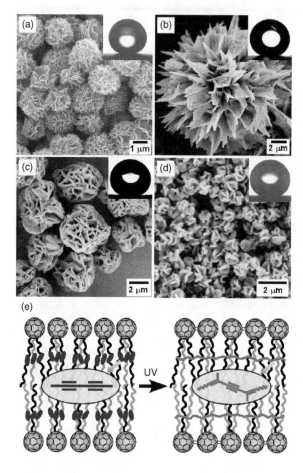

Figure 1-2 SEM images of flower-shaped supramolecular assemblies of **10** from (a) 1,4-dioxane and (b) *n*-dodecane, (c) **11** from diethoxyethane, (d) **12** from THF/MeOH mixture. Insets show the photographs of a water droplet on the surface with static water contact angles (a) 152°, (b) 164°, (c) 148°, and (d) 146°. (e) Schematic representation of the photo-cross-linking process in the bilayer structural subunit of **12**. *Source:* Reprinted with permission [42–44].

1.4 FUNCTIONS ORIGINATED FROM THREE-DIMENSIONAL FLAKELIKE MICROPARTICLES

One of the challenging tasks in the case of supramolecular architectures created by the self-assembly approach is to find out suitable applications. As a new strategy, reports from our group have shown how best the assemblies can be effectively utilized. Two unique approaches have been implemented to diversify the applications.

1.4.1 Supramolecular Molding Method

The remarkable and reliable method to develop well-designed hard matter using molecularly assembled supramolecular soft matter is by metallization. Interestingly, supramolecular architectures with structural and morphological diversity can deliver a broad range of high-definition templates for the design and synthesis of unusual functional inorganic nanoarchitectures. In this direction, transcription of C_{60}-based microparticles having flakelike outer surface features obtained through self-assembly into various metals was demonstrated [45]. The sputtering of the desired metal (Au, Pt, Ti, and Ni) directly onto a thin film of the supramolecular assemblies, followed by the removal of the C_{60} template using organic solvents, enabled successfully transferring the self-assembled features directly on to the metal surfaces (Fig. 1-3a). The advantage of this transcription method is the possibility of recovering and reusing the template, making the entire process sustainable. In order to understand the features of the nanostructured metal surfaces, it has been applied in surface-enhanced Raman experiments. For example, surface-enhanced Raman scattering (SERS) of the resulting metal Au nanoflake exhibited an enhancement factor on an order of 10^5. This study has revealed a simple method for the transcription of the morphological features

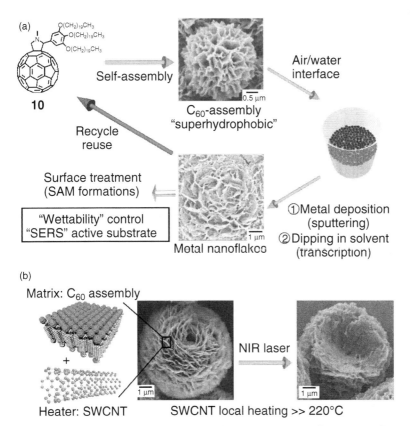

Figure 1-3   (a) Schematic representation for the preparation of metal nanoflake surfaces by using supramolecular architectures of **10** as templates. (b) Scheme of the coassembly of **10**-SWCNT for evaluating photothermal conversion of SWCNT; SEM images before (left) and after (right) illumination by an NIR laser (50 mW). *Source:* Reprinted with permission [45,48].

of a self-assembled object directly to different active metal surfaces, which has useful applications [46].

1.4.2   Thermal Indicator for NIR-Induced Local Heating of Carbon Nanotube

A promising strategy for assessing temperature rise during photothermal conversion of single-walled carbon nanotube (SWCNT) on NIR irradiation [47] was demonstrated by microparticles obtained by the coassembly of **10** and SWCNT (Fig. 1-3b) [48]. The flake-shaped microparticles of **10**-SWCNT were prepared by heating 1,4-dioxane solution of **10** and SWCNT to 70°C with ultrasonication followed by cooling to room temperature. On irradiation with NIR laser of lower intensity (50 mW), the microparticles started to deform (Fig. 1-3b), and when the laser intensity was increased to 90 mW, they were destroyed immediately. The flake-shaped surfaces of **10**-SWCNT assemblies have a mesomorphic-to-isotropic transition (melting point) at 191.8°C. Interestingly, microdisc (**9**-SWCNT, 2–8 μm) and flake-shaped microparticles (**13**-SWCNT, 2–4 μm) with melting points of 217.2 and 223.0°C, respectively, were also deformed on NIR laser illumination. This study demonstrates that NIR irradiation of C_{60}-SWCNT assembly can reach a local heating temperature in excess of around 220°C. The advantage of this system is the possibility of *in situ* visualization of deformation during photothermal conversion by means of an optical microscope and the tuning of assembly melting point by selecting an appropriate C_{60} derivative with suitable length or number of alkyl chains. More importantly, considering that SWCNTs are widely used in biology for local heating with operations conducted around body temperature, our results serve as a reminder that NIR irradiation of carbon nanotubes can induce an extreme temperature rise.

1.5   PHOTOCONDUCTIVE SOFT MATERIALS

The covalent functionalization of C_{60} has resulted in the formation of various functional assemblies, including organogels and liquid crystals (LCs). Nakamura *et al.* have reported organogels of alkyl-conjugated C_{60} derivative (**14**, Scheme 1-2) and developed self-assembled nanowires of C_{60} using the Langmuir Blodgett method [49]. The control over solubility of C_{60} obtained by derivatization using an L-glutamide moiety (**15–17**, Scheme 1-2) has led to the formation of organogels, especially in mixed organic solvents [50]. In recent years, there has been considerable interest in the design and synthesis of C_{60}-based LC assemblies because of the unique properties resulting from the ordering of different LCs. Deschenaux [51] and Felder-Flesch [52] are the leading contributors in the area of C_{60}-based LCs. The reports from their groups have demonstrated how synthetic design strategies can be utilized to develop plenty of C_{60}-based LCs and to impart anisotropic liquid crystalline character to completely isotropic fullerenes. One of the disadvantages of these systems is that the content of C_{60} part in the liquid crystalline C_{60} derivatives is relatively low, and hence the optoelectronic properties corresponding to C_{60} part will be limited.

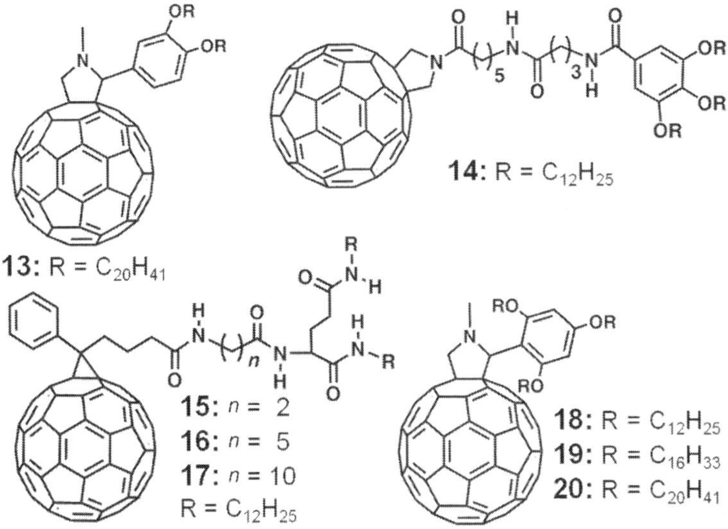

13: R = C$_{20}$H$_{41}$

14: R = C$_{12}$H$_{25}$

15: n = 2
16: n = 5
17: n = 10
R = C$_{12}$H$_{25}$

18: R = C$_{12}$H$_{25}$
19: R = C$_{16}$H$_{33}$
20: R = C$_{20}$H$_{41}$

Scheme 1-2 Chemical structures of fulleropyrrolidine functionalized with 3,4-alkyloxyphenyl groups (**13**), 3,4,5-tris(dodecyloxy)benzamide-linked C$_{60}$ organogelator (**14**), C$_{60}$ gelator with L-glutamide moiety (**15–17**), fulleropyrrolidine functionalized with 2,4,6-alkyloxyphenyl groups (**18–20**).

1.5.1 C$_{60}$-Rich Thermotropic Liquid Crystals

Since C$_{60}$-based LCs find direct applications in organic photovoltaic devices because of their predetermined, controllable organization at the molecular level, it is extremely important to select the necessary anisotropic building blocks along with C$_{60}$ to attain the preferred ordering. Recently, alkylated C$_{60}$ derivatives, that is, **9, 10, 13** (Schemes 1-1 and 1-2) with a high C$_{60}$ content (up to 50%) and high carrier mobility in the highly ordered mesophase were reported [53]. Polarized optical microscopic (POM) studies of **10** showed a birefringent optical texture (Fig. 1-4a) comparable to smectic phases with a fluid nature in the wide temperature range between 62 and 193°C. An unusual long-range ordered lamellar mesophase in which molecules are assumed to arrange their long axis, on average, perpendicular to the plane of the layers with the C$_{60}$ moieties in a head-to-head configuration was confirmed by the presence of multiple Bragg peaks in the LC state (Fig. 1-4b) [54]. The mesomorphic fullerenes retain reversible electrochemistry as cast films and exhibit electron mobility of $\sim$3 × 10^{-3} cm^2/Vs, which is the largest photoconductivity value for C$_{60}$-containing LCs. The dense packing of fullerenes carrying the charges is responsible for the high electron mobility observed in the mesophase.

1.5.2 Room Temperature Fullerene Liquids

The advantages of synthetic organic chemistry have been extended to design some room temperature liquid fullerenes. The controlled aggregation of fullerenes at

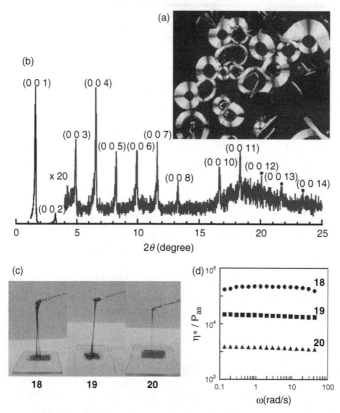

Figure 1-4 (a) Polarized optical micrographic texture of mesophase of **10** at 190°C on cooling from the isotropic phase at a rate of 0.1°C/min and (b) XRD patterns of **10** at 185°C. (c) Photographs of the liquid fullerenes (**18–20**) at room temperature with (d) the corresponding viscosity values. *Source:* Reprinted with permission [53,55].

the molecular level by attaching a 2,4,6-tri(alkyloxy)phenyl group to the fullero-pyrrolidine (**18–20**, Scheme 1-2) resulted in a new kind of nanocarbon fluid matter (Fig. 1-4c) [55]. The substitution pattern and length of alkyl chains were carefully chosen to serve as an effective steric stabilizer, preventing C_{60} aggregation. The lack of perfect molecular ordering in the liquid material is evidenced by very broad peaks in the XRD pattern of the liquid form. The higher loss modulus (G'') value than the storage modulus (G') has confirmed the liquid character of these derivatives at room temperature, and the viscosity of liquid fullerenes can be effectively controlled by changing the alkyl chain length (Fig. 1-4d). An important feature is that liquids with higher alkyl chains show lower viscosity, which is a completely opposite trend to that seen in alkanes. The liquid C_{60} compounds retain the characteristic electrochemical features of C_{60} and carrier mobility of $\sim 3 \times 10^{-2}$ cm^2/Vs (**20** at 20°C). These features make it an extremely attractive novel carbon material for future applications because of the absence of structural defects and quite high C_{60} content.

1.6   CONCLUSIONS

The rational design of covalently functionalized C_{60} derivatives has enlightened the self-assembly of fullerenes. Hence the remarkable achievement in crafting supramorphologies with molecular-level precision using the simple spherical $\pi-\pi$ interaction of fullerenes has gathered much attention. The important criterion in the design of C_{60} derivatives to obtain functional assemblies with a highly ordered C_{60} arrangement and having a high C_{60} content has been realized by alkyl-conjugated fullerenes. In this way, a concept of hydrophobic amphiphilicity through the delicate balance between hydrophobic interactions of the aromatic and aliphatic moieties has been established. The self-organized supramolecular objects exhibited superhydrophobicity, and further robustness of these structures were enhanced through polymerization of the diacetylene unit. Self-assembled microparticles have been used to transfer the nanomorphology to different metal surfaces, enabling the development of metal nanoflakes as highly sensitive SERS surfaces. The incorporation of carbon nanotubes into self-assembly enabled monitoring the temperature rise during photothermal conversion on NIR irradiation. In addition to superhydrophobic characteristics, these molecules revealed the presence of mesophases, with the largest electron mobility reported for C_{60}-containing LCs. The synthetic manipulation of the alkylated fullerenes has ended up with new fluid nanocarbon materials with good photoconductivity. Recent developments in our group have delivered new C_{60} derivatives with better crystallinity and higher C_{60} content. Photoconductive flowerlike supramolecular architectures were developed via self-assembly of a C_{60} derivative (C_{60} content of 84%) bearing a pyridine substituent [56]. Arene–perfluoroarene interaction has been used to develop transparent millimeter-sized flat crystalline C_{60} sheets with anisotropic photoconductivity through 1:1 coassembly of phenyl- and perfluorophenyl-substituted fullerenes [57]. We believe that the control over the morphology of π-conjugated compounds with a high π-core content would be used to diversify their variety of applications in organic electronics.

REFERENCES

1. Lehn, J. M. *Supramolecular Chemistry: Concepts and Perspectives*. VCH, Weinheim, Germany, 1995.
2. Reinhoudt, D. N., Crego-Calama, M. (2002). Synthesis beyond the molecule. *Science*, *295*, 2403–2407.
3. Whitesides, G. M., Grzybowski, B. (2002). Self-assembly at all scales. *Science*, *295*, 2418–2421.
4. Hoeben, F. J. M., Jonkheijm, P., Meijer, E. W., Schenning, A. P. H. J. (2005). About supramolecular assemblies of π-conjugated systems. *Chem. Rev.*, *105*, 1491–1546.
5. de Greef, T. F. A., Meijer, E. W. (2008). Supramolecular polymers. *Nature*, *453*, 171–173.
6. Schenning, A. P. H. J., Jonkheijm, P., Hoeben, F. J. M., van Herrikhuyzen, J., Meskers, S. C. J., Meijer, E. W., Herz, L. M., Daniel, C., Silva, C., Phillips, R. T., Friend, R. H., Beljonn, D., Miura, A., Feyter, S. D., Zdanowska, M., Uji-i, H., Schryver, F. C. D., Chen, Z., Würthner, F., Mas-Torrent, M., den Boer, D., Durkut, M., Hadley, P. (2004). Towards supramolecular electronics. *Synth. Met.*, *147*, 43–48.

7. Zhang, J., Hoeben, F. J. M., Pouderoijen, M. J., Schenning, A. P. H. J., Meijer, E. W., Schryver, F. C. D., Feyter, S. D. (2006). Hydrogen-bonded oligo(p-phenylenevinylene) functionalized with perylene bisimide: self-assembly and energy transfer. *Chem.—Eur. J.*, *12*, 9046–9055.

8. Katsonis, N., Xu, H., Haak, R. M., Kudernac, T., Tomović, Ž., George, S., Van der Auweraer, M., Schenning, A. P. H. J., Meijer, E. W., Feringa, B. L., Feyter, S. D. (2008). Emerging solvent-induced homochirality by the confinement of achiral molecules against a solid surface. *Angew. Chem. Int. Ed.*, *47*, 4997–5001.

9. Ajayaghosh, A., Praveen, V. K. (2007). π-Organogels of self-assembled p-phenylenevinylenes: soft materials with distinct size, shape, and functions. *Acc. Chem. Res.*, *40*, 644–656.

10. Ajayaghosh, A., Praveen, V. K., Vijayakumar, C. (2008) Organogels as scaffolds for excitation energy transfer and light harvesting. *Chem. Soc. Rev.*, *37*, 109–122.

11. Würthner, F. (2004). Perylene bisimide dyes as versatile building blocks for functional supramolecular architectures. *Chem. Commun.*, *14*, 1564–1579.

12. Kronenberg, N. M., Deppisch, M., Würthner, F., Lademann, H. W. A., Deinga, K., Meerholz, K. (2008). Bulk heterojunction organic solar cells based on merocyanine colorants. *Chem. Commun.*, 6489–6491.

13. Hill, J. P., Jin, W., Kosaka, A., Fukushima, T., Ichihara, H., Shimomura, T., Ito, K., Hashizume, T., Ishii, N., Aida, T. (2004). Self-assembled hexa-*peri*-hexabenzocoronene graphitic nanotube. *Science*, *304*, 1481–1483.

14. Tashiro, K., Aida, T. (2007). Metalloporphyrin hosts for supramolecular chemistry of fullerenes. *Chem. Soc. Rev.*, *36*, 189–197.

15. Kato, T., Mizoshita, N., Kishimoto, K. (2006). Functional liquid-crystalline assemblies: self-organized soft materials. *Angew. Chem. Int. Ed.*, *45*, 38–68.

16. Babu, S. S., Möhwald, H., Nakanishi, T. (2010). Recent progress in morphology control of supramolecular fullerene assemblies and its applications. *Chem. Soc. Rev.*, *39*, 4021–4035.

17. Segura, J. L., Martín, N., Guldi, D. M. (2005). Materials for organic solar cells: the C_{60}/π-conjugated oligomer approach. *Chem. Soc. Rev.*, *34*, 31–47.

18. López, A. M., Mateo-Alonsobc, A., Prato, M. (2011). Materials chemistry of fullerene C_{60} derivatives. *J. Mater. Chem.*, *21*, 1305–1318.

19. Asanuma, H., Li, H., Nakanishi, T., Möhwald, H. (2010). Fullerene derivatives that bear aliphatic chains as unusual surfactants: hierarchical self-organization, diverse morphologies, and functions. *Chem.—Eur. J.*, *16*, 9330–9338.

20. Israelachvili, J., *Intermolecular and Surface Forces*. Academic Press, London, 1991.

21. Fuhrhop, J.-H., Helfrich, W. (1993). Fluid and solid fibers made of lipid molecular bilayers. *Chem. Rev.*, *93*, 1565–1582.

22. Hafkamp, R. J. H., Feiters, M. C., Nolte, R. J. M. (1994). Tunable supramolecular structures from a gluconamide containing imidazole. *Angew. Chem. Int. Ed. Engl.*, *33*, 986–987.

23. Löwik, D. W. P. M., van Hest, J. C. M. (2004). Peptide based amphiphiles. *Chem. Soc. Rev.*, *33*, 234–245.

24. Elemans, J. A. A. W., Rowan, A. E., Nolte, R. J. M. (2003). Mastering molecular matter. Supramolecular architectures by hierarchical self-assembly. *J. Mater. Chem.*, *13*, 2661–2670.

25. Zhou, S., Burger, C., Chu, B., Sawamura, M., Nagahama, N., Toganoh, M., Hackler, U. E., Isobe, H., Nakamura, E. (2001). Spherical bilayer vesicles of fullerene-based surfactants in water: a laser light scattering study. *Science*, *291*, 1944–1947.

26. Zhang, G., Liu, Y., Liang, D., Gan, L., Li, Y. (2010). Facile synthesis of isomerically pure fullerenols and formation of spherical aggregates from $C_{60}(OH)_8$. *Angew. Chem. Int. Ed.*, *49*, 5293–5295.

27. Cassell, A. M., Asplund, C. L., Tour, J. M. (1999). Self-assembling supramolecular nanostructures from a C60 derivative: nanorods and vesicles. *Angew. Chem. Int. Ed.*, *38*, 2403–2405.

28. Shiga, T., Motohiro, T. (2007). Supramolecular structures formed by the self assembly of ionic fullerenes in binary liquid mixtures. *J. Mater. Res.*, *22*, 3029–3035.

29. Brough, P., Bonifazi, D., Prato, M. (2006). Self-organization of amphiphilic [60]fullerene derivatives in nanorod-like morphologies. *Tetrahedron*, *62*, 2110–2114.

30. Nakanishi, T., Schmitt, W., Michinobu, T., Kurth, D. G., Ariga, K. (2005). Hierarchical supramolecular fullerene architectures with controlled dimensionality. *Chem. Commun.*, 5982–5984.

31. Murakami, H., Watanabe, Y., Nakashima, N. (1996). Fullerene lipid chemistry: self-organised multilayer films of a C60-bearing lipid with main and subphase transitions. *J. Am. Chem. Soc.*, *118*, 4484–4485.

32. Nakanishi, T., Morita, M., Murakami, H., Sagara, T., Nakashima, N. (2002). Structure and electrochemistry of self-organized fullerene-lipid bilayer films. *Chem.—Eur. J.*, *8*, 1641–1648.

33. Mouri, E., Nakanishi, T., Nakashima, N., Matsuoka, H. (2002). Nanostructure of fullerene-bearing artificial lipid monolayer on water surface by *in situ* X-ray reflectometry. *Langmuir*, *18*, 10042–10045.

34. Gayathri, S. S., Agarwal, A. K., Suresh, K. A., Patnaik, A. (2005). Structure and synamics in solvent-polarity-induced aggregates from a C_{60} fullerene-based dyad. *Langmuir*, *21*, 12139–12145.

35. Gayathri, S. S., Patnaik, A. (2007). Aggregation of a C_{60}-didodecyloxybenzene dyad: structure, dynamics, and mechanism of vesicle growth. *Langmuir*, *23*, 4800–4808.

36. Nakanishi, T. (2010). Supramolecular soft and hard materials based on self-assembly algorithms of alkyl-conjugated fullerenes. *Chem. Commun.*, *46*, 3425–3436.

37. Nakanishi, T., Ariga, K., Michinobu, T., Yoshida, K., Takahashi, H., Teranishi, T., Möhwald, H., Kurth, D. G. (2007). Flower-shaped supramolecular assemblies: hierarchical organization of a fullerene bearing long aliphatic chains. *Small*, *3*, 2019–2023.

38. Nakanishi, T., Wang, J., Möhwald, H., Kurth, D. G., Michinobu, T., Takeuchi, M., Ariga, K. (2009). Supramolecular shape shifter: polymorphs of self-organised fullerene assemblies. *J. Nanosci. Nanotechnol.*, *9*, 550–556.

39. Nakanishi, T., Miyashita, N., Michinobu, T., Wakayama, Y., Tsuruoka, T., Ariga, K., Kurth, D. G. (2006). Perfectly straight nanowires of fullerenes bearing long alkyl chains on graphite. *J. Am. Chem. Soc.*, *128*, 6328–6329.

40. Nakanishi, T., Takahashi, H., Michinobu, T., Takeuchi, M., Teranishi, T., Ariga, K. (2008). Fullerene nanowires on graphite: epitaxial self-organisations of a fullerene bearing double long-aliphatic chains. *Colloids Surf., A: Physicochem. Eng. Aspects*, *321*, 99–105.

41. Li, X.-M., Reinhoudt, D., Crego-Calama, M. (2007). What do we need for a superhydrophobic surface? A review on the recent progress in the preparation of superhydrophobic surfaces. *Chem. Soc. Rev.*, *36*, 1350–1368.

42. Nakanishi, T., Michinobu, T., Yoshida, K., Shirahata, N., Ariga, K., Möhwald, H., Kurth, D. G. (2008). Nanocarbon superhydrophobic surfaces created from fullerene-based hierarchical supramolecular assemblies. *Adv. Mater.*, *20*, 443–446.

43. Nakanishi, T., Shen, Y., Wang, J., Li, H., Fernandes, P., Yoshida, K., Yagai, S., Takeuchi, M., Ariga, K., Kurth, D. G., Möhwald, H. (2010). Superstructures and superhydrophobic property in hierarchical organized architectures of fullerenes bearing long alkyl tails. *J. Mater. Chem.*, *10*, 1253–1260.

44. Wang, J., Shen, Y., Kessel, S., Fernandes, P., Yoshida, K., Yagai, S., Kurth, D. G., Möhwald, H., Nakanishi, T. (2009). Self-assembly made durable: water-repellent materials formed by cross-linking fullerene derivatives. *Angew. Chem. Int. Ed.*, *48*, 2166–2170.

45. Shen, Y., Wang, J., Kuhlmann, U., Hildebrandt, P., Ariga, K., Möhwald, H., Kurth, D. G., Nakanishi, T. (2009). Supramolecular templates for nanoflake-metal surfaces. *Chem.—Eur. J.*, *15*, 2763–2767.

46. Sezer, M., Feng, J.-J., Ly, H. K., Shen, Y., Nakanishi, T., Kuhlmann, U., Hildebrandt, P., Möhwald, H., Weidinger, I. M. (2010). Multi-layer electron transfer across nanostructured Ag-SAM-Au-SAM junctions probed by surface enhanced Raman spectroscopy. *Phys. Chem. Chem. Phys.*, *12*, 9822–9829.

47. Singh, P., Campidelli, S., Giordani, S., Bonifazi, D., Biancoa, A., Prato, M. (2009). Organic functionalisation and characterisation of single-walled carbon nanotubes. *Chem. Soc. Rev.*, *38*, 2214–2230.

48. Shen, Y., Skirtach, A. G., Seki, T., Yagai, S., Li, H., Möhwald, H., Nakanishi, T. (2010). Assembly of fullerene-carbon nanotubes: temperature indicator for photothermal conversion. *J. Am. Chem. Soc.*, *132*, 8566–8568.

49. Tsunashima, R., Noro, S.-I., Akutagawa, T., Nakamura, T., Kawakami, H., Toma, K. (2008). Fullerene nanowires: self-assembled structures of a low-molecular-weight organogelator fabricated by the Langmuir-Blodgett method. *Chem.—Eur. J.*, *14*, 8169–8176.

50. Watanabe, N., Jintoku, H., Sagawa, T., Takafuji, M., Sawada, T., Ihara, H. (2009). Self-assembling fullerene derivatives for energy transfer in molecular gel system. *J. Phys.: Conf. Ser.*, *159*, 012016.

51. Campidelli, S., Bourgun, P., Guintchin, B., Furrer, J., Stoeckli-Evans, H., Saez, I. M., Goodby, J. W., Deschenaux, R. (2010). Diastereoisomerically pure fulleropyrrolidines as chiral platforms for the design of optically active liquid crystals. *J. Am. Chem. Soc.*, *132*, 3574–3581.

52. Mamlouk, H., Heinrich, B., Bourgogne, C., Donnio, B., Guillon, D., Felder-Flesch, D. (2007). A nematic [60]fullerene supermolecule: when polyaddition leads to supramolecular self-organization at room temperature. *J. Mater. Chem.*, *17*, 2199–2205.

53. Nakanishi, T., Shen, Y., Wang, J., Yagai, S., Funahashi, M., Kato, T., Fernandes, P., Möhwald, H., Kurth, D. G. (2008). Electron transport and electrochemistry of mesomorphic fullerenes with long-range ordered lamellae. *J. Am. Chem. Soc.*, *130*, 9236–9237.

54. Fernandes, P. A. L., Yagai, S., Möhwald, H., Nakanishi, T. (2010). Molecular arrangement of alkylated fullerenes in the liquid crystalline phase studied with X-ray diffraction. *Langmuir*, *26*, 4339–4345.

55. Michinobu, T., Nakanishi, T., Hill, J. P., Funahashi, M., Ariga, K. (2006). Room temperature liquid fullerenes: an uncommon morphology of C_{60} derivatives. *J. Am. Chem. Soc.*, *128*, 10384–10385.

56. Zhang, X., Nakanishi, T., Ogawa, T., Saeki, A., Seki, S., Shen, Y., Yamauchi, Y., Takeuchi, M. (2010). Flowerlike supramolecular architectures assembled from C_{60} equipped with a pyridine substituent. *Chem. Commun.*, *46*, 8752–8754.

57. Babu, S. S., Saeki, A., Seki, S., Möhwald, H., Nakanishi, T. (2011). Millimeter-sized flat crystalline sheet architectures of fullerene assemblies with anisotropic photoconductivity. *Phys. Chem. Chem. Phys.*, *13*, 4830–4834.

CHAPTER 2

TUNING AMPHIPHILICITY OF BUILDING BLOCKS FOR CONTROLLED SELF-ASSEMBLY AND DISASSEMBLY: A WAY FOR FABRICATION OF FUNCTIONAL SUPRAMOLECULAR MATERIALS

Huaping Xu and Xi Zhang

Department of Chemistry, Tsinghua University, Beijing, China

2.1 INTRODUCTION

Supramolecular chemistry aims at developing highly complex chemical systems from components interacting by means of noncovalent intermolecular forces. As initiated by Lehn, the field was and is the basis for most of the essential biochemical processes of life. It has grown over 20 years into a major domain of modern teaching, research, and technology. Self-assembly is one of the key issues in supramolecular science and refers to the autonomous organization of components into patterns or structures. In nature, the components of cells and organisms self-assemble spontaneously, leading to hierarchically organized structures that allow life to exist. In science, the self-assembly of small molecules into large functional nanostructures has led to the construction of supramolecular systems with defined dimensions, both in solution and on solid substrates, which display unique properties through collective interactions, much like natural systems [1,2].

Self-assembly processes are mainly classified as static self-assembly and dynamic self-assembly. While the current understanding of self-assembly comes from the examination of static systems, many mechanistic details of dynamic self-assembly and disassembly are poorly understood [3,4]. Although studies

Supramolecular Soft Matter: Applications in Materials and Organic Electronics, First Edition.
Edited by Takashi Nakanishi.
© 2011 John Wiley & Sons, Inc. Published 2011 by John Wiley & Sons, Inc.

on self-assembly have been pursued for more than three decades, making components and letting them assemble "correctly" is still far from being routine [5]. The control of the subtle balance between weak interactions (e.g., hydrogen bonding, polar attractions, van der Waals forces, hydrophilic–hydrophobic interactions, and charge-transfer interaction) and the use of cooperative effects to influence self-assembly and disassembly processes in order to generate artificial functional systems thus remains a great challenge [6].

Amphiphilicity is one of the molecular bases for self-assembly. An amphiphile, a molecule that contains both hydrophilic and hydrophobic parts, can self-assemble in solution or at interface to form diversified molecular assemblies, such as micelles, reversed micelles, lyotropic mesophase, monolayers, and vesicles [7,8]. The hydrophilic part of the amphiphile is preferentially immersed in water, while the hydrophobic part preferentially resides in air or in the nonpolar solvent. The amphiphiles are aggregated to form different molecular assemblies by the repelling and coordinating actions between the hydrophilic and hydrophobic parts and the surrounding environment [9]. If we can tune the amphiphilicity of the building blocks, we can control the process of the self-assembly to some extent. The tuning of the amphiphilicity of the building blocks, including small surfactants and amphiphilic copolymers, can be used for controlling self-assembly and disassembly [10].

There is no doubt that molecular chemistry remains an important tool for creating new matter and materials. However, self-assembly provides an alternative route in this regard, which involves controlling spatiotemporal structures for the design and fabrication of functional supramolecular assemblies and materials [1–6]. The self-assembly of amphiphiles has shown significant importance in many research fields, for example, as building blocks for the fabrication of novel organic nanotubes or nanofibers toward electric and medical devices [11–15], candidates for drug delivery [16–19], templates for processing well-defined materials [20–22], nano- or microreactors for carrying out reactions in aqueous solutions and even for artificial enzyme mimicking [23–27], and stabilizer for incorporating emulsion used for cleaning and green organic reactions [28,29]. Moreover, the introduction of stimuli responsiveness into the building blocks can lead to the fabrication of smart supramolecular materials that are responsive to external stimuli, such as light irradiation, heat, pH change, and so on.

This chapter discusses different methods for tuning the amphiphilicity of building blocks for controlled self-assembly and disassembly, as shown in Fig. 2-1. It discusses irreversible methods that convert amphiphilic building blocks to either hydrophilic or hydrophobic by chemical approaches. On transforming to nonamphiphilic, the molecular assemblies formed by the amphiphilic building blocks can be collapsed irreversibly. There are also methods that can be used to tune the amphiphilicity reversibly. In this respect, reversible stimuli-responsive and supramolecular chemical methods are involved. The alternation between the hydrophilic and hydrophobic parameters allows for the reversible self-assembly and disassembly of the building blocks.

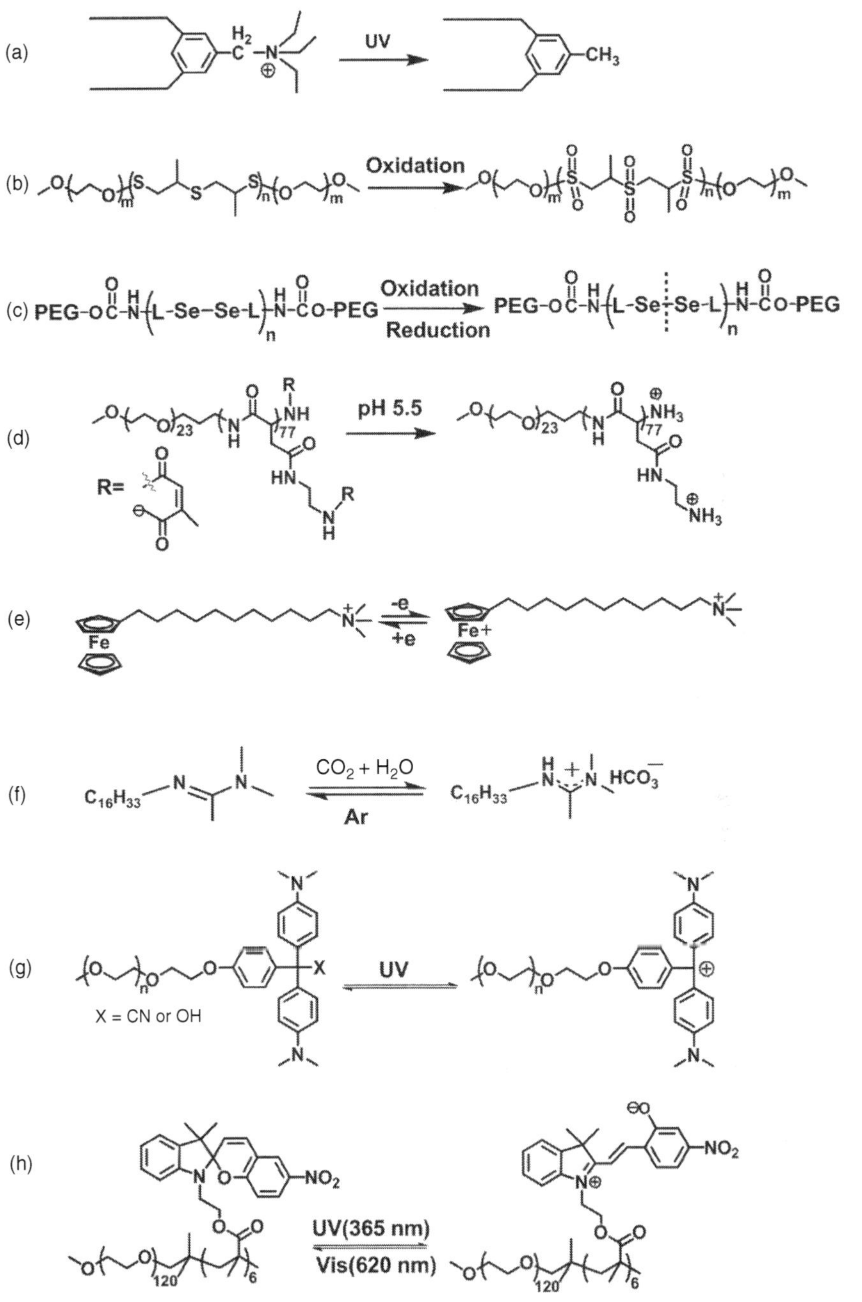

Figure 2-1 Schematic illustration of the irreversible ((a) UV irradiation [30], (b) oxidation [31], (c) redox switch [32], (d) pH variation [33]) and reversible ((e) redox switch [34], (f) reversible combination with carbon dioxide [35], and (g) and (h) photo-irradiation [36]) methods for tuning the amphiphilicity of building blocks.

2.2   IRREVERSIBLE METHODS TO TUNE THE AMPHIPHILICITY OF BUILDING BLOCKS

Normally, there are two kinds of irreversible reactions used to tune the molecular amphiphilicity: the *in situ* polarity variation of the special groups on the building blocks and the detachment of the labile groups from the building blocks. In both cases, the polarity variation induced by those irreversible reactions can change the molecular amphiphilicity concomitantly, which will endow the building blocks with new properties for application in materials science. In this section, we discuss how to use photochemical, oxidation–reduction, and pH-stimuli reactions for tuning the amphiphilicity.

2.2.1   Photo-Irradiated Irreversible Methods

Without the need for additional substances, light is one of the most desirable stimuli for tuning the molecular amphiphilicity, thus providing methods for clean and rapid control of critical micelle concentrations (CMCs), surface tension, aggregation behavior, and types of aggregates. Ringsdorf *et al.* employed two different photoreactions to tune the amphiphilicity of the building blocks, which strongly influenced the stability of the liposomal structure formed by the amphiphiles [30]. One is to tune the amphiphilicity through *in situ* photoreaction on the molecular skeleton. The amphiphile bearing a photosensitive head of pyridinioamidates can form liposomes by self-assembly in water. The photosensitive positively charged pyridinioamidate on the surface of the liposome was triggered to form neutral diazepin on UV light irradiation. The transformation of the head groups from polar to nonpolar induced the liposome to be metastable and even to collapse. This work opened an avenue to tune the amphiphilicity of the self-assembling systems, such as micelles and liposomes, surface wettability, and so on. The other photostimuli method to vary the amphiphilicity is the detachment of photolabile groups from the amphiphiles. As shown in Fig. 2-1a, the hydrophilic head of the amphiphilic 3,5-dialkoxybenzylammonium salt was detached through UV light irradiation. Therefore, the liposome formed by such amphiphiles was deformed significantly [37,38]. It should be pointed out that this line of research provides a series of photodegradative surfactants for application in soft lithography and separation technology [39,40].

2.2.2   Redox Response

Redox responsive polymers have attracted wide interest for their promising applications in controllable encapsulation and delivery in physiological environments, where the redox process is constantly and widely present. It has been reported that tumor cells exhibit a more oxidative atmosphere intracellularly than healthy cells. As an unconventional irreversible method, oxidation reactions were also well developed to tune molecular amphiphilicity for its potential use in drug delivery in the oxidative environment of extracellular fluids, physiologically and pathophysiologically. Usually, it is easy to understand that the oxidative reaction, with the oxygen atom involved, can enhance the polarity of targeted molecules.

One simple example is the surface treatment by oxygen plasma bombardment: the surface hydrophilicity can be improved significantly by the oxidation of surface molecules. If a similar oxidation reaction is introduced to change the polarity of some neutral or charged groups on the amphiphile, it is hoped that the hydrophilic–lipophilic balance of the amphiphile will be destroyed.

Recently, Hubbell and his coworkers designed and synthesized an ABA-type triblock copolymer as the candidate amphiphile to control molecular self-assembly by oxidation reaction. The ABA-type triblock copolymer has hydrophilic poly(ethylene glycol) (PEG) as A-part and hydrophobic poly(propylene sulfide) (PPS) as B-part, in which the PPS segment is responsive to oxidative chemicals, such as H_2O_2, and can be converted from hydrophobic to hydrophilic on oxidation. Compared to other poloxamer macroamphiphiles [31], these copolymers favored vesicle formation at room temperature without a cosolvent because of the greater hydrophobicity of PPS, which also had a low T_g [43]. As shown in Fig. 2-1b, in the presence of adequate H_2O_2, the PPS segment was oxidized into hydrophilic sulfones and the vesicle-like aggregates were converted to long wormlike micelles. It is known that the oxidatively destabilized copolymers can be removed from the body by glomerular filtration in the kidneys, which makes this oxidative method attractive for clinical use [44]. Later on, a very good example was proposed by encapsulating oxidant-generating enzyme glucose oxidase (GOx) into PEG–PPS–PEG triblock polymer vesicles. H_2O_2 produced by GOx-catalyzed conversion of glucose into gluconolactone could oxidize $-S-$ to $-SO_2-$ in the PPS segment, inducing the conversion of the amphiphilic triblock copolymer to a hydrophilic polymer, which resulted in the breaking of the vesicle [45].

Similar to sulfur-containing compounds, selenium-containing compounds also exhibit a very good redox responsive property. Selenium-containing compounds have been widely used in pharmacochemistry as antioxidants for the well-known glutathione peroxidase (GPx) activity, among which diselenide is a promising candidate for a dual redox response owing to its good activity in the presence of either oxidants or reductants [46–48]. Normally, Se–Se bonds are cleaved and oxidized to seleninic acid in the presence of oxidants and reduced to selenol in a reducing environment [49,50]. Thus, the amphiphilicity of the block copolymer will be destroyed. Recently in our study, a diselenide-containing block copolymer was synthesized and its dual redox responsive disassembly on addition of oxidants or reductants investigated, as shown in Fig. 2-1c [32]. The results show that the redox-responsive PEG–PUSeSe–PEG micelles were quite stable under ambient conditions but could exhibit very good sensitivity to external redox stimuli. The incorporated species could be released from the micelles when some effective oxidants or reductants were added in a mild environment. Considering the active nature of the Se–Se bonds existing in the block copolymer, we are attempting to use some other stimuli, such as γ rays, to change the amphiphilicity of the block copolymer and release the loaded species. It is fully expected that this kind of multiresponsive block copolymer aggregate may function as a controlled drug delivery system, enabling a combination of chemotherapeutics and actinotherapy.

2.2.3 pH-Stimuli Methods

Although many new stimuli methods are being explored to tune the molecular amphiphilicity of the building blocks, pH stimuli is still one of the most important methods because of the possibility of being used in real clinical drug delivery [51]. The key reason for the latent use of pH-stimuli building blocks is due to the different pH conditions in normal organs and tumors. In tumor tissues *in vivo*, pH value drops evidently from normal extracellular physiological environment (pH 7.4) to pH 6.0–5.0, and to around pH 4–5 in primary and secondary lysosomes. This significant pH variation provides great opportunities for us to develop pH-responsive systems for drug delivery. Inspired by the "Ringsdorf model," in which a labile linker is introduced between the drug and the biocompatible polymeric backbone [52,53], many models have been created with the aim of attaching acid- or base-cleavable groups on amphiphiles [54–56]. Some pH-responsive groups have been well reviewed to fabricate controllable surfactants or amphiphilic polymers [57–62].

Some new and important progresses on tuning the molecular amphiphilicity and material properties by pH-stimuli need to be highlighted. For example, Kataoka and his coworkers recently reported a pH-responsive block copolymer that can perform a charge conversion on one block by external pH stimuli [33]. As shown in Fig. 2-1d, PEG-pAsp (EDACit) was a double-hydrophilic block copolymer. The side comb-type citraconic amide groups endowed the pAsp block with negative charges. Therefore, this block copolymer was associated with positively charged lysozyme, based on electrostatic interaction to form poly ionic complex (PIC)-type micelles. Moreover, the citraconic amide groups attached on pAsp chain were hydrolyzed through a slight change in pH. After the hydrolysis of citraconic amide under mild condition at pH 5.5, the charge of the block copolymer changed from negative to positive. In this case, lysozyme encapsulated in the micelle was repulsed and released. This active and prompt pH-stimuli response in tuning the molecular amphiphilicity is promising for effective clinical drug delivery.

2.3 REVERSIBLE STIMULI-RESPONSIVE METHODS

There is an increasing interest in developing reversible methods, and by these methods, the molecular amphiphilicity is expected to be reversibly tuned by stimuli-responsive groups attached on the amphiphiles. By employing suitable external stimuli, those stimuli-responsive groups may change their polarity and hence the molecular amphiphilicity but without detachment of the groups from the amphiphiles. In addition, when the change is reversible, the molecular amphiphilicity can be varied between amphiphilicity and hydrophilicity or between amphiphilicity and hydrophobicity, providing ways to realize the control of self-assembly and disassembly.

2.3.1   Redox Switches

Since the early report on ferrocene-containing surfactant by Saji *et al.* [63], diversified redox-switched groups are attracting more and more attention on their use for preparing controllable building blocks. Among those groups, ferrocene is widely used for its easy conversion between oxidized and reduced forms by electrochemistry or chemicals. Ferrocenyl surfactants, including single- or double-tailed cationic surfactant and bola-amphiphile, permit the control of the formation and disruption of self-assemblies such as micelles and vesicles by redox reactions. In addition, oxidized ferrocene is stable enough to be monitored by NMR (nuclear magnetic resonance) and other spectroscopic methods. Therefore, the amphiphilic molecules with ferrocene groups can be well investigated during the redox process.

The surfactant with a ferrocene head group showed aggregation behavior at concentrations above its CMC. After the ferrocene groups were oxidized by an oxidant, such as $Ce(SO_4)_2$, as shown in Fig. 2-1e, the charged ferrocene groups brought better hydrophilicity to the surfactant, leading to the breakage of the micelles. The property of such surfactant response to redox switch allows the surfactants to reversibly load small hydrophobic species into its micelle aggregates [64]. Later on, it was demonstrated that ferrocene-head-containing surfactants could incorporate small hydrophobic molecules to prepare an organic film on an electrode surface. The mechanism was that the micelles formed by surfactants with a ferrocene moiety could load or disperse some sparingly water-soluble molecules onto the electrode surfaces. By electrochemical technique, the small molecules were released from the micelles onto the electrode surface, resulting in film formation [65,66]. Similar to this work, another surfactant with ferrocene as the hydrophobic end group was exploited as the container to separate two different drug-like substrates. The solubilization and deposition on solid surface of the poor water-soluble molecules could be realized by assembly and disassembly of redox-active surfactants. This reversible process provided a proof of concept for the separation of two sparingly water-soluble molecules through multiple redox switches [34].

Besides ferrocene-containing small amphiphiles, the introduction of ferrocene into copolymers can lead to the fabrication of supramolecular materials and devices. One of the well-known ferrocene-containing polymers is poly(ferrocenylsilanes) (PFS), synthesized first by Manners and his coworkers [67]. The self-organization of PFS copolymer in the solid state or in solution yields well-defined supramolecular, nanoscaled organometallic architectures [68,69]. PFS also displays a stimuli-responsive behavior because the ferrocene groups can be oxidized and reduced reversibly. This performance allows PFS to serve as a candidate for the realization of a single macromolecular motor [70]. For example, ethylene sulfide end-capped PFS could be immobilized on gold substrate, and the other terminal was physically adsorbed on Atomic Force Microscope (AFM) tip, establishing a molecular bridge between the AFM tip and the gold substrate. On application of electrochemical potentials, PFS could be oxidized to the cationic form and the single chain of PFS was lengthened because of the electrostatic repulsion between the oxidized ferrocene centers along the chain. After reduction, the oxidized PFS chain recovered its neutral form, leading to the contraction of the single chain of PFS. The closed mechanoelectrochemical cycles of single PFS chain allowed the conversion from electrochemical energy to mechanical energy. A maximum efficiency of 26% was observed for this single macromolecular motor. Since the efficiency depended on the stretching ratio, a higher efficiency was expected if the single chain of PFS could be chemically immobilized on both substrate and tip, guaranteeing a higher stretching ratio.

2.3.2 Tuning the Amphiphilicity by Reversible Combination of Carbon Dioxide

A new approach was explored by Jessop and his coworkers to tune the molecular amphiphilicity by the combination of carbon dioxide and amidine groups. Before being exposed to carbon dioxide, the mixture of two liquids, namely, DBU (1,8-diazabicyclo-[5.4.0]-undec-7-ene) and 1-hexanol, was a nonionic solvent. After the reaction with carbon dioxide, DBU was converted to the cationic form, DBUH. The mixture became an ionic liquid concomitantly. On the other hand, by exposing to enough inert gases, such as nitrogen or argon, the carbon dioxide bound in the mixture could be compelled out and the liquid mixture was reversed to the nonionic state [71–73]. Very recently, based on the polarity conversion, this kind of solvent was employed to dissolve the reagents for a chemical synthesis and then precipitate the product under control [74]. For example, the monomer styrene could be polymerized to polystyrene when dissolved well in a nonpolar solvent. On exposure of the system to carbon dioxide, the polymer could not be dissolved in the polar solvent anymore, and instead, it deposited on the bottom of the container. In addition, the solvent could be used repeatedly by removing carbon dioxide using nitrogen or argon. Therefore, this method, involving carbon dioxide, is considered as a green concept to tune the amphiphilicity because only harmless and inexpensive gases are used in the system.

On the basis of a similar concept, as shown in Fig. 2-1f, Jessop *et al.* designed and synthesized other kinds of compounds bearing carbon dioxide-responsive long-chain alkyl amidine groups that could be reversibly transformed into charged surfactants by exposure to an atmosphere of carbon dioxide [35]. The cationic surfactants were proved to be effectively used for stabilizing water/alkane or oil emulsions and for microsuspension polymerization. Taking crude oil as an example, the mixture of water and oil could form a fairly stable emulsion in the presence of cationic surfactants, being separated into two layers after enough argon treatment. Carbon dioxide is easily obtained from the atmosphere; thus, the amphiphilicity control of building blocks by exposure to carbon dioxide has potential industrial applications, such as the separation of oil from oil sands, the transportation of oil emulsion through pipelines, and then the release of oil when needed.

2.3.3   Photocontrolled Methods

The photocontrolled method is an attractive alternative since it provides a very broad range of tunable parameters, for example, wavelength, duration, and intensity. Some photosensitive moieties, such as azobenzene, spiropyran (SP), stilbene, and malachite green, show significant reversible conversion between nonpolar and polar states through external photostimuli. Herein, we discuss how to tune the molecular amphiphilicity reversibly by introducing photoresponsive groups onto the backbone of amphiphiles.

Among the various photosensitive moieties mentioned above, malachite green is of unique significance because it is hydrophobic in its neutral form but becomes cationic and hydrophilic on UV light irradiation. Therefore, the introduction of malachite green into the molecular building blocks for self-assembly may lead to the development of photosensitive supramolecular systems. Recently, our group has designed and synthesized some building blocks containing photoresponsive malachite green groups [36]. By tuning the molecular amphiphilicity using the photoreaction of malachite, controllable assembly and disassembly of vesicles were achieved. As shown in Fig. 2-1g, the PEG-terminated malachite green derivative shows amphiphilic property because malachite green moiety was hydrophobic in its neutral form and the PEG segment is hydrophilic. As expected, this kind of amphiphilic molecules formed vesicle-like aggregates. On UV light irradiation, the color of the solution changed to deep green, indicating that the photochromic moiety of the amphiphile was ionized to its corresponding cationic form, which led to the disassembly of these vesicles. Interestingly, the cationic form could thermally recover to its electrically neutral form, and the disassembled species could re-form amphiphiles on the basis of a thermal reverse reaction, and vesicles will be reassembled spontaneously.

By using the reversible change of the amphiphilicity, the malachite green derivative can be applied to disperse single-walled carbon nanotubes in aqueous solution in a controllable way [75]. First, single-walled carbon nanotubes were well dispersed in an aqueous solution by mixing with a certain amount of malachite green on the basis of hydrophobic and $\pi-\pi$ interactions between the nanotube surface and amphiphilic malachite green derivative. Second, the carbon nanotubes could be released and precipitated in the solution after the deneutralization of malachite green groups. Here, the malachite green derivative behaved like a molecular claw, wrapping carbon nanotubes and liberating them by photostimuli when needed. This process has a great potential use for further application of carbon materials (carbon nanotubes, graphene, etc.) in new electronic devices and catalytic fields.

Moreover, the amphiphilicity change of malachite green by the photoreaction can endow the supramolecular entities with great macroscopic motion. Recently, we described the facile reversible UV-controlled and fast transition from emulsion to gel by using a photoresponsive polymer with a malachite green group [76]. The photoresponsive polymer with the hydrophobic malachite green group could be used for the formation of an oil-in-water emulsion. However, on UV irradiation for 5 min, the photochromic malachite green group could be ionized to its corresponding cation, leading to the transformation from emulsion to gel. On shaking, such gel could recover the emulsion state, and further UV irradiation could turn the emulsion into gel again. Such a transition from emulsion to gel by photochemical reaction and reverse transition by shaking treatment can be repeated several times.

SP, another photoresponsive group, can also be used for such transitions, since it can be switched between neutral and zwitterionic forms reversibly by alternating UV and visible light irradiation. The photochromatism of SP can be used to control the molecular amphiphilicity of the surfactants attached with SP. In that case, the controlled self-assembly modulation of the building blocks with SP groups can be realized by light irradiation [77,78]. Even more, the hydrophilic–lipophilic balance of such surfactants could be changed after photoreaction that led to the change of surface tension [79,80]. Recently, Matyjaszewski *et al*. synthesized an amphiphilic block copolymer with poly(ethylene oxide) (PEO) as one block and poly(SP) as another block by the Atom Transfer Radical Polymerization (ATRP) method. As shown in Fig. 2-1h, by alternating UV and visible light irradiation, the photochromic SP could be changed between the neutral SP form and zwitterionic merocyanine (ME). It means that the amphiphilic block copolymer PEO-*b*-SP could be converted to double-hydrophilic block copolymer PEO-*b*-ME. The micelle-like aggregates formed by PEO-*b*-SP were destroyed through the conversion of SP to ME. Moreover, the micelles could be re-formed after the recovery of SP. It was also shown that the reversible micelle formation could be used to incorporate hydrophobic molecules under control [81].

The conversion of the two isomers of azobenzene, trans and cis forms, can take place reversibly by alternating UV and visible light irradiation. Azobenzene with the photoresponsive property has been applied well in material science. The change of dipole between *trans*-azobenzene and *cis*-azobenzene is estimated as 4.4 D, responsible for a slight polarity change between *trans*-azobenzene and *cis*-azobenzene. This polarity change of azobenzene between the trans and cis forms can be amplified by attaching azobenzene groups on the polymers. By the photoisomerization of azobenzene, the polymers or the particles with azobenzene are explored as smart photoresponsive materials, such as surface relief gratings and light-triggered bending polymer films [82–92].

Besides single stimuli, multistimuli responsive methods are also being explored for tuning the amphiphilicity of the building blocks [93–97]. For example, we have demonstrated that the self-assembled monolayer (SAM) of mercapto-terminated malachite green derivative on a rough gold substrate gave rise to a dual pH-stimuli- and photo-irradiation-responsive surface. Interestingly, the amphiphilicity conversion of malachite green could change the surface wettability macroscopically from superhydrophobic to superhydrophilic. This dual-tunable on/off switch broadens the insights for fabrication of new wetting-controlled stimuli materials [98].

2.4   SUPRAMOLECULAR METHODS

Supramolecular chemistry concerns molecular assemblies and noncovalent interactions. The reversibility of noncovalent interactions used in supramolecular chemistry, such as electrostatic interaction, hydrogen-binding, $\pi-\pi$ interaction, van der Waals force, and host–guest interaction, endows the assemblies with stimuli-responsive ability. Herein, the main idea is how to construct supramolecular amphiphiles that are formed on the basis of noncovalent interactions. The supramolecular amphiphiles refer to supramolecules of two component systems: one is a host amphiphile noncovalently binding with a guest molecule; the other is a nonamphiphilic building block binding with a hydrophilic guest. By changing the noncovalent interactions, the amphiphilicity of those supramolecular amphiphiles can be tunable between more hydrophilic and less hydrophobic or between more hydrophobic and less hydrophilic. Therefore, the supramolecular amphiphiles are hoped to display controlled aggregation behaviors and new functions.

2.4.1   Electrostatic Interaction

A new concept of designing a photocontrollable supramolecular polymeric amphiphile through the electrostatic association between an azobenzene-containing surfactant (AzoC10) and a double-hydrophilic block ionomer, poly(ethylene glycol)-b-poly(acrylic acid) (PEG_{43}-PAA_{153}), has been recently developed by our group, as is shown in Fig. 2-2a [41]. PEG_{43}-PAA_{153} is a double-hydrophilic block copolymer; however, on complexation with azobenzene-containing surfactant based on electrostatic interaction, the complex changes to amphiphilic. The amphiphilicity variation makes the block ionomer complex self-assemble in aqueous solution and form vesicle-like aggregates. The photoisomerization of azobenzene moieties in the block ionomer complex can reversibly tune the amphiphilicity of the surfactants, inducing the disassembly of the vesicles. Such block ionomer complex vesicles are further evaluated as nanocontainers capable of encapsulating and releasing guest solutes on demand, controlled by light irradiation. For example, vesicles encapsulating the fluorescein sodium display clear spherical images observed by fluorescence microscopy. However, such fluorescence-marked images disappear after releasing the solute from the vesicles, triggered by UV light. Such novel materials are of both basic and practical significance, especially as prospective nanocontainers for cargo delivery.

2.4.2   Hydrogen Bonding Method

The use of complementary hydrogen bonding for fabricating supramolecular amphiphiles dates back to the early work of Kunitake *et al.*, who employed substituted melamines as hydrogen acceptors and isocyanuric acid derivatives

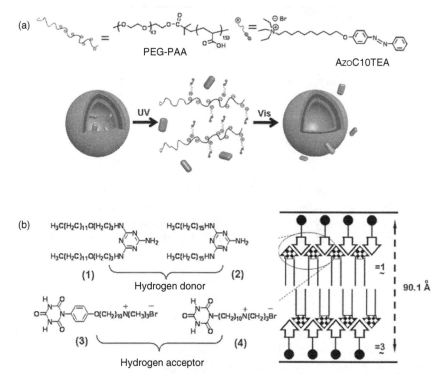

Figure 2-2   Schematic illustration of (a) photocontrollable supramolecular polymeric amphiphile formed by self-assembly of the block ionomer complex [41] and (b) hydrogen-bonding-directed supramolecular amphiphile [42].

as hydrogen donors [42]. As shown in Fig. 2-2b, sparingly water-soluble (1) and (2) with hydrogen acceptor groups were used as guest building blocks. They both can bind with the host hydrogen-donor-containing amphiphiles (3) and (4) on the basis of hydrogen bonding interaction to form supramolecular amphiphiles. By combining with a hydrogen acceptor, the amphiphilicity of the amphiphilic hydrogen donor can be changed greatly because the hydrophobic segment was elongated. However, the hydrogen bonding interaction was greatly weakened in polar water, and the supramolecular amphiphile could not be formed by mixing the hydrogen donor and acceptor directly. To prepare the supramolecular amphiphiles, the authors dissolved the hydrogen donor and acceptor in a cosolvent ethanol and removed the solvent to form a solid-state mixture. After repeating this cycle three times, the final product was dissolved in water. Unlike the single amphiphiles, the supramolecular amphiphiles could form disklike aggregates, which were confirmed to be bilayer structures. By changing the alkyl length of hydrogen acceptors, the supramolecular amphiphiles displayed a different aggregation behavior. Since the strength of hydrogen bonding interaction depended on the applied temperature, the bilayer structure of the aggregates formed by the supramolecular amphiphiles could be thermally dissociated reversibly [99–101]. The concept of this family of hydrogen bonding-based supramolecular amphiphiles opens an avenue to fabricate new supramolecular amphiphiles on the basis of the hydrogen bond.

2.4.3 Host–Guest Modulation Employing Cyclodextrin as Host

Among so many supramolecular host–guest systems, some host molecules as building blocks can implant other building blocks that are employed as the guest in their bodies. Till now, some hosts are well investigated, including cyclodextrins (CDs), crown ethers, cucurbiturils, calixarenes, and so on. In this section, we present some examples of successfully tuning the amphiphilicity by the introduction of CD.

CD, which is divided into α-CD, β-CD, and γ-CD, is mainly formed by the bridge-linking of glucose units. CD can be dissolved in water very well because of the water-soluble nature of the outer surface, and it provides a hydrophobic cavity resulting from the water-insoluble interior. This series of host molecules can precisely bind with many kinds of small hydrophobic molecules depending on the complementarity of size and shape. For this reason, CD has been explored well as a water-soluble nanocontainer and reactor [102,103]. Stoddart, Harada, and Wenz *et al.*, one after the other, reviewed some works about the utilization of CDs as a building ring to fabricate rotaxane-like supramolecular assemblies through the self-assembly of CDs with amphiphiles [104–106]. Usually, two possibilities are expected when CD binds with a guest molecule. If the guest is sparingly water soluble, it will become a supramolecule amphiphile by binding with CD because the CD can function as a hydrophilic head. For a normal amphiphile, it can bind well with CD by the interaction between CD and the alkyl chain. This host–guest interaction can increase the solubility of the hydrophobic chain and change the amphiphilicity, which results in the disruption of the aggregates of the amphiphiles [107–111].

It is well known that the host–guest interaction between azobenzene and α-CD or β-CD can be controlled by the photoisomerization of azobenzene. Recently, our group has taken this unique feature and presented a concept of combining host–guest chemistry and photochemistry to tune the amphiphilicity. In our work, an azobenzene-containing surfactant was used, which could form vesicle-like aggregates in aqueous solution. The enhanced water solubility by the binding of α-CD with azobenzene disrupted the vesicles remarkably. After the photoisomerization of azobenzene from trans to cis form by UV light irradiation, α-CD could not bind with azobenzene any more and slid onto the alkyl chain, as shown in Fig. 2-3a. This movement of α-CD changed the amphiphilicity of the supramolecular amphiphile, and new aggregates were formed subsequently, whose size was smaller than that formed by pure amphiphiles. When azobenzene recovered its trans form by visible light irradiation, α-CD moved back onto the azobenzene group and induced the disruption of the aggregates again. An interesting finding was that the shuttle-like movement of α-CD on the amphiphile could tune the aggregation behavior reversibly by photoisomerization of azobenzene [112].

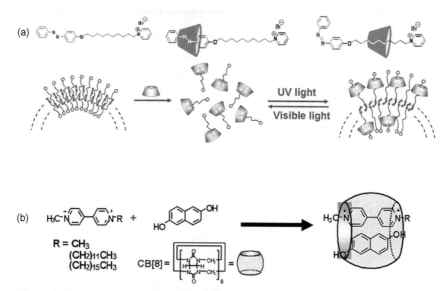

Figure 2-3 Schematic illustration of (a) photocontrolled reversible supramolecular amphiphile of azobenzene-containing surfactant with α-CD [112] and (b) supramolecular amphiphile formed by charge-transfer complex in a host [113].

As mentioned above, the movement of CD can induce the amphiphilicity change of the guest molecule; thus, we have reason to believe that the surface modified with such rotaxane-like assemblies can display different wettability on the movement of CD. Very recently, we fabricated this kind of molecular shuttle onto a rough gold substrate by SAM, whose wettability could respond to the photostimuli [114]. In this work, an azobenzene-containing building block with a mercapto group at the end preassembled with α-CD in water, based on the interaction between azobenzene and α-CD, forming a sort of supramolecular amphiphile. This supramolecular amphiphile could form a mixed SAM with n-butylthiol on gold substrate. The purpose of fabrication of the mixed SAMs was to ensure that there was enough free space for the movement of the molecular shuttle. Before UV light irradiation, α-CD stayed on the top of the surface and the SAM displayed hydrophilic property. After UV light irradiation, α-CD slid down onto the alkyl chain and the SAM became hydrophobic. Nearly $50°$ change in the surface contact angle with a water droplet as an indicator was found before and after UV light irradiation. This great change of wettability could be cycled several times by alternating UV and visible light irradiation. The shuttle movement mechanism of α-CD on this SAM was well supported by cyclic voltammetric studies.

Furthermore, based on a similar idea, we showed the photocontrolled switching between PAA-g-CD-attached Azo SAM and Azo SAM using the light-driven reversible host–guest interaction of azobenzene and CD to reversibly adsorb and release electroactive Cyt c. The PAA-g-CD-attached Azo SAM also possesses pH sensitivity, which can be used to reversibly immobilize electroactive Cyt c triggered by environmental pH [115]. The integration of the pH sensitivity and the photoactivated switching to form dual-responsive reactivated biointerfaces for reversible sorption and release of electroactive Cyt c meets the modern requirements of developing bioscience and biotechnology and is anticipated to provide an excellent platform for potentially wide-ranging applications in biomimetics, biomembranes, controlled bioseparation, stimuli-responsive biomedical technologies, biosensing, and optobioelectronic devices.

2.4.4 Host–Guest Modulation Employing Cucurbituril as Host

Cucurbituril is another interesting host molecule, with a similar hydrophobic cavity to CD, which can be used to incorporate guest molecules. It has been studied very well as building blocks for supramolecular assembly, both in solutions and on solid surfaces [116,117]. Kim *et al.* presented an early example by combining the host–guest chemistry with charge-transfer interaction to form a surpamolecular amphiphile in aqueous solution. The surfactants with bipyridium groups themselves formed micelles in aqueous solution. Interestingly, in the presence of cucurbituril, the electron acceptor of the bipyridium group and the electron donor of dihydroxynaphthalene could form a pair in the cavity of cucurbituril, based on the charge-transfer interaction, as is shown in Fig. 2-3b. This supramolecular amphiphile displayed new aggregation behavior in water and formed vesicle-like aggregates as observed by TEM (transmission electron microscopy). On addition of oxidative chemicals, such as cerium(IV) ammonium nitrate, the electron donor of dihydroxynaphthalene was oxidized and the supramolecular assemblies based on charge-transfer interaction could be collapsed [113].

2.4.5 Charge-Transfer Interaction

Charge-transfer interactions between electron donors and acceptors have been developed as a new noncovalent force for the fabrication of supramolecular assemblies. The formation of charge-transfer complexes is always driven by a combination of charge-transfer interactions and other noncovalent interactions, such as hydrophobic interactions, hydrogen bonding, and so on. One characteristic of charge-transfer complexes is their high charge-carrier densities, which lead to their potential application as electronic and optoelectronic materials [118–122]. Very recently, our group established a series of supramolecular systems for the modulation of self-assembly by the weak charge-transfer interaction [123–125]. As shown in Fig. 2-4a, an amphiphilic electron donor PYR and a hydrophobic electron acceptor DNB were involved in the formation of a charge-transfer-interaction-based supramolecular amphiphile. The amphiphilic PYR itself could form tubelike aggregates at a concentration more than its CMC. By the method mentioned in Kunitake's work, dissolving and drying three times, a mixture of PYR with DNB could be obtained. Interestingly, this supramolecular amphiphile formed stable vesicles in water. It was evident

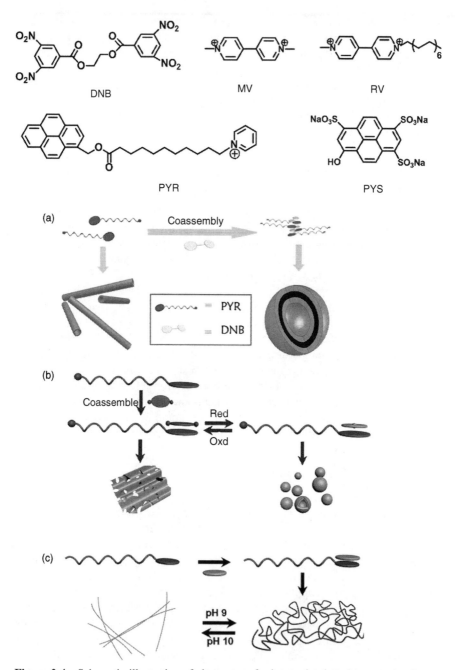

Figure 2-4   Schematic illustration of charge-transfer-interaction-based supramolecular amphiphiles: (a) transformation from tubes to vesicles [123], (b) redox responsive reversible transformation between irregular continuous aggregates and vesicles [124], and (c) formation of pH-responsive nanofibers [125].

that the introduction of a charge-transfer interaction into the supramolecular assembly could induce the transformation of the aggregates from tubelike to vesicle-like. This morphology transformation is understood by the curvature change of the aggregates by the introduction of charge-transfer interaction. This finding provides a simple model that affords further understanding of the membrane deformation and micropatterning details in various cellular events.

Furthermore, supramolecular amphiphile can also be formed by the complexation of electron donor PYR and redox-responsive electron acceptor MV driven by charge-transfer interactions. The self-assembly architecture of the supramolecular amphiphile can change reversibly between irregular continuous aggregates and organized vesicles in response to redox stimuli, as shown in Fig. 2-4b. Compared with the classic stimuli-responsive amphiphile system, where the responsive group is covalently linked to the building blocks, the supramolecular amphiphile has the advantage of good reversibility and facile preparation. It is hoped that these findings will potentially further extend the scope of supramolecular strategy into fabricating nanoscopic intelligent materials [124].

The concept of "supramolecular amphiphiles" was further extended to fabricate one-dimensional ultralong nanofibers on the basis of the water-soluble charge-transfer complex formation between viologen derivatives (RV) and the 8-hydroxypyrene-1,3,6-trisulfonic acid trisodium salt (PYS). As can be seen in Fig. 2-4c, ultralong nanofibers are obtained from a supramolecular amphiphile based on a water-soluble charge-transfer complex with a 1:1 PYS/MV stoichiometry. The formation of the complex is driven by combined interactions including coulombic attractions, charge-transfer interactions, and the hydrophobic effect. The straightness of the nanofiber can be tuned by changing the pH of the system [125]. Compared with conventional materials based on charge-transfer complexes, the easy and environmentally friendly preparation of the water-soluble charge-transfer complex is advantageous. It is hoped that such a method may be extended to fabricate tunable self-assembling materials for electronic and optoelectronic devices. Moreover, it could be used as a model system for understanding amyloid fibril formation and neurodegenerative diseases.

2.5 CONCLUSION AND OUTLOOK

We have tried to provide a balanced overview about how to tune the amphiphilicity of building blocks, leading to controlled self-assembly and disassembly. There is no doubt that biological self-assembly often happens in a controlled manner, which is responsible for the different functions of life. Inspired by the perfect biological self-assembly, there is an increasing interest to learn from nature for the realization of controlled self-assembly and disassembly in artificial systems.

Reproducibility and ease of scale-up are basic criteria for fabrication of new functional supramolecular materials with high performance; however, they are difficult to achieve without controlled self-assembly approaches. The control of amphiphilicity of building blocks opens an avenue for controlled self-assembly;

however, the responsive speed needs to be improved sometimes. In addition, new systems that are responsive to mild and different types of stimuli need to be developed. Biocompatibility and toxicity should be considered in designing smart supramolecular systems for clinical purposes.

It should be pointed out that there are different levels of self-assembly, and a high level of self-assembling methods is highly demanded. The high level of self-assembly refers to hierarchical self-assembly and even dynamic self-assembly, which are crucial for fabricating complicated functional supramolecular materials and devices that are adaptive, self-replicative, and self-healing. Biomacromolecules include various functions that are so far not achievable by synthetic materials. It is hoped that combining biomacromolecules with inorganic species in hybrid assemblies may broaden the range of materials with spatiotemporal functions. In summary, the control of self-assembly and disassembly is a field that embodies the creative power of not only chemists but also scientists in materials and chemical biology.

ACKNOWLEDGMENTS

The authors thank the National Basic Research Program (2007CB808000), National Natural Science Foundation of China (20904028, 50973051, 20974059), Tsinghua University Initiative Scientific Research Program (2009THZ02-2), the NSFC-DFG joint grant (TRR 61) for financial support.

REFERENCES

1. Lehn, J.-M. (1988). Supramolecular chemistry—scope and perspectives: molecules—dupermolecules—molecular devices. *J. Inclusion Phenom. Macrocycl. Chem.*, *6*, 351–396.
2. Lehn, J.-M. (1995). *Supramolecular Chemistry: Concepts and Perspectives*. VCH, Weinheim.
3. Whitesides, G. M., Grzbowski, B. (2002). Self-assembly at all scales. *Science*, *295*, 2418–2421.
4. Service, R. F. (2005). How far can we push chemical self-assembly? *Science*, *309*, 95.
5. Prins, L. J., Reinhoudt, D. N., Timmerman, P. (2001). Noncovalent synthesis using hydrogen bonding. *Angew. Chem. Int. Ed.*, *40*, 2382–2426.
6. Barth, J. V., Constantini, G., Kern, K. (2005). Engineering atomic and molecular nanostructures at surfaces. *Nature*, *437*, 671–679.
7. Isrealachvilli, J. N., Mitchell, D. J., Ninham, B. W. (1976). Theory of self-assembly of hydrocarbon amphiphiles into micelles and bilayers. *J. Chem. Soc., Faraday Trans. 2*, *72*, 1525–1568.
8. Israelachivili, J. N. (1985). *Intermolecular and Surface Forces*. Academic Press, New York.
9. Vögtle, F. (1993). *Supramolecular Chemistry*. John Wiley & Sons, Ltd., England.
10. Wang, Y., Xu, H., Zhang, X. (2009). Tuning the amphiphilicity of building blocks: controlled self-assembly and disassembly for functional supramolecular materials. *Adv. Mater.*, *21*, 2849–2864.
11. Shimizu, T., Masuda, M., Minamikawa, H. (2005). Supramolecular nanotube architectures based on amphiphilic molecules. *Chem. Rev.*, *105*, 1401–1444.
12. Hill, J. P., Jin, W., Kosaka, A., Fukushima, T., Ichihara, H., Shimomura, T., Ito, K., Hashizume, T., Ishii, N., Aida, T. (2004). Self-assembled hexa-peri-hexabenzocoronene graphitic nanotube. *Science*, *304*, 1481–1483.
13. Yamamoto, Y., Fukushima, T., Jin, W., Kosaka, A., Hara, T., Nakamura, T., Saeki, A., Seki, S., Tagawa, S., Aida, T. (2006). Charge and anion ordering in (TMTTF)2X: quasi-one-dimensional conductors. *Adv. Mater.*, *18*, 1297–1300.

14. Claussen, R. C., Rabatic, B. M., Stupp, S. I. (2003). Aqueous self-assembly of unsymmetric peptide bolaamphiphiles into nanofibers with hydrophilic cores and surfaces. *J. Am. Chem. Soc.*, *125*, 12680–12681.

15. Vemula, P. K., John, G. (2008). Crops: a green approach toward self-assembled soft materials. *Acc. Chem. Res.*, *41*, 769–782.

16. Kabanov, A. V., Kabanov, V. A. (1998). Interpolyelectrolyte and block ionomer complexes for gene delivery: physico-chemical aspects. *Adv. Drug Delivery Rev.*, *30*, 49–60.

17. Allen, C., Maysinger, D., Eisenberg, A. (1999). Nano-engineering block copolymer aggregates for drug delivery. *Colloids Surf., B: Biointerf.*, *16*, 3–27.

18. Kwon, G. S., Kataoka, K. (1995). Block copolymer micelles as long circulating drug vehicles. *Adv. Drug Delivery Rev.*, *16*, 295–309.

19. Rosler, A., Vandermeulen, G. W. M., Klok, H.-A. (2001). Advanced drug delivery devices via self-assembly of amphiphilic block copolymers. *Adv. Drug Delivery Rev.*, *53*, 95–108.

20. Sierra, L., Lopez, B., Gil, H., Guth, J.-L. (1999). Synthesis of mesoporous silica from sodium silica solutions and a poly(ethylene oxide)-based surfactant. *Adv. Mater.*, *11*, 307–311.

21. Ruiz-Hitzky, E., Letaïef, S., Préot, V. (2002). Novel organic–inorganic mesophases: self-templating synthesis and intratubular swelling. *Adv. Mater.*, *14*, 439–443.

22. Zhang, Q., Ariga, K., Okabe, A., Aida, T. (2004). A condensable amphiphile with a cleavable tail as a "Lizard" template for the sol-gel synthesis of functionalized mesoporous silica. *J. Am. Chem. Soc.*, *126*, 988–989.

23. Walde, P., Ichikawa, S. (2001). Enzymes inside lipid vesicles: preparation, reactivity and applications. *Biomol. Eng.*, *18*, 143–177.

24. Vriezema, D. M., Aragonés, M. C., Elemans, J. A. A. W., Cornelissen, J. J. L. M., Rowan, A. E., Nolte, R. J. M. (2005). Self-assembled nanoreactors. *Chem. Rev.*, *105*, 1445–1490.

25. Vriezema, D. M., Garcia, P. M. L., Oltra, N. S., Hatzakis, N. S., Kuiper, S. M., Nolte, R. J. M., Rowan, A. E., van Hest, J. C. M. (2007). Positional assembly of enzymes in polymersome nanoreactors for cascade reactions. *Angew. Chem. Int. Ed.*, *46*, 7378–7382.

26. Kishimura, A., Koide, A., Osada, K., Yamasaki, Y., Kataoka, K. (2007). Encapsulation of myoglobin in PEGylated polyion complex vesicles made from a pair of oppositely charged block ionomers: a physiologically available oxygen carrier. *Angew. Chem. Int. Ed.*, *46*, 6085–6088.

27. Wang, Y., Xu, H., Ma, N., Wang, Z., Zhang, X., Liu, J., Shen, J. (2006). Block copolymer micelles as matrixes for incorporating diselenide compound: water-soluble glutathione peroxidase mimic fine-tuned by ionic strength. *Langmuir*, *22*, 5552–5555.

28. Hager, M., Currie, F., Holmberg, K. (2003). Organic reactions in microemulsions. *Top. Curr. Chem.*, *227*, 53–74.

29. Morikawa, M., Yoshihara, M., Endo, T., Kimizuka, N., (2005). α-Helical polypeptide microcapsules formed by emulsion-templated self-assembly. *Chem.—Eur. J.*, *11*, 1514–1518.

30. Haubs, M., Ringsdorf, H. (1985). Photoreactions of *N*-(1-pyridinio)amidates in monolayers and liposomes. *Angew. Chem. Int. Ed. Engl.*, *24*, 882–883.

31. Schillén, K., Bryskhe, K., Mel'nikova, Y. S. (1999). Vesicles formed from a poly(ethylene oxide)-poly(propylene oxide)-poly(ethylene oxide) triblock copolymer in dilute aqueous solution. *Macromolecules*, *32*, 6885–6888.

32. Ma, N., Li, Y., Xu, H., Wang, Z., Zhang, X. (2010). Dual redox responsive assemblies formed from diselenide block copolymers. *J. Am. Chem. Soc.*, *132*, 442–443.

33. Lee, Y., Fukushima, S., Bae, Y., Hiki, S., Ishii, T., Kataoka, K. (2007). A protein nanocarrier from charge-conversion polymer in response to endosomal pH. *J. Am. Chem. Soc.*, *129*, 5362–5363.

34. Rosslee, C. A., Abbott, N. L. (2001). Principles for microscale separations based on redox-active surfactants and electrochemical methods. *Anal. Chem.*, 73, 4808–4814.

35. Liu, Y., Jessop, P. G., Cunningham, M., Eckert, C. A., Liotta, C. L. (2006). Switchable surfactants. *Science*, *313*, 958–960.

36. Jiang, Y., Wang, Y., Ma, N., Wang, Z., Smet, M., Zhang, X. (2007). Self-organization of a UV-responsive PEG-terminated malachite green derivative: vesicle formation and photoinduced disassembly. *Langmuir*, *23*, 4029–4034.

37. Haubs, M., Ringsdorf, H. (1987). Photosensitive monolayers, bilayer membranes and polymers. *Nouv. J. Chim.*, *11*, 151–156.

38. Ringsdorf, H., Schlarb, B., Venzmer, J. (1988). Molecular architecture and function of polymeric oriented systems: models for the study of organization, surface recognition, and dynamics of biomembranes. *Angew. Chem. Int. Ed. Engl.*, *27*, 113–158.

39. Itoh, Y., Horiuchi, S., Yamamoto, K. (2005). Photodegradative surfactants: photolyses of p-alkylbenzyltrimethylammonium and alkylbenzyldimethylammonium halides in aqueous solution. *Photochem. Photobiol. Sci.*, *4*, 835–839.

40. Itoh, Y., Yamamoto, K., Shirai, H. (2003). Photodegradative surfactants: photolysis of p-dodecylbenzyltrimethylammonium bromide in aqueous solution. *Chem. Lett.*, *32*, 8–9.

41. Wang, Y., Han, P., Xu, H., Wang, Z., Zhang, X., Kabanov, A. V. (2010). Photocontrolled self-assembly and disassembly of block ionomer complex vesicles: A facile approach toward supramolecular polymer nanocontainers. *Langmuir*, *26*, 709–715.

42. Kimizuka, N., Kawasaki, T., Kunitake, T. (1993). Self-organization of bilayer membranes from amphiphilic networks of complementary hydrogen bonds. *J. Am. Chem. Soc.*, *115*, 4387–4388.

43. Napoli, A., Tirelli, N., Wehrli, E., Hubbell, J. A. (2002). Lyotropic behavior in water of amphiphilic ABA triblock copolymers based on poly(propylene sulfide) and poly(ethylene glycol). *Langmuir*, *18*, 8324–8329.

44. Napoli, A., Valentini, M., Tirelli, N., Müller, M., Hubbell, J. A. (2004). Oxidation-responsive polymeric vesicles. *Nat. Mater.*, *3*, 183–189.

45. Napoli, A., Boerakker, M. J., Tirelli, N., Nolte, R. J. M., Sommerdijk, N. A. J. M., Hubbell, J. A. (2004). Glucose-oxidase based self-destructing polymeric vesicles. *Langmuir*, *20*, 3487–3491.

46. Rotruck, J. T., Pope, A. L., Ganther, H. E., Swanson, A. B., Hafeman, D. G., Hoekstra, W. G. (1973). Selenium: biochemical role as a component of glutathione peroxidase. *Science*, *179*, 588–590.

47. Zhang, X., Xu, H., Dong, Z., Wang, Y., Liu, J., Shen, J. (2004). Highly efficient dendrimer-based mimic of glutathione peroxidase. *J. Am. Chem. Soc.*, *126*, 10556–10557.

48. Xu, H., Gao, J., Wang, Y., Smet, M., Dehaen, W., Zhang, X. (2006). Hyperbranched polyselenides as glutathione peroxidase mimics. *Chem. Commun.*, 796–798.

49. Fredga, A. (1972). Organic selenium chemistry. *Ann. N.Y. Acad. Sci.*, *192*, 1–9.

50. Trahanovsky, W. S. (1978). *Oxidation in Organic Chemistry, Part C*. Academic Press, New York.

51. Haag, R. (2004). Supramolecular drug-delivery systems based on polymeric core-shell architectures. *Angew. Chem. Int. Ed.*, *43*, 278–282.

52. Ringsdorf, H. (1975). Structure and properties of pharmacologically active polymers. *J. Polym. Sci., Polym. Symp.*, *51*, 135–153.

53. Gros, L., Ringsdorf, H., Schupp, H. (1981). Polymeric antitumor agents on a molecular and on a cellular level? *Angew. Chem. Int. Ed. Engl.*, *20*, 305–325.

54. Gillies, E. R., Fréchet, J. M. J. (2003). A new approach towards acid sensitive copolymer micelles for drug delivery. *Chem. Commun.*, *14*, 1640–1641.

55. Krämer, M., Stumbé, J.-F., Türk, H., Krause, S., Komp, A., Delineau, L., Prokhorova, S., Kautz, H., Haag, R. (2002). pH-Responsive molecular nanocarriers based on dendritic core-shell architectures. *Angew. Chem. Int. Ed.*, *41*, 4252–4256.

56. Bae, Y., Fukushima, S., Harada, A., Kataoka, K. (2003). Design of environment-sensitive supramolecular assemblies for intracellular drug delivery: polymeric micelles that are responsive to intracellular pH change. *Angew. Chem. Int. Ed.*, *42*, 4640–4643.

57. Hellberg, P.-E., Bergström, K., Holmberg, K. (2000). Cleavable surfactants. *J. Surfact. Deterg.*, *3*, 81–91.

58. Cordes, E. H., Bull, H. G. (1974). Mechanism and catalysis for hydrolysis of acetals, ketals, and ortho esters. *Chem. Rev.*, *74*, 581–603.

59. Liu, F., Eisenberg, A. (2003). Preparation and pH triggered inversion of vesicles from poly(acrylic acid)-*block*-polystyrene-*block*-poly(4-vinyl pyridine). *J. Am. Chem. Soc.*, *125*, 15059–15064.

60. Jaeger, D. A., Li, B., Clark, Jr., T. (1996). Cleavable double-chain surfactants with one cationic and one anionic head group that form vesicles. *Langmuir*, *12*, 4314–4316.

61. Zhu, J., Munn, R. J., Nantz, M. H. (2000). Self-cleaving ortho ester lipids: a new class of pH-vulnerable amphiphiles. *J. Am. Chem. Soc.*, *122*, 2645–2646.

62. Boomer, J. A., Thompson, D. H. (1999). Synthesis of acid-labile diplasmenyl lipids for drug and gene delivery applications. *Chem. Phys. Lipids*, *99*, 145–153.

63. Saji, T., Hoshino, K., Aoyagui, S. (1985). Reversible formation and disruption of micelles by control of the redox state of the surfactant tail group. *J. Chem. Soc., Chem. Commun.*, 865–866.

64. Saji, T., Hoshino, K., Aoyagui, S. (1985). Reversible formation and disruption of micelles by control of the redox states of the head group. *J. Am. Chem. Soc.*, *107*, 6865–6868.

65. Hoshino, K., Saji, T. (1987). Electrochemical formation of an organic thin film by disruption of micelles. *J. Am. Chem. Soc.*, *109*, 5881–5883.

66. Saji, T., Hoshino, K., Ishii, Y., Goto, M. (1991). Formation of organic thin films by electrolysis of surfactants with the ferrocenyl moiety. *J. Am. Chem. Soc.*, *113*, 450–456.

67. Foucher, D. A., Tang, B.-Z., Manners, I. (1992). Ring-opening polymerization of strained, ring-tilted ferrocenophanes: a route to high-molecular-weight poly(ferrocenylsilanes). *J. Am. Chem. Soc.*, *114*, 6246–6248.

68. Manners, I. (1999). Poly(ferrocenylsilanes): novel organometallic plastics. *Chem. Commun.*, 857–865.

69. Wang, X., Guerin, G., Wang, H., Wang, Y., Manners, I., Winnik, M. A. (2007). Cylindrical block copolymer micelles and co-micelles of controlled length and architecture. *Science*, *317*, 644–647.

70. Shi, W., Giannotti, M. I., Zhang, X., Hempenius, M. A., Schönherr, H., Vancso, G. J. (2007). Closed mechanoelectrochemical cycles of individual single-chain macromolecular motors by AFM. *Angew. Chem. Int. Ed.*, *46*, 8400–8404.

71. Jessop, P. G., Heldebrant, D. J., Li, X., Eckert, C. A., Liotta, C. L. (2005). Reversible nonpolar-to-polar solvent. *Nature*, *436*, 1102.

72. Yamada, T., Lukac, P. J., George, M., Weiss, R. G. (2007). Reversible, room-temperature ionic liquids. amidinium carbamates derived from amidines and aliphatic primary amines with carbon dioxide. *Chem. Mater.*, *19*, 967–969.

73. Phan, L., Andreatta, J. R., Horvey, L. K., Edie, C. F., Luco, A.-L., Mirchandani, A., Darensbourg, D. J., Jessop, P. G. (2008). Switchable-polarity solvents prepared from a single liquid component. *J. Org. Chem.*, *73*, 127–132.

74. Phan, L., Chiu, D., Heldebrant, D. J., Huttenhower, H., John, E., Li, X., Pollet, P., Wang, R., Eckert, C. A., Liotta, C. L., Jessop, P. G. (2008). Switchable solvents consisting of amidine/alcohol or guanidine/alcohol mixtures. *Ind. Eng. Chem. Res.*, *47*, 539–545.

75. Chen, S., Jiang, Y., Wang, Z., Zhang, X., Dai, L., Smet, M. (2008). Light- controlled single-walled carbon nanotube dispersions in aqueous solution. *Langmuir*, *24*, 9233–9236.

76. Jiang, Y., Wan, P., Xu, H., Wang, Z., Zhang, X., Smet, M. (2009). Facile reversible UV-controlled and fast transition from emulsion to gel by using a photoresponsive polymer with a malachite green group. *Langmuir*, *25*, 10134–10138.

77. Sanchez, C., Lebeau, B., Chaput, F., Boilot, J.-P. (2003). Optical properties of functional hybrid organic-inorganic nanocomposites. *Adv. Mater.*, *15*, 1969–1994.

78. Berkovic, G., Krongauz, V., Weiss, V. (2000). Spiropyrans and spirooxazines for memories and switches. *Chem. Rev.*, *100*, 1741–1754.

79. Tazuke, S., Kurihara, S., Yamaguchi, H., Ikeda, T. (1987). Photochemically triggered physical amplification of photoresponsiveness. *J. Phys. Chem.*, *91*, 249–251.

80. Drummond, C. J., Albers, S., Furlong, D. N., Wells, D. (1991). Photocontrol of surface activity and self-assembly with a spirobenzopyran surfactant. *Langmuir*, *7*, 2409.

81. Lee, H., Wu, W., Oh, J. K., Mueller, L., Sherwood, G., Peteanu, L., Kowalewski, T., Matyjaszewski, K. (2007). Light-induced reversible formation of polymeric micelles. *Angew. Chem. Int. Ed.*, *46*, 2453–2457.

82. Gao, J., He, Y., Xu, H., Song, B., Zhang, X., Wang, Z., Wang, X. (2007). Azobenzene-containing supramolecular polymer films for laser-induced surface relief gratings. *Chem. Mater.*, *19*, 14–17.

83. Yu, Y., Nakano, M., Ikeda, T. (2003). Directed bending of a polymer film by light. *Nature*, *425*, 145.

84. Klajn, R., Bishop, K. J. M., Fialkowski, M., Paszewski, M., Campbell, C. J., Gray, T. P., Grzybowski, B. A. (2007). Plastic and moldable metals by self-assembly of sticky nanoparticle aggregates. *Science*, *316*, 261–264.

85. Kunitake, T., Nakashima, N., Shimomura, M., Okahata, Y. (1980). Unique properties of chromophore-containing bilayer aggregates: enhanced chirality and photochemically induced morphological change. *J. Am. Chem. Soc.*, *102*, 6642–6644.

86. Hayashita, T., Kurosawa, T., Miyata, T., Tanaka, K., Igawa, M. (1994). Effect of structural variation within cationic azo-surfactant upon photoresponsive function in aqueous solution. *Colloid Polym. Sci.*, *272*, 1611–1619.

87. Eastoe, J., Vesperinas, A. (2005). Self-assembly of light-sensitive surfactants. *Soft Matter*, *1*, 338–347.

88. Orihara, Y., Matsumura, A., Sito, Y., Ogawa, N., Saji, T., Yamaguchi, A., Sakai, H., Abe, M. (2001). Reversible release control of an oily substance using photoresponsive micelles. *Langmuir*, *17*, 6072–6076.

89. Lim, H. S., Han, J. T., Kwak, D., Jin, M., Cho, K. (2006). Photoreversibly switchable superhydrophobic surface with erasable and rewritable pattern. *J. Am. Chem. Soc.*, *128*, 14458–14459.

90. Ichimura, K., Oh, S.-K., Nakagawa, M. (2000). Light-driven motion of liquids on a photoresponsive surface. *Science*, *288*, 1624–1626.

91. Wang, G., Tong, X., Zhao, Y. (2004). Preparation of azobenzene-containing amphiphilic diblock copolymers for light-responsive micellar aggregates. *Macromolecules*, *37*, 8911–8917.

92. Tong, X., Wang, G., Soldera, A., Zhao, Y. (2005). How can azobenzene block copolymer vesicles be dissociated and reformed by light? *J. Phys. Chem. B*, *109*, 20281–20287.

93. Zhang, W., Shi, L., Ma, R., An, Y., Xu, Y., Wu, K. (2005). Micellization of thermo- and pH-responsive triblock copolymer of poly(ethyleneglycol) -*b*-poly(4-vinylpyridine)-*b*-poly(*N*-isopropylacrylamide). *Macromolecules*, *38*, 8850–8852.

94. Sumaru, K., Kameda, M., Kanamori, T., Shinbo, T. (2004). Characteristic phase transition of aqueous solution of poly(*N*-isopropylacrylamide) functionalized with spirobenzopyran. *Macromolecules*, *37*, 4949–4955.

95. Willet, N., Gohy, J.-F., Lei, L., Heinrich, M., Auvray, L., Varshney, S., Jérôme, R., Leyh, B. (2007). Fast multiresponsive micellar gels from a smart ABC triblock copolymer. *Angew. Chem. Int. Ed.*, *46*, 7988–7992.

96. Zeng, J., Shi, K., Zhang, Y., Sun, X., Zhang, B. (2008). Construction and micellization of a noncovalent double hydrophilic block copolymer. *Chem. Commun.*, 3753–3755.

97. Schilli, C. M., Zhang, M., Rizzardo, E., Thang, S. H., Chong, Y. K., Edwards, K., Karlsson, G., Müller, A. H. E. (2004). A new double-responsive block copolymer synthesized via RAFT polymerization: poly(*N*-isopropylacrylamide)- *block*-poly(acrylic acid). *Macromolecules*, *37*, 7861–7866.

98. Jiang, Y., Wan, P., Smet, M., Wang, Z., Zhang, X. (2008). Self-assembled monolayers of a malachite green derivative: surfaces with pH- and UV-responsive wetting properties. *Adv. Mater.*, *20*, 1972–1977.

99. Kimizuka, N., Kawasaki, T., Hirata, K., Kunitake, T. (1998). Supramolecular membrane: spontaneous assembly of aqueous bilayer membrane via formation of hydrogen-bonded pairs of melamine and cyanuric acid derivatives. *J. Am. Chem. Soc.*, *120*, 4094–4104.

100. Kimizuka, N., Kawasaki, T., Kunitake, T. (1994). Spectral characteristics and molecular orientation of azobenzene-containing hydrogen-bond-mediated bilayer membranes. *Chem. Lett.*, 1399–1402.

101. Kimizuka, N., Kawasaki, T., Kunitake, T. (1994). Thermal stability and specific dye binding of a hydrogen-bond-mediated bilayer membrane. *Chem. Lett.*, 33–36.

102. Breslow, R., Dong, S. D. (1998). Biomimetic reactions catalyzed by cyclodextrins and their derivatives. *Chem. Rev.*, *98*, 1997–2012.

103. Takahashi, K. (1998). Organic reactions mediated by cyclodextrins. *Chem. Rev.*, *98*, 2013–2034.

104. Nepogodiev, S. A., Stoddart, J. F. (1998). Cyclodextrin-based catenanes and rotaxanes. *Chem. Rev.*, *98*, 1959–1976.

105. Harada, A. (2001). Cyclodextrin-based molecular machines. *Acc. Chem. Res.*, *34*, 456–464.

106. Wenz, G., Han, B., Müller, A. (2006). Cyclodextrin rotaxanes and polyrotaxanes. *Chem. Rev.*, *106*, 782–817.

107. Diaz, A., Quintela, P. A., Schuette, J. M., Kaifer, A. E. (1988). Complexation of redox-active surfactants by cyclodextrins. *J. Phys. Chem.*, *92*, 3537–3542.

108. Dharmawardana, U. R., Christian, S. D., Tucker, E. E., Taylor, R. W., Scamehorn, J. F. (1993). A surface tension method for determining binding constants for cyclodextrin inclusion complexes of ionic surfactants. *Langmuir*, *9*, 2258–2263.

109. Mwakibete, H., Bloor, D. M., Wyn-Jones, E. (1994). Determination of the complexation constants between alkylpyridinium bromide and. alpha.- and. beta.-cyclodextrins using electromotive force methods. *Langmuir*, *10*, 3328–3331.

110. Junquera, E., Tardajos, G., Aicart, E. (1993). Effect of the presence of. beta.-cyclodextrin on the micellization process of sodium dodecyl sulfate or sodium perfluorooctanoate in water. *Langmuir*, *9*, 1213–1219.

111. Zou, J., Tao, F., Jiang, M. (2007). Optical switching of self-assembly and disassembly of noncovalently connected amphiphiles. *Langmuir*, *23*, 12791–12794.

112. Wang, Y., Ma, N., Wang, Z., Zhang, X. (2007). Photo-controlled reversible supramolecular assembly of an azobenzene-containing surfactant with α-cyclodextrin. *Angew. Chem. Int. Ed.*, *46*, 2823–2826.

113. Jeon, Y. J., Bharadwaj, P. K., Choi, S. W., Lee, J. W., Kim, K. (2002). Supramolecular amphiphiles: spontaneous formation of vesicles triggered by formation of a charge-transfer complex in a host. *Angew. Chem. Int. Ed.*, *41*, 4474–4476.

114. Wan, P., Jiang, Y., Wang, Z., Zhang, X. (2008). Tuning surface wettability through photocontrolled reversible molecular shuttle. *Chem. Commun.*, *44*, 5710–5712.

115. Wan, P., Wang, Y., Jiang, Y., Xu, H., Zhang, X. (2009). Fabrication of reactivated biointerface for dual-controlled reversible immobilization of cytochrome c. *Adv. Mater.*, *21*, 4362–4365.

116. Ko, Y. H., Kim, E., Hwang, I., Kim, K. (2007). Supramolecular assemblies built with host-stabilized charge-transfer interactions. *Chem. Commun.*, 1305–1315.

117. Kim, K., Jeon, W. S., Kang, J.-K., Lee, J. W., Jon, S. Y., Kim, T., Kim, K. (2003). A pseudorotaxane on gold: formation of self-assembled monolayers, reversible dethreading and rethreading of the ring, and ion-gating behavior. *Angew. Chem. Int. Ed.*, *42*, 2293–2296.

118. Gabriel, G. J., Sorey, S., Iverson, B. L. (2005). Altering the folding patterns of naphthyl trimers. *J. Am. Chem. Soc.*, *127*, 2637–2640.

119. Okabe, A., Fukushima, T., Ariga, K., Aida, T. (2002). Color-tunable transparent mesoporous silica films: immobilization of one-dimensional columnar charge-transfer assemblies in aligned silicate nanochannels. *Angew. Chem. Int. Ed.*, *41*, 3414–3417.

120. Pisula, W., Kastler, M., Wasserfallen, D., Robertson, J. W. F., Nolde, F., Kohl, C., Müllen, K. (2006). Pronounced supramolecular order in discotic donor–acceptor mixtures. *Angew. Chem. Int. Ed.*, *45*, 819–823.

121. Ringsdorf, H., Bengs, H., Karthaus, O., Wüsterfeld, R., Ebert, M., Wendorff, J. H., Kohne, B., Praefcke, K. (1990). Induction and variation of discotic columnar phases through doping with electron acceptors. *Adv. Mater.*, *2*, 141–144.

122. Green, M. M., Ringsdorf, H., Wagner, J., Wüstefeld, R. (1990). Induction and variation of chirality in discotic liquid crystalline polymers. *Angew. Chem. Int. Ed.*, *29*, 1478–1481.

123. Wang, C., Yin, S., Chen, S., Xu, H., Wang, Z., Zhang, X. (2008). Controlled self-assembly manipulated by charge-transfer interactions: from tubes to vesicles. *Angew. Chem. Int. Ed.*, *47*, 9049–9053.

124. Wang, C., Guo, Y., Wang, Y., Xu, H., Zhang, X. (2009). Redox responsive supramolecular amphiphiles based on reversible charge transfer interactions. *Chem. Commun.*, 5380–5382.

125. Wang, C., Guo, Y., Wang, Y., Xu, H., Wang, R., Zhang, X. (2009). Supramolecular amphiphiles based on water-soluble charge transfer complex: fabrication of ultra-long nanofiber with tunable straightness. *Angew. Chem. Int. Ed.*, *48*, 8962–8965.

CHAPTER 3

ORGANIC–INORGANIC SUPRAMOLECULAR MATERIALS

Katsuhiko Ariga, Jonathan P. Hill, and Qingmin Ji

World Premier International (WPI), Research Center for Materials
Nanoarchitectonics (MANA), National Institute for Materials Science (NIMS),
Tsukuba, Ibaraki, Japan
Japan Science and Technology Agency (JST), Core Research of Evolutional
Science and Technology (CREST), Tsukuba, Ibaraki, Japan

3.1 INTRODUCTION

Nanotechnology using soft organic materials is an attractive challenge despite the fact that most of current nanotechnology has been developed by using top-down microfabrication techniques of hard inorganic materials such as that seen in silicon-based technologies. Moore's law predicts that the current rate of minia-turization of silicon-based technology will be affected by the physical limits of device dimensions in the very near future [1,2]. Therefore, bottom-up processes based on spontaneous assembly of small building blocks, where supramolecular chemistry of noncovalent molecular associations plays a crucial role, will become more important in nanotechnology [3,4]. Soft organic supramolecular assemblies also possess flexible stimulus-response properties that are not commonly seen in rigid inorganic devices. One of the ultimate goals of nanotechnology would be to emulate physically the biochemical mechanisms of living systems, which can be also regarded as cooperative device systems of supramolecular soft matter.

Supramolecular soft matter is organized by noncovalent interactions including electrostatics, hydrogen bonding, and metal coordination, resulting in materials with more flexible characteristics. Although these supramolecular approaches mostly use organic and biological components for assembly, inorganic nanomaterials are also useful for supramolecular structures based on the same molecular interactions. Recent developments in nanomaterials science allow us to use inorganic substances with nanometric dimensions and controllable morphologies in supramolecular assemblies. Mixed supramolecular assemblies using organic, biological, or inorganic materials have a great potential in the creation of novel

Supramolecular Soft Matter: Applications in Materials and Organic Electronics, First Edition.
Edited by Takashi Nakanishi.
© 2011 John Wiley & Sons, Inc. Published 2011 by John Wiley & Sons, Inc.

types of supramolecular soft matter. In particular, organic–inorganic supramolecular hybrids can achieve soft flexible functions of organic components, keeping the mechanical stability and structural precision originating from the inorganic moieties.

In this chapter, we describe the challenges in the construction of organic–inorganic supramolecular materials. Examples described in this chapter are mainly cited from previous research of one of the authors (Katsuhiko Ariga) as well as related reports from other groups. Topics described in this chapter are categorized according to morphological types such as (i) film-type supramolecular hybrids, (ii) endo-type mesoporous supramolecular hybrids, and (iii) exo-type mesoporous supramolecular hybrids.

3.2 FILM-TYPE SUPRAMOLECULAR HYBRIDS

Fabrication of supramolecular materials into thin films is a useful potential method for integration of supramolecular functions into artificial device structures. Of the various methods available for thin film preparation, the methodology for self-assembled monolayers (SAMs) is an efficient method to directly immobilize supramolecular elements on device surfaces [5,6]. Apart from thiol compounds on gold surfaces, organosilane compounds are often used for the formation of organic–inorganic hybrid structures on the surface of an oxide such as glass in the form of a SAM. The thin-film-type hybrids formed also provide media for immobilization of functional elements, including dyes, proteins, and nucleic acids, on inorganic device structures such as electrodes and field-effect transistors. This feature enables us to use SAM hybrids for material separation and sensing.

Okahata and coworkers demonstrated control of permeation of materials through the thinnest possible lipid film by combining the Langmuir–Blodgett (LB) technique and the SAM method [7,8]. In their method, an organosilane monolayer was first prepared and polymerized at the air–water interface and then transferred onto a solid inorganic porous substrate for immobilization (Fig. 3-1a). Prior to immobilization of the monolayer onto a solid substrate, a monolayer of the organosilane compounds was spread at the air–water interface. The pH condition of the subphase is a crucial factor for the monolayer state, that is, a well-packed monolayer can be formed under acidic conditions (pH 2) because of stabilization of the monolayer by acid-catalyzed Si–O–Si condensation. Temperature dependence of the molecular area at a given surface pressure, such as 20 mN/m, indicated that obvious changes in thermal expansion occurred at around 45°C. This temperature corresponds to the phase transition temperature, from the crystalline phase to the liquid crystalline phase of the polymerized organosilane monolayer, as similarly confirmed by differential scanning calorimetric (DSC) measurement. Mild heat treatment ensured immobilization of the monolayer covalently onto the surface of the porous glass plate (5-nm pore) after LB transfer of the monolayer. Permeation behaviors of water-soluble fluorescent probe molecules (naphthalene derivative) through the monolayer immobilized on

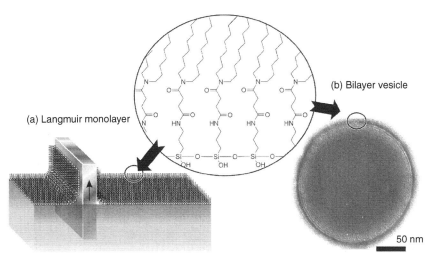

Figure 3-1   Silica-supported lipid assemblies: (a) monolayer-type and (b) vesicle-type (cerasome).

the porous glass plate was evaluated spectroscopically by monitoring changes in fluorescence intensity. Permeation coefficients at different temperatures exhibited a discontinuity at the phase transition temperature of the polymerized monolayer, that is, permeation of the probe molecules through the monolayer is apparently changed between the two phases, crystalline state and liquid crystalline state. This was the first example of permeation control using single monolayers with a thickness of only 2 nm. This corresponded to the thinnest biomembrane-mimicking film operating as a permeation valve on the surface of an inorganic porous material.

The same research group immobilized a similar monolayer onto SnO_2 electrodes by forming covalent linkages through organic and inorganic moieties [9], where permeability of an electrochemical probe ($Fe(CN)_6^{4-}/Fe(CN)_6^{3-}$) through the monolayer was investigated by monitoring changes in the oxidation peak current of the cyclic voltammograms (CVs). The effect of temperature on the permeability of the probe molecule showed a drastic increase of material permeability near the phase transition temperature. Alkyl chain disorder during the unstable coexistence of crystalline and liquid crystalline states of the monolayer probably induced enhanced permeation of probe molecules over a particular temperature range. The permeability of the probe molecule can also be modulated by the addition of some kinds of alcohols. Interestingly, some small alcohols efficiently blocked probe permeation, while bulky alcohols had less effect. Small alcohols are preferentially inserted into the organosilane monolayer. These results suggest the potential use of monolayer-hybrid electrodes for shape-selective discrimination of alcohol guests.

In a more bio-oriented system, Ariga, Hisaeda, and coworkers developed electrodes modified with a vitamin B_{12} derivative with the aid of organosilane monolayers [10]. The heptapropyl and heptaoctyl esters of vitamin B_{12} containing

Co(II) or Co(III) were used for this application and could be stably incorporated into the matrix organosilane monolayer. The monolayer containing vitamin B_{12} derivative was then transferred onto an indium-tin-oxide (ITO) electrode. A CV measurement revealed that this electrode in aqueous solution exhibited a Co(II)/Co(I) redox couple at −0.65 V versus Ag/AgCl, corresponding to that of the vitamin B_{12} derivative dissolved in methanol. Actual material conversion by the function of vitamin B_{12} with the related electrode was demonstrated using thicker organic–inorganic hybrid films obtained by the sol–gel process using the vitamin B_{12} derivative, heptapropyl cobyrinate perchlorate [11]. The controlled-potential electrolysis of benzyl bromide using the sol–gel modified electrode at −1.20 V versus Ag/AgCl in aqueous solution containing 0.1 M KCl afforded dehalogenated products, bibenzyl and toluene, with a total turnover number of >1000 per hour.

Thin-film-type organosilica hybrids are not limited to sheet-like structures. Because such amphiphile structures are capable of forming spherical structures such as micelles and vesicles, cell-membrane-like organic–inorganic hybrid supermolecules can be constructed using the related strategy. Lipid vesicles are known as standard models of a spherical cell membrane, but their limited mechanical stability is often disadvantageous for some kinds of practical applications. If such structures are supported by an inorganic framework through hybridization, they would become more useful materials. Organic–inorganic hybrids with cell-like structures have been investigated using organosilane compounds and/or related materials. Pinnavaia and coworkers reported vesicle-type mesostructured silica materials using gemini-type surfactants as templates with silica precursors [12]. The vesicular silica has a shell structure consisting of one or more undulated silica sheets of about 3-nm thickness. This vesicle-type silica can be used in catalysis and molecular separations. Caruso, Möhwald, and coworkers reported preparation of hollow silica vesicles through layer-by-layer (LbL) assembling techniques using colloidal nanoparticles as a template [13]. Silica particles and organic polyelectrolytes were initially adsorbed on a colloidal core in an LbL manner. Dissolution of the colloidal core resulted in a hollow silica–polymer hybrid vesicle. Calcination of the hybrid vesicles left a hollow vesicle. As Frey and coworkers demonstrated, aggregation of the uncondensed amphiphilic spherosilsesquioxane derivative and cross-linking at high pH led to the formation of liposome-like silica particles [14]. Sommerdijk and coworkers used an amphiphilic block copolymer consisting of hydrophilic poly(ethyleneoxide) blocks and hydrophobic poly(methylphenylsilane) segments for preparation of vesicle aggregates [15]. Their UV-sensitive nature can be used for controlled release systems.

Katagiri et al. did pioneering work on the preparation of organic–inorganic hybrid lipid vesicles as a strengthened cell-membrane model (Fig. 3-1b) [16,17]. Their hybrid vesicles have a siloxane network covalently attached to the bilayer membrane surface. Therefore, they are called cerasome, a term that is derived from combining "cera" (from "ceramics") and "soma." Alkoxysilane-bearing amphiphiles were dispersed in a weakly acidic aqueous solution using a vortex mixer, resulting in a stable dispersion at room temperature. Observation by transmission electron microscopy (TEM) with the aid of hexaammonium hep-tamolybdate tetrahydrate confirmed formation of the vesicular structures, where the multilamellar cerasomes with a bilayer thickness of about 4 nm and vesic-ular diameter of 150 nm were clearly visible. Some cerasome objects formed aggregates that maintained the original spherical vesicle structure without caus-ing collapse and fusion of the cerasome, probably due to the formation of the intra- and intermembrane siloxane network.

This fact suggests possible formation of multicellular mimics using the cerasome strategy. The possibilities are not limited to unintentional aggregation, and subjecting the cerasome structure to LbL techniques can result in multi-cellular mimics in a predesigned fashion. LbL assemblies between the cationic polyelectrolyte (poly(diallyldimethylammonium chloride), PDDA) and anionic vesicles (conventional lipid vesicles and anionic cerasome) on solid supports were investigated using a quartz crystal microbalance (QCM) upon sensitive mass detection on the surface [18]. The LbL assembly between conventional anionic vesicles composed of dihexadecyl phosphate and cationic PDDA on a QCM plate resulted in frequency shifts that corresponded to flat lipid bilayer for-mation on the surface, indicating that the vesicles of dihexadecyl phosphate were collapsed during the LbL assembling process. In contrast, LbL assembly between the prepared anionic cerasome and cationic PDDA exhibited significantly large frequency shifts of the QCM response that were in good agreement with the cerasome deposition in accordance with their spherical structure. Maintenance of the spherical morphology of the assembled structure was further confirmed by atomic force microscopy (AFM). The highly stable nature of organic–inorganic cerasome structures enables this layered multivesicle assembly. By changing the structure of the organosilane amphiphile, smaller cationic cerasomes were sim-ilarly synthesized. Using both the anionic and cationic cerasomes, direct LbL assembly of cerasome structures in the absence of polyelectrolyte counterions became possible [19]. The presence of closely packed cerasome particles, similar to a stone pavement, in both layers was clearly confirmed by AFM observa-tions. These assembled structures can be recognized as a multicellular mimic, and further functionalization of the cerasome surface with functional bio units such as enzymes and antibodies potentially creates various kinds of biomimetic nanohybrids.

Because the LbL method is useful for assembling various nanocomponents, including organic polymers, biomaterials, molecular assemblies, and inorganic substances [20–24], various kinds of film-type organic–inorganic supramolecu-lar hybrids can be constructed using this technique. Typical examples of LbL films containing inorganic building blocks are obtained from the assembly processes of nanoparticles such as silica nanoparticles and counterionic polyelectrolytes. With these assemblies, thin films containing inorganic nanoparticles can be easily obtained. Appropriate selection of solid supports for the LbL assembly enables us to prepare freestanding films through selective etching of the solid support (sac-rificial layer). For example, Tsukruk and coworkers demonstrated the fabrication of freely suspended, multilayered nanocomposite membranes containing gold nanoparticles [25]. The LbL films of organic–inorganic hybrids are even useful for more advanced microfabrication. Lvov and coworkers reported the fabrication of microcantilevers consisting of clay/polymer nanocomposites, where sequenced procedures including patterning, phototreatment, etching, and LbL assembly were thoughtfully combined [26].

3.3   ENDO-TYPE MESOPOROUS SUPRAMOLECULAR HYBRIDS

Recently, much attention has been paid to mesoporous materials such as mesoporous silica and mesoporous carbon [27–29]. They have regular pores and huge surface area and pore volumes. The latter features are especially attractive for some kinds of applications such as catalysis, materials separation, and drug release. In addition, the well-designed size and dimension of the pores can be attractive media for immobilization of organic supramolecular structures in confined environments. Therefore, hybrids between inorganic mesoporous materials and organic functional substances will became large targets in the related research areas. In this chapter, we classify these hybrids into two categories, endo-type mesoporous supramolecular hybrids and exo-type mesoporous supramolecular hybrids. The former assemble organic supramolecular structures within inorganic mesopore channels, while the latter assemble organic components at the external surfaces of inorganic mesoporous materials.

Because mesoporous silica materials have reactive silanol groups, introduction of functional groups at the interiors or exteriors of their pores is possible [30]. Modification of mesoporous silica at its pore inlet with organic functional groups leads to mesoporous materials capable of controlled release of materials from the mesopore interior to the outside. For example, Fujiwara *et al.* prepared MCM-41 mesoporous silica functionalized with photoactive coumarin, which was grafted only at the pore outlet. Coumarin moieties were dimerized by irradiation with UV light [31], causing stable trapping of the guest drug (cholestane) in the mesopores. The dimerized coumarin was cleaved on irradiation at a different UV wavelength, resulting in the release of the trapped cholestane. A similar approach involved the use of various stimuli-responsive functional groups, including supramolecular complexes, polymer chains, and colloidal stoppers [32].

In addition to pore inlet modification, introduction of organic functions at the surfaces of the pore interior has been a research target. One of the unique approaches is shown in Fig. 3-2a, where dense functionalization at the pore interior and high accessibility of external guests were achieved using a condensable amphiphile with a cleavable alkyl tail as a template surfactant [33]. A dialkoxysilane functionality as a part of the head group makes covalent connection with silica at the internal surfaces of mesopores on sol–gel reactions with tetraethyl orthosilicate, resulting in mesoporous silica channels that are filled with organic units. Cleavage and removal of the alkyl tail by selective hydrolysis of the ester at the C-terminal leaves open pores with surfaces covalently functionalized by alanine residues. Because of the head biting and tail removal, this strategy is called the lizard template method. Regular mesoporous structures

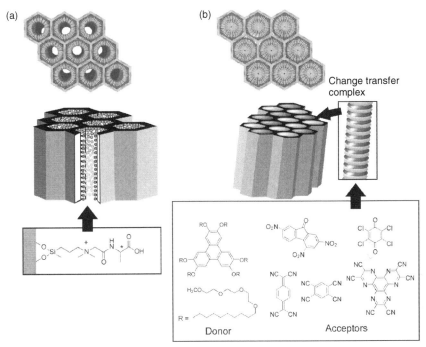

Figure 3-2 Endo-type mesoporous supramolecular hybrids: (a) covalently immobilized peptide assembly on mesopore and (b) charge-transfer complex column in mesopore.

functionalized by amino acid groups were confirmed with X-ray diffraction (XRD), TEM, Fourier transform infrared spectroscopy (FT-IR), and thermogravimetric analysis (TGA). Temperature-programmed desorption (TPD) analysis with NH_3 as a basic guest also confirmed the exposure of the alanine C-terminal in the silicate channel. The immobilization of biological residues by this strategy could be used for separation and sensing of biological substances. For example, reactor applications of these mesoporous hybrids were demonstrated by Aida and coworkers, who demonstrated the catalytic capability of unhydrolyzed materials on the acetalization of ketones, such as cyclohexanone, in ethanol under mild conditions [34]. Its amphiphilic core–shell architecture of the immobilized rod-like micelle within silica channels provides a medium that can simultaneously incorporate hydrophobic and hydrophilic reactants. These reactants can be activated by hydrogen bonding interactions with the peptidic functionalities located at the core–shell interface.

Instead of covalent hybridization, confining organic functional moieties within size-defined mesopores should result in unique organic–inorganic supramolecular hybrid materials, because organic functional substances often show unusual properties at size- and motion-restricted media. For example, Schwartz and coworkers incorporated a poly(phenylenevinylene) derivative from solution into a calcined mesoporous silica and demonstrated that energy migration along the incorporated polymer is slower than Förster energy transfer between the polymer chains [35]. Ozin and coworkers reported ring-opening polymerization of [1]silaferrocenophane in MCM-41 to provide poly(ferrocenylsilanes) [36]. Pyrolysis of the composite resulted in the formation of iron nanoparticles with superparamagnetic properties. These postloading approaches may include uncertainty in the presence of organic moieties within mesopores. In order to avoid such problems, Aida and Tajima used hexadecadiynyltrimethylammonium bromide as a template to synthesize mesoporous silica containing poly(diacetylene) as microfibers [37]. Red-shifted features of the electronic spectra of the silica–polymer hybrids suggest confinement of the poly(diacetylene) and an elongated effective conjugation within the silica nanochannel. The same research group similarly developed poly(pyrrole)-containing mesoporous silica films [38]. The polypyrrole chains are highly constrained and insulated when incorporated within hexagonal nanoscopic channels, where recombination of polarons into bipolarons is significantly suppressed. In contrast, the two-dimensional lamellar phase affords spatial freedom for electron recombination.

This approach is not limited to the confinement of polymer chains within mesopore channels. Confinement of supramolecular assemblies within size-defined mesopore channels also provides attractive organic–inorganic supramolecular hybrids. Aida and coworkers reported the immobilization of a one-dimensional columnar charge-transfer (CT) assembly in a mesoporous silica film by the sol–gel reaction with CT complexes of an amphiphilic triphenylene donor and acceptors (Fig. 3-2b) [39]. The films obtained are basically transparent, and their colors can be tuned by appropriate selection of acceptors. As a confinement effect, an extension of the apparent conjugation length was suggested from their spectral shifts. In addition, confining CT columns within the mesopore channel significantly increased their stability, that is, neither solvatochromism nor guest exchange activity was observed. For these hybrid structures, tunable photoconductive properties are also expected, because the donor/acceptor molar ratio can be varied over a wide range from 1:1 to 9:1.

Confinement of biomolecular assemblies was also investigated by confining assemblies of the peptide segments within highly organized mesopores to prepare a novel bio-silica hybrid, called proteosilica [40,41]. For preparation of proteosilioca, tetramethyl orthosilicate in aqueous methanol was first partially gelated in the presence of an appropriate amount of HCl as acid catalyst and then further reacted after the addition of the amphiphilic peptides. Spin coating of the transparent solution obtained on a cover glass resulted in the formation of proteosilica as a transparent film. The peptide assembly of proteosilica provides a chiral environment where photochromic dye molecules such as spiropyran can be doped. Asymmetric photoreaction of spiropyran dopant was demonstrated to show reversible circular dichromic activities. Such hybrid systems can be applied in devices for nondestructive memory recording and reading.

In a unique approach for endo-type mesoporous supramolecular hybrids, Caruso and coworkers have developed bioinorganic hybrids using LbL assembly within and at the surface of mesopores [42–44]. They first immobilized various proteins (catalase, peroxidase, cytochrome c, lysozyme, transferrin, urease, and bovine serum albumin (BSA)) within mesoporous channels of bimodal meso-porous silica (BMS) spheres followed by LbL assembly to cover the BMS sphere surface with a multilayer shell that was assembled to encapsulate the protein in the nanoporous particles. Destruction of the mesoporous sphere results in hollow capsules that included proteins that could also be used as biomimetic reactors. As a more advanced attempt, polyelectrolytes have been infiltrated into the meso-pores of silica spheres preloaded with proteins to cross-link these biopolymers. In this case, removal of the mesoporous silica template leads to self-standing nanoporous protein-based particles.

3.4 EXO-TYPE MESOPOROUS SUPRAMOLECULAR HYBRIDS

In contrast to the above-mentioned approaches, mesoporous substances themselves can be assembled within supramolecular assemblies using organic binders and connectors to provide hierarchic organic–inorganic supramolecular hybrids. This kind of mesoporous hybrids can be recognized as exo-type mesoporous supramolecular hybrids. Use of mesostructured materials as film components in the LbL assembly can result in the preparation of functional hierarchic organic–inorganic nanohybrids. Because the LbL technique is based mainly on electrostatic interactions, this method is widely applicable to charged inorganic substances. For example, LbL films of mesoporous carbon can have controlled numbers of layers, within which well-defined carbon nanospaces are extended. Such structures can be used for adsorption and sensing of particular guest molecules and their assembly based on their size matching to carbon mesopores [45]. For this purpose, surface-oxidized mesoporous carbon (CMK-3) as a charged component for LbL assembly was synthesized [46]. LbL films of charged CMK-3 and polyelectrolyte prepared on a QCM plate were subjected to *in situ* measurement of nonionic aqueous guests (Fig. 3-3). The prepared mesoporous carbon LbL films exhibited novel features, including

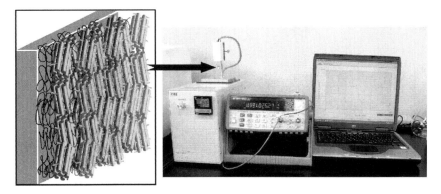

Figure 3-3 LbL assembly of mesoporous carbon for sensor application.

highly selective sensing size matching between guests and carbon nanospace and serendipitous discovery of the capillary-condensation-like highly cooperative adsorption in the liquid phase.

Another type of hierarchic mesoporous structure can be constructed by LbL assembly of mesoporous silica capsules with the aid of coassembled silica nanoparticles and counterionic polyelectrolytes. Mesoporous silica capsules have hierarchic pore structures, that is, both an empty core and small mesopores at the silica wall. The assembled LbL films, called mesoporous compartment films [47,48], have one more hierarchic structure, regularly layered structures. Therefore, hierarchic silica capsule structures with the main cavity inside and the mesopore channels at the walls are further assembled into layered structures in a single mesoporous compartment film (Fig. 3-4a). The nanocompartment films obtained can stably accommodate liquid guest substances such as water and liquid fragrances. Interestingly, release of the trapped guest molecules follows the stepped release mode because of the nonequilibrated concurrent evaporation of material from the pore channels to the exterior and capillary penetration from the interior into the mesoporous channels. The water release profiles can be tuned by various structural parameters of the films, such as the number of layers and sizes of the coadduct particles, ambient temperature, and physical properties of the guest species. This can be regarded as a stimulus-free controlled release system that was hardly observed in any other currently available controlled release systems. This new system is of great utility for the development of clean stimulus-free controlled drug release applications.

Modification of procedures in mesoporous silica capsule also provides carbon-based mesoporous capsules. LbL assembly of mesoporous carbon capsule with polyelectrolyte was similarly conducted by precoating of noncharged carbon capsule with charged surfactants (Fig. 3-4b) [49]. The LbL film of dual-pore carbon capsules obtained exhibited excellent adsorption capabilities for volatile guests, such as aromatic hydrocarbons. Enhanced adsorption of the aromatic guests, as compared with aliphatic guests, into the layered mesoporous carbon film, might originate from strong $\pi-\pi$ interactions. Interestingly, impregnation

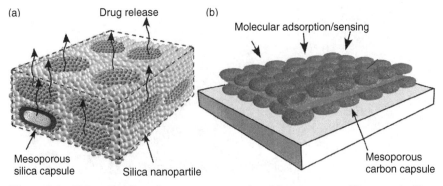

Figure 3-4 LbL assemblies of mesoporous capsules: (a) mesoporous silica capsule film for drug release and (b) mesoporous carbon capsule film for material adsorption/sensing.

of additional recognition components into the carbon capsules permits further control over adsorption selectivity between aromatic and nonaromatic substances and between acids and bases. Therefore, the layered carbon capsule films will find application in sensing and separation science because of their designable selectivity.

3.5 CONCLUSIONS

In this chapter, various types of organic–inorganic supramolecular hybrids, mainly including film-type supramolecular hybrids, endo-type mesoporous supramolecular hybrids, and exo-type mesoporous supramolecular hybrids, are briefly described. Concepts on how to fabricate these hybrids would be effective to construct nanotechnological devices using supramolecular soft matter. Judging from the examples described in this chapter, design and control of interfacial structures are crucial keys to construct organic–inorganic supramolecular hybrids with high functions. Just mixing organic and inorganic components is not a successful fabrication method for high-performance hybrids because fine control of interfacial structure is necessary. Therefore, methodologies and concepts regarding design and fabrication of interfacial structures in supermolecular hybrids is critically important for future technologies as well as in the development of analytical methods of such interfacial structures.

ACKNOWLEDGMENT

This work was partly supported by World Premier International Research Center Initiative (WPI Initiative), MEXT, Japan, and Core Research for Evolutional Science and Technology (CREST) program of Japan Science and Technology Agency (JST), Japan. Q. J. is especially grateful for support from the Japan Society for the Promotion of Science (JSPS) fellowship program.

REFERENCES

1. Lundstrom, M. (2003). Moore's law forever? *Science*, *299*, 210–211.
2. Thompson, S. E., Parthasarathy, S. (2006). Moore's law: the future of Si microelectronics. *Mater. Today*, *9*, 20–25.
3. Ariga, K., Nakanishi, T., Hill, J. P. (2007). Self-assembled microstructures of functional molecules. *Curr. Opin. Colloid Interface Sci.*, *12*, 106–120.
4. Ariga, K., Hill, J. P., Lee, M. V., Vinu, A., Charvet, R., Acharya, S. (2008). Challenges and breakthroughs in recent research on self-assembly. *Sci. Technol. Adv. Mater.*, *9*, 014109.
5. Yang, L., Lua, Y.-Y., Lee, M. V., Linford, M. R. (2005). Chemomechanical functionalization and patterning of silicon. *Acc. Chem. Res.*, *38*, 933–942.
6. DiBenedetto, S. A., Facchetti, A., Ratner, M. A., Marks, T. J. (2009). Molecular self-assembled monolayers and multilayers for organic and unconventional inorganic thin-film transistor applications. *Adv. Mater.*, *21*, 1407–1433.

7. Okahata, Y., Ariga, K., Nakahara, H., Fukuda, K. (1986). Permeation control by a phase transition of the dialkylsilane monolayer immobilized on a porous glass plate. *J. Chem. Soc., Chem. Commun.*, 1069–1071.

8. Ariga, K., Okahata, Y. (1989). Polymerized monolayers of single-, double-, and triple-chain silane amphiphiles and permeation control through the monolayer-immobilized porous glass plate in an aqueous solution. *J. Am. Chem. Soc.*, *111*, 5618–5622.

9. Okahata, Y., Yokobori, M., Ebara, Y., Ebato, H., Ariga, K. (1990). Electrochemical properties of covalently bonded silane amphiphile monolayers on a tin dioxide electrode. *Langmuir*, *6*, 1148–1153.

10. Ariga, K., Tanaka, K., Katagiri, K., Kikuchi, J., Shimakoshi, H., Ohshima, E., Hisaeda, Y. (2001). Langmuir monolayer of organoalkoxysilane for vitamin B_{12}-modified electrode. *Phys. Chem. Chem. Phys.*, *3*, 3442–3446.

11. Shimakoshi, H., Nakazato, A., Tokunaga, M., Katagiri, K., Ariga, K., Kikuchi, J., Hisaeda, Y. (2003). Preparation of a sol-gel modified electrode trapped with a vitamin B_{12} derivative and its photoelectrochemical reactivity. *Dalton Trans.*, 2308–2312.

12. Kim, S. S., Zhang, W., Pinnavaia, T. J. (1998). Ultrastable mesostructured silica vesicles. *Science*, *282*, 1302–1305.

13. Caruso, F., Caruso, R. A., Möhwald, H. (1998). Nanoengineering of inorganic and hybrid hollow spheres by colloidal templating. *Science*, *282*, 1111–1114.

14. Knischka, R., Dietsche, F., Hanselmann, R., Frey, H., Mulhaupt, R. (1999). Silsesquioxane-based amphiphiles. *Langmuir*, *15*, 4752–4756.

15. Kros, A., Jansen, J. A., Holder, S. J., Nolte, R. J. M., Sommerdijk, N. A. J. M. (2002). Silane-based hybrids for biomedical applications. *J. Adhes. Sci. Technol.*, *16*, 143–155.

16. Katagiri, K., Ariga, K., Kikuchi, J. (1999). Preparation of organic-inorganic hybrid vesicle "cerasome" derived from artificial lipid with alkoxysilyl head. *Chem. Lett.*, 661–662.

17. Katagiri, K., Hashizume, M., Ariga, K., Terashima, T., Kikuchi, J. (2007). Preparation and characterization of a novel organic-inorganic nanohybrid "cerasome" formed with liposomal membrane and silicate surface. *Chem.—Eur. J.*, *13*, 5272–5281.

18. Katagiri, K., Hamasaki, R., Ariga, K., Kikuchi, J. (2002). Layer-by-layer self-assembling of liposomal nanohybrid "cerasome" on substrates. *Langmuir*, *18*, 6709–6711.

19. Katagiri, K., Hamasaki, R., Ariga, K., Kikuchi, J. (2002). Layered paving of vesicular nanoparticles formed with cerasome as a bioinspired organic-inorganic hybrid. *J. Am. Chem. Soc.*, *124*, 7892–7893.

20. Ariga, K., Hill, J. P., Ji, Q. (2007). Layer-by-layer assembly as a versatile bottom-up nanofabrication technique for exploratory research and realistic application. *Phys. Chem. Chem. Phys.*, *9*, 2319–2340.

21. Srivastava, S., Kotov, N. A. (2008). Composite layer-by-layer (LBL) assembly with inorganic nanoparticles and nanowires. *Acc. Chem. Res.*, *41*, 1831–1841.

22. Ariga, K., Hill, J. P., Ji, Q. (2008). Biomaterials and biofunctionality in layered macromolecular assemblies. *Macromol. Biosci.*, *8*, 981–990.

23. De Geest, B. G., De Koker, S., Sukhorukov, G. B., Kreft, O., Parak, W. J., Skirtach, A. G., Demeester, J., De Smedt, S. C., Hennink, W. E. (2009). Polyelectrolyte microcapsules for biomedical applications. *Soft Matter*, *5*, 282–291.

24. Ariga, K., Ji, Q., Hill, J. P., Kawazoe, N., Chen, G. (2009). Supramolecular approaches to biological therapy. *Expert Opin. Biol. Ther.*, *9*, 307–320.

25. Jiang, C., Tsukruk, V. V. (2006). Freestanding nanostructures via layer-by-layer assembly. *Adv. Mater.*, *18*, 829–840.

26. Hua, F., Cui, T., Lvov, Y. M. (2004). Ultrathin cantilevers based on polymer-ceramic nanocomposite assembled through layer-by-layer adsorption. *Nano Lett.*, *4*, 823–825.

27. Vinu, A., Mori, T., Ariga, K. (2006). New families of mesoporous materials. *Scie. Technol. Adv. Mater.*, *7*, 753–771.

28. Manzano, M., Colilla, M., Vallet-Regi, M. (2009). Drug delivery from ordered mesoporous matrices. *Expert Opin. Drug Delivery*, *6*, 1383–1400.

29. Yang, Q., Liu, J., Zhang, L., Li, C. (2009). Functionalized periodic mesoporous organosilicas for catalysis. *J. Mater. Chem.*, *19*, 1945–1955.

30. Vinu, A., Hossain, K. Z., Ariga, K. (2005). Recent advances in functionalization of mesoporous silica. *J. Nanosci. Nanotechnol.*, *5*, 347–371.
31. Mal, N. K., Fujiwara, M., Tanaka, Y. (2003). Photocontrolled reversible release of guest molecules from coumarin-modified mesoporous silica. *Nature*, *421*, 350–353.
32. Ariga, K., Vinu, A., Hill, J. P., Mori, T. (2007). Coordination chemistry and supramolecular chemistry in mesoporous nanospace. *Coord. Chem. Rev.*, *251*, 2562–2591.
33. Zhang, Q., Ariga, K., Okabe, A., Aida, T. (2004). Condensable amphiphile with cleavable tail as 'lizard' template for sol-gel synthesis of functionalized mesoporous silica. *J. Am. Chem. Soc.*, *126*, 988–989.
34. Otani, W., Kinbara, K., Zhang, Q., Ariga, K., Aida, T. (2007). Catalysis of a peptidic micellar assembly covalently immobilized within mesoporous silica channels: importance of amphiphilic spatial design. *Chem.—Eur. J.*, *13*, 1731–1736.
35. Nguyen, T.-Q., Wu, J., Doan, V., Schwartz, B. J., Tolbert, S. H. (2000). Control of energy transfer in oriented conjugated polymer-mesoporous silica composites. *Science*, *288*, 652–656.
36. MacLachlan, M. J., Ginzburg, M., Coombs, N., Raju, N. P., Greedan, J. E., Ozin, G. A., Manners, I. (2000). Superparamagnetic ceramic nanocomposites: synthesis and pyrolysis of ring-opened poly(ferrocenylsilanes) inside periodic mesoporous silica. *J. Am. Chem. Soc.*, *122*, 3878–3891.
37. Aida, T., Tajima, K. (2001). Photoluminescent silicate microsticks containing aligned nanodomains of conjugated polymers by sol-gel-based *in situ* polymerization. *Angew. Chem. Int. Ed.*, *40*, 3803–3806.
38. Ikegame, M., Tajima, K., Aida, T. (2003). Template synthesis of polypyrrole nanofibers insulated within one-dimensional silicate channels: Hexagonal versus lamellar for recombination of polarons into bipolarons. *Angew. Chem. Int. Ed.*, *42*, 2154–2157.
39. Okabe, A., Fukushima, T., Ariga, K., Aida, T. (2002). Color-tunable transparent mesoporous silica films: immobilization of one-dimensional columnar charge-transfer assemblies in aligned silicate nanochannels. *Angew. Chem. Int. Ed.*, *41*, 3414–3417.
40. Ariga, K. (2004). Silica-supported biomimetic membranes. *Chem. Rec.*, *3*, 297–307.
41. Ariga, K., Aimiya, T., Zhang, Q., Okabe, A., Niki, M., Aida, T. (2002). "Proteosilica" a novel nanocomposite with peptide assemblies in silica nanospace: photoisomerization of spiropyran doped in chiral environment. *Int. J. Nanosci.*, *1*, 521–525.
42. Wang, Y., Caruso, F. (2004). Enzyme encapsulation in nanoporous silica spheres. *Chem. Commun.*, 1528–1529.
43. Wang, Y., Caruso, F. (2005). Mesoporous silica spheres as supports for enzyme immobilization and encapsulation. *Chem. Mater.*, *17*, 953–961.
44. Wang, Y., Caruso, F. (2006). Nanoporous protein particles through templating mesoporous silica spheres. *Adv. Mater.*, *18*, 795–800.
45. Ariga, K., Vinu, A., Ji, Q., Ohmori, O., Hill, J. P., Acharya, S., Koke, J., Shiratori, S. (2008). A layered mesoporous carbon sensor based on nanopore-filling cooperative adsorption in the liquid phase. *Angew. Chem. Int. Ed.*, *47*, 7254–7257.
46. Vinu, A., Hossian, K. Z., Srinivasu, P., Miyahara, M., Anandan, S., Gokulakrishnan, N., Mori, T., Ariga, K., Balasubramanian, V. V. (2007). Carboxy-mesoporous carbon and its excellent adsorption capability for proteins. *J. Mater. Chem.*, *17*, 1819–1825.
47. Ji, Q., Miyahara, M., Hill, J. P., Acharya, S., Vinu, A., Yoon, S. B., Yu, J.-S., Sakamoto, K., Ariga, K. (2008). Stimuli-free auto-modulated material release from mesoporous nanocompartment films. *J. Am. Chem. Soc.*, *130*, 2376–2377.
48. Ji, Q., Acharya, S., Hill, J. P., Vinu, A., Yoon, S. B., Yu, J.-S., Sakamoto, K., Ariga, K. (2009). Hierarchic nanostructure for auto-modulation of material release: mesoporous nanocompartment films. *Adv. Funct. Mater.*, *19*, 1792–1799.
49. Ji, Q., Yoon, S. B., Hill, J. P., Vinu, A., Yu, J.-S., Ariga, K. (2009). Layer-by-layer films of dual-pore carbon capsules with designable selectivity of gas adsorption. *J. Am. Chem. Soc.*, *131*, 4220–4221.

STIMULI RESPONSIVE DYE ORGANIZED SOFT MATERIALS

CHAPTER 4

FUNCTIONAL MATERIALS FROM SUPRAMOLECULAR AZOBENZENE DYE ARCHITECTURES

Charl F. J. Faul

School of Chemistry, University of Bristol, Bristol, UK
Bristol Centre for Nanoscience and Quantum Information, University of Bristol, Bristol, UK

4.1 INTRODUCTION

Research in supramolecular chemistry, and the closely related field of nano-technology, has, since Lehn, Pederson, and Cram received the Nobel Prize in 1987, reached a level of maturity where a number of questions are directed at researchers active in the area. These questions have their origin in the many sweeping statements made in published papers claiming a plethora of possible applications of their materials, which have mainly been left unanswered. In a recent paper, and closely followed by a Faraday Discussion Meeting (on soft nanotechnology), the questions were formalized by Ozin and Cademartiri [1] and Whitesides and Lipomi [2]. In his concluding remarks, Whitesides challenged researchers to move away from synthesis and structure and focus their vision and efforts toward the areas of function and application and the grand challenges currently facing scientists internationally.

It is within this above-outlined context that this chapter on supramolecular azobenzene dye architectures is written. In-depth reviews have been written at the beginning of the 2000s concerning optical properties of photochromic materials [3], simple polymer host–guest, and covalently modified polymer-based azobenzene systems [4]. More recently, Barrett *et al.* reviewed the photomechanical effects in azobenzene-containing soft materials [5], whereas Evans *et al.* reviewed (polymeric) photoresponsive systems and biomaterials [6]. It is therefore not the aim of this chapter to provide an in-depth treatment of these areas of

Supramolecular Soft Matter: Applications in Materials and Organic Electronics, First Edition.
Edited by Takashi Nakanishi.
© 2011 John Wiley & Sons, Inc. Published 2011 by John Wiley & Sons, Inc.

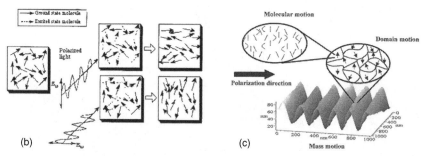

Scheme 4-1 (a) Generalised structure and photoactivity of an azobenzene dye. (b) Optical generation of anisotropy. *Source*: Reprinted with permission from Ref. [3]. Copyright 2000 American Chemical Society. (c) Overview of the three levels of motion induced by light. *Source:* Reprinted with permission from Ref. [4]. Copyright 2002 American Chemical Society.

very high background relevance, but rather refer the reader to these (and many other) reviews for in-depth treatments of those subject areas.

The scope of this chapter, as reflected in the title, is to explore the developments in the new area of supramolecular azobenzene dye architectures that have emerged, mainly since 2005. After a brief discussion of various supramolecular strategies currently available and the most important issues around the azobenzene functionality relevant to this discussion, the subject is divided into two widely used approaches for the production of such dye architectures, namely, (i) ionic self-assembled and (ii) hydrogen-bonded supramolecular approaches. These discussions are followed by a short summary, conclusions, and future areas of interest.

4.2   AZOBENZENE DYES FOR FUNCTIONAL MATERIALS

Azobenzene dyes are well-known chromophores with a long history in the chemical industry. They contain an azo group ($-N=N-$) as a linker between two (substituted) phenyl groups (Scheme 4-1a). Azobenzenes show strong absorption in the UV and visible spectrum owing to the presence of the π-conjugated system, which can easily be tuned through changes in the substitution patterns on the two phenyl rings. Azobenzenes are typically divided into three classes, as suggested by Rao [7], according to the relative energies of their $n-\pi^*$ and

$\pi - \pi^*$ transitions: (i) azobenzene-type molecules (i.e., similar in properties to unsubstituted azobenzenes) with a weak $n - \pi^*$ band in the visible region and a strong $\pi - \pi^*$ band in the UV region, (ii) aminoazobenzene-type molecules (ortho- or para-substituted with an electron-donating moiety) with overlapping $n - \pi^*$ and $\pi - \pi^*$ bands, and (iii) pseudo-stilbene-type molecules (with electron-withdrawing and electron-donating groups at the 4 and 4' positions), where the order of the energies of the transitions is reversed.

However, by far the most interesting aspect of azobenzene moieties is the possibility of reversibly addressing the molecular conformation between the trans (E) and cis (Z) states using a light source, so-called photoisomerization (Scheme 4-1a). The exact mechanism of this photoinduced molecular motion, and the consequent change in molecular structure, has not been conclusively finalized, and two mechanisms (rotation or inversion mechanisms or a combination thereof) have been proposed [8] and intensely investigated.

However, the most important factor to consider here is the fact that this molecular motion has a number of effects [4] on at least three different length scales:

The first is on the molecular level, as described above. Of particular interest is the fact that when the photoisomerization is carried out under illumination with polarized light, molecular motion will take place only as long as the azobenzene chromophore's transition dipole moment has a component parallel to the polarization of the light. The consequence of such illumination is that the chromophores will undergo molecular motion until they are aligned perpendicular to the direction of polarization (i.e., this orientation will be enriched relative to others), thus leading to the induction of anisotropic optical properties in the materials (Scheme 4-1b). Photoinduced anisotropy (PIA), or the induction of dichroic and birefringent properties in materials after irradiation by polarized light, has been known since the early 1920s (the so-called Weigert effect [9,10]). In the case of azobenzene dyes, this effect was observed for the first time in the late 1950s [11] and is exploited in the forms of photoinduced birefringence and dichroism for optical data storage, optical waveguides, and other optical elements.

On the second level, which corresponds to the domain or nanoscale level, motion in domains (liquid-crystalline or crystalline) is observed. Natansohn and Rochon state that the chromophore is required to be bound to the polymer matrix or be part of tightly organized structures (e.g., liquid-crystalline, Langmuir–Blodgett, or monolayer films). Although, at the time of writing, only covalent and simple (mixed) guest–host systems were known, supramolecular interactions would fall into the category of "tightly bound," as is discussed later in this chapter.

The third level involves motion of the chromophore-containing substrate on a macroscopic scale and can be described as a photomechanical effect (Scheme 4-1c). This motion produces all-optical surface patterns on the micrometer scale and was only discovered relatively recently [12,13]. This effect is extensively exploited in the production of surface relief gratings (SRGs) for holographic storage, optical filters, subwavelength gratings, and resonant couplers.

From the above short discussion, the usefulness and potential for application of the azobenzene chromophore should hopefully be obvious. The question that remains is why new approaches and strategies for producing chromophore-containing materials would be necessary. From a practical point of view, tedious organic synthesis (e.g., for chromophores covalently bound to a polymeric backbone) and undesired chromophore aggregation are two important issues that need to be addressed—new routes to alleviate/avoid these issues would be welcomed by researchers (who might not necessarily be organic synthetic chemists) and those interested in finding commercial applications based on these materials. These, and a number of further issues, are discussed in the following section on supramolecular approaches to functional materials production.

4.3   STRATEGIES FOR THE PRODUCTION OF FUNCTIONAL SUPRAMOLECULAR MATERIALS

The production of supramolecular materials has traditionally been based on the following approaches: (i) hydrogen-bonded assemblies (low-molecular-weight [14] as well as high-molecular-weight assemblies [15]), (ii) metal coordination, (iii) electrostatic/ionic interactions (e.g., layer-by-layer (LbL) [16] or ionic self-assembly [17] approaches), (iv) stacking interactions [18], and (v) van der Waals weak noncovalent interactions. The general advantages of these approaches, when compared to traditional guest–host and covalent strategies, are that they provide new routes for the production of nanostructured functional materials that

- do not (necessarily) require complicated organic synthetic efforts; cost-effectiveness and low complexity are therefore underlying driving factors;
- provide possibilities to tune phase behavior, aggregation behavior, and active component (e.g., azobenzene dye) content in a facile manner;
- allow for the choice and consequent exploitation of either cooperative or noncooperative aggregation behavior;
- provide for the facile inclusion of functional components (e.g., azobenzene dyes or even combinations with other functional moieties) into supramolecular architectures for applications;
- enable the inclusion and tuning of dynamic behavior, thus providing a measure of control over the temporal and spatial stability of such materials according to the required application;
- provide opportunities for reproducible and scalable production of larger quantities of materials for device fabrication.

At the moment, it seems as if not any one specific strategy would provide access to all of these advantages at the same time. However, two of these strategies seem to show the most promise and are critically discussed below in terms of their specific advantages and disadvantages for the production of functional supramolecular azobenzene dye architectures. In both cases, a number of possible applications are envisaged in the areas of optics and communication (wave guides, signal processing, frequency conversion), data storage, and liquid crystal (LC) technologies (as LC alignment layers).

4.4   IONIC SELF-ASSEMBLY

4.4.1   Polyelectrolyte-Based Materials

Although the use of ionic interactions to generate materials was already reported in 1955 by Mysels *et al.* (during investigations into the determination of critical micelle concentrations) [19], it was not until 1994 that this strategy became more widely publicized and used for the generation of nanostructured materials [20]. This was followed by an intense period of activity by a number of researchers, covering areas as diverse as polypeptide–surfactant [21], (also for drug delivery [22]), DNA–surfactant [23], polysiloxane–fluorosurfactants (for low-surface-energy applications) [24], and polymer-hexabenzocoronenes (for organic light-emitting diode (OLED) applications) [25]. In 2000, a study was published on the use of a polyelectrolyte with a diazosulfonate side chain and fluorinated surfactants. However, no mention was made of any optical characterization of any kind. In 2002, Thünemann published an extensive review covering these (and many more) aspects of early polyelectrolyte–surfactant research activities [26].

It is, however, interesting to note that Wegner *et al.* already published a study in 1991 (which has gone largely unnoticed in the literature) on the production of a polyelectrolyte glass with linear and nonlinear optical properties [27]. They made use of a statistical copolymer polycation and the azobenzene dye ethyl orange (EO) to prepare noncrystalline films of high optical quality. Only the linear and nonlinear optical properties of these materials were investigated, and the unique properties of the azobenzene chromophore were not explored in any particular way. It is useful to note that macroscopic photoinduced motion (e.g., production of SRGs) was not reported until 1995 [12,13].

The first instance the author could find of ionically self-assembled polyelectrolyte–azobenzene materials being investigated specifically for their optical functionality was the investigation by Stumpe *et al.* [28], followed closely by the published study of Lu *et al.* [29]. Both these followed the studies on low-molecular-weight materials published by Faul and Stumpe in 2006 [30]. The motivation for the first investigation was the previously reported low efficiencies of SRG inscription in an LbL approach [31], as well as the very time-consuming preparation steps required for the LbL approach. Stumpe *et al.* varied the azobenzene dye structure and the polyelectrolytes charge type and densities in order to investigate different strengths of ionic interaction (Scheme 4-2, poly(diallyldimethylammonium chloride) (**pDADMAC**), poly(ethylene imine) (**PEI**), dyes **A**, **B**, and **C**). Induced optical anisotropy was found to be unstable and relaxed according to an exponential decay function. SRG recording, on the other hand, was far more successful in most cases, with grating heights of up to 1.8 μm recorded with high efficiencies. More importantly, these inscribed gratings could be overwritten (for more complex structures), as well as erased, while at the same time exhibiting very high thermal stability (up to 150°C).

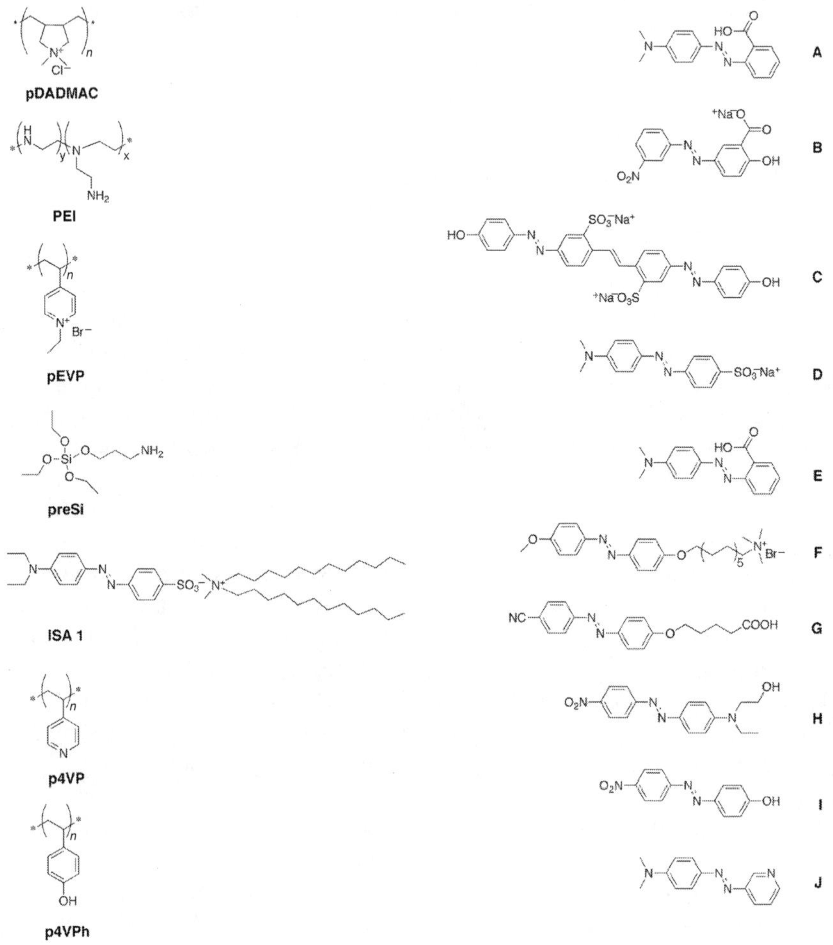

Scheme 4-2 Structures of components of various supramolecular dye complexes.

No mention is made of any supramolecular order in the films, as produced from solution casting methods.

Lu *et al.* reported [29] on the very simple system of a poly(ethylvinylpyridinium) (**pEVP**) bromide polyelectrolyte and methyl orange (Scheme 4-2, **pEVP**, dye **D**). The authors clearly showed the presence of a lamellar mesophase from their X-ray investigations. Photoinduced birefringence was, after a slight initial decay, stable. They also ascertained that anisotropic surface arrangements could be induced for LC alignment (a so-called command surface) at low irradiation doses (3.6 mJ/cm^2), with films thermally stable up to 120°C. At higher irradiation doses, periodic structures were induced on the surface, which also acted as LC alignment layers. No mention is made in either of the two studies of the influence of humidity and moisture uptake into the prepared films.

After the initial report, a number of papers by Stumpe/Goldenberg *et al.*, Bazuin *et al.*, and various other groups appeared, utilizing the ionic self-assembly strategy with various modifications. After the initial publications in 2007, Stumpe's group published a number of papers in quick succession, all focusing on various aspects of the generation of microstructures/SRGs. They proposed an innovative approach by combining well-known sol–gel chemistry precursors and photoaddressable azobenzene dyes to produce an *in situ* cross-linkable photoactive material, where no aggregation of the chromophores was observed (Scheme 4-2, **preSi**) [32]. The conditions had to be tuned so that no or minimal cross-linking took place before the SRG structures were induced, as polysiloxane network formation would certainly hinder any photoinduced mass transport. This phenomenon was, at that stage, not fully characterised or quantified. However, under the right conditions and correct choice of azobenzene dye molecular structure, they were successful in producing efficient grating structures that also exhibited temporal and thermal stability (up to four weeks and 150°C). The issues of residual solvent, thermal pretreatment, and variation of dye content was studied in a follow-on study (Scheme 4-2, dye **E**) [33]. However, no attempt was made to chemically quantify the degree of cross-linking in the prepared films.

In a separate, but parallel, series of investigations, Bazuin *et al.* reported on a number of very defined and systematic studies into ionic self-assembled azobenzene-containing systems. Their initial communication [34], where photoinduced functionality was explicitly investigated, focused on a system very similar to that of Lu *et al.* [29], except for the fact that the polyelectrolyte used here was a methylpyridinium derivative (rather than the ethyl version used by Lu *et al.*). Not surprisingly, they also found a lamellar/smectic-like mesoscale arrangement of the complex. This complex was, however, slightly more stable than the ethyl derivative, showing less relaxation, even at higher temperatures (even at 180°C, where residual birefringence was still evident). In comparison, this supramolecular complex showed photoinduced birefringence values similar or superior to that of covalent polymeric systems. This high stability (both temporal and thermal) was ascribed not only to the strong ionic interactions but also to the absence of a flexible spacer between the chromophore and the polymeric backbone. An extensive study into materials without a spacer was then published shortly after the initial report, where the variables investigated were (i) the character and length of the terminal/tail portion of the azobenzene chromophore and (ii) addition of a flexible spacer into the polyelectrolyte backbone [35]. The outcome of this study can be summarized as follows: generally speaking, the changes in molecular architecture had no or little influence on the structure or thermal properties. However, it was found that the more rigid the overall structure of the complex, the better the optical properties, the higher the photoinduced birefringence and temporal and thermal stability, and the greater the SRG amplitude. It was also found that the "tail groups" of the azobenzene chromophores with the possibility for hydrogen-bonding capabilities (e.g.,–OH groups) showed higher diffraction efficiencies and SRG amplitudes.

Lu *et al.* then followed up on their initial study by investigating the effect of spacer length on PIA in poly(acrylic acid)-based complexes [36]. They showed no SRG generation or photoinduced birefringence decay studies and only reported on the order parameter S from simple absorption studies and the ability of their materials to act as LC alignment layer. They concluded that the complex with the longest spacer (12 CH_2 units, Scheme 4-2, dye **F**) produced the highest PIA. From their UV-absorption studies, they also clearly found an aggregation of the chromophores in the supramolecular complexes.

Interestingly, in a study [37] published shortly after that of Lu *et al.*, Bazuin *et al.* investigated the influence of spacer lengths in a number of systems where the charges were inverted (i.e., cationic azobenzene moieties and anionic polystyrenesulfonate) to what they had investigated before. Special attention was paid to the preparation of complexes and presence of solvents (both from casting, as well as absorbed H_2O from the atmosphere). The conclusions concerning the optical properties echoed their other study—more rigid architectures provided better optical properties—but was in direct contrast to that found by Lu *et al.* The contradictory findings of the two groups had, up to the point of writing, not been explored any further and would certainly provide a helpful insight into a problem that will have significant influence on the possible applications of such materials. However, the more extensive studies by Bazuin *et al.* would certainly indicate that their assessment has more general applicability.

Bazuin *et al.* also explored the supramolecular organization and material properties of these materials in detail and found that, for example, the degradation temperatures (T_g) showed some correlation with both the dye spacer lengths as well as the dipole moments of the mesogenic units. Various options for internal chromophore arrangements were also explored to explain the lamellar repeat distances.

A small number of further studies, related to polymeric materials, are noteworthy in this section. Marcos *et al.* [38] made use of complexes of poly(propylene imine), poly(amidoamine), random hyperbranched PEI, and a methylated PEI with a cyanofunctionalized azobenzene with a short spacer (four methylene units) and a carboxylic acid moiety (Scheme 4-2, dye **G**). They clearly showed stable induced birefringence (with slight increases after irradiation was stopped), as well as reaching in-plane order parameters up to 0.67, mostly in nonoptimised irradiation conditions. No expansion of these initial studies has been published up to date. Wang *et al.* [39] reported on the nonlinear optical properties and PIA of a complex, methyl orange and poly(butylvinylpyridinium), which is closely related to those found in the early work of Lu *et al.* Finally, Jiang *et al.* published [40] an interesting approach of combinations of chromo- and lumophores and combinations of interaction strategies (ionic and hydrogen bonding) to produce fluorescent micropatterns.

4.4.2   Oligoelectrolyte-Based Materials

Faul *et al.* published a series of papers from 2002 on the preparation of highly ordered (crystalline and liquid-crystalline) materials from the combination of charged oligoelectrolytes and oppositely charged surfactants [41]. Their investigations initially focused on the use of multicharged azobenzene dyes as oligoelectrolytic structural units [42,43] but then branched out to include a wide variety of other oligoelectrolytic species [44–46]. These early studies did not investigate the functionality of the azobenzene-containing complexes.

In 2006, Faul and Stumpe published a paper investigating the induction of optical anisotropy in a number of low-molecular-weight azobenzene dye–double-tailed ammonium surfactant ionically self-assembled (ISA) complexes (see Scheme 4-2, **ISA1**) [30]. Good quality films (from spin or solution casting) with mosaic textures were obtained for complexes based on EO, with strong aggregation of the complexes evident from the dramatic blue shift observed in the absorption maximum (in both the solution and solid state). The presence of aggregates were ascribed to the very strong cooperative zipperlike binding nature of the ISA process [42]. This cooperative behavior is in stark contrast to the materials formed utilizing a hydrogen-bonding strategy (see discussion later). Irradiation of EO-based complexes lead to PIA and impressive dichroic ratios (ca. 50, see Fig. 4-1a and b) in such low-molecular-weight materials—much higher than those achieved for polymer-based systems. It was furthermore possible to pattern these materials by irradiation through masks (Fig. 4-1c and d), and to use these photoaligned materials to induce optical phase gratings (Fig. 4-1e), and as LC alignment layers. These materials were temporally and thermally stable and even showed increased induced anisotropy after thermal annealing at $120°C$.

In a second, detailed follow-up paper, the photoinduction processes and their correlation to the mesophases of the materials were investigated [47]. This study focused exclusively on the EO–didodecyldimethyl ammonium ISA complex (Scheme 4-2, **ISA1**) and used temperature-dependent UV and X-ray (both small-angle and wide-angle) studies to provide insight into the photo-orientation processes.

However, two issues, very specific to these low-molecular-weight materials, should be noted here: (i) no SRG features could be induced in such low-molecular-weight complexes; this seems to be a particular property of these materials, rather than a general property for materials prepared by ISA (as efficient SRG induction was reported in polyelectrolyte-based systems, as discussed above) and (ii) the levels of induced anisotropy are proportional to the irradiation dose (rather than intensity). Practically, this was showed to not be problematic for lower levels of induced anisotropy (e.g., for dichroic ratios below 10, an irradiation dose of 1 J/cm^2 was found to be sufficient). However, to achieve the very high levels of PIA reported, extremely high doses (approaching 1 kJ/cm^2, see Fig. 4-1b) were required. Such requirements would be impractical for applications. Further unpublished studies have shown that simple changes in the azobenzene dye spacer length has little or no effect on the power input requirements as found in the initially investigated systems [48]. It is noteworthy that the study by Goldenberg et al. [32], where a low-molecular-weight system was initially prepared and then cross-linked at the appropriate stage to produce the (required) high-molecular-weight polymeric system, seemed to provide a good and beneficial mixture of these two ionic approaches.

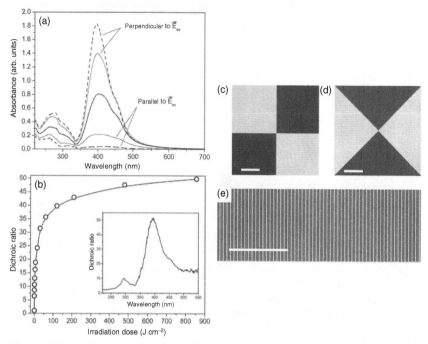

Figure 4-1 (a) Changes of polarized absorbance spectra in a film of the ISA complex under irradiation with linearly polarized light (solid line, initial spectrum; dotted line, after irradiation dose of 0.25 J cm^{-2}; dashed line, after irradiation dose of 850 J cm^{-2}). (b) Kinetics of induced dichroic ratio in a film of the complex calculated at $k_{test} = 400$ nm; inset graph: spectral dependence of induced dichroic ratio after an irradiation dose of 850 J cm^{-2}. The figures on the right show various films after a second irradiation step (c and d), and (e) after irradiation with two circularly polarized interfering beams. *Source:* Reproduced with permission from Ref. [30]. Copyright Wiley-VCH Verlag GmbH & Co. KGaA.

In summary and overview of all the aforementioned investigations into ionic self-assembled complexes of azobenzene chromophores, the following points are worth highlighting:

1. Stable PIA optical properties and SRG have been successfully induced in supramolecular complexes produced using the ISA strategy.

2. Seen in the light of the more detailed investigation by Bazuin *et al.*, shorter spacer lengths seem to lead to higher PIA stability. This issue could easily be further explored and the hypothesis tested against further variations in the types of materials used.

3. The low-molecular-weight ISA complexes seem not to be easily utilized for applications owing to the extremely high energy input required to reach high PIA levels.

4. The complexes are formed in a cooperative manner, which leaves little room for the preparation of nonstoichiometric complexes (see below).

4.5 HYDROGEN-BONDED POLYMERIC ASSEMBLIES

4.5.1 Polymer-Based Materials

Although the production of hydrogen-bonded supramolecular side-chain polymeric systems have been reported some time ago [49], the production of intricate and hierarchically organized switchable nanostructured materials using a hydrogen-bonding approach was initiated by the work of Ikkala and ten Brinke [50]. It was, however, only in 2005 that activity in the field of hydrogen-bonded azobenzene systems was initiated by the work of Ikkala and Priimagi [51]. They compared three systems—one with no interactions between the polymer (polystyrene) and the azobenzene chromophore, a hydrogen-bonded poly-4-vinylphenol-based system (with the commonly used azobenzene dye Disperse Red 1, dye **H** in Scheme 4-2), and an ionic polystyrenesulfonic acid-based system. This investigation followed on a large body of work by Ikkala's research laboratories on various hydrogen-bonded systems based on polyvinylpyridines and various phenolic low-molecular-weight materials [15,52] and seemed to be a natural progression into functional hydrogen-bonded systems. Interestingly enough, they showed that the ionically self-assembled system could accommodate a far larger percentage of chromophore (63 wt% doping, corresponding to a 1:1 charge-neutralized complex) when compared to the other systems. Second harmonic generation (SHG), in conjunction with UV/Vis spectroscopy, was used to probe the aggregation behavior of the chromophore. No further investigation into the use of the azobenzene functionality was reported in this study. However, the main advantages of the noncovalent approach to the preparation of functional architectures were highlighted clearly in their study—the possibility to include high concentrations of chromophores and the ability to prevent aggregation at such high desired concentrations.

A number of studies by Priimagi *et al.* followed in quick succession, in which photoinduced birefringence was demonstrated in 2007 [53] and then investigated in further detail in 2008 [54]. They found that the use of hydrogen bonds reduced the mobility of "guest" azobenzene dyes, thus ensuring higher values as well as better preserved photoinduced birefringence than classic guest–host systems. The values were of comparable magnitude to that of covalently modified polymeric systems. A further benefit was the possibility of using higher concentrations of active chromophore when compared to classic azobenzene-containing systems. The 2008 study focused on the concentration-dependent behavior of 4-nitro-4′-hydroxyazobenzene hydrogen bonded to poly(4-vinylpyridine) (Scheme 4-2, **p4VP** and dye **I**). It was found that two concentration-dependent regimes for the noncooperative hydrogen bonding exist: the first at lower chromophore concentrations (up to a 33% complexation of the backbone, i.e., where every third binding site is occupied) and second, at higher chromophore concentrations, where behavior is influenced (but not dominated) by chromophore–chromophore interactions. The latter leads to significantly stabilized and enhanced photoinduced birefringence.

In a final paper in 2009, SRG formation in further hydrogen-bonded systems was studied [55]. A number of different hydrogen-bonding combinations were explored during these investigations, based on complexes of poly(4-vinylphenol) and varying content of hydrogen-bonding azobenzene chromophore (Scheme 4-2, **p4VPh** and dye **J**). Very efficient SRG formation was found for these systems, with an increase in diffraction efficiency with a decrease in the Mw (molecular weight) of the polymer. Thermal stability was fair (up to 100°C), but with little dependence on the Mw of the polymer. This combination of properties would be useful in applications, as it indicates that lower Mw materials (easy to process, good inscription efficiencies) can be used with little loss of thermal stability.

A simple host–guest system (consisting of noninteracting polystyrene as the polymeric component) was employed for (very useful) comparison in the first two studies (but not in the final study). Generally, it was found that the supramolecular systems showed comparable values for long-term stability of induced birefringence to that of their covalent counterparts [53] and that the noncooperative binding behavior of hydrogen-bonded systems provides a significant advantage (over host–guest systems) when trying to avoid aggregation of chromophores at high concentrations [54]. Furthermore, these systems exhibited enhanced temporal stability of induced birefringence, clearly making them very attractive for applications.

After Ikkala's first study appeared in 2005, a number of papers appeared exploring various aspects of hydrogen-bonded systems. Zhang *et al.* were, to the author's knowledge, the first group to report on efficient and stable SRG induction in hydrogen-bonded systems [56]. The system investigated comprised a carboxylic acid azobenzene derivative forming hydrogen bonds with P4VP (although, in some cases, this mode of binding is seen as proton transfer and thus more closely related to ionic interactions). They found that an increase in dye content yielded increased modulation depths in the inscribed SRGs and that tuning of the peripheral moieties around the azobenzene core could play a role in the SRG-forming process. It was found that azobenzene dyes containing weaker electron-donating substituents (such as methoxy groups) showed very little induction of SRG (with modulation depths of only 10 nm recorded), while amine substituents seemed to produce much more efficient SRG formation. However, this effect was not explored in any detail. They furthermore found that the produced SRGs were stable up to at least 120°C, with little or no change in modulation depth. Heating to 150°C lead to the erasure of the SRG structures.

Two other studies are also of interest here: the first, by the same authors, presented an interesting approach by utilizing the formation of a main-chain supramolecular azobenzene polymer for the formation of photoactive materials [57]. They used the quadruple hydrogen-bonding strategy developed by Meijer and Sijbesma [58] to provide stable materials for photoinduced dichroism and SRGs. However, they found that the inscription depth (15 nm) for the SRG structures was very inefficient and not sufficient for application.

Seki's group published the second study in 2008 [59]. Although not the first to show SRG induction, they made use of the less well-known hydrogen-bonding motif of an imidazole moiety and a polycarboxylic acid polymer (Scheme 4-3).

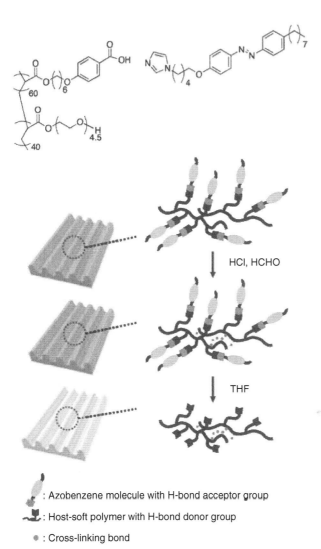

: Azobenzene molecule with H-bond acceptor group

: Host-soft polymer with H-bond donor group

: Cross-linking bond

Scheme 4-3 The approach by Seki *et al.*, where the strongly absorbing azobenzene chromophore is removed once the SRG has been prepared. *Source*: Reproduced with permission from Ref. [59]. Copyright Wiley-VCH Verlag GmbH & Co. KGaA.

Most importantly, they showed the possibility of removing the strongly absorbing azobenzene chromophore by solvent extraction after the induction of the SRG structures. Although this strategy has not proved to be extremely successful in preserving the desired SRG functionality (because of a drastic reduction in the feature sizes), it could certainly lead to interesting developments through postremoval functionalization with other hydrogen-bonding moieties. A further point of interest is the fact that a block copolymer was for the first time utilized in supramolecular azobenzene architectures.

In summary, hydrogen-bonded azobenzene supramolecular architectures have led to equally applicable PIA and SRG (as was found for ISA-based supramolecular complexes). Most of the complexes prepared had the hydrogen-bonding moiety very close to the central azobenzene core; thus, no similar arguments concerning stability of photoinduced properties have been raised so far (as was the case for the ISA complexes). The noncooperative manner of assembly provides advantages when it comes to careful control of the chromophore concentration at nonstoichiometric ratios.

4.6 SUMMARY AND CONCLUSIONS

Two approaches, only recently applied for the production of supramolecular azobenzene architectures, were discussed in this chapter: ionic and hydrogen-bonding routes to self-assembly. The presented results showed that these newly applied noncovalent approaches clearly have promise for the production of functional materials and devices. The obvious advantages of both of these approaches are that (i) a wide range of materials is accessible utilizing facile preparation methods (i.e., without major synthetic efforts); (ii) facile inclusion of high concentrations of chromophores with no or very little undesirable aggregation is possible; (iii) PIA and SRG formation levels are similar to or, in the case of the low-molecular-weight ISA complexes, higher than the covalent counterparts; (iv) removal of the strongly absorbing chromophore has been shown to be possible in the case of SRG formation in hydrogen-bonding systems, and (v) noncooperative hydrogen-bonding provides advantages when it comes to careful control of the chromophore concentration at nonstoichiometric ratios.

In terms of future outlook, and possible areas of further and more detailed investigations, the following options are either still unexplored or fairly new and are worth further investigation:

- tuning of the optical properties of the used chromophores (e.g., the addition of a further azobenzene moiety, addition and tuning of substituents with strong electron-withdrawing or electron-donating effects);
- inclusion of further functionalities (possibly through hydrogen bonding to azobenzene chromophores, as shown by Jiang *et al.* [40].) and possible application in a biological context;
- further exploration of the use of block copolymers to induce hierarchical structures and further (orthogonal) functionalities;
- clarification of the influence and role the specific phase structures of such formed complexes play during photoaddressing processes;
- investigation into low-molecular-weight hydrogen-bonded systems is still an area open to investigation.

Finally, unlike many other vague claims about the application and use of functional nanostructured materials, it is expected that the routes discussed and opportunities presented here will lead to the application of functional materials from supramolecular azobenzene architectures in a range of devices in the very near future.

REFERENCES

1. Ozin, G. A., Cademartiri, L. (2009). Nanochemistry: what is next? *Small*, *5*, 1240–1244.
2. Whitesides, G. M., Lipomi, D. J. (2009). Soft nanotechnology: "structure" vs. "function". *Faraday Discuss.*, *143*, 373–384.
3. Delaire, J. A., Nakatani, K. (2000). Linear and nonlinear optical properties of photochromic molecules and materials. *Chem. Rev.*, *100*, 1817–1845.
4. Natansohn, A., Rochon, P. (2002). Photoinduced motions in azo-containing polymers. *Chem. Rev.*, *102*, 4139–4175.
5. Barrett, C. J., Mamiya, J. I., Yager, K. G., Ikeda, T. (2007). Photo-mechanical effects in azobenzene-containing soft materials. *Soft Matter*, *3*, 1249–1261.
6. Ercole, F., Davis, T. P., Evans, R. A. (2010). Photo-responsive systems and biomaterials: photochromic polymers, light-triggered self-assembly, surface modification, fluorescence modulation and beyond. *Polym. Chem.*, *1*, 37–54.
7. Rau, H. (1990). in *Photoisomerization of Azobenzenes* (Eds.: Rebek, J.), CRC Press, Boca Raton, FL.
8. Rau, H., Luddecke, E. (1982). On the rotation-inversion controversy on photo-isomerization of azobenzenes. Experimental proof of inversion. *J. Am. Chem. Soc.*, *104*, 1616–1620.
9. Weigert, F. (1920). On the specific effect of polarised radiation. *Ann. Phys.*, *63*, 681–725.
10. Weigert, F. (1920). A new effect of radiation. *Z. Phys.*, *2*, 1–12.
11. Teitel, A. (1957). Ueber eine besondere mechanische wirkung des polarisierten lichts. *Naturwissenschaften*, *44*, 370–371.
12. Rochon, P., Batalla, E., Natansohn, A. (1995). Optically induced surface gratings on azoaromatic polymer films. *Appl. Phys. Lett.*, *66*, 136–138.
13. Kim, D. Y., Tripathy, S. K., Li, L., Kumar, J. (1995). Laser-induced holographic surface-relief gratings on nonlinear-optical polymer films. *Appl. Phys. Lett.*, *66*, 1166–1168.
14. MacDonald, J. C., Whitesides, G. M. (1994). Solid-state structures of hydrogen-bonded tapes based on cyclic secondary diamides. *Chem. Rev.*, *94*, 2383–2420.
15. Ikkala, O., ten Brinke, G. (2002). Functional materials based on self-assembly of polymeric supramolecules. *Science*, *295*, 2407–2409.
16. Johnston, A. P. R., Cortez, C., Angelatos, A. S., Caruso, F. (2006). Layer-by-layer engineered capsules and their applications. *Curr. Opin. Colloid Interface Sci.*, *11*, 203–209.
17. Faul, C. F. J., Antonietti, M. (2003). Ionic self-assembly: facile synthesis of supramolecular materials. *Adv. Mater.*, *15*, 673–683.
18. Engelkamp, H., Middelbeek, S., Nolte, R. J. M. (1999). Self-assembly of disk-shaped molecules to coiled-coil aggregates with tunable helicity. *Science*, *284*, 785–788.
19. Mukerjee, P., Mysels, K. J. (1955). A re-evaluation of the spectral change method of determining critical micelle concentration. *J. Am. Chem. Soc.*, *77*, 2937–2943.
20. Antonietti, M., Conrad, J. (1994). Synthesis of very highly ordered liquid-crystalline phases by complex-formation of polyacrylic-acid with cationic surfactants. *Angew. Chem. Int. Ed. Engl.*, *33*, 1869–1870.
21. Ponomarenko, E. A., Waddon, A. J., Bakeev, K. N., Tirrell, D. A., MacKnight, W. J. (1996). Self-assembled complexes of synthetic polypeptides and oppositely charged low molecular weight surfactants. Solid-state properties. *Macromolecules*, *29*, 4340–4345.
22. Thunemann, A. F., General, S. (2001). Nanoparticles of a polyelectrolyte-fatty acid complex: carriers for Q(10) and triiodothyronine. *J. Controlled Release*, *75*, 237–247.
23. Radler, J. O., Koltover, I., Salditt, T., Safinya, C. R. (1997). Structure of DNA-cationic liposome complexes: DNA intercalation in multilamellar membranes in distinct interhelical packing regimes. *Science*, *275*, 810–814.

24. Thunemann, A. F. (2000). Nano-structured materials with low surface energies formed by polyelectrolytes and fluorinated amphiphiles (PEFA). *Polym. Int.*, *49*, 636–644.

25. Thunemann, A. F., Ruppelt, D., Burger, C., Mullen, K. (2000). Long-range ordered columns of a hexabenzo bc,ef,hi,kl,no,qr coronene-polysiloxane complex: towards molecular nanowires. *J. Mater. Chem.*, *10*, 1325–1329.

26. Thunemann, A. F. (2002). Polyelectrolyte-surfactant complexes (synthesis, structure and materials aspects). *Prog. Polym. Sci.*, *27*, 1473–1572.

27. Meyer, W. H., Pecherz, J., Mathy, A., Wegner, G. (1991). Polyelectrolyte glasses with linear and nonlinear optical properties. *Adv. Mater.*, *3*, 153–156.

28. Kulikovska, O., Goldenberg, L. M., Stumpe, J. (2007). Supramolecular azobenzene-based materials for optical generation of microstructures. *Chem. Mater.*, *19*, 3343–3348.

29. Xiao, S., Lu, X., Lu, Q. (2007). Photosensitive polymer from ionic self-assembly of azobenzene dye and poly(ionic liquid) and its alignment characteristic toward liquid crystal molecules. *Macromolecules*, *40*, 7944–7950.

30. Zakrevskyy, Y., Stumpe, J., Faul, C. F. J. (2006). A supramolecular approach to optically anisotropic materials: photosensitive ionic self-assembly complexes. *Adv. Mater.*, *18*, 2133–2136.

31. Zucolotto, V., Mendonca, C. R., dos Santos, D. S., Balogh, D. T., Zilio, S. C., Oliveira, O. N., Constantino, C. J. L., Aroca, R. F. (2002). The influence of electrostatic and H-bonding interactions on the optical storage of layer-by-layer films of an azopolymer. *Polymer*, *43*, 4645–4650.

32. Kulikovska, O., Goldenberg, L. M., Kulikovsky, L., Stumpe, J. (2008). Smart ionic sol-gel-based azobenzene materials for optical generation of microstructures. *Chem. Mater.*, *20*, 3528–3534.

33. Kulikovsky, L., Kulikovska, O., Goldenberg, L. M., Stumpe, J. (2009). Phenomenology of photoinduced processes in the ionic sol-gel-based azobenzene materials. *Acs Appl. Mater. Interfaces*, *1*, 1739–1746.

34. Zhang, Q., Bazuin, C. G., Barrett, C. J. (2008). Simple spacer-free dye-polyelectrolyte ionic complex: side-chain liquid crystal order with high and stable photoinduced birefringence. *Chem. Mater.*, *20*, 29–31.

35. Zhang, Q., Wang, X., Barrett, C. J., Bazuin, C. G. (2009). Spacer-free ionic dye-polyelectrolyte complexes: influence of molecular structure on liquid crystal order and photoinduced motion. *Chem. Mater.*, *21*, 3216–3227.

36. Xiao, S. F., Lu, X. M., Lu, Q. H., Su, B. (2008). Photosensitive liquid-crystalline supramolecules self-assembled from ionic liquid crystal and polyelectrolyte for laser-induced optical anisotropy. *Macromolecules*, *41*, 3884–3892.

37. Zhang, Q., Bazuin, C. G. (2009). Liquid crystallinity and other properties in complexes of cationic azo-containing surfactomesogens with poly(styrenesulfonate). *Macromolecules*, *42*, 4775–4786.

38. Marcos, M., Alcala, R., Barbera, J., Romero, P., Sanchez, C., Serrano, J. L. (2008). Photosensitive ionic nematic liquid crystalline complexes based on dendrimers and hyperbranched polymers and a cyanoazobenzene carboxylic acid. *Chem. Mater.*, *20*, 5209–5217.

39. Zhang, X. Q., Wang, C. S., Pan, X., Xiao, S. F., Zeng, Y., He, T. C., Lu, X. M. (2010). Nonlinear optical properties and photoinduced anisotropy of an azobenzene ionic liquid-crystalline polymer. *Opt. Commun.*, *283*, 146–150.

40. Chen, X. B., Liu, B. J., Zhang, H. B., Guan, S. W., Zhang, J. J., Zhang, W. Y., Chen, Q. D., Jiang, Z. H., Guiver, M. D. (2009). Fabrication of fluorescent holographic micropatterns based on azobenzene-containing host-guest complexes. *Langmuir*, *25*, 10444–10446.

41. Faul, C. F. J. (2006). Liquid-crystalline materials by the ionic self-assembly route. *Mol. Cryst. Liq. Cryst.*, *450*, 255–265.

42. Faul, C. F. J., Antonietti, M. (2002). Facile synthesis of optically functional, highly organized nanostructures: dye-surfactant complexes. *Chem. Eur. J.*, *8*, 2764–2768.

43. Guan, Y., Antonietti, M., Faul, C. F. J. (2002). Ionic self-assembly of dye-surfactant complexes: influence of tail lengths and dye architecture on the phase morphology. *Langmuir*, *18*, 5939–5945.

44. Wei, Z. X., Laitinen, T., Smarsly, B., Ikkala, O., Faul, C. F. J. (2005). Self-assembly and electrical conductivity transitions in conjugated oligoaniline-surfactant complexes. *Angew. Chem. Int. Ed.*, *44*, 751–756.

45. Guan, Y., Zakrevskyy, Y., Stumpe, J., Antonietti, M., Faul, C. F. J. (2003). Perylenediimide-surfactant complexes: thermotropic liquid-crystalline materials via ionic self-assembly. *Chem. Commun.*, 894–895.

46. Camerel, F., Strauch, P., Antonietti, M., Faul, C. F. J. (2003). Copper-metallomesogen structures obtained by ionic self-assembly (ISA): molecular electro-mechanical switching driven by cooperativity. *Chem. Eur. J.*, *9*, 3764–3771.

47. Zakrevskyy, Y., Stumpe, J., Smarsly, B., Faul, C. F. J. (2007). Photo-induction of optical anisotropy in an azobenzene-containing ionic self-assembly liquid-crystalline material. *Phys. Rev. E*, *75*, 031703.

48. Oakley, R. J., Fischer, T., Stumpe, J., Faul, C. F. J.unpublished results.

49. Kato, T., Frechet, J. M. J. (1989). Stabilization of a liquid-crystalline phase through noncovalent interaction with a polymer side chain. *Macromolecules*, *22*, 3818–3819.

50. Ruokolainen, J., Makinen, R., Torkkeli, M., Makela, T., Serimaa, R., ten Brinke, G., Ikkala, O. (1998). Switching supramolecular polymeric materials with multiple length scales. *Science*, *280*, 557–560.

51. Priimagi, A., Cattaneo, S., Ras, R. H. A., Valkama, S., Ikkala, O., Kauranen, M. (2005). Polymer-dye complexes: a facile method for high doping level and aggregation control of dye molecules. *Chem. Mater.*, *17*, 5798–5802.

52. Ikkala, O., ten Brinke, G. (2004). Hierarchical self-assembly in polymeric complexes: towards functional materials. *Chem. Commun.*, 2131–2137.

53. Priimagi, A., Kaivola, M., Rodriguez, F. J., Kauranen, M. (2007). Enhanced photoinduced birefringence in polymer-dye complexes: hydrogen bonding makes a difference. *Appl. Phys. Lett.*, *90*, 121103-1-3.

54. Priimagi, A., Vapaavuori, J., Rodriguez, F. J., Faul, C. F. J., Heino, M. T., Ikkala, I., Kauranen, M., Kaivola, M. (2008). Hydrogen-bonded polymer–azobenzene complexes: enhanced photoinduced birefringence with high temporal stability through interplay of intermolecular interactions. *Chem. Mater.*, *20*, 6358–6363.

55. Priimagi, A., Lindfors, W., Kaivola, M., Rochon, P. (2009). Efficient surface-relief gratings in hydrogen-bonded polymer-azobenzene complexes. *Acs Appl. Mater. Interfaces*, *1*, 1183–1189.

56. Gao, J., He, Y. N., Liu, F., Zhang, X., Wang, Z. Q., Wang, X. G. (2007). Azobenzene-containing supramolecular side-chain polymer films for laser-induced surface relief gratings. *Chem. Mater.*, *19*, 3877–3881.

57. Gao, J., He, Y. N., Xu, H. P., Song, B., Zhang, X., Wang, Z. Q., Wang, X. G. (2007). Azobenzene-containing supramolecular polymer films for laser-induced surface relief gratings. *Chem. Mater.*, *19*, 14–17.

58. Sijbesma, R. P., Beijer, F. H., Brunsveld, L., Folmer, B. J. B., Hirschberg, J., Lange, R. F. M., Lowe, J. K. L., Meijer, E. W. (1997). Reversible polymers formed from self-complementary monomers using quadruple hydrogen bonding. *Science*, *278*, 1601–1604.

59. Zettsu, N., Ogasawara, T., Mizoshita, N., Nagano, S., Seki, T. (2008). Photo-triggered surface relief grating formation in supramolecular liquid crystalline polymer systems with detachable azobenzene units. *Adv. Mater.*, *20*, 516–521.

CHAPTER **5**

STIMULI-RESPONSIVE SUPRAMOLECULAR DYE ASSEMBLIES

Shiki Yagai

Department of Applied Chemistry and Biotechnology, Graduate School of Engineering, Chiba University, Chiba, Japan

5.1 INTRODUCTION

Artificial dye assemblies constructed based on a supramolecular approach [1] have attracted tremendous attention as new organic materials featuring complex self-organized architectures [2–6]. By using developed protocols of specific noncovalent interactions, the control of chromophore packing arrangement and/or dimensionality of self-assembled architectures is enabled, which could open up new avenues toward realization of organic materials with novel optical and electronic properties arising from well-organized multichromophoric architectures. Furthermore, the concerted action of various noncovalent interactions often results in the formation of dye assemblies with intriguing photochemical and electronic properties, as a result of unique chromophore packing and/or complex self-organized structures. Salient examples of such programmed self-assembling systems of functional dyes are circular [7,8] and tubular [9,10] multichromophoric architectures of chlorophyll pigments in the light-harvesting antenna complexes of photosynthetic bacteria. These complexes provide the most convincing demonstration of the important relationships between the high-performance photochemical and electronic functions (harvesting of light and transportation of energy and electron) and highly organized multichromophoric architectures of dyes (chromophore packing arrangements and overall nanoarchitectures). Furthermore, several supramolecularly programmed dye nanostructures have proven to act as efficient media for the transportation of energy and charges, enabling their dimension-controlled transportation required for the realization of nanoscale electronics [4].

With the accumulation of the construction protocols of well-defined supramolecular dye assemblies, a new functional aspect has been incorporated,

Supramolecular Soft Matter: Applications in Materials and Organic Electronics, First Edition.
Edited by Takashi Nakanishi.
© 2011 John Wiley & Sons, Inc. Published 2011 by John Wiley & Sons, Inc.

that is, response to external stimuli. Stimuli-responsive dye assemblies are a class of "smart" self-assemblies, which exhibit a dramatic change in their self-assembled architectures and/or physical properties in response to the application of an environmental stimulus, such as temperature, light, pH, solvent polarity, mechanical stress, chemical additives, and so on. Since most molecular assemblies dissolved in solvent undergo thermal transition between monomeric and aggregated states, we do not involve such systems in the category of stimuli-responsive dye assemblies. However, if one can successfully endow such systems with other functionalities that can be modulated by aggregation, we can categorize them as thermoresponsive dye assemblies. Stimuli-responsive dye assemblies have potential applications as intelligent optical and electronic materials, functional properties of which can be manipulated by external stimuli.

A large number of stimuli-responsive dye assemblies have been reported so far, and this chapter does not cover all of them. For example, the use of photochromic dyes such as azobenzenes and diarylethenes as supramolecular building blocks enables the construction of diverse photoresponsive supramolecular dye assemblies from discrete supramolecular complexes to vesicular assemblies [11,12]. Dyes with an ionic character can be employed for the construction of anion- or cation-responsive dye assemblies with transformable nanostructures [13–16]. Recently, mechanochromic supramolecular dye assemblies have been reported as liquid crystalline functional soft materials [17,18]. The liquid crystalline phase provides an effective platform for the demonstration of stimuli-responsive dye assemblies because of coexisting ordered and dynamic properties [19]. Furthermore, the absence of any solvents in this soft phase is advantageous over solution-based systems in view of practical use. Dye-based organogels [20–24] are one of the most exciting frontiers of stimuli-responsive dye assemblies. The presence of solvent molecules in the gel phase seems to provide a unique opportunity to construct dye assemblies showing a more dynamic response to external stimuli compared to liquid crystalline systems. For these intriguing stimuli-responsive dye assemblies, readers are directed to excellent reviews and other chapters of this book. This chapter focuses on the recent examples of stimuli-responsive supramolecular dye assemblies featuring extended π-conjugated systems that can be applied to organic (opto)electronic devices.

5.2   SUPRAMOLECULAR DYE ASSEMBLIES WITH STIMULI-RESPONSIVE OPTICAL PROPERTIES

A sustained challenge in the research field of functional dyes is the control of stacking arrangements of chromophores, which greatly affect their optical (e.g., absorption and fluorescence) and electronic properties (e.g., transportation of excitation energy and charges) through electronic interactions. If one can modulate the stacking arrangements of chromophores in a self-assembled structure by external stimuli, such an assembly may show changes in optical and electronic properties in response to the stimuli. Perylene bisimides (PBIs; perylene-3,4:9,10-bis(dicarboxiimide)) are among the most attractive functional dyes owing to their outstanding optical and electronic properties that are favorable for application as optoelectronic functional materials, together with the relative ease of synthetic modification [25–27]. Furthermore, their self-assembling capabilities arising from strong $\pi-\pi$ stacking interactions enable the construction of supramolecular functional materials with a defined size and shape.

The stacking arrangements of PBIs in the crystalline state are strongly influenced by the substituents at the imide positions, and can be discussed by invoking longitudinal, transverse, and rotational displacements between neighboring molecules. Empirical [28] and theoretical [29] studies have shown that such diverse displacements between stacked PBI molecules impact absorption properties of the crystalline PBI derivatives (so-called crystallochromy) through electronic interaction (excitonic coupling). If packing arrangements of PBIs are in a kinetically trapped state, they might shift to a thermodynamically stable state by some kind of external stimulus. Gregg showed that a dramatic change occurs in the absorption spectrum of N,N'-bis(2-phenylethyl) perylene bisimide (PBI-1, Fig. 5-1a) on exposure to solvent vapor [30]. The films of PBI-1 prepared by evaporation show absorption maxima at $\sim$540 and 500 nm, exhibiting a red shade. On exposure to solvent vapor such as dichloromethane, the films turn into black materials with absorption maxima at $\sim$610 and 410 nm. It is considered that the solvent vapors permeate the films and impart enough mobility to the molecules so that they can crystallize into a more ordered state. Mizuguchi also reported that the amorphous films of PBI-1 can be transformed into a crystalline state not only by exposing to acetone vapor but also by thermal treatment [31]. These amorphous–crystalline transitions can be related to the transition from H- to J-type packing arrangements of PBIs by taking characteristic absorption changes into account [32].

Despite remarkable variations in the packing arrangements of PBIs in the crystalline state, reported UV–vis spectra of the aggregates of "core-unsubstituted" PBIs in solution display similar spectral shapes that are ascribable to H-aggregates by a hypsochromically shifted absorption maximum at $\sim$490 nm, independent of the imide substituents. While H-aggregated PBIs exhibit high n-type charge carrier mobilities as desired in electronic devices [33–36], the extended exciton mobility of J-aggregated PBIs could be exploited in organic solar cells. J-aggregation of PBIs has been achieved so far by introducing bulky substituents in the bay area (1,6,7, and 12 positions) of perylene core that induce a large longitudinal offset of stacked π faces, resulting in a remarkable red-shift of the absorption band [37–41]. However, several molecularly engineered PBI derivatives have recently been shown to form soluble J-type aggregates in organic solvents. While PBI-2 (Fig. 5-1b) possessing tridodecyloxyphenyl (TDP) wedges self-assembles into gel-forming helical H-aggregates in apolar solvents [34], the introduction of the bulky 3,7-dimethyloctyl group on the wedge as in PBI-3 (Fig. 5-1b) results in the formation of soluble J-type aggregates that are more efficiently gelate apolar solvents with low concentrations [42]. The fabrication of photovoltaic devices has been achieved by blending the fibrous J-aggregates with hole-transporting polymers [43].

PBI-4 equipped with a monotopic melamine hydrogen-bonding unit was also found to form J-aggregates in the apolar solvent upon complexing with complementary ditopic cyanuric acid derivatives (Fig. 5-1c) [44]. A unique feature of this system is the reversible transformation between H- and J-aggregated states with the stoichiometry of complementary guests and temperature control. In the presence of the 0.5 equivalent of dodecylated cyanuric acid (dCA), PBI-4 forms a stable H-aggregate (H-dimer) in methylcyclohexane (MCH) as a result of cooperative multiple hydrogen bonding and $\pi-\pi$ stacking interactions (see "H-dimer" in Fig. 6-1c of Chapter 6). The absorption maximum of the H-dimer locates at 495 nm, blue-shifted from that of monomeric PBI-4 (515 nm). The H-dimer is fairly stable: the melting temperature at which half of the aggregates dissociate is 68°C in MCH at [**1**] = 14 μM. Interestingly, further addition of

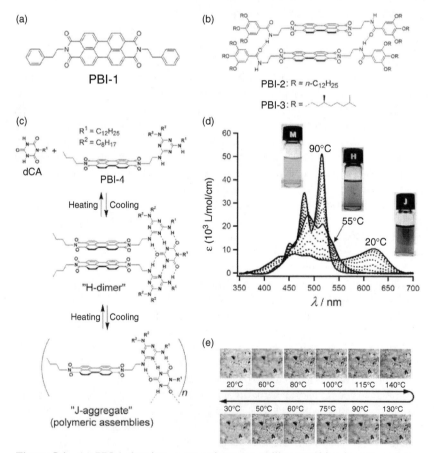

Figure 5-1   (a) PBI-1 showing an amorphous–crystalline transition by exposure to a solvent vapor. (b) Aggregation motif of PBI-2 and PBI-3. (c) Aggregation scheme of PBI-4 and dCA. (d) Temperature-dependent absorption spectra of the 1:1 mixture of PBI-4 and dCA in MCH. (e) Thermoresponsive color change of the J-aggregated film of the 1:1 mixture of PBI-4 and dCA. (*See insert for color representation of the figure.*)

the 0.5 equivalent of dCA (i.e., PBI-4:dCA = 1:1) resulted in the formation of J-aggregates having an absorption maximum at 620 nm. Atomic force microscopy (AFM) measurements of the spin-coated solution showed fibrous aggregates, indicating the evolution of an extended supramolecular network from the ditopic cyanurates and monotopic melamines possessing the extended π-system. Cyanuric acid and its alkylated derivatives are known to form supramolecular lattices scaffolded by single and/or double hydrogen-bonding interactions between neighboring molecules [45]. Hence, it can be considered that one acceptor–donor–acceptor (imide type) triple hydrogen-bonding array of dCA is occupied by the donor–acceptor–donor triple hydrogen-bonding array of the melamine moiety of **1**, whereas another array is used for extended aggregation (see "J-aggregate" in Fig. 5-1c). J-type aggregation of PBI moieties might be achieved in either an intra- or an intersupramolecular manner.

On increasing the temperature of the above J-aggregate solution, the UV–vis spectral change indicating the J-to-H transition occurs below 55°C, and above this temperature the H-to-M (monomer) transition exclusively takes place (Fig. 5-1d). Thus, the solution exhibits a prominent color change of green (J) → red (H) → orange (M) on heating. It is believed that thermodynamically stable species are selectively formed at applied temperatures. More surprisingly, thin films prepared from the J-aggregated solutions also show a reversible green–red thermochromism between 20 and 140°C (Fig. 5-1e). The mechanism of this thermoresponsive chromophore packing is apparently different from that of the thermal interconversion between J-aggregates and H-dimers in solution, because such a large structural transition may not be feasible in the condensed state. It is likely that the thermally inducible disordering and reorganization of the supramolecular polymer network affects the local chromophore packing arrangement in the solid state. The fully reversible thermal interconversion between H- and J-type chromophore packing in solution and in solid state is an intriguing property that might be utilized for a supramolecular thermometer [46] and a new type of information storage material.

One of the most challenging tasks in the supramolecular dye chemistry is the construction of highly organized dye assemblies that are capable of performing efficient light harvesting and subsequent directional energy and electron transfer, as observed in the light-harvesting antenna systems of photosynthetic organisms [47]. Exploitation of such functions by using artificial dye assemblies [23,48–53] may realize new optoelectronic soft devices. If one can successfully impart stimuli-responsive properties to such functional dye assemblies, intelligent dye assemblies might be realized where specific functions can be controlled by external stimuli. Ajayaghosh et al. reported a thermally gated fluorescence resonance energy transfer (FRET) system using a self-assembling oligo(p-phenylenevinylene) (OPV–COOH) as donor scaffolds and Rhodamine B as an acceptor (Fig. 5-2a) [54]. At room temperature, OPV–COOH self-assembles to form π-stacked nanotapes in a dodecane–chloroform mixture, entrapping the coexisting acceptors. Upon exciting OPV–COOH at 460 nm, the emission from the self-assembled nanotapes is quenched, and the emission from the entrapped Rhodamine B acceptor is observed at ~600–650 nm as a result of FRET.

On increasing the temperature, the dissociation of the OPV nanotapes occurs, thus switching the FRET off. Since the emission of monomeric OPV–COOH matches with the absorption of Rhodamine B, the results clearly demonstrate that the $\pi-\pi$ stacked architecture is crucial for the FRET.

Recently, Würthner *et al.* reported an interesting stimuli-responsive FRET system by using bilayer vesicles composed of amphiphilic PBIs (Fig. 5-2b). Mixing of micelle-forming amphiphilic PBI-5 with PBI-6 (molar ratio = 20:1) affords bilayer vesicles with diameters of $\sim$20 nm as a result of the lowered spontaneous curvature of stacked PBI-5 by symmetrical PBI-6 [55]. The resulting vesicular assemblies can be stabilized by the photopolymerization of the terminal acrylate group attached to PBI-5 in the presence of a photoinitiator. The authors successfully applied this "nanocapsules" to establish fluorescent pH-sensing systems using FRET between the PBI vesicles and encapsulated guests [56]. When the PBI vesicles are prepared in the presence of the bispyrene (BP) derivative as a FRET donor, encapsulation of the donor occurs. BP has a pH-dependent dual fluorescence property: under a basic condition, it emits green light (460–540 nm) as a result of excimer formation from a folded conformation, whereas under an acidic condition purple-blue emission (370–420 nm) of the monomeric pyrene is observed because the electrostatic repulsion between protonated amino groups hampers the folding. Thus, excitation of the donor-encapsulated vesicles at 363 nm (the absorption band of the donor) results in a pH-dependent FRET from the donor to the acceptor PBI bilayers absorbing at $\sim$450–600 nm. Because the PBI bilayer emits red light (600–750 nm), characteristic of PBI excimers, this system exhibits fluorescence covering the whole visible light range upon changing pH. At pH = 9.0, all the above three emissions, that is, 370- to 420-nm emission from the unfolded BP, 460- to 540-nm emission from the folded BP, and 600–750 nm emission from the PBI bilayers are mixed, showing elusive white emission.

5.3 SUPRAMOLECULAR DYE ASSEMBLIES WITH STIMULI-RESPONSIVE NANOSTRUCTURES

A rational design of functional molecular building blocks leading to desired nanostructures with size and shape control, is crucial for the success of organic electronics on the nanoscale [57]. While many functional dyes have an intrinsic ability to self-assemble into extended one-dimensional nanostructures through $\pi-\pi$ stacking interactions, more precise control of their dimensionalities might be possible by using highly directional noncovalent interactions such as multiple hydrogen bonds [58–60]. The validity of multiple hydrogen-bonding interactions to control the self-assembly of functional building blocks has been demonstrated for OPVs [61–67], azobenzenes [68–73], porphyrins [74–77], merocyanines [78–86], and PBIs [78,87–91], and is receiving increasing attention as a promising tool for the realization of nanoscale electronics and optics [2,6].

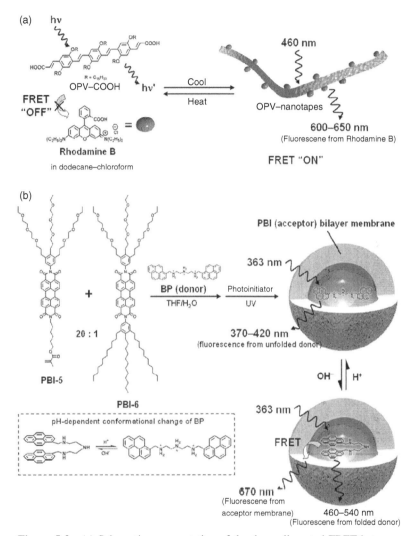

Figure 5-2 (a) Schematic representation of the thermally gated FRET between OPV–COOH and Rhodamine B. (b) Schematic representation of the formation of the bilayer membrane vesicle composed of amphiphilic perylene bisimides PBI-5 and PBI-6, and its fluorescent pH-sensing system using FRET with encapsulated bispyrene (BP).

As already described in the discussion on PBI-4, complementary multiple hydrogen-bonding interactions between melamines and barbituric acid or cyanuric acid derivatives are useful molecular glues for the creation of well-defined supramolecular dye assemblies [92–94]. The use of this type of complementary, two-component systems is highly attractive from the following viewpoints. (i) Two (or more) different functional building blocks can be homogeneously mixed without macroscopic phase separation, but on the molecular level they can be phase separated, which imparts a new functionality to supramolecular dye assemblies such as supramolecular p/n heterojunction [40,95]. (ii) Diverse supramolecular architectures can be obtained by structural modification of either one of the two components [96]. (iii) If either one of the two components could be designed as it self-aggregates to form a well-defined nanostructure, such a self-assembled nanostructure can be transformed into other structures by the addition of the complementary components. The third point could be utilized for the construction of stimuli-responsive supramolecular assemblies where the addition of a specific guest induces transformation of nanostructures [97]. In the cases of melamines–barbiturates/cyanurates, however, most of the reported assemblies are based on the complexation of these complementary components through triple hydrogen-bonding interactions, and there have been only few examples of self-aggregation of the individual components. On the basis of scanning tunneling microscopic investigations on the liquid–solid interface, the research groups of Meijer and DeFeyter revealed that OPVs equipped with a ditopic melamine hydrogen-bonding unit can form hydrogen-bonded hexamers (rosettes) through double hydrogen bonding. In a nonpolar solvent, well-defined columnar nanostructures were visualized by AFM, indicating the hierarchical organization of the OPV rosettes mainly driven by solvophobic and $\pi-\pi$ stacking interactions [98]. Thus, an appropriate molecular design of functional building blocks utilizing melamines, barbiturates, and cyanurates might enable the construction of stimuli-responsive functional assemblies, nanoscale morphologies of which can be transformed on addition of complementary guests.

Yagai *et al.* synthesized a xylylene-linked OPV dimer (XOPV), the OPV units of which are capped on their end by a monotopic melamine hydrogen-bonding unit (Fig. 5-3a) [67]. XOPV self-aggregates in MCH to form curled, flexible, and fibrous nanostructures as visualized by AFM. The capability to form such flexible quasi-one-dimensional nanostructures endows this compound with the ability to gelate the solvent at millimolar concentrations ($c > 2$ mM), on cooling hot homogeneous solutions to room temperature. When 1 equivalent of a guest cyanurate (e.g., dCA) was added to the gels of XOPV followed by heating, the formation of gel-like materials, even viscous liquids, was not observed after cooling. This clearly indicates the formation of new assemblies as a result of 1:1 host–guest complexation between XOPV and dCA. AFM revealed that the nanostructures are transformed into rigid nanostrips, the lengths of which are significantly shorter than those of the self-aggregated XOPV. The solution of XOPV·dCA could be reconverted to the gel state by adding 1 equivalent of BMx [99] after the heating–cooling procedure. Since BMx is a host that can strongly bind with cyanurates compared to XOPV for entropic reasons, dCA selectively complex with BMx and thereby gel-forming flexible nanostructures are reproduced through the self-aggregation of XOPV.

Similar to melamines, complementary barbiturates and cyanurates are considered to be useful nanostructure-directing units by taking into consideration their diverse hydrogen-bonded patterns in the crystal structures [100]. However, the examples demonstrating their capability to lead functional molecules into well-defined nanostructures by themselves are scant. Liu *et al.* have shown that

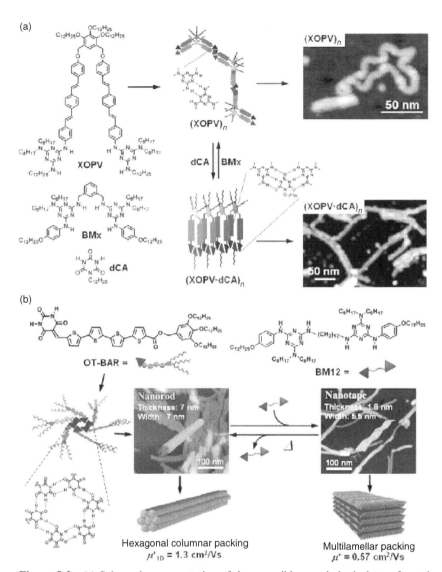

Figure 5-3 (a) Schematic representation of the reversible morphological transformation of self-assembled nanostructures of XOPV induced by the addition of guest (dCA) and competing host (BM*x*) and the corresponding AFM images. (b) Schematic representation of the reversible morphological transformation of self-assembled OT-BAR nanostructures induced by the addition of a host (BM12) and thermal stimulus and the corresponding AFM images.

a barbituric acid merocyanine dye possessing long alkyl chains forms nanospirals in the Langmuir–Blodgett films [101]. The role of the hydrogen-bonding interaction between barbituric acid moieties was confirmed by Fourier transform infrared spectroscopy (FT-IR) measurements. Bassani *et al.* reported the synthesis of oligothiophene and fullerene derivatives conjugated with melamine or barbituric acid group, and investigated photovoltaic properties of their coaggregates [102]. Photovoltaic devices of the hydrogen-bonded coaggregates gave a 2.5-fold enhancement in light energy to electrical energy conversion when compared to the control system using the intact C_{60}.

Recently, the self-assembly of several π-conjugated oligomers equipped with barbituric acid on one end and TDP aliphatic wedge on the other have been reported by Yagai *et al.* AFM studies showed that an OT-BAR possessing a quaterthiophene core forms well-defined nanorods with uniform diameters of 7 nm in MCH (Fig. 5-3b) [103]. UV–vis spectra of the nanorod dispersions showed the absorption maximum at 439 nm, which was strongly blue-shifted from that of the molecularly dissolved state in dichloromethane (518 nm). Thus, oligothiophene moieties in the nanorods are closely packed into H-aggregates. The nanorods are fairly stable as they remain intact on heating to 80°C in MCH. X-ray diffraction (XRD) measurements of the solid thin films prepared from the nanorod dispersions showed the existence of a hexagonal columnar packing structure with a lattice constant of 6.6 nm. None of the hydrogen-bonding patterns in the reported crystal structures of barbituric acid derivatives satisfied the observed columnar packing. Assuming a density of 1 g/cm^3, the number of OT-BAR molecules per columnar stratum of 3.5 Å thickness (typical $\pi-\pi$ stacking distance) is 6.8. Therefore, the OT-BAR most likely forms hexameric or heptameric rosettes through double hydrogen bonding between barbituric acid moieties, which hierarchically organize into nanorods. As a result of closely H-aggregated oligothiophene moieties within the nanorods, the films of the isotropically oriented nanorods showed a remarkably high charge carrier mobility of 1.3 cm^2/Vs as evaluated by flash-photolysis time-resolved microwave conductivity (FP-TRMC) measurements.

On adding 1 equivalent of BM12 [83], which is a flexible ditopic bismelamine, to the nanorod dispersions in MCH, flat, tapelike nanostructures (nanotapes) with the thickness of 1.7 nm and the width of ca 5.5 nm are generated (Fig. 5-3b), demonstrating that the self-aggregates of OT-BAR are converted to coaggregates of OT-BAR and BM12. In the coaggregates, oligothiophene moieties are no longer packed in an H-aggregated state because their dispersions show absorption maximum at 510 nm close to that of the monomeric species. Nevertheless, multilamellar structures formed by the hierarchical organization of the nanotapes showed a high hole mobility of 0.57 cm^2/Vs, probably owing to the interchain $\pi-\pi$ stacking of the oligothiophene moieties.

Interestingly, the nanotapes of OT-BAR and BM12 can be thermally reconverted to the nanorods of OT-BAR on heating in MCH. This was demonstrated by variable temperature UV–vis measurements of the nanotape dispersions showing the formation of H-aggregates on increasing the temperature (510 nm at 20°C to

439 nm at 70°C). The heating-induced H-aggregation is unusual behavior for dye assemblies because most dye aggregations are enthalpically favored processes. AFM images of the samples prepared from the nanotape dispersion heated at 80°C indeed showed the formation of rodlike nanostructures. Released BM12 molecules, on heating, are thus considered to exist in a molecularly dissolved state. The mixture of nanorods of OT-BAR and monomeric BM12 regenerated nanotapes upon aging at an ambient temperature. Although the formation of nanorods from the molecularly dissolved OT-BAR is an enthalpy-driven process, their heating-induced formation from coaggregated OT-BAR and BM12 is apparently an entropy-driven process in which the release of closely packed "flexible" BM12 compensates the entropic loss on the reorganization of OT-BAR [104].

Further exciting stimuli-responsive assemblies have been constructed by OPV-BAR (Fig. 5-4a) [105]. Despite the analogous molecular structure, the nanostructures formed by OPV-BAR are surprisingly different from the nanorods formed by OT-BAR. When MCH solutions of OPV-BAR were drop cast or spin coated onto highly oriented pyrolytic graphite (HOPG), a variety of nanostructures was visualized by AFM, the morphologies of which were dependent on the concentrations of the precast solutions. The nanostructures obtained from solutions with concentrations less than 2×10^{-5} M predominantly show closed, ring-shaped morphologies (nanorings [106,107]) with relatively uniform diameters and thicknesses of ca 38 and 2.6 nm, respectively (AFM image i). Spontaneous formation of nanorings in solution without the aid of the dewetting process on the substrates is supported by dynamic light-scattering analysis of the precast solution as well as partially overlapping parts between neighboring nanorings in AFM images. AFM and transmission electron microscopic (TEM) observation by using different substrates further support the spontaneous formation of nanorings. Surprisingly, when the concentration was increased to 4×10^{-5} M, a considerable number of open-ended, curved nanostructures emerged (AFM image ii). The formation of these curved nanostructures suggests the existence of a spontaneous curvature encoded in the self-assembly of OPV-BAR. A further increase in the concentration to 1×10^{-4} M resulted in the evolution of rodlike nanostructures reminiscent of the tobacco mosaic virus (AFM image iii).

On heating the solvent-free films of OPV-BAR to 200°C, a liquid crystalline mesophase was formed, which showed an XRD pattern assignable to a rectangular columnar packing. Thus, it is strongly suggested that the hexameric rosette is also the primary supramolecular species for this compound but here the rosettes stack on top of each other with an offset, giving rise to the column with an ellipsoidal cross section. In aliphatic solvents, such columns are isolated by solvation and behave as tapelike nanofibers. At concentrations below 2×10^{-5} M, where the lengths of columns are moderate (≈90 nm as judged from the circumferences of nanorings), intrachain end-to-end interaction occur to form rings. It is believed that translational and rotational offsets between rosettes are responsible for the spontaneous curvature of columns. At concentrations above 1×10^{-4} M, columns are further elongated to over 200 nm, which eventually leads to their coiling into nanorods to minimize the surface-free energy [109].

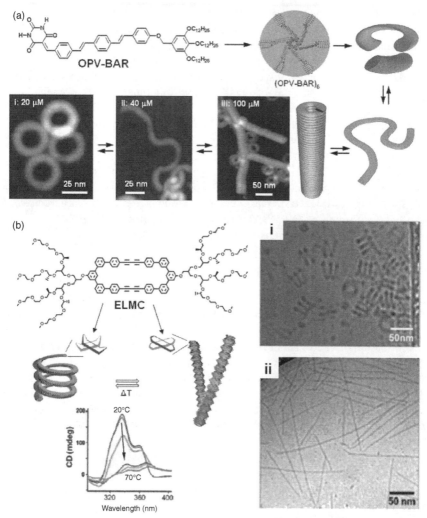

Figure 5-4   (a) Schematic representation of the concentration-dependent reversible morphological transition (ring-tape-coil) of self-assembled OPV-BAR nanostructures and the corresponding AFM images. (b) Schematic representation of the thermoreversible morphological transition (coil rod) of self-assembled amphiphilic elliptical macrocyclic compound (ELMC) and the corresponding TEM images. *Source:* Reproduced with permission from Ref. [108].

The exotic stimuli response on the self-assembled nanostructures of OPV-BAR is reminiscent of the dynamic self-organization processes of several proteins. The tobacco mosaic virus coat protein (capsomer) is an excellent example where the ring (disk)-to-helical coil transition could be regulated by pH or ionic strength even in the absence of RNA [110]. The β protein of the bacteriophage λ self-assembles into rings, which are transformed into helically elongated filaments by the action of DNA [111]. Such complex self-organization processes of proteins might be a result of reversible multivalent interactions between higher order structures of proteins. It is thus surprising that such a small molecule as OPV-BAR shows organization behavior that is similar to the proteins. Apart from their complexity, perfect morphological features of the nanoring and the nanorod are particularly attractive as nanomaterials with unique electronic, magnetic, as well as optical properties, and the authors are currently modifying the molecular structures of the building blocks such as the conjugated length of π-core and the bulkiness of aliphatic chains, and exploring the effect of external environment such as solvent and temperature in order to selectively construct a single nanostructure. Furthermore, the investigation of morphology-dependent semiconductive properties is also under way.

The research group of Lee recently reported elliptical macrocycles that show thermoreversible transformation of helical coils and straight rods (Fig. 6-4b of Chapter 6) [108]. In aqueous media, the amphiphilic elliptical macrocyclic compound (ELMC) predominantly self-assembles into helical coils with a uniform diameter of $\sim$30 nm and a regular pitch of $\sim$10 nm, as shown by cryo-TEM observation (TEM image i). The cross-sectional diameter of the coils is $\sim$3 nm, which is consistent with the estimated length of the macrocycle of ELMC. Therefore, it is suggested that the macrocycles stack on top of one another with one-handed mutual rotation, which was also confirmed by a pronounced Cotton effect in the circular dichroism (CD) spectrum.

A dramatic decrease of the CD signal was observed at 50°C on increasing the temperature (Fig. 5-4b). The cryo-TEM image of the solution at 50°C showed the formation of straight rodlike micelles with a uniform diameter of $\sim$3 nm (TEM image ii). When the solution was cooled to room temperature, the straight rods started to undulate and eventually recovered the original helical coils after three days accompanied by the recovery of the Cotton effect. The authors explain this interesting nanostructural transition by invoking the lower critical solution temperature (LCST) of the ethylene oxide chains in aqueous media [112]. The ethylene oxide chains are hydrated well at room temperature, thus taking a random coil conformation and behave as sterically demanding side chains. Above LCST, they are dehydrated and thereby the macrocycles can be more closely packed to form straight columnar structures.

Reversible extension and contraction of polymers or supramolecular coordination polymer strands induced by a change in the external environment have been reported for oligo(phenylene ethynylene)s by Moore *et al.* (induced by solvent) [113], polyphenylacetylenes by Yashima *et al.* (induced by temperature) [114], coordination supramolecular polymers by Lee *et al.* (induced by temperature) [115], dendronized polyphenylacetylenes by Percec *et al.* (induced by temperature) [116], and tripeptide amphiphiles by Stupp *et al.* (induced by light) [117]. However, such systems based on $\pi-\pi$ stacked small molecular building blocks are rare because external stimulus often results in the dissociation of assemblies into monomeric species. Thus, the reversible morphological changes shown by OPV-BAR and ELMC may be a new type of stimuli response, mimicking the naturally occurring dynamic self-organization process of proteins.

5.4 CONCLUSIONS

Supramolecular construction of functional dye assemblies is currently regarded as a unique methodology toward (i) the realization of nanoscale organic electronics and optics; (ii) the creation of functional soft materials with complex self-organized architectures; and (iii) the exploration of new optical and electronic properties of the employed dyes that is not accessible through the conventional dye chemistry. Furthermore, complexity of self-organized structures supported by various cooperative noncovalent interactions often endows supramolecular dye assemblies with extreme sensitivity to external stimuli, imparting a fourth aim to the supramolecular construction of functional dye assemblies. Stimuli-responsive dye assemblies constructed in aqueous systems can be directly applied to the biological and analytical fields. Even though many stimuli-responsive dye assemblies reported so far have been achieved in organic solution, we hope that the conception outlined in this chapter illustrates how we can translate information given as external stimuli into functional outputs through dye assemblies. Furthermore, the construction protocols of stimuli-responsive dye assemblies in solution will be also applicable to specific soft systems where dynamic motions of molecules are allowed, such as gels [21] and liquid crystalline systems [19].

REFERENCES

1. Lehn, J.-M. (1995). *Supramolecular Chemistry: Concepts and Perspectives*. VCH, Weinheim.
2. Schenning, A. P. H. J., Meijer, E. W. (2005). Supramolecular electronics; nanowires from self-assembled π-conjugated systems. *Chem. Commun.*, 3245–3258.
3. Schenning, A. P. H. J., *et al.* (2004). Towards supramolecular electronics. *Synth. Met.*, *147*, 43–48.
4. Meijer, E. W., Schenning, A. P. H. J. (2002). Chemistry: material marriage in electronics. *Nature*, *419*, 353–354.
5. Würthner, F., ed. (2005). *Supramolecular Dye Chemistry*, Topics in Current Chemistry, Vol. 258. Springer-Verlag, New York.
6. Hoeben, F. J. M., Jonkheijm, P., Meijer, E. W., Schenning, A. P. H. J. (2005). About supramolecular assemblies of π-conjugated systems. *Chem. Rev.*, *105*, 1491–1546.
7. McDermott, G., Prince, S. M., Freer, A. A., Hawthornthwaite-Lawless, A. M., Papiz, M. Z., Cogdell, R. J., Isaacs, N. W. (1995). Crystal structure of an integral membrane light-harvesting complex from photosynthetic bacteria. *Nature*, *374*, 517–521.
8. Pullerits, T., Sundström, V. (1996). Photosynthetic light-harvesting pigment-protein complexes: toward understanding how and why. *Acc. Chem. Res.*, *29*, 381–389.
9. Holzwarth, A. R., Griebenow, K., Schaffner, K. (1992). Chlorosomes, photosynthetic antennae with novel self-organized pigment structures. *J. Photochem. Photobiol., A*, *65*, 61–71.
10. Tamiaki, H. (1996). Supramolecular structure in extramembraneous antennae of green photosynthetic bacteria. *Coord. Chem. Rev.*, *148*, 183–197.
11. Yagai, S., Karatsu, T., Kitamura, A. (2005). Photocontrollable self-assembly. *Chem.—Eur. J.*, *11*, 4054–4063.
12. Yagai, S., Kitamura, A. (2008). Recent advances on photoresponsive supramolecular self-assemblies. *Chem. Soc. Rev.*, *37*, 1520–1529.
13. Engelkamp, H., Middelbeek, S., Nolte, R. J. M. (1999). Self-assembly of disk-shaped molecules to coiled-coil aggregates with tunable helicity. *Science*, *284*, 785–788.

14. Ajayaghosh, A., Chithra, P., Varghese, R. (2007). Self-assembly of tripodal squaraines: cation-assisted expression of molecular chirality and change from spherical to helical morphology. *Angew. Chem. Int. Ed.*, *46*, 230–233.

15. Maeda, H. (2008). Anion-responsive supramolecular gels. *Chem.—Eur. J.*, *14*, 11274–11282.

16. Kishimura, A., Yamashita, T., Aida, T. (2005). Phosphorescent organogels via "metallophilic" interactions for reversible RGB-color switching. *J. Am. Chem. Soc.*, *127*, 179–183.

17. Sagara, Y., Kato, T. (2008). Stimuli-responsive luminescent liquid crystals: change of photo-luminescent colors triggered by a shear-induced phase transition. *Angew. Chem. Int. Ed.*, *47*, 5175–5178.

18. Sagara, Y., Kato, T. (2009). Mechanically induced luminescence changes in molecular assemblies. *Nat. Chem.*, *1*, 605–610.

19. Kato, T., Mizoshita, N., Kishimoto, K. (2006). Functional liquid-crystalline assemblies: Self-organized soft materials. *Angew. Chem. Int. Ed.*, *45*, 38–68.

20. Terech, P., Weiss, R. G. (1997). Low-molecular mass gelators of organic liquids and the properties of their gels. *Chem. Rev.*, *97*, 3133–3159.

21. Ishi-i, T., Shinkai, S. (2005). Dye-based organogels: stimuli-responsive soft materials based on one-dimensional self-assembling aromatic dyes. *Top. Curr. Chem.*, *258*, 119–160.

22. Bhattacharya, S., Samanta, S. K. (2009). Soft functional materials induced by fibrillar networks of small molecular photochromic gelators. *Langmuir*, *25*, 8378–8381.

23. Ajayaghosh, A., Praveen, V. K., Vijayakumar, C. (2008). Organogels as scaffolds for excitation energy transfer and light harvesting. *Chem. Soc. Rev.*, *37*, 109–122.

24. Fages, F., ed. (2005). *Low Molecular Mass Gelators, Design, Self-Assembly, Function*, Topics in Current Chemistry, Vol. 256. Springer-Verlag, New York.

25. Würthner, F. (2004). Perylene bisimide dyes as versatile building blocks for functional supramolecular architectures. *Chem. Commun.*, 1564–1579.

26. Langhals, H. (2005). Control of the interactions in multichromophores: novel concepts. Perylene bisimides as components for larger functional units. *Helv. Chim. Acta*, *88*, 1309–1343.

27. Wasielewski, M. R. (2009). Self-assembly strategies for integrating light harvesting and charge separation in artificial photosynthetic systems. *Acc. Chem. Res.*, *42*, 1910–1921.

28. Klebe, G., Graser, F., Hadicke, E., Berndt, J. (1989). Crystallochromy as a solid-state effect: correlation of molecular conformation, crystal packing and colour in perylene-3,4:9,10-bis(dicarboximide) pigments. *Acta Crystallogr., Sect. B*, *45*, 69–77.

29. Kazmaier, P. M., Hoffmann, R. (1994). A theoretical study of crystallochromy. Quantum interference effects in the spectra of perylene pigments. *J. Am. Chem. Soc.*, *116*, 9684–9691.

30. Gregg, B. A. (1996). Evolution of photophysical and photovoltaic properties of perylene bis(phenylethylimide) films upon solvent vapor annealing. *J. Phys. Chem.*, *100*, 852–859.

31. Mizuguchi, J. (1998). Electronic characterization of *N*,*N*-bis(2-phenylethyl)perylene-3,4:9,10-bis(dicarboxyimide) and its application to optical disks. *J. Appl. Phys.*, *84*, 4479–4486.

32. Kasha, M., Rawls, H. R., El-Bayoumi, M. A. (1965). Exciton model in molecular spectroscopy. *Pure Appl. Chem.*, *11*, 371–392.

33. Struijk, C. W., Sieval, A. B., Dakhorst, J. E. J., van Dijk, M., Kimkes, P., Koehorst, R. B. M., Donker, H., Schaafsma, T. J., Picken, S. J., van de Craats, A. M., Warman, J. M., Zuilhof, H., Sudholter, E. J. R. (2000). Liquid crystalline perylene diimides: architecture and charge carrier mobilities. *J. Am. Chem. Soc.*, *122*, 11057–11066.

34. Li, X.-Q., Stepanenko, V., Chen, Z., Prins, P., Siebbeles Laurens, D. A., Würthner, F. (2006). Functional organogels from highly efficient organogelator based on perylene bisimide semiconductor. *Chem. Commun.*, 3871–3873.

35. Che, Y., Datar, A., Balakrishnan, K., Zang, L. (2007). Ultralong nanobelts self-assembled from an asymmetric perylene tetracarboxylic diimide. *J. Am. Chem. Soc.*, *129*, 7234–7235.

36. Briseno, A. L., Mannsfeld, S. C. B., Reese, C., Hancock, J. M., Xiong, Y., Jenekhe, S. A., Bao, Z., Xia, Y. (2007). Perylenediimide nanowires and their use in fabricating field-effect transistors and complementary inverters. *Nano Lett.*, *7*, 2847–2853.

37. Würthner, F. (2006). Bay-substituted perylene bisimides: twisted fluorophores for supramolecular chemistry. *Pure Appl. Chem.*, *78*, 2341–2349.

38. Kaiser, T. E., Wang, H., Stepanenko, V., Würthner, F. (2007). Supramolecular construction of fluorescenct J-aggregates. *Angew. Chem. Int. Ed.*, *46*, 5541–5544.

39. Schenning, A. P. H. J., Jeroen, V. H., Jonkheijm, P., Chen, Z., Würthner, F., Meijer, E. W. (2002). Photoinduced electron transfer in hydrogen-bonded oligo(p-phenylene vinylene)-perylene bisimide chiral assemblies. *J. Am. Chem. Soc.*, *124*, 10252–10253.

40. Würthner, F., Chen, Z., Hoeben, F. J. M., Osswald, P., You, C.-C., Jonkheijm, P., Herrikhuyzen, J. V., Schenning, A. P. H. J., van der Schoot, P. P. A. M., Meijer, E. W., Beckers, E. H. A., Meskers, S. C. J., Janssen, R. A. J. (2004). Supramolecular p-n-heterojunctions by co-self-organization of oligo(p-phenylene vinylene) and perylene bisimide dyes. *J. Am. Chem. Soc.*, *126*, 10611–10618.

41. Yagai, S., Hamamura, S., Wang, H., Stepanenko, V., Seki, T., Unoike, K., Kikkawa, Y., Karatsu, T., Kitamura, A., Würthner, F. (2009). Unconventional hydrogen-bond-directed hierarchical co-assembly between perylene bisimide and azobenzene-functionalized melamine. *Org. Biomol. Chem.*, *7*, 3926–3929.

42. Würthner, F., Bauer, C., Stepanenko, V., Yagai, S. (2008). A black perylene bisimide super gelator with an unexpected J-type absorption band. *Adv. Mater.*, *20*, 1695–1698.

43. Wicklein, A., Ghosh, S., Sommer, M., Würthner, F., Thelakkat, M. (2009). Self-assembly of semiconductor organogelator nanowires for photoinduced charge separation. *ACS Nano*, *3*, 1107–1114.

44. Yagai, S., Seki, T., Karatsu, T., Kitamura, A., Würthner, F. (2008). Transformation from H- to J-aggregated perylene bisimide dyes by complexation with cyanurates. *Angew. Chem. Int. Ed.*, *47*, 3367–3371.

45. Wiebenga, E. H. (1952). Crystal structure of cyanuric acid. *J. Am. Chem. Soc.*, *74*, 6156–6157.

46. Tsuda, A., Sakamoto, S., Yamaguchi, K., Aida, T. (2003). A novel supramolecular multicolor thermometer by self-Assembly of a π-extended zinc porphyrin complex. *J. Am. Chem. Soc.*, *125*, 15722–15723.

47. Green, B. R., Parson, W.W., eds. (2003). *Light-Harvesting Antennas in Photosynthesis*, *Advances in Photosynthesis and Respiration series*, Vol. 13. Kluwer Academic Publishers, Dordrecht, The Netherlands.

48. Hoeben, F. J. M., Shklyarevskiy, I. O., Pouderoijen, M. J., Engelkamp, H., Schenning, A. P. H. J., Christianen, P. C. M., Maan, J. C., Meijer, E. W. (2006). Direct visualization of efficient energy transfer in single oligo(p-phenylene vinylene) vesicles. *Angew. Chem. Int. Ed.*, *45*, 1232–1236.

49. Sugiyasu, K., Fujita, N., Shinkai, S. (2004). Visible-light-harvesting organogel composed of cholesterol-based perylene derivatives. *Angew. Chem. Int. Ed.*, *43*, 1229–1233.

50. Del Guerzo, A., Olive, A. G. L., Reichwagen, J., Hopf, H., Desvergne, J.-P. (2005). Energy transfer in self-assembled [n]-acene fibers involving≥100 donors per acceptor. *J. Am. Chem. Soc.*, *127*, 17984–17985.

51. Hoeben, F. J. M., Herz, L. M., Daniel, C., Jonkeijm, P., Schenning, A. P. H. J., Silva, C., Meskers, S. C. J., Beljonne, D., Phillips, R. T., Friend, R. H., Meijer, E. W. (2004). Efficient energy transfer in mixed columnar stacks of hydrogen-bonded oligo(p-phenylene vinylene)s in solution. *Angew. Chem. Int. Ed.*, *43*, 1976–1979.

52. Wolffs, M., Hoeben, F. J. M., Beckers, E. H. A., Schenning, A. P. H. J., Meijer, E. W. (2005). Sequential energy and electron transfer in aggregates of tetrakis[oligo(p-phenylene vinylene)] porphyrins and C60 in water. *J. Am. Chem. Soc.*, *127*, 13484–13485.

53. Tamiaki, H., Miyatake, T., Tanikaga, R., Holzwarth, A. R., Schaffner, K. (1996). Self-assembly of an artificial light-harvesting antenna: energy transfer from a zinc chlorin to a bacteriochlorin in a supramolecular aggregate. *Angew. Chem. Int. Ed. Engl.*, *35*, 772–774.

54. Praveen, V. K., George, S., J., Varghese, R., Vijayakumar, C., Ajayaghosh, A. (2006). Self-assembled π-nanotapes as donor scaffolds for selective and thermally gated fluorescence resonance energy transfer (FRET). *J. Am. Chem. Soc.*, *128*, 7542–7550.

55. Zhang, X., Chen, Z., Wuerthner, F. (2007). Morphology control of fluorescent nanoaggregates by co-self-assembly of wedge- and dumbbell-shaped amphiphilic perylene bisimides. *J. Am. Chem. Soc.*, *129*, 4886–4887.

56. Zhang, X., Rehm, S., Safont-Sempere, M. M., Würthner, F. (2009). Vesicular perylene dye nanocapsules as supramolecular fluorescent pH sensor system. *Nat. Chem.*, *1*, 623–629.

57. Hill, J. P., Jin, W., Kosaka, A., Fukushima, T., Ichihara, H., Shimomura, T., Ito, K., Hashizume, T., Ishii, N., Aida, T. (2004). Self-assembled hexa-peri-hexabenzocoronene graphitic nanotube. *Science*, *304*, 1481–1483.

58. Zimmerman, S. C., Corbin, P. S. (2000). Heteroaromatic modules for self-assembly using multiple hydrogen bonds. *Struct. Bond.*, *96*, 63–94.

59. Sijbesma, R. P., Meijer, E. W. (2003). Quadruple hydrogen bonded systems. *Chem. Commun.*, 5–16.

60. Binder, W., ed. (2007). *Hydrogen-Bonded Polymers*, *Advances in Polymer Science*, Vol. 207. Springer-Verlag, Heidelberg.

61. El-ghayoury, A., Peeters, E., Schenning, A. P. H. J., Meijer, E. W. (2000). Quadruple hydrogen bonded oligo(p-phenylene vinylene) dimers. *Chem. Commun.*, 1969–1970.

62. Schenning, A. P. H. J., Jonkheijm, P., Peeters, E., Meijer, E. W. (2001). Hierarchical order in supramolecular assemblies of hydrogen-bonded oligo(p-phenylene vinylene)s. *J. Am. Chem. Soc.*, *123*, 409–416.

63. Jonkheijm, P., Hoeben, F. J. M., Kleppinger, R., Van Herrikhuyzen, J., Schenning, A. P. H. J., Meijer, E. W. (2003). Transfer of π-conjugated columnar stacks from solution to surfaces. *J. Am. Chem. Soc.*, *125*, 15941–15949.

64. Jonkheijm, P., van der Schoot, P. P. A. M., Schenning, A. P. H. J., Meijer, E. W. (2006). Probing the solvent-assisted nucleation pathway in chemical self-assembly. *Science*, *313*, 80–83.

65. George, S. J., Tomovic, Z., Smulders, M. M. J., Greef, T. F. A., Leclère, P. E. L. G., Meijer, E. W., Schenning, A. P. H. J. (2007). Helicity induction and amplification in an oligo(p-phenylenevinylene) assembly through hydrogen-bonded chiral acids. *Angew. Chem. Int. Ed.*, *46*, 8206–8211.

66. Yagai, S., Kubota, S., Unoike, K., Karatsu, T., Kitamura, A. (2008). Cyanurate-guided self-assembly of a melamine-capped oligo(p-phenylenevinylene). *Chem. Commun.*, 4466–4468.

67. Tazawa, T., Yagai, S., Kikkawa, Y., Karatsu, T., Kitamura, A., Ajayaghosh, A. (2010). A complementary guest induced morphology transition in a two-component multiple H-bonding self-assembly. *Chem. Commun.*, *46*, 1076–1078.

68. Kimizuka, N., Kawasaki, T., Hirata, K., Kunitake, T. (1998). Supramolecular membranes. Spontaneous assembly of aqueous bilayer membrane via formation of hydrogen bonded pairs of melamine and cyanuric acid derivatives. *J. Am. Chem. Soc.*, *120*, 4094–4104.

69. Kawasaki, T., Tokuhiro, M., Kimizuka, N., Kunitake, T. (2001). Hierarchical self-assembly of chiral complementary hydrogen-bond networks in water: reconstitution of supramolecular membranes. *J. Am. Chem. Soc.*, *123*, 6792–6800.

70. Yagai, S., Karatsu, T., Kitamura, A. (2003). Photoresponsive melamine barbiturate hydrogen-bonded assembly. *Chem. Commun.*, 1844–1845.

71. Yagai, S., Nakajima, T., Karatsu, T., Saitow, K., Kitamura, A. (2004). Phototriggered self-assembly of hydrogen-bonded rosette. *J. Am. Chem. Soc.*, *126*, 11500–11508.

72. Yagai, S., Nakajima, T., Kishikawa, K., Kohmoto, S., Karatsu, T., Kitamura, A. (2005). Hierarchical organization of photoresponsive hydrogen-bonded rosettes. *J. Am. Chem. Soc.*, *127*, 11134–11139.

73. Yagai, S., Iwashima, T., Kishikawa, K., Nakahara, S., Karatsu, T., Kitamura, A. (2006). Photoresponsive self-assembly and self-organization of hydrogen-bonded supramolecular tapes. *Chem.—Eur. J.*, *12*, 3984–3994.

74. Drain, C. M., Fischer, R., Nolen, E. G., Lehn, J. M. (1993). Self-assembly of a bisporphyrin supramolecular cage induced by molecular recognition between complementary hydrogen bonding sites. *J. Chem. Soc., Chem. Commun.*, 243–235.

75. Drain, C. M., Russell, K. C., Lehn, J.-M. (1996). Self-assembly of a multi-porphyrin supramolecular macrocycle by hydrogen bond molecular recognition. *Chem. Commun.*, 337–338.

76. Drain, C. M., Shi, X., Milic, T., Nifiatis, F. (2001). Self-assembled multiporphyrin arrays mediated by self-complementary quadruple hydrogen bond motifs. *Chem. Commun.*, 287–288.

77. Ohkawa, H., Takayama, A., Nakajima, S., Nishide, H. (2006). Cyclic tetramer of a metalloporphyrin based on a quadruple hydrogen bond. *Org. Lett.*, *8*, 2225–2228.

78. Zhu, P., Kang, H., Facchetti, A., Evmenenko, G., Dutta, P., Marks, T. J. (2003). Vapor phase self-assembly of electrooptic thin films via triple hydrogen bonds. *J. Am. Chem. Soc.*, *125*, 11496–11497.

79. Würthner, F., Yao, S., Heise, B., Tschierske, C. (2001). Hydrogen bond directed formation of liquid-crystalline merocyanine dye assemblies. *Chem. Commun.*, 2260–2261.

80. Würthner, F., Schmidt, J., Stolte, M., Wortmann, R. (2006). Hydrogen-bond-directed head-to-tail orientation of dipolar merocyanine dyes: a strategy for the design of electrooptical materials. *Angew. Chem. Int. Ed.*, *45*, 3842–3846.

81. Prins, L. J., Thalacker, C., Würthner, F., Timmerman, P., Reinhoudt, D. N. (2001). Chiral exciton coupling of merocyanine dyes within a well defined hydrogen-bonded assembly. *Proc. Natl. Acad. Sci. U.S.A.*, *98*, 10042–10045.

82. Würthner, F., Yao, S. (2003). Merocyanine dyes containing imide functional groups: synthesis and studies on hydrogen bonding to melamine receptors. *J. Org. Chem.*, *68*, 8943–8949.

83. Yagai, S., Higashi, M., Karatsu, T., Kitamura, A. (2004). Binary supramolecular gels based on bismelamine·cyanurate/Barbiturate noncovalent polymers. *Chem. Mater.*, *16*, 3582–3585.

84. Yagai, S., Higashi, M., Karatsu, T., Kitamura, A. (2005). Dye-assisted structural modulation of hydrogen-bonded binary supramolecular polymers. *Chem. Mater.*, *17*, 4392–4398.

85. Yagai, S., Higashi, M., Karatsu, T., Kitamura, A. (2006). Tunable interchromophore electronic interaction of a merocyanine dye in hydrogen-bonded supramolecular assemblies scaffolded by bismelamine receptors. *Chem. Commun.*, 1500–1502.

86. Saadeh, H., Wang, L., Yu, L. (2000). Supramolecular solid-state assemblies exhibiting electrooptic effects. *J. Am. Chem. Soc.*, *122*, 546–547.

87. Würthner, F., Thalacker, C., Sautter, A. (1999). Hierarchical organization of functional perylene chromophores to mesoscopic superstructures by hydrogen bonding and $\pi-\pi$ interactions. *Adv. Mater.*, *11*, 754–758.

88. Thalacker, C., Wurthner, F. (2002). Chiral perylene bisimide-melamine assemblies: hydrogen bond-directed growth of helically stacked dyes with chiroptical properties. *Adv. Funct. Mater.*, *12*, 209–218.

89. Yagai, S., Monma, Y., Kawauchi, N., Karatsu, T., Kitamura, A. (2007). Supramolecular nanoribbons and nanoropes generated from hydrogen-bonded supramolecular polymers containing perylene bisimide chromophores. *Org. Lett.*, *9*, 1137–1140.

90. Seki, T., Yagai, S., Karatsu, T., Kitamura, A. (2008). Formation of supramolecular polymers and discrete dimers of perylene bisimide dyes based on melamine-cyanurates hydrogen-bonding interactions. *J. Org. Chem.*, *73*, 3328–3335.

91. Seki, T., Yagai, S., Karatsu, T., Kiamura, A. (2008). Miniaturization of nanofibers composed of melamine-appended perylene bisimides and cyanurates. *Chem. Lett.*, *37*, 764–765.

92. Whitesides, G. M., Simanek, E. E., Mathias, J. P., Seto, C. T., Chin, D., Mammen, M., Gordon, D. M. (1995). Noncovalent synthesis: using physical-organic chemistry to make aggregates. *Acc. Chem. Res.*, *28*, 37–44.

93. Prins, L. J., Reinhoudt, D. N., Timmerman, P. (2001). Noncovalent synthesis using hydrogen bonding. *Angew. Chem. Int. Ed.*, *40*, 2382–2426.

94. Yagai, S. (2006). Supramolecular complexes of functional chromophores based on multiple hydrogen-bonding interactions. *J. Photochem. Photobiol., C*, *7*, 164–182.

95. Schenning, A. P. H. J., van Herrikhuyzen, J., Jonkheijm, P., Chen, Z., Wuerthner, F., Meijer, E. W. (2002). Photoinduced electron transfer in hydrogen-bonded oligo(p-phenylene vinylene)-perylene bisimide chiral assemblies. *J. Am. Chem. Soc.*, *124*, 10252–10253.

96. Yagai, S., Kinoshita, T., Higashi, M., Kishikawa, K., Nakanishi, T., Karatsu, T., Kitamura, A. (2007). Diversification of self-organized architectures in supramolecular dye assemblies. *J. Am. Chem. Soc.*, *129*, 13277–13287.

97. Hirst, A. R., Smith, D. K., Feiters, M. C., Geurts, H. P. M., Wright, A. C. (2003). Two-component dendritic gels: easily tunable materials. *J. Am. Chem. Soc.*, 125, 9010–9011.

98. Jonkheijm, P., Miura, A., Zdanowska, M., Hoeben, F. J. M., De Feyter, S., Schenning, A. P. H. J., De Schryver, F. C., Meijer, E. W. (2004). π-Conjugated oligo(*p*-phenylenevinylene) rosettes and their tubular self-assembly. *Angew. Chem. Int. Ed.*, *43*, 74–78.

99. Bielejewska, A. G., Marjo, C. E., Prins, L. J., Timmerman, P., de Jong, F., Reinhoudt, D. N. (2001). Thermodynamic stabilities of linear and crinkled tapes and cyclic rosettes in melamine-cyanurate assemblies: a model description. *J. Am. Chem. Soc.*, *123*, 7518–7533.

100. MacDonald, J. C., Whitesides, G. M. (1994). Solid-state structures of hydrogen-bonded tapes based on cyclic secondary diamides. *Chem. Rev.*, *94*, 2383–2420.

101. Huang, X., Li, C., Jiang, S., Wang, X., Zhang, B., Liu, M. (2004). Self-assembled spiral nanoarchitecture and supramolecular chirality in Langmuir-Blodgett films of an achiral amphiphilic barbituric acid. *J. Am. Chem. Soc.*, *126*, 1322–1323.

102. Huang, C.-H., McClenaghan, N. D., Kuhn, A., Hofstraat, J. W., Bassani, D. M. (2005). Enhanced photovoltaic response in hydrogen-bonded all-organic devices. *Org. Lett.*, *7*, 3409–3412.

103. Yagai, S., Kinoshita, T., Kikkawa, Y., Karatsu, T., Kitamura, A., Honsho, Y., Seki, S. (2009). Interconvertible oligothiophene nanorods and nanotapes with high charge carrier mobilities. *Chem.—Eur. J.*, *15*, 9320–9324.

104. Kang, J., Rebek, J. (1996). Entropically driven binding in a self-assembling molecular capsule. *Nature*, *382*, 239–241

105. Yagai, S., Kubota, S., Saito, H., Unoike, K., Karatsua, T., Kitamura, A., Ajayaghosh, A., Kanesato, M., Kikkawa, Y. (2009). Reversible transformation between rings and coils in a dynamic hydrogen-bonded self-assembly. *J. Am. Chem. Soc.*, *131*, 5408–5410.

106. Yagai, S., Mahesh, S., Kikkawa, Y., Unoike, K., Karatsu, T., Kitamura, A., Ajayaghosh, A. (2008). Toroidal nanoobjects from rosette assemblies of melamine-linked oligo(*p*-phenyleneethynylene)s and cyanurates. *Angew. Chem. Int. Ed.*, *47*, 4691–4694.

107. Kim, J.-K., Lee, E., Huang, Z., Lee, M. (2006). Nanorings from the self-assembly of amphiphilic molecular dumbbells. *J. Am. Chem. Soc.*, *128*, 14022–14023.

108. Kim, J.-K., Lee, E., Kim, M.-C., Sim, E., Lee, M. (2009). Reversible transformation of helical coils and straight rods in cylindrical assembly of elliptical macrocycles. *J. Am. Chem. Soc.*, *131*, 17768–17770.

109. Hill, D. J., Mio, M. J., Prince, R. B., Hughes, T. S., Moore, J. S. (2001). A field guide to foldamers. *Chem. Rev.*, *101*, 3893–4011.

110. Klug, A. (1983). From macromolecule to biological molecular assembly. *Angew. Chem. Int. Ed. Engl.*, *95*, 579–636.

111. Passy, S. I., Yu, X., Li, Z., Radding, C. M., Egelman, E. H. (1999). Rings and filaments of β protein from bacteriophage λ suggest a superfamily of recombination proteins. *Proc. Natl. Acad. Sci. U.S.A.*, *96*, 4279–4284.

112. Dormidontova, E. E. (2002). Role of competitive PEO-water and water-water hydrogen bonding in aqueous solution PEO behavior. *Macromolecules*, *35*, 987–1001.

113. Prince, R. B., Saven, J. G., Wolynes, P. G., Moore, J. S. (1999). Cooperative conformational transitions in phenylene ethynylene oligomers: chain-length dependence. *J. Am. Chem. Soc.*, *121*, 3114–3121.

114. Yashima, E., Maeda, K., Sato, O. (2001). Switching of a macromolecular helicity for visual distinction of molecular recognition events. *J. Am. Chem. Soc.*, *123*, 8159–8160.

115. Kim, H.-J., Lee, E., Park, H.-S., Lee, M. (2007). Dynamic extension-contraction motion in supramolecular springs. *J. Am. Chem. Soc.*, *129*, 10994–10995.

116. Percec, V., Rudick, J. G., Peterca, M., Heiney, P. A. (2008). Nanomechanical function from self-organizable dendronized helical polyphenylacetylenes. *J. Am. Chem. Soc.*, *130*, 7503–7508.

117. Li, L.-S., Jiang, H., Messmore, B. W., Bull, S. R., Stupp, S. I. (2007). A torsional strain mechanism to tune pitch in supramolecular helices. *Angew. Chem. Int. Ed.*, *46*, 5873–5876.

CHAPTER 6

ANION-RESPONSIVE SUPRAMOLECULAR DYE CHEMISTRY

Hiromitsu Maeda

College of Pharmaceutical Sciences, Institute of Science and Engineering, Ritsumeikan University, Kusatsu, Japan
PRESTO, Japan Science and Technology Agency (JST), Kawaguchi, Japan

6.1 INTRODUCTION

Recently, soft materials have been attracting increasing attention as transformable functional materials because of their moderate mobility and flexibility, which readily enable them to change their bulk shape and properties depending on the ambient conditions [1]. In particular, soft materials comprising π-conjugated molecules exhibit attractive properties and have the potential to be used for the fabrication of electronic and optical devices, the properties of which can be tuned according to the environment. The introduction of building subunits that are responsive to chemical stimuli would result not only in the supramolecular structures exhibiting stimuli-responsive behaviors but also in the formation of various materials consisting of multiple components. Therefore, the combination of these building units and additives (chemical stimuli) provides numerous soft materials having potential utility under appropriate conditions.

Gels are soft materials that are fairly less mobile agglomerates than the solution state; these agglomerates exhibit mechanical properties that are characteristic of a solid. Some of the gels incorporate solvent molecules into a 3-D entangled network of dimensionally controlled fibril- and tape-like organized aggregates consisting of gelators. In contrast to polymer gels [2], supramolecular gels are dimensionally controlled assemblies that consist of low-molecular-weight (LMW) molecules held together by noncovalent interactions such as hydrogen bonding, metal coordination, van der Waals interaction, and π–π stacking. Gels derived from molecular assemblies, the components of which can be readily

Supramolecular Soft Matter: Applications in Materials and Organic Electronics, First Edition.
Edited by Takashi Nakanishi.

replaced with alternatives, may provide promising materials systems that can be used in drug delivery and tissue engineering [3]. The size of the organized structures in the supramolecular gels is limited to the scale range from micrometers to nanometers. Smaller scale materials allow the assemblies to disperse in solutions, whereas larger scale materials afford precipitates. Crystals are fairly ordered organized structures as compared to dimensionally controlled ones. The initial step in the formation of a gel involves obtaining appropriate-scale (width and length) gelator molecules that can form the fibril- and tape-like structures that compose gels. Such molecular assemblies can be influenced by external conditions, including the solvent, temperature, concentration, and additives.

Chemical stimuli could afford versatile supramolecular gels that change their states depending on the interactions between the additives and gelator molecules. For example, Shinkai *et al.* reported that cholesterol-substituted phenanthroline derivatives gelate various alcohols, polar solvents, and nonpolar solvents and that propanol gel exhibits tunable electronic states and emission behaviors when trifluoroacetic acid (TFA) is added [4]. Ghoussoub and Lehn produced pH-dependent hydrogels made of amide-bridged guanine–polyethylene glycol (PEG) hybrids in the presence of potassium ions in the form of KCl and [2.2.2]cryptand [5]. Furthermore, Xu *et al.* reported the synthesis of a 9-fluorenylmethyloxycarbonyl (Fmoc)-substituted D-alanine dimer that forms a hydrogel via interactions between its fluorenyl moieties and by hydrogen bonding between its amide units. The hydrogel transforms into a solution after the addition of vancomycin [6]. In these cases, the additives (chemical stimuli) appear to behave as inhibitors of gel formation by interacting at the gelating sites of the building components, which are crucial for molecular assembly. If gelators and additives are appropriately selected, the combination of supramolecular gels responsive to charged species can form functional materials and devices, which consist of cations and anions in an ordered arrangement.

Anions play essential roles in controlling the gel states in biotic systems. For example, iodide (I^-) causes the depolymerization of actin filaments (F-actin) and the corresponding transition from a gel-like state to a solution state by influencing the hydration shell that surrounds and stabilizes the F-actin polymers [7]. In this case, the anion affects biotic macromolecular systems such as protein and gel-like states. Moreover, various artificial anion receptors have been actively synthesized in the past two decades [8]. However, only a few reports currently exist on anion-controlled supramolecular gels consisting of gelator molecules with an anion-binding ability. Furthermore, most of the anion-responsive gels reported thus far comprise anion receptors bearing amide or urea units, which also act as hydrogen-bonding sites that support molecular assembly. If the hydrogen-bonding donor NH sites in the amide or urea units interact with anions, the gelators may not form stable assemblies, resulting in the transformation of the gels, for example, into a solution. In addition, the multimodal interactions of the gelators with anions would afford various types of transformations. Of course, there are examples where gelator (receptor)–anion complexes form more ordered states, such as crystals and gels. Such transitions have also been observed between crystals and monomers in solution. In any case, the solvent is a crucial factor that dominates the properties of gels. This solvent-based definition of gels conveys diversified and new features of the chemistry of supramolecular assembly and anion binding.

In this chapter, the focus is on supramolecular gels of which the structures and properties are tunable by electrostatic hydrogen bonding between anion stimuli and the gelator-building components [9]. To be exact, anion-responsive gel materials may include supramolecular gels consisting of anionic gelators, cationic gelators with counter anions [10], and systems derived from ionic liquids [10b,11], which have been and will be reviewed elsewhere [3c]. Anion-responsive supramolecular gels are not simply materials that exhibit state transitions in response to anion stimuli but are promising "gel salts" that contain small amounts of ionic species because of their structural design and the modifications to their building components (gelators, anions, and cations).

6.2 HYDROGEN-BONDING-BASED ANION-RESPONSIVE SUPRAMOLECULAR GELS

NH units are well-known anion-binding sites owing to the polarization of π-conjugation with the neighboring carbonyl moiety [8]. For example, Thordarson et al. reported an alkyl-substituted pyromellite tetraamide derivative (**1**, Fig. 6-1a) that gelates cyclohexane, hexane, diethyl ether, and toluene as a result of hydrogen bonding between the gelators. Tetraamide **1** binds to anions that are added as tetrabutylammonium (TBA) salts through conformation changes of its amide units in dilute solutions. Therefore, the cyclohexane gel of **1** is transformed to a solution by the addition of 0.25 equivalent of TBA salts (Cl$^-$, Br$^-$, I$^-$, AcO$^-$, and NO$_3{}^-$) due to the disruption in hydrogen bonding between the gelator molecules and anions. The transformation from a gel to a solution on adding TBA salts on the gel surface takes several seconds to minutes, depending on the affinity of **1** for the anions (Cl$^-$ > AcO$^-$ > Br$^-$ > NO$_3{}^-$ > I$^-$). In contrast to organic salts, NaI does not induce such a transition to the solution state [12]. Žinić et al. reported an oxalamide-substituted anthraquinone (**2**, Fig. 6-1b), which gelates alcohols (ethanol and 1-butanol) and aromatic solvents (benzene, toluene, and p-xylene) through $\pi-\pi$ stacking between the gelator planes. In sharp contrast to compound **2**, the regioisomeric 2-substituted derivative only affords precipitates. This result suggests that the positions of the amide-substituents significantly influence the gelation abilities. Addition of 10 equivalents of tetrabutylammonium fluoride (TBAF) transforms the p-xylene gel of **2** into a solution with an accompanying color change. This is due to the interactions of F$^-$ with the amide groups, which interferes with the hydrogen bonding between the gelator molecules. In contrast, the ethanol gel of **2** changes color only after F$^-$ is added, which does not entirely disrupt the intermolecular interaction between the gelators due to the solvation of F$^-$ by the protic ethanol molecules [13].

Wei et al. reported an acylhydrazone molecule (**3**, Fig. 6-1c), which forms stable yellow-colored organogels from DMF (dimethylformamide) and DMSO (dimethyl sulfoxide) but not from hydrocarbon solvents, ethers, ethanol, and tetrahydrofuran (THF) because of the poor solubility of **3**. The molecular

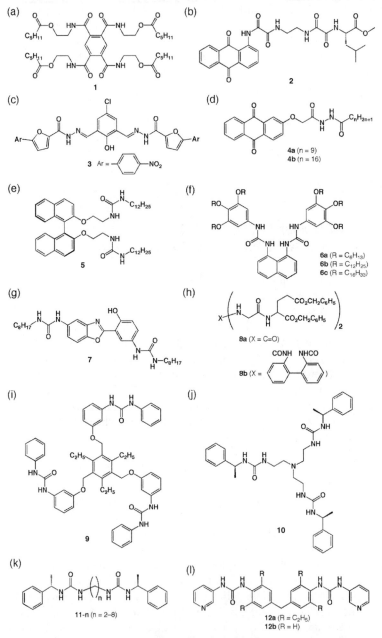

Figure 6-1   (a) Pyromellite tetraamide derivative **1**; (b) oxalamide-substituted anthraquinone derivative **2**; (c) phenol-based acylhydrazone **3**; (d) hydrazide-substituted anthraquinone derivative **4a,b**; (e) urea-substituted binaphthyl derivative **5**; (f) urea-substituted naphthalene derivatives **6a–c**; (g) urea-pendant 2-(2-hydroxyphenyl)benzoxazole **7**; (h) glycine-glutamic acid-based bisurea derivatives **8a,b**; (i) benzene-coupled triurea **9**; (j) triurea **10**; (k) bisurea gelators **11-n** ($n = 2 - 8$); (l) pyridyl-substituted bisureas **12a,b**.

structure and phenol OH group are found to play a critical role in the gelation process. Addition of F^-, AcO^-, and $H_2PO_4^-$ (5 equivalents) as solid TBA salts to the DMF gel of **3** at 20°C causes the gradual decomposition of the gelated state to wine-colored (F^-) and red (AcO^- and $H_2PO_4^-$) solutions. Conversely, the addition of Cl^-, Br^-, I^-, HSO_4^-, and ClO_4^- as TBA salts does not lead to gel decomposition. From ^1H NMR (nuclear magnetic resonance) analyses, anion-induced deprotonation at OH and NH causes gel–sol transition. Furthermore, addition of $HClO_4$, H_2O, or methanol to the F^--disintegrated solution causes the gels to re-form [14]. Chen and Ma reported anthraquinone-based hydrazide derivatives (**4a,b**, Fig. 6-1d) that form gels from various solvents such as toluene, 1,2-dichloroethane, $CHCl_3$, chlorobenzene, bromobenzene, 1,2-dichlorobenzene, ethanol, 2-propanol, 1-butanol, CCl_4, cyclohexanone, acetone, ethyl acetate, methyl benzoate, THF, and o-xylene. Various spectra such as ^1H NMR, infrared (IR), UV/vis absorption, and powder X-ray diffraction (XRD) suggest that the $\pi-\pi$ interactions of anthraquinone and hydrogen bonding at the NH sites are essential for organized structures. Anion-responsive gel–sol transitions were observed on addition of F^-, AcO^-, and $H_2PO_4^-$ (40 equivalents) as TBA salts to the $CHCl_3$ gel of **4a** and were possibly due to the deprotonation of hydrazide NH. In contrast to these anions, addition of Cl^-, Br^-, and I^- preserved the gel state [15].

Urea moieties are also used as anion recognition sites because of their polarized NH units [8]. Wang *et al.* synthesized a urea-substituted binaphthalene derivative (**5**, Fig. 6-1e), which produces transparent supramolecular gels (6 mg/mL) from cyclohexane through both hydrogen bonding and $\pi-\pi$ stacking interaction and has a transition temperature of 20°C. Addition of TBAF (1 equivalent) interferes with the hydrogen bonding between the gelator molecules and thereby affords a solution. In this case, significant absorption spectral changes caused by F^- binding were not observed. This result is due to σ-bonding between the hydrogen-bonding amide units and the chromophore binaphthyl unit [16]. On the other hand, Yi and Li *et al.* synthesized urea-bridged hybrids of alkyl-substituted aryl units and the naphthyl moiety (**6a–c**, Fig. 6-1f), which use hydrogen bonding and $\pi-\pi$ stacking interaction to form emissive supramolecular gels from various solvents. These gels are decomposed to solutions by the addition of F^- but re-form with the addition of H^+. However, the re-formed gels exhibit emission behaviors and organized structures different from those of the original gels [17]. Lee *et al.* reported gelation-induced fluorescence enhancement and F^--responsive behavior of urea-pendant 2-(2-hydroxyphenyl)benzoxazole (HPB) (**7**, Fig. 6-1g). A powdered form of **7** was completely dissolved in hot DMF, and cold toluene was added drop by drop to the solution (0.5 wt%, DMF:toluene = 1:9 (v/v)), giving a stable and highly fluorescent gel. The reference molecule that lacked a phenol OH form did not form a gel because the molecular stacking between the HPB units was disrupted due to lack of planarity. Therefore, the planarity of HPB from intramolecular hydrogen bonding is one of the driving forces for gelation. Placing F^- on top of the DMF/toluene gel immediately produced a gel-to-sol transition with a color change from a translucent colorless gel to a solution with a strong greenish emission, although the disruption of the gel structure appeared in the presence of other anions (Cl⁻, Br⁻, and AcO⁻) [18]. Furthermore, Teng *et al.* prepared glycine-glutamic acid-based bisurea derivatives (**8a,b**, Fig. 6-1h), which can be used to form gels of aromatic solvents (**8a**) and alcohol solvents (**8b**). These gelator molecules contained both urea and amide moieties. The addition of F⁻ as a TBA salt to a toluene gel of **8a** resulted in a rapid transition from a translucent gel to a homogeneous solution because of the disruption of hydrogen bonding between gelator molecules. Conversely, the gel of **8a** could be preserved after the addition of a small amount of Cl⁻, Br⁻, and I⁻. In sharp contrast, the ethanol gel of **8b** was retained without any changes, even after the addition of F⁻ (20 equivalents) as a TBA salt. In this case, hydrogen-bonding interactions were not significantly influenced by the addition of F⁻ [19].

Yamanaka *et al.* synthesized a triurea-coupled benzene (**9**, Fig. 6-1i) that affords opaque supramolecular gels from acetone after sonication. However, cooling the thermally dissolved acetone solution resulted in precipitation rather than gel formation. Compound **9** gelates in diethyl phthalate, methanol, and THF but not in nonpolar solvents such as hexane, toluene, and CH_2Cl_2. IR measurements demonstrated that hydrogen bonding at the urea units is an essential factor for gel formation. This result is also supported by the gelation behaviors of derivatives lacking these interaction sites. Similar to other examples, acetone gel is transformed to solution by the addition of TBAF and TBACl. Required amounts for the transitions increase with the ionic radii of the anions, and are correlated to the binding affinities of **9** for the corresponding anions. Furthermore, addition of $BF_3 \cdot OEt_2$ to the F⁻-mediated solution state enables the recovery of the gel state, whereas addition of $ZnBr_2$ mediated by anions such as halides (F⁻, Cl⁻, Br⁻, and I⁻) and AcO⁻ also affords supramolecular gels [20].

Here, gel systems that contain anion-binding receptors have been introduced, which exhibit state transitions without direct interactions with anions. Steed *et al.* synthesized a triurea-substituted molecule (**10**, Fig. 6-1j) that exhibits Cl⁻ affinity (ca 10^3 M⁻¹ in DMSO) and gelates in DMSO–H_2O (1:1 (v/v), 0.5 wt%). Addition of 1 equivalent of NaCl to the gel impedes the gelation process and decreases the amount of solvents included in the gel, thereby resulting in the formation of both gel and crystals. Single-crystal diffraction analysis revealed that the crystals that formed in the absence of Cl⁻ salts comprised only the gelator molecules that were bound through intramolecular and intermolecular hydrogen bonding between the urea units [21]. Steed *et al.* also reported alkyl-bridged bisurea compounds (**11-n**, Fig. 6-1k), which form gels when they encounter an even number of methylene units ($n = 2, 4, 6,$ and 8) but do not gel in the presence of an odd number of methylene units ($n = 3, 5,$ and 7) in CH_3CN, $CHCl_3$, MeOH, toluene, and solvent mixtures such as DMSO–H_2O and MeOH–H_2O (1 wt%). These differences in the even- and odd-number effect of the spacing of the methylene units are correlated with the orientation of the urea units at the interaction sites, which is also corroborated by XRD analysis. Addition of anions (0.1 equivalent as TBA salts) to the gels of **11-n** affects the storage modulus, which is augmented in the order corresponding to the addition of AcO$^-$ (small), Cl$^-$, Br$^-$, BF$_4$$^-$, and the absence of anions (large), as a function of oscillation stress. The degrees to which anions inhibit gelation are, to some extent, correlated with the anion–gelator binding affinities [22].

Apart from gel-to-solution transitions caused by the dispersion of the building units, the formation of new aggregates or crystals has also been reported. Steed *et al.* synthesized a pyridyl-substituted bisurea derivative (**12a**, Fig. 6-1l) that gelates CHCl$_3$/MeOH (1:1 (v/v), 10 mM) via hydrogen bonding. Bisurea **12a** remains as a gel with metal coordination polymers after the addition of AgPF$_6$ and AgNO$_3$. The organized structures observed by scanning electron microscopy (SEM) are distinct and depend on the coexisting anionic species. In contrast, bisurea **12b**, which does not have alkyl substituents, affords crystals by itself and gelates THF/H$_2$O (2:1 (v/v)) in the presence of AgBF$_4$. In contrast to **12a**, this compound (**12b**) forms crystals when AgNO$_3$ is added. Single-crystal X-ray analysis has revealed that metal coordination is essential for aggregation and that interactions between urea units and anions affect the assembled structures. NO$_3$$^-$ associates fairly well with urea sites to form crystals, whereas BF$_4$$^-$ allows the formation of a more soluble state and thus supramolecular gels [23,24].

6.3 METAL-COORDINATED GELS RESPONSIVE TO ANIONS

As observed in the supramolecular gels in Ref. [23], metal complexation of the gelator molecules is also an essential factor that is required to stabilize molecular assemblies in both discrete and dispersed coordination forms. Metal-coordinated supramolecular gels can exhibit state transformation by the anion-stimulated "collapse" of the building units, which are the metal complexes in this case. Wu *et al.* generated the amide-bridged hybrid of an alkyl-substituted aryl moiety and a pyridyl unit, which affords a Ag$^+$-bridged dimer complex. The triflate salt of the Ag$^+$ complex (**13**, Fig. 6-2a) gelates toluene/ethanol (10:1 (v/v)) via hydrogen bonding and aromatic interactions between the Ag^{I} complexes. In contrast, Co^{2+}/Cu^{2+}/Ni^{2+} complexes do not produce gels because of their nonlinear coordination behaviors, which are not effective for the formation of organized structures. By adding KI in ethanol, the Ag$^+$ complex, as a triflate salt (**13**), provides a free ligand, and, as a result, a solution containing AgI precipitates is generated. Addition of excess Ag$^+$ salts re-forms the gel state. In addition to I$^-$, other anions such as Br$^-$ and Cl$^-$ also produce transitions from gels to solutions by the formation of their Ag$^+$ salts [25].

The anion-driven liberation of metal ions has been shown to control emission behavior. Aida *et al.* reported an alkyl-substituted pyrazole–Au$^+$ trinuclear complex (**14**, Fig. 6-2b(i)) that forms a red-emissive supramolecular gel (λ_{em} = 640 nm, λ_{ex} = 284 nm) from hexane. On addition of AgOTf, the gel is transformed into a blue-emissive gel (λ_{em} = 458 nm, λ_{ex} = 370 nm), which re-forms into a red-emissive gel when Cl$^-$ is added as a cetyltrimethylammonium salt (Fig. 6-2b(ii)). In this system, the Ag$^+$ ions, which are associated with Au$^+$ ions through metal–metal interactions, are partially inserted into the stacking

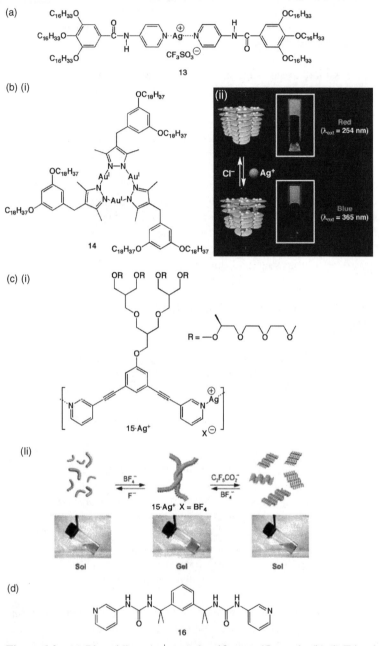

Figure 6-2 (a) Bispyridine–Ag$^+$ complex **13** as a triflate salt. (b) (i) Triazole–Au$^+$ complex **14** and (ii) emission control of supramolecular gels by Ag$^+$/Cl$^-$ addition. *Source*: Redrawn from Ref. [26]. Copyright (2005) ACS. (c) (i) Ag$^+$ complex of dendrimer-like, oxyethylene-substituted bispyridine derivative **15**. Ag$^+$ and (ii) reversible transitions between morphologies of coordination polymers and the corresponding gel/solution states by anion exchanges. *Source*: Redrawn from Ref. [27b]. Copyright (2005) Wiley. (d) Pyridyl-attached bisurea molecule **16**.

planes of the trinuclear complexes and thereby create a space between the planes. It is this Ag–Au interaction in the blue-emissive gel that is disrupted by the added Cl^- ions, which associate with the Ag^+ ions to re-form the red-emissive gel. Control of red, green, and blue (RGB) colors by physical and chemical stimuli has been achieved using a green-emissive solution ($\lambda_{em} = 510$ nm, $\lambda_{ex} = 370$ nm), which is obtained by heating a blue-emissive gel [26].

Anions can indeed control the state of a gel without the need for specific interactions with the gelators (molecules or complexes). For example, Lee *et al.* prepared a dendrimer-like, oxyethylene-substituted bispyridine derivative (**15**, Fig. 6-2c(i)) that forms coordination polymers or oligomers by Ag^+ complexation, such as dispersed helical structures and discrete cyclic conformations. The morphologies of the organized structures depend on the coexisting counter anions such as NO_3^-, BF_4^-, and $CF_3SO_3^-$. The assembled structures can be modified by electrostatic interactions between **15**·Ag^+ and the anions, whose ionic radii are crucial in determining the morphologies. These dispersed and discrete coordination assemblies from **15**·Ag^+ gelate methanol, suggesting the formation of soft materials derived from different topologies of coordination polymers and oligomers. Addition of F^- to the BF_4^- salt of **15**·Ag^+, which has an entangled helical conformation in the gel state, transforms the gel to the solution state, which includes free monomers of **15**, by the formation of AgF salts. Furthermore, the addition of $C_2F_5CO_2^-$ also transforms the helical structures of the gel to fairly dispersed assemblies of linear polymers, thereby affording the solution state. In all of these cases, a reversible transition between the solution and gel states has been observed (Fig. 6-2c(ii)) [27]. In addition, Clarke and Steed *et al.* reported a pyridyl-attached bisurea molecule (**16**, Fig. 6-2d), which was found to gelate a methanol solution at 1 wt% in the presence of 0.2–0.5 equivalent of $CuCl_2$, even though **16** does not form a gel by itself. Addition of more than ca 0.5 equivalent of $CuCl_2$ results in the precipitation of a new crystalline phase with a greenish color and the disappearance of the gel. The crystallization of **16** with $Cu(NO_3)_2$ does not result in gelation but crystallization, suggesting that hydrogen bonding and the replacement of NO_3^- with coordinating Cl^- could result in more flexible chains, leading to gelation instead of crystallization. Effects of the addition of Cl^- and AcO^- as TBA salts on the rheological properties were also observed [28].

6.4 PYRROLE-BASED, ANION-RESPONSIVE π-CONJUGATED MOLECULES THAT FORM SUPRAMOLECULAR ASSEMBLIES

Thus far, some examples of anion-responsive supramolecular gels have been illustrated. However, in order to systematically examine anion-controllable gels systems, the synthesis of a series of anion-responsive gelator molecules that can be modified to form appropriate structures for desired functional materials is discussed. However, current investigations on such anion-responsive supramolecular gels appear to be quite primitive. Therefore, it is essential to design and synthesize fairly planar π-conjugated systems that can efficiently bind anions. Focusing on the components, pyrrole is one building block for planar anion receptors that is well known as a π-conjugated aromatic heterocyclic molecule [29]. Pyrrole is found not only in biotic dyes such as heme and chlorophyll but also in artificial porphyrin derivatives [30]. In contrast to the relatively inert benzene, pyrrole has more electrons (six) than atoms (five) in its framework, and therefore, pyrrole moieties are reactive and stabilized by, for example, incorporation into aromatic macrocycles such as porphyrins or substitution by electron-withdrawing moieties. Pyrrole exhibits "duality" at its nitrogen moiety, which behaves both as a hydrogen-bonding acceptor or a metal coordination ligand at the N site and a hydrogen-bonding donor due to the NH site. Furthermore, since the π-planes of the pyrrole unit also enable stacking and formation of π-ligands for metal complexing, pyrrole rings can act as potential building blocks for nanoscale supramolecular structures. However, pyrrole rings are often in preorganized macrocycles such as porphyrins, wherein their N sites are located on the inside of fairly rigid closed structures, and consequently, pyrroles cannot use all of their potential as interaction sites. Therefore, new aspects of pyrrole rings may be revealed by exploring novel *acyclic* oligopyrrolic systems, which may find application in functional supramolecular materials that exploit these new properties and phenomena. In particular, supramolecular assemblies built from planar, pyrrole-based anion receptors would show anion-responsive behavior and, under appropriate conditions, could form functional organized structures comprising cationic and anionic species. In this section, a brief introduction and recent progress of pyrrole-based anion receptors that can form soft materials such as supramolecular gels is given.

Recently, Maeda *et al.* began focusing on the synthesis of acyclic π-conjugated molecules with hydrogen-bonding donor pyrrole NH site(s) [31–34]. In 2005, Maeda *et al.* reported the first dipyrrolyldiketone BF$_2$ complex **17** (Fig. 6-3a), consisting of two pyrrole rings and a boron-bridged 1,3-propanedione moiety, as a candidate π-conjugated acyclic anion receptor with potential to form stacking structures. Pyrrole rings, which exist even in acyclic structure, are stabilized by the neighboring electron-withdrawing carbonyl unit. Therefore, the skeleton structure could be appropriate for various uses such as sensors, functional materials, and so on. The BF$_2$ complex **17** does not form a preorganized conformation because the two pyrrole NH are not located at the appropriate positions for anion binding. On addition of anions, receptor **17**, as a "molecular flipper," exhibits inversions of the two pyrrole rings and binds anions using the pyrrole NH and bridging CH to form a planar receptor–anion complex (Fig. 6-3a). N–H$\cdots$X$^-$ and bridging C–H$\cdots$X$^-$ interactions are implicated from the ^{1}H NMR chemical shifts of other molecular flippers on addition of anions as TBA salts. Furthermore, discrete resonances of the two species—the free receptor and the anion complex—suggest that the equilibrium between these states is too slow to be detected on the NMR timescale, possibly because of the requirement for pyrrole to invert before anion binding [32a]. The available library of pyrrole derivatives reported so far shows that the introduction of substituents to the receptor framework of

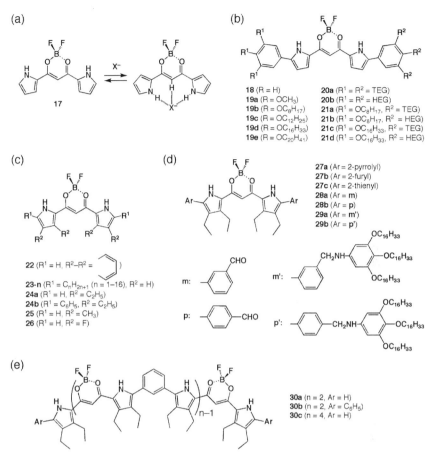

Figure 6-3 (a) Acyclic anion receptor **17** and anion-binding mode; (b–e) various derivatives of dipyrrolyldiketone BF₂ complexes.

17 yields a variety of molecular flippers (Fig. 6-3b–e). In fact, Maeda *et al.* prepared the α-aryl-substituted **18–21** [33a,d,e,g] and β-benzo-fused **22** [33h] along with α-alkyl-substituted **23-n** [32a,e], β-alkyl- and β-fluorine-substituted **24a**, **25**, and **26** [32c,d,f], and (partially) binding-site-blocked receptors [32b] from the corresponding pyrrole derivatives. β-Substituted receptors **24a**, **25**, and **26** can act as building blocks for π-extended derivatives and covalently linked oligomers because of their free α positions. Evidently, selective iodination at the α-pyrrole positions of β-substituted **24a**, **25**, and **26** by treatment with *N*-iodosuccinimide (NIS) in CH₂Cl₂ is found to be a key step in synthesizing iodinated derivatives [33b,l], which are essential starting materials in coupling reactions used for various functional molecules. For example, Suzuki cross coupling of iodinated derivatives and arylboronic acids afford bisaryl-substituted **24b,c**, **27a–c**, and **28a,b** [33b,c,f]. Furthermore, covalently linked dimers and tetramer **30a–c** prepared by similar procedures were found to exhibit anion-driven helical structures [33b,i].

Electronic properties of single molecules could be applied to potential electronic and optical materials such as supramolecular assemblies. For example, the absorption maxima (λ_{max}) of **18**, **19a** [33a], and **21** [33h] in CH_2Cl_2 appear at 500, 516, and 528 nm in CH_2Cl_2, respectively, which are redshifted as compared to **17** (432 nm) [32a], **23-16** (457 nm) [32e], **24a** (452 nm), **25a** (449 nm), and **26a** (421 nm). UV/vis absorption bands of **24b** and **27a–c** in CH_2Cl_2 are observed at 499, 551, 538, and 527 nm, respectively, suggesting redshifts that are comparable to that of α-unsubstituted **24a** (452 nm) [33b,c]. Those of formyl substituted **28a,b** in CH_2Cl_2 are observed at 495 and 510 nm, respectively [33f]. Furthermore, oligomers **30a–c** show λ_{max} values at 489, 514, and 478 nm because of incomplete π-conjugation at the *meta*-phenylene linkage [33b,j]. Absorption and emission spectra of anion receptors change in the presence of anions as TBA salts, suggesting their potential as colorimetric anion sensors. From experimental and theoretical data, affinity for anions can be determined from the following factors: (i) electronic effects of peripheral substituents, (ii) steric effects of α-substituents, and (iii) relative stabilities of the preorganized conformation. UV/vis absorption spectral changes of, for example, **17** in CH_2Cl_2 indicate K_a values of 15,000, 2100, 930,000, and 270,000 M^{-1} for binding with Cl^-, Br^-, AcO^-, and $H_2PO_4^-$, respectively [32a,e]. Compared with unsubstituted **17**, K_a values of α-phenyl **18** were greater, especially for halides, possibly due to pentacoordination in contrast to oxoanion binding. Furthermore, compared with **18**, β-ethyl **24b** shows similar and smaller K_a values for oxoanions and halides, respectively [33a–c]. ^{1}H NMR spectral changes of α-aryl-substituted receptors on anion binding provide valuable insights into the binding behaviors of pyrrole NH, bridging CH, and aryl-*o*-CH. In addition, at lower temperatures with a small amount of Cl^-, a new NH signal appears between that of the free receptor and the [1+1] complex. This is attributed to the [2+1] binding complex **18**$_2\cdot Cl^-$, and the formation of the [2+1] complex is also supported by diffusion-ordered spectroscopy (DOSY) NMR and electrospray ionization time-of-flight mass spectrometry (ESI-TOF-MS) [33a]. However, even in $CHCl_3$ containing 0.5% EtOH, pyrrolyl **27a** shows considerably larger K_a values, $>10^6$ M^{-1} for Cl^-, $H_2PO_4^-$, and AcO^-, than those of **24b**, **27b**, and **27c** and α-unsubstituted **24a** because of the presence of multiple polarized NH sites [33c]. Note that smaller K_a values are expected with an increase in the number of phenyl substituents in **24a,b** because of the steric hindrance between the α-phenyl and β-ethyl moieties.

Solid-state assemblies of, for example, **17** [32a,33h], **18** [33a], **19a** [33a], **22** [33h], **23-1,2,4** [32d], **24a** [32f], **24b** [33b], **25** [32f], **26** [32c], **27a–c** [33c], and **28a,b** [33f] have been revealed by single-crystal X-ray analyses. In most cases, pyrrole nitrogens are on opposite sides of the molecule in these fairly planar receptor molecules, which form stacking structures with an offset arrangement. Interactions between NH and BF, which are essential for specific assembling modes such as hydrogen-bonding dimers, are observed in these receptors. Interestingly, parent **17** affords polymorphs of single crystals according to crystallization conditions [33h]. In the solid state, the receptors **24b**, **27a–c**, and **28a,b** form stacking dimers or highly ordered stacking structures due to β-ethyl substituents [33b,c,f]. In addition, solid-state structures of the receptor–anion complexes **17**·Cl$^-$ [32a], **17**·Br$^-$ [33h], **18**·Cl$^-$ [33a], **18**·Br$^-$ [33h], **19a**·Cl$^-$ [33a], **22**·Br$^-$ [33h], **23-2**·Cl$^-$ [32e], **26**·Cl$^-$ [32c], **27a**·Cl$^-$ [33c], **30a**·Cl$^-$ [33i], **30b$_2$**·Cl$_2$$^-$ [33i], and **30c**·Cl$_2$$^-$ [33i], prepared from TBA or tetrapropylammonium (TPA) salts, as revealed by single-crystal X-ray analyses, show anion-bridged 1-D chains (**17**, **22**, and **26**), regular [1 + 1] complexes (**18**, **19a**, **23-2**, and **27a**), and helical structures (**30a–c**) [35]. In the packing diagram, receptor–anion complexes behave as the building components of the electrostatically mediated, alternately stacking structures consisting of "planar anions (receptor–anion complexes)" and tetraalkylammonium cations. These "charge-by-charge" columnar assemblies, which were observed in the crystal states as 3-D organized structures, allow the formation of dimension-controlled organic salts under appropriate conditions as discussed in the next section.

Further, Maeda *et al.* investigated the solid-state electronic and optical properties of **17** (three crystals: **17r**, **17y**, and **17v**), **18**, and **22** and their Br$^-$ complexes with the aim of discovering the unique properties based on the ordered assembly of π-conjugated molecules. Excitation maxima (λ_{max}) are distinct for **17r** (522 nm), **17y** (340 and 494 nm), **17v** (540 nm), **18** (579 nm), and **22** (589 nm), and emissions reflect these values: λ_{em} = 630 nm (**17r**), 524 nm (**17y**), 604 nm (**17v**), 610 nm (**18**), and 685 nm (**22**), respectively, when excited at 500 nm (**17r**, **17v**, **18**, and **22**) and 400 nm (**17y**). In fact, redshift values of the excitation and emission spectra in the solid state compared to the absorption and emissions from dispersed monomers in the solution state (CH$_2$Cl$_2$) are 90 and 156 nm for **17r**, 62 and 50 nm for **17y**, 108 and 130 nm for **17v**, 79 and 81 nm for **18**, and 61 and 145 nm for **22**, respectively. These shifts correlate mainly with the arrangements of the π-conjugated molecules in the solid state. On the other hand, the excitation and emission maxima of crystalline Br$^-$-binding complexes are observed at 380, 450, and 468 nm (λ_{max}) and 484 and 583 nm (λ_{em}) (**17**·Br$^-$-TBA$^+$), 406 and 560 nm (λ_{max}) and 580 nm (λ_{em}) (**18**·Br$^-$-TPA$^+$), and 581 nm (λ_{max}) and 612 nm (λ_{em}) (**22**·Br$^-$-TBA$^+$), respectively, suggesting that TBA or TPA cations are located between receptor–anion complexes and therefore interfere with $\pi-\pi$ interactions of π-plane chromophores. Blue shifts, which are moderate as compared to free receptors, are derived from the insertion of aliphatic cations between chromophores (receptor–anion complexes), which weakens their exciton coupling. However, redshifts (compared to the solution state) are presumably due to the edge-to-edge interactions of receptor–anion complexes. Ordered stacking structures of π-conjugated molecules, as observed in crystal structures of anion receptors and their anion complexes, are suitable for use as charge-conductive materials. Flash photolysis time-resolved microwave conductivity (FP-TRMC) measurements [36] allow the behavior of mobile charge carriers along all the axes of the crystals to be estimated. When a 355-nm laser pulse was applied at 25°C, anisotropy in the mobilities was observed in **17y** (0.8, 0.2, and 0.07 cm^2/Vs for *a*, *b*, and *c* axes) and **22** (1, 0.2, and 0.06 cm^2/Vs) in contrast to **17r** (0.05, 0.02, and 0.03 cm^2/Vs), **17v** (0.7, 0.8, and 0.6 cm^2/Vs), and **18** (0.02, 0.03, and 0.05 cm^2/Vs). Apparently, slipped parallel stacking of π-planes leads to high hole mobility and anistropic properties along the stacking direction. Further, Br^- complexes show smaller conductivities of 7×10^{-3}, 8×10^{-3}, and 3×10^{-4} cm^2/Vs (**17**·Br^--TBA^+) and 4×10^{-3}, 4×10^{-3}, and 7×10^{-4} cm^2/Vs (**18**·Br^--TPA^+) for a, b, and c axes of the crystals, and 8×10^{-3} and 7×10^{-3} cm^2/Vs (**22**·Br^--TBA^+) for a and c axes of the crystal, respectively. Therefore, crystals of Br^- complexes of tetraalkylammonium salts give smaller conductivities than those of anion-free crystals [33h].

Peripheral modifications to receptor molecules allow stabilization of stacking structures not only in the solid state but also in soft materials. The introduction of aliphatic chains to the pyrrole-based, π-conjugated molecules affords soft materials on the basis of two interactions between π-units and alkyl moieties. Differential scanning calorimetry (DSC) measurements of **19b–e** reveal mesophases at room temperature to 191.0°C, room temperature to 183.3°C, 36.7–172.5°C, and 65.6–156.0°C, respectively. Polarizing optical microscopy (POM) images of mesophases of **19b–d** show ribbon-like textures, whereas that of **19e** shows a mosaic-like texture. XRD analyses of **19b–d** in mesophases suggest that thermotropic liquid crystalline states are formed due to discotic hexagonal columnar (Col$_h$) phases that are, in turn, composed of dimeric assemblies (Z = ca 2 for $\rho = 1$) as building units. For **19b–d**, unit parameters a equal 31.2, 36.7, 39.8, and 46.0 Å, respectively, and unit parameters c equal 4.45, 4.52, 4.45, and 4.50 Å, respectively. The formation of stacking units comprising two rod-like molecules is rare but can be achieved by weak interactions such as hydrogen-bonding and dipole–dipole interactions between the pyrrole NH and BF$_2$ moieties. FP-TRMC measurements at 25°C provided conductivities of liquid crystal materials **19b** and **19c** at 0.093 and 0.22 cm^2/Vs, respectively. These values are fairly high and comparable to that of the crystal state such as **19d** (0.25 cm^2/Vs). Next, the effects of pyrrole N sites are examined using hexadecyloxy-substituted **19d′** and **19d″**, which have one and two methyl substituents at pyrrole N sites, respectively. DSC measurements reveal the occurrence of mesophases at 44.4–102.9°C (**19d′**) and 50.0–79.0°C (**19d″**). Compared to the dendritic POM texture and the weak XRD diffraction pattern of **19d′**, which suggests the formation of Col$_h$ phases, a flake-like POM texture and XRD analysis of the doubly N-blocked **19d″** reveals a rectangular columnar (Col$_r$ ($P2_1/a$)) phase of $a = 64.7$ Å, $b = 30.2$ Å, and $c = 4.5$ Å without the formation of a dimer structure (Z = ca 2 for $\rho = 0.8$). Results in **19d′** and **19d″** are observed to be correlated with the number of pyrrole NH sites [33g].

Organized structures can be formed in aqueous solutions by exploiting the interactions between hydrophobic moieties inside the assemblies and the association of hydrophilic sites with water molecules [1]. Whereas amphiphiles PEG-substituted amphiphilic π-conjugated acyclic oligopyrroles **21c,d** precipitate from water, derivatives **20a,b** and **21a,b** are soluble. Their λ_{max} values are observed at 496, 506, 462, and 481 nm (1×10^{-5} M), respectively, in water, and these values are blueshifted as compared with those in MeOH (510, 512, 510, and 512 nm), which suggests the formation of H-aggregates in aqueous solutions. Fluorescence spectra of **20a,b** and **21a,b** in aqueous solutions are observed as each monomer's fluorescence peaks at 571, 572, 672, and 671 nm with low-emission quantum yields (Φ_F, determined at λ_{ex} values that are equal to the respective λ_{max} values) of 0.01, 0.09, 0.02, and 0.02, respectively, which are characteristic aspects of H-aggregates. Solid films cast from aqueous solutions of **20a,b** and **21a,b** exhibit almost the same redshifted UV/vis absorption profiles ($\lambda_{max} = 540$, 530, 548, and 531 nm, respectively) as those cast from CH_2Cl_2 solutions. This result suggests that the removal of water molecules by slow evaporation at room temperature or by freeze drying disrupts H-aggregate formations, which are supported by water molecules, and instead leads to other assembled structures (J-type aggregates). Cryo-TEM analysis of **21a** (1×10^{-5} M) shows vesicular structures [37] with diameters of 30–80 nm, which are consistent with the results of dynamic light scattering (DLS) measurements. The wall thickness of the capsules is estimated to be ca 5 nm, which reflects hydrophobic segments consisting of bilayers of amphiphilic molecules (ca 3.9 nm from AM1 calculations). In contrast to amphiphiles containing only hydrophilic chains, **21a** possibly forms bilayers such as biotic lipids by exploiting the hydrophobic interactions of aliphatic chains and locating the hydrophilic triethylene glycol (TEG) chains on the outside to form water-soluble vesicles [33e].

6.5 CHARGE-BY-CHARGE ASSEMBLIES FROM ANION-RESPONSIVE SUPRAMOLECULAR GELS

Anion receptors **19b–d** gelate octane (10 mg/mL), and transition temperatures between the gel and solution states are −8.5 (**19b**), 4.5 (**19c**), and 27.5°C (**19d**), suggesting that the longer alkyl chains afford more stable gels [33a]. In contrast, icosyloxy-substituted **19e** forms precipitates due to fairly strong van der Waals interactions between its longer alkyl chains [33g]. The octane gel of hexadecyloxy-substituted **19d** (10 mg/mL) exhibits split absorption bands with absorption maxima at 525 and 555 nm along with a shoulder at 470 nm due to the formation of stacking structures. This is in contrast to the single peak at 493 nm exhibited in a diluted solution containing dispersed monomers. This supramolecular organogel formation is achieved via noncovalent interactions between π-conjugated moieties and their substituents, and this hypothesis is supported by atomic force microscopy (AFM) (Fig. 6-4a(i)), SEM, and XRD observations. Addition of anions (10 equivalents) in the solid form (TBA salts) to the fluorescent octane gel results in a transition to the solution state; the gels are gradually transformed into solutions beginning from areas close to where solid salts have been added (Fig. 6-4a(ii)). In this process, once the receptor (gelator) molecules in the gel bind to anions, the counter TBA cations concertedly approach the receptor–anion complexes to form soluble ion pairs, thereby producing the octane solution (Fig. 6-4a(iii)). In the case of the gel of **19d**, these transitions are quite distinct from the crystal (e.g., **19a**·Cl⁻) because of the insolubility of the TBA salt of **19a**·Cl⁻ in apolar hydrocarbon solvents [33a]. Furthermore, the singly N-blocked **19d′** affords a solution, whereas the doubly N-blocked **19d″** precipitates with ambient cooling and gels after rapid cooling following heating at 40.0°C to a solution state. Compounds **19d′** and **19d″** in concentrated states

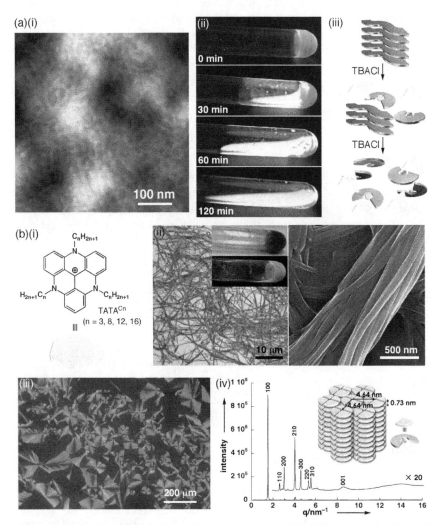

Figure 6-4 (a) (i) AFM 2-D image in a tapping mode of **19d** (from octane gel) cast by spin coating on a silicon substrate, (ii) transition of supramolecular organogel of **19d** in octane (10 mg/mL) at 20°C on addition of Cl⁻ (10 equivalents) added as a solid TBA salt (under UV (365 nm) light), and (iii) possible anion-responsive process of gel-to-solution transition; (b) (i) structure of TATA cations, (ii) OM (left) and SEM (right) measurements of an octane xerogel of **19d**·Cl⁻-(TATA^{C3})⁺ at 20°C and photographs of the gel (inset: under visible (top) and UV$_{365 \, nm}$ (bottom) light), (iii) POM image of **19d**·Cl⁻-(TATA^{C3})⁺ at 87°C by cooling at 2°C/min from Iso, and (iv) XRD pattern of **19d**·Cl⁻-(TATA^{C3})⁺ in mesophase at 70°C after cooling from Iso and proposed assembling model (inset).

(10 mg/mL) exhibit broad absorption bands with λ_{max} at 470 and 413 nm, respectively, and when excited at each λ_{max}, **19d′** and **19d″** exhibit broad emission bands with λ_{em} at 649 and 640 nm, respectively, suggesting that stacking structures of **19d′** are formed in a solution state and those of **19d″** are formed in a gel state. In contrast to **19d″**, which shows no anion-responsive behaviors owing to the absence of NH sites, **19d′** shows changes in emission and color of the solution when 1 equivalent of TBACl is added to the solution. On the basis of these results, organized structures in the solution and gelated states with ca 1 wt% concentrations are affected by the geometries of π-conjugated molecules [33g]. As β-substituted receptors bearing aliphatic chains, **29a,b**, prepared by the condensation of the formyl-substituted **28a,b** with 3,4,5-trihexadecyloxyaniline followed by reduction with NaBH(OAc)$_3$, appear to be fairly dispersed, and non-gel-like states are formed at room temperature from octane (10 mg/mL). When the solutions are cooled, **29a,b** form opaque solutions below ca 1 and ca $-10°$C, respectively, and form gel-like materials below ca -10 and ca $-30°$C, respectively. UV/vis and NMR analyses and DLS support the formation of assembled structures. In these systems, the connection of three π-conjugated moieties—a core π-plane and two side aryl units—by sp^3 methylene bridges is the basis for the formation of supramolecular assemblies that are not very rigid but yield soft materials by ordered organization under appropriate conditions. Addition of 1 equivalent of Cl$^-$ as a TBA salt can modulate the nature of the assembled structures, along with their optical and electronic properties [33f].

Although intermolecular N–H$\cdots$F–B hydrogen-bonding and dipole–dipole interactions may also be essential in other states of assemblies, the systems discussed in the previous paragraph are prototypical examples of supramolecular gels that use π–π interactions as the main force for aggregation. Thus, anions as additives may not always act as inhibitors but may sometimes act as building units of soft materials. From this point of view, structural modifications of anion receptors and the choice of appropriate combinations of anions, cations, receptors, and solvents have been investigated in order to harness the fascinating properties of supramolecular gels that are sensitive to chemical stimuli. In particular, counter cations are essential in determining the states of supramolecular gels even though they do not directly interact with gelator molecules.

In order to achieve dimension-controlled assemblies comprising alternately stacking positively and negatively charged species, the introduction of planar cations instead of sterical cations was attempted. In fact, we combined receptors **18** and **19d** with anion salts of planar 4,8,12-trialkyl-4,8,12-triazatriangulenium (TATA) cations [38], TATA^{Cn+} (n = 3, 8, 12, and 16) (Fig. 6-4b(i)). Single-crystal X-ray analysis of equivalent mixtures of **18** and TATA^{C3}·X (X = Cl and Br) exhibited alternately stacking charge-by-charge structures comprising planar receptor–anion complexes and TATA cations, wherein the distance between receptor–anion complexes in, for example, **18**·Cl$^-$-(TATA^{C3})$^+$ is 6.85 Å, which is 0.44 Å shorter than that of **18**·Cl$^-$-TPA$^+$ (7.29 Å). On the other hand, ion pairs

19d·Cl⁻-(TATA^{Cn})⁺ (10 mg/mL of **19d** with 1 equivalent of TATA salts) afforded opaque gels from **19d**·Cl⁻-(TATA^{C3})⁺ (inset of Fig. 6-4b(ii)) and precipitated from **19d**·Cl⁻-(TATA^{Cn})⁺ ($n = 8$, 12, and 16) in octane at 20°C after heating to over 40°C to solution. The gel of **19d**·Cl⁻-(TATA^{C3})⁺ is transformed to solution at 35°C, which is higher at ca 8°C than the gel of **19d**, also suggesting that appropriate TATA salts can stabilize gels. Optical microscopy (OM), AFM, and SEM analyses showed submicrometer-scale morphologies of xerogels or precipitates of **19d**·Cl⁻-(TATA^{Cn})⁺: **19d**·Cl⁻-(TATA^{Cn})⁺ ($n = 3$, 8, and 12) formed entangled fibril structures (Fig. 6-4b(ii)), whereas **19d**·Cl⁻-(TATA^{C16})⁺ afforded irregular-shaped structures. Longer alkyl chains of receptor–anion complexes help stabilize the dimension-controlled stacking assemblies, whereas those of TATA units provide less ordered morphologies. Gel or precipitates of **19d**·Cl⁻-(TATA^{Cn})⁺ ($n = 3$, 8, 12, and 16) without solvent removal showed blueshifts in UV/vis absorption and emission spectra by longer alkyl chains in TATA cations, assuming that longer alkyl chains might loosen the charge-by-charge stacking columns and, as a result, disturb the column bundling to affect their electronic and optical properties. Further, synchrotron radiation X-ray scattering analysis (BL40B2 at SPring-8) of **19d**·Cl⁻-(TATA^{C3})⁺ fibers as a xerogel revealed the structural details as a Col$_h$ phase with $a = 4.25$ nm and $c = 0.73$ nm based on a trimeric assembly ($Z = 3.00$ for $\rho = 1$). The c value (0.73 nm) corresponds to the distance between the stacking structure of both a receptor–anion complex and a TATA cation, and the circular trimeric assemblies are also consistent with the fan-like geometry of the pyrrole-inverted receptor–anion complex. On the contrary, a dried sample of **19d**·Cl⁻-(TATA^{C16})⁺ exhibited peaks of a discotic lamellar structure with an intercolumnar distance of ca 5 nm, suggesting that longer alkyl chains at the periphery of a TATA cation may disturb the formation and alignment of a hexagonally ordered stacking columnar structure [39].

DSC of an octane xerogel of **19d**·Cl⁻-(TATA^{C3})⁺ (5°C/min) suggested phase transitions at 75 and 96°C during the first heating, 88 and 42°C during the first cooling, and 44 and 96°C during the second heating. Slow cooling at 2°C/min from the isotropic liquid (Iso) phase afforded larger focal conic domains in POM images along with greater areas of dark domains (Fig. 6-4b(iii)) than fast cooling at 5°C/min, suggesting that discotic columnar structures are well aligned perpendicularly to substrates. At 70°C after cooling from Iso, synchrotron radiation X-ray scattering analysis showed relatively sharp peaks of a Col$_h$ phase with $a = 4.64$ nm and $c = 0.73$ nm based on a tetrameric assembly ($Z = 3.58$ for $\rho = 1$) (Fig. 6-4b(iv)). Further, the shear-driven alignment of the organized structure, which was prepared by shearing around 90°C and cooling to room temperature, also suggests the formation of liquid crystal in the mesophase. An optical response in POM images was preliminarily observed on application of an electric field. Since no significant decompositions of building components appeared after repeated transitions between the mesophase and the isotropic liquid states, the potential stability of these charge-by-charge assemblies is worth noting [39].

6.6 CONCLUSIONS

The use of an appropriate molecular design and fine synthetic procedures for building subunits can allow the formation of various anion-responsive soft materials such as supramolecular gels. Are anions always inhibitors for soft materials comprising anion-responsive molecules? The answer is no. The planarity of π-conjugated dye molecules can be used as the main driving force to construct various organized structures, the stabilities of which are not significantly affected by anion binding. In this context, under appropriate conditions, various anions can be incorporated into organized structures as versatile negative-charged building subunits with counter cationic components [40]. In fact, planar receptor–anion complexes as negatively charged components readily assemble with planar cations through electrostatic and $\pi-\pi$ interactions. Columnar assemblies based on these charge-by-charge stacking structures were found to form supramolecular organogels and a liquid crystal columnar phase, showing a property to readily form oriented structures due to the existence of charge-by-charge columns. The creation of a library of charge-by-charge assemblies, covering various combinations of planar ionic species, would provide promising functional soft materials as oriented salts for future application.

REFERENCES

1. (a) Israelachvili, J. N. (1992). *Intermolecular and Surface Forces*. Academic Press, London, p. 450; (b) Hamley, I. W. (2000). *Introduction to Soft Matter—Polymers, Colloids, Amphiphiles and Liquid Crystals*. John Wiley & Sons, Chichester, p. 342.
2. (a) Flory, P. J. (1974). Gels and gelling process. *Faraday Discuss. Chem. Soc.*, *58*, 7–18; (b) Osada, Y., Khokhlov, A. R., eds. (2001). *Polymer Gels and Networks*. Marcel Dekker, New York; (c) Siegel, R. A., ed. (2005). *Fundamentals and Applications of Polymer Gels*. Wiley-VCH, Weinheim.
3. (a) Fages, F., ed. (2005). *Low Molecular Mass Gelators, Topics in Current Chemistry*. Springer-Verlag, Berlin, Vol. 256, p. 283; (b) Ishi-i, T., Shinkai, S. (2005). Dye-based organogels: Stimuli responsive soft materials based on one-dimensional self-assembling aromatic dyes. in *Supramolecular Dye Chemistry: Topics in Current Chemistry* (Ed.: Würthner, F.), Springer-Verlag, Berlin, Vol. *258*, pp. 119–160; (c) Weiss, R. G., Terech, P., eds. (2006). *Molecular Gels*. Springer, Dordrecht, p. 978; (d) Terech, P., Weiss, R. G. (1997). Low molecular mass gelators of organic liquids and the properties of their gels. *Chem. Rev.*, *97*, 3133–3159; (e) Abdallah, D. J., Weiss, R. G. (2000). Organogels and low molecular mass organic gelators. *Adv. Mater.*, *12*, 1237–1247; (f) van Esch, J. H., Feringa, B. L. (2000). New functional materials based on self-assembling organogels: From serendipity towards design. *Angew. Chem. Int. Ed.*, *39*, 2263–2266.
4. Sugiyasu, K., Fujita, N., Takeuchi, M., Yamada, S., Shinkai, S. (2003). Proton-sensitive fluorescent organogels. *Org. Biomol. Chem.*, *1*, 895–899.
5. Ghoussoub, A., Lehn, J.-M. (2005). Dynamic sol–gel interconversion by reversible cation binding and release in G-quartet-based supramolecular polymers. *Chem. Commun.*, 5763–5765.
6. Zang, Y., Gu, H., Yang, Z., Xu, B. (2003). Supramolecular hydrogels respond to ligand-receptor interaction. *J. Am. Chem. Soc.*, *125*, 13680–13681.
7. Kabir, S. R., Yokoyama, K., Mihashi, K., Kodama, T., Suzuki, M. (2003). Hyper-mobile water is induced around actin filaments. *Biophys. J.*, *85*, 3154–3161.
8. Books on anion binding: (a) Bianchi, A., Bowman-James, K., García-España, E., eds. (1997). *Supramolecular Chemistry of Anions*. Wiley-VCH, New York; (b) Singh, R. P., Moyer, B. A., eds.

(2004). *Fundamentals and Applications of Anion Separation*. Kluwer Academic/Plenum Publishers, New York; (c) Stibor, I., eds. (2005). *Anion Sensing, Topics in Current Chemistry*. Springer-Verlag, Berlin, Vol. 255, p. 238; (d) Sessler, J. L., Gale, P. A., Cho, W.-S. (2006). *Anion Receptor Chemistry*. RSC, Cambridge; (e) Vilar, R., ed. (2008). *Recognition of Anions, Structure and Bonding*. Springer-Verlag, Berlin.

9. Reviews on anion-responsive supramolecular gels: (a) Maeda, H. (2008). Anion-responsive supramolecular gels. *Chem.—Eur. J.*, *14*, 11274–11282; (b) Lloyd, G. O., Steed, J. W. (2009). Anion-tuning of supramolecular gel properties. *Nat. Chem.*, *1*, 437–442; (c) Piepenbrock, M.-O. M., Lloyd, G. O., Clarke, N., Steed, J. W. (2010). Metal- and anion-binding supramolecular gels. *Chem. Rev.*, *110*, 1960–2004; (d) Steed, J. W. (2010). Anion-tuned supramolecular gels: a natural evolution from urea supramolecular chemistry. *Chem. Soc. Rev.*, *39*, 3689–3699.

10. For example: (a) Oda, R., Huc, I., Candau, S. J. (1998). Gemini surfactants as new, low molecular weight gelators of organic solvents and water. *Angew. Chem. Int. Ed. Engl.*, *37*, 2689–2691; (b) Yoshida, M., Koumura, N., Misawa, Y., Tamaoki, N., Matsumoto, H., Kawanami, H., Kazaoui, S., Minami, N. (2007). Oligomeric electrolyte as a multifunctional gelator. *J. Am. Chem. Soc.*, *129*, 11039–11041; (c) Tam, A. Y.-Y., Wong, K. M.-C., Wang, G., Yam, V. W.-W. (2007). Luminescent metallogels of platinum(II) terpyridyl complexes: interplay of metal···metal, $\pi-\pi$ and hydrophobic–hydrophobic interactions on gel formation. *Chem. Commun.*, 2028–2030; (d) Tam, A. Y.-Y., Wong, K. M.-C., Zhu, N., Wang, G., Yam, V. W.-W. (2009). Luminescent alkynylplatinum(II) terpyridyl metallogels stabilized by Pt···Pt, $\pi-\pi$, and hydrophobic-hydrophobic interactions. *Langmuir*, *25*, 8685–8695.

11. For example: Kimizuka, N., Nakashima, T. (2001). Spontaneous self-assembly of glycolipid bilayer membranes in sugar-philic ionic liquids and formation of ionogels. *Langmuir*, *17*, 6759–6761.

12. Webb, J. E. A., Crossley, M. J., Turner, P., Thordarson, P. (2007). Pyromellitamide aggregates and their response to anion stimuli. *J. Am. Chem. Soc.*, *129*, 7155–7162.

13. Dzolic, Z., Cametti, M., Cort, A. D., Mandolini, L., Žinić, M. (2007). Fluoride-responsive organogelator based on oxalamide-derived anthraquinone. *Chem. Commun.*, *34*, 3535–3537.

14. Zhang, Y.-M., Lin, Q., Wei, T.-B., Qin, X.-P., Li, Y. (2009). A novel smart organogel which could allow a two channel anion response by proton controlled reversible sol–gel transition and color changes. *Chem. Commun.*, 6074–6076.

15. Liu, J.-W., Yang, Y., Chen, C.-F., Ma, J.-T. (2010). Novel anion-tuning supramolecular gels with dual-channel response: reversible sol-gel transition and color changes. *Langmuir*, *26*, 9040–9044.

16. Wang, C., Zhang, D., Zhu, D. (2007). A chiral low-molecular-weight gelator based on binaphthalene with two urea moieties: modulation of the CD spectrum after gel formation. *Langmuir*, *23*, 1478–1482.

17. Yang, H., Yi, T., Zhou, Z., Zhou, Y., Wu, J., Xu, M., Li, F., Huang, C. (2007). Switchable fluorescent organogels and mesomorphic superstructure based on naphthalene derivatives. *Langmuir*, *23*, 8224–8230.

18. Kim, T. H., Choi, M. S., Sohn, B.-H., Park, S.-Y., Lyoo, W. S., Lee, T. S. (2008). Gelation-induced fluorescence enhancement of benzoxazole-based organogel and its naked-eye fluoride detection. *Chem. Commun.*, 2364–2366.

19. Teng, M., Kuang, G., Jia, X., Gao, M., Li, Y., Wei, Y. (2009). Glycine-glutamic-acid-based organogelators and their fluoride anion responsive properties. *J. Mater. Chem.*, *19*, 5648–5664.

20. Yamanaka, M., Nakamura, T., Nakagawa, T., Itagaki, H. (2007). Reversible sol–gel transition of a tris–urea gelator that responds to chemical stimuli. *Tetrahedron Lett.*, *48*, 8990–8993.

21. Stanley, C. E., Clarke, N., Anderson, K. M., Elder, J. A., Lenthall, J. T., Steed, J. W. (2006). Anion binding inhibition of the formation of a helical organogel. *Chem. Commun.*, 3199–3201.

22. Piepenbrock, M.-O. M., Lloyd, G. O., Clarke, N., Steed, J. W. (2008). Gelation is crucially dependent on functional group orientation and may be tuned by anion binding. *Chem. Commun.*, 2644–2646.

23. Applegarth, L., Clark, N., Richardson, A. C., Parker, A. D. M., Radosavljevic-Evans, I., Goeta, A. E., Howard, J. A. K., Steed, J. W. (2005). Modular nanometer-scale structuring of gel fibres by sequential self-organization. *Chem. Commun.*, 5423–5425.

24. Proline-functionalized calix[4]arene exhibited anion-triggered hydrogels. The gelating molecule is not a π-conjugated system, but consists of a phenyl moiety connected by sp^3 carbon, and therefore,

the details are omitted in this chapter. Becker, T., Goh, C. Y., Jones, F., McIldowie, M. J., Mocerino, M., Ogden, M. I. (2008). Proline-functionalised calix[4]arene: an anion-triggered hydrogelator. *Chem. Commun.*, 3900–3902.

25. Li, Q., Wang, Y., Li, W., Wu, L. (2007). Structural characterization and chemical response of a Ag-coordinated supramolecular gel. *Langmuir*, *23*, 8217–8223.

26. Kishimura, A., Yamashita, T., Aida, T. (2005). Phosphorescent organogels via "metallophilic" interactions for reversible RGB-color switching. *J. Am. Chem. Soc.*, *127*, 179–183.

27. (a) Kim, H.-J., Zin, W.-C., Lee, M. (2004). Anion-directed self-assembly of coordination polymer into tunable secondary structure. *J. Am. Chem. Soc.*, *126*, 7009–7014; (b) Kim, H.-J., Lee, J.-H., Lee, M. (2005). Stimuli-responsive gels from reversible coordination polymers. *Angew. Chem. Int. Ed.*, *44*, 5810–5814.

28. Piepenbrock, M.-O. M., Clarke, N., Steed, J. W. (2009). Metal ion and anion-based "tuning" of a supramolecular metallogel. *Langmuir*, *25*, 8451–8456.

29. Fischer, H., Orth, H. (1934). *Die Chemie des Pyrrols*. Akademische Verlagsgesellschaft M.B.H., Leipzig.

30. (a) Kadish, K. M., Smith, K. M., Guilard, R., eds. (2000). *The Porphyrin Handbook*. Academic Press, San Diego, CA,; (b) Kadish, K. M., Smith, K. M., Guilard, R., eds. (2010). *Handbook of Porphyrin Science*. World Scientific, New Jersey.

31. (a) Maeda, H. (2007). Supramolecular chemistry of acyclic oligopyrroles. *Eur. J. Org. Chem.*, 5313–5325; (b) Maeda, H. (2009). Acyclic oligopyrroles as building blocks of supramolecular assemblies. *J. Inclusion Phenom.*, *64*, 193–214; (c) Maeda, H. (2010). Supramolecular chemistry of pyrrole-based π-conjugated acyclic anion receptors. in *Handbook of Porphyrin Science* (Eds.: Kadish, K. M., Smith, K. M., Guilard, R.), World Scientific, New Jersey, Vol. *8*, Chapter 38; (d) Maeda, H. (2010). Acyclic oligopyrrolic anion receptors. in *Anion Complexation in Supramolecular Chemistry, Topics in Heterocyclic Chemistry* (Eds.: Gale, P. A., Dehaen, W.), Springer-Verlag, Berlin, Vol. *24*, pp. 103–144; (e) Maeda, H., Haketa, Y. (2011). Charge-by-charge assemblies based on planar anion receptors. *Pure Appl. Chem.*, *83*, 189–199.

32. (a) Maeda, H., Kusunose, Y. (2005). Dipyrrolyldiketone difluoroboron complexes: novel anion sensors with C–H⋯X⁻ interactions. *Chem.—Eur. J.*, *11*, 5661–5666; (b) Fujimoto, C., Kusunose, Y., Maeda, H. (2006). CH⋯anion interaction in BF₂ complexes of C₃-bridged oligopyrroles. *J. Org. Chem.*, *71*, 2389–2394; (c) Maeda, H., Ito, Y., (2006). BF₂ complex of fluorinated dipyrrolyldiketone: a new class of efficient receptor for acetate anions. *Inorg. Chem.*, *45*, 8205–8210; (d) Maeda, H., Kusunose, Y., Mihashi, Y., Mizoguchi, T. (2007). BF₂ Complexes of β-tetraethyl-substituted dipyrrolyldiketones as anion receptors: potential building subunits for oligomeric systems. *J. Org. Chem.*, *72*, 2612–2616; (e) Maeda, H., Terasaki, M., Haketa, Y., Mihashi, Y., Kusunose, Y. (2008). BF₂ complexes of α-alkyl-substituted dipyrrolyldiketones as acyclic anion receptors. *Org. Biomol. Chem.*, *6*, 433–436; (f) Maeda, H., Haketa, Y., Bando, Y., Sakamoto, S. (2009). Synthesis, properties, and solid-state assemblies of β-alkyl-substituted dipyrrolyldiketone BF₂ complexes. *Synth. Met.*, *159*, 792–796.

33. (a) Maeda, H., Haketa, Y., Nakanishi, T. (2007). Aryl-substituted C₃-bridged oligopyrroles as anion receptors for formation of supramolecular organogels. *J. Am. Chem. Soc.*, *129*, 13661–13674; (b) Maeda, H., Haketa, Y. (2008). Selective iodinated dipyrrolyldiketone BF₂ complexes as potential building units for oligomeric systems. *Org. Biomol. Chem.*, *6*, 3091–3095; (c) Maeda, H., Mihashi, Y., Haketa, Y. (2008). Heteroaryl-substituted C₃-bridged oligopyrroles: potential building subunits of anion-responsive π-conjugated oligomers. *Org. Lett.*, *10*, 3179–3182; (d) Maeda, H., Eifuku, N. (2009). Alkoxy-substituted derivatives of π-conjugated acyclic anion receptors: effects of substituted positions. *Chem. Lett.*, *38*, 208–209; (e) Maeda, H., Ito, Y., Haketa, Y., Eifuku, N., Lee, E., Lee, M., Hashishin, T., Kaneko, K. (2009). Solvent-assisted organized structures based on amphiphilic anion-responsive π-conjugated systems. *Chem.—Eur. J.*, *15*, 3709–3716; (f) Maeda, H., Fujii, R., Haketa, Y. (2010). Supramolecular assemblies derived from formyl-substituted π-conjugated acyclic anion receptors. *Eur. J. Org. Chem.*, 1469–1482; (g) Maeda, H., Terashima, Y., Haketa, Y., Asano, A., Honsho, Y., Seki, S., Shimizu, M., Mukai, H., Ohta, K. (2010). Discotic columnar mesophases derived from 'rod-like' π-conjugated anion-responsive acyclic oligopyrroles. *Chem. Commun.*, *46*, 4559–4561; (h) Maeda, H., Bando, Y., Haketa, Y., Honsho, Y., Seki,

S., Nakajima, H., Tohnai, N. (2010). Electronic and optical properties in the solid-state molecular assemblies of anion-responsive pyrrole-based π-conjugated systems. *Chem.—Eur. J.*, *16*, 11653–11661;   (i) Haketa, Y., Maeda, H. (2011). From helix to macrocycle: Anion-driven conformation control of π-conjugated acyclic oligopyrroles. *Chem.—Eur. J.*, *17*, 1485–1492;   (j) Maeda, H., Eifuku, N., Haketa, Y., Ito, Y., Lee, E., Lee, M. (2011). Water-supported organized structures based on wedge-shaped amphiphilic derivatives of dipyrrolyldiketone boron complexes. *Phys. Chem. Chem. Phys.*, *13*, 3843–3850;   (k) Maeda, H., Terashima, Y. Solvent-dependent supramolecular assemblies of π-conjugated anion-responsive acyclic oligopyrroles. *Chem. Commun.*, in press. DOI: 10.1039/C1CC12827B;   (l) Haketa, Y., Sakamoto, S., Chigusa, K., Nakanishi, T., Maeda, H. Synthesis, Crystal Structures, and Supramolecular Assemblies of Pyrrole-Based Anion Receptors Bearing Modified Pyrrole β-Substituents. *J. Org. Chem.*, in press. DOI: 10.1021/jo2008687;   (m) Maeda, H., Kinoshita, K., Naritani, K., Bando, Y. Self-sorting self-complementary assemblies of π-conjugated acyclic anion receptors. *Chem. Commun.*, in press. DOI: 10.1039/C1CC12120K   (n) Maeda, H., Kitaguchi, K., Haketa, Y. Anion-responsive covalently linked and metal-bridged oligomers. Submitted.

34. Modifications at a boron unit, substitution by diols and aryl moieties, have also been reported: (a) Maeda, H., Fujii, Y., Mihashi, Y. (2008). Diol-substituted boron complexes of dipyrrolyldiketones as anion receptors and covalently linked 'pivotal' dimers. *Chem. Commun.*, 4285–4287; (b) Maeda, H., Takayama, M., Kobayashi, K., Shinmori, H. (2010). Modification at boron unit: tuning electronic and optical properties of π-conjugated acyclic anion receptors. *Org. Biomol. Chem.*, *8*, 4308–4315; (c) Maeda, H., Bando, Y., Shimomura, K., Yamada, I., Naito, M., Nobusawa, K., Tsumatori, H., Kawai, K. (2011). Chemical-stimuli-controllable circularly polarized luminescence from anion-responsive π-conjugated molecules. *J. Am. Chem. Soc.*, *133*, 9266–9269.

35. (a) Molecular structures along with the assembled modes are also found to be controlled by counter cations: manuscript in preparation;   (b) Single-crystal structures of boron-modified derivatives and their anion complexes were also reported. See ref 34(b) and (c).

36. (a) Yamamoto, Y., Fukushima, T., Suna, Y., Ishii, N., Saeki, A., Seki, S., Tagawa, S.; Taniguchi, M., Kawai, T., Aida, T. (2006). Photoconductive coaxial nanotubes of molecularly connected electron donor and acceptor layers. *Science*, *314*, 1761–1764;   (b) Yamamoto, Y., Zhang, G., Jin, W., Fukushima, T., Minari, T., Ishii, N., Saeki, A., Seki, S. Tagawa, S., Minari, T. Tsukagoshi, K. Aida, T. (2009). Ambipolar-transporting coaxial nanotubes with a tailored molecular graphene–fullerene heterojunction. *Proc. Natl. Acad. Sci. U.S.A*, *50*, 21051–21056;   (c) Umeyama, T., Tezuka, N., Seki, S., Matano, Y., Nishi, M., Hirao, K., Lehtivuori, H., Tkachenko, V. N., Lemmetyinen, H., Nakao, Y., Sakaki, S., Imahori, H. (2010). Selective formation and efficient photocurrent generation of [70]fullerene-single-walled carbon nanotube composites. *Adv. Mater.*, *22*, 1767–1770.

37. Luisi, P. L., Walde, P., eds. (2000). *Giant Vesicles*. Wiley-VCH, Chichester.

38. (a) Laursen, B. W., Krebs, F. C. (2000). Synthesis of a triazatriangulenium salt. *Angew. Chem. Int. Ed.*, *39*, 3432–3434;   (b) Laursen, B. W., Krebs, F. C. (2001). Synthesis, structure, and properties of azatriangulenium salts. *Chem.—Eur. J.*, *7*, 1773–1783.

39. Haketa, Y., Sasaki, S., Ohta, N., Masunaga, H., Ogawa, H., Araoka, F., Takezoe, H., Maeda, H. (2010). Oriented salts: dimension-controlled charge-by-charge assemblies from planar receptor–anion complexes. *Angew. Chem. Int. Ed.*, *49*, 10079–10083.

40. Modified anions as TBA salts also provide assembled structures by complexation with planar anion receptors: Maeda, H., Naritani, K., Honsho, Y., Seki, S. (2011). Anion Modules: Building blocks of supramolecular assemblies by combination with π-conjugated anion receptors. *J. Am. Chem. Soc.*, *133*, 8896–8899.

DIMENSION CONTROLLED ORGANIC FRAMEWORKS

CHAPTER 7

POLYMERIC FRAMEWORKS: TOWARD POROUS SEMICONDUCTORS

Jens Weber,[1] *Michael J. Bojdys,*[2] *and Arne Thomas*[2]

[1]Max Planck Institute of Colloids and Interfaces, Potsdam, Germany
[2]Technische Universität Berlin, Berlin, Germany

7.1 INTRODUCTION

Porous materials are solids with high and accessible surface areas, making them ideal candidates for applications in catalysis, adsorption, and storage and for purification purposes. Besides the well-known inorganic examples, such as porous metal oxides, zeolites, or activated carbons, new materials of entirely organic composition line up to extend the range of properties and applications of this class of materials.

In recent years, considerable efforts have been made toward the synthesis, characterization, and application of micro- and mesoporous polymers and covalent organic frameworks (COFs). Such materials have been prepared using different methods, mainly templating and scaffolding approaches. The resulting materials can be distinguished from the well-known macroporous polymers (e.g., polystyrene or polyurethane foams) by their much smaller pore size and consequently much higher surface area. Microporous materials generally have pore diameters below 2 nm. Examples are activated carbons, zeolites, or metal–organic frameworks (MOFs). Typical surface areas that are observed for these materials are ~ 1000 m^2/g, but materials that exceed this value have also been reported. Mesoporous materials have pores in the region between 2 and 50 nm. Characteristic examples are mesoporous aluminum oxides or silicas from the Mobil Composition of Matter (MCM) [1] and Santa Barbara Amorphous (SBA) [2] family and mesoporous carbons (e.g., Carbon Mesoporous Korea (CMK)) [3] derived therefrom. Here the typical surface areas are in the range of $100-1000$ m^2/g.

This short summary of typical micro- and mesoporous materials shows that known and applied materials with pores smaller than 10 nm are mainly inorganic

Supramolecular Soft Matter: Applications in Materials and Organic Electronics, First Edition.
Edited by Takashi Nakanishi.
© 2011 John Wiley & Sons, Inc. Published 2011 by John Wiley & Sons, Inc.

in nature. This can be explained by the large capillary forces and surface energies that generally occur when very small pores and very high surface areas are introduced into a material. To maintain porosity on this scale, the pore wall should be composed of a rigid and stable compound, preferably connected in three dimensions. This, however, excludes most organic building blocks, which allow translational and rotational degrees of freedom around bonds or bond angles so that the porous structure is prone to collapse. Nevertheless, some special organic compositions are quite inflexible and thus have a better chance to withstand capillary pressures. Examples are the so-called "high-performance" polymers such as aromatic polyimides or polybenzimidazoles. These polymers have a low degree of rotational freedom, and their structure is often described as "rigid rods" [4]. Through cross-linking, i.e. polymerization into 3D networks, these polymers become sufficiently stable and rigid to support even the highest surface areas. Another class of polymers, exhibiting comparable mechanical and structural stability, is the class of conjugated polymers such as polyparaphenylenes or polythiophenes [5,6]. Indeed, recent work, which will be discussed in the following chapters, has shown that meso- and even microporosity can be introduced into conjugated polymer networks. Hence, the design-principle for porous semiconducting polymers ideally combines accessible porosity with dimensions in the nanoscale and high surface areas with a pore wall composition featuring organic, π-conjugated, semiconducting building blocks. This combination gives rise to a fascinating field of applications, from organic electronics to photocatalytic applications.

Thus, a preformed porous system could be filled with a second phase, for example, a dye or an electron or a hole conductor. Recently, materials with comparable architecture have become the object of many research activities, although they have so far been produced by a different technique. Organic bulk heterojunction solar cells are composed of one hole- and one electron-conducting phases [7–11], with probably the most prominent example of [6,6]-phenyl-C_{61}-butyric acid methyl ester (PCBM) (electron acceptor) and poly-3-hexylthiophene (P3HT) (electron donor) [12–14]. Such electron donor/acceptor (D/A) composites are assembled in interpenetrating networks with high D/A interfaces, which allow for an efficient separation of the excitons formed during light irradiation. Since the lifetime of the exciton is short, its diffusion length in organic materials is only about 10–20 nm [15]. The donor and acceptor phases should therefore be assembled in such a way that nanodomains with dimensions comparable to the exciton diffusion length are formed. Once the exciton is formed and dissociated, the hole and the electron must drift to the electrodes within the span of their lifetimes [16]. All these requirements are crucially dependent on the nanomorphology of the photoactive layer, and thus such as solvent influence, composition dependence, thin-film preparation conditions, crystallization, and post-treatment have been used to influence this morphology [11]. A new approach would be the incorporation of one of the functional compounds into a porous network of the other. Tuning of pore size and pore channels thus would allow the a priori design of interpenetrating networks, which are important not only for the generation of organic solar cells but also for organic light-emitting diodes (OLEDs) and organic field effect transistors (OFETs).

In the following, we describe pathways for the preparation of such porous polymeric semiconductors. Several examples that show high prospects for the mentioned applications have been reported and will be the subject of discussion in the following chapters.

In the first section, an overview of the general methods for the synthesis of meso- and microporous polymers is given. Several examples are shown exemplifying the different approaches. The second section focuses on porous networks with pore walls composed of conjugated, semiconducting polymers. Here, the relation between the development of 3D organic semiconductors and porous semiconducting polymers is discussed. The section also gives a comprehensive overview of porous semiconducting polymers. Besides optoelectronic applications, these polymers have shown other interesting applications, for example, as materials for gas storage or as catalyst support. The third section describes a particular types of porous conjugated polymers, namely, COFs. In the last section, a special type of organic semiconductor, graphitic carbon nitride, is discussed in detail, as the semiconducting properties of this material have been recently extensively investigated and used for important applications such as photocatalysis for the production of hydrogen from water.

7.2 GENERAL SYNTHETIC AND ANALYTICAL METHODS FOR POROUS POLYMERS

7.2.1 Synthetic Routes toward Mesoporous Polymers

The formation of mesoporous polymers relies in most cases on a structure-directing template, that is, the replication of one nanostructure into another. Usually, this yields a 3D negative of the respective template structure. A general discrimination can be made between the use of organic soft matter (e.g., block copolymer or surfactant micelles) and inorganic hard matter (e.g., mesoporous silica or silica nanoparticles). Details of these techniques have been discussed in recent reviews [17–19].

In the case of soft-templating, most of the examples make use of the block-copolymer approach [17]. Block copolymers that consist of two immiscible but covalently linked polymer chains tend to undergo microphase separation in the solid state. Depending on the chemical nature and the block length ratio, the accessible microphases manifest as spheres, cylinders, lamellae, and so on. If one block is removed selectively after the microphase separation has occurred, a mesoporous polymer is obtained. The pore size is dictated by the size of the self-assembled mesophase, which itself depends on the chemical identity of the block copolymer, that is, the molecular weight of the blocks. Often, cross-linking of the remaining block has to be performed before etching of the sacrificial block in order to ensure sufficient stability against pore closure.

The use of surfactant self-assembly (e.g., polymeric surfactants) in order to induce polymer growth in a well-defined mesophase was shown to be applicable only for a limited number of systems such as ordered, mesoporous phenolic and melamine resins [20–22]. While this approach is broadly applicable for the synthesis of ordered, mesoporous, inorganic materials, it is hard to control in the case of mesoporous polymer synthesis because of thermodynamic reasons: usually, the two polymer phases demix macroscopically, yielding only macroporous materials.

As mentioned above, mesoscopically ordered, organic matter can be used successfully in the synthesis of mesoporous, inorganic materials, and vice versa. Nanostructured inorganic materials, mostly mesoporous silica or silica nanoparticles, have been frequently used as hard templates for the synthesis of mesoporous polymers [18,23–26]. The template is mixed together with a monomer melt or solution, yielding either a suspension of the template in the monomer ("endotemplates") or the incorporation of the monomers in the pores of the template ("exotemplate") [27]. Polymerization of the monomers can be subsequently carried out by thermal, chemical, electrochemical, or photochemical polymerization (or a combination thereof). Etching of the hard-template finally yields the mesoporous polymer. As described above, cross-linking is usually a prerequisite for stable mesopores.

For a complete picture, we should mention that some reports also exist on the successful synthesis of mesoporous polymers by well-controlled phase-separation processes [28,29]. These processes may also be regarded as variants of soft-templating as they involve certain polymeric additives.

7.2.2 Synthetic Schemes for Microporous Polymers

An additional hurdle needs to be overcome should the desired product also exhibit permanent microporosity. Capillary pressure and high surface energy tend to close small pores by simple deformation of the framework [18,25], and the absence of microporosity is the downside of the "soft" character commonly associated with organic materials. The classical way of dealing with undesired pore collapse is to work with a high degree of cross-linking, effectively yielding hard, rigid organic materials. A more recent approach introduces stiff, bulky, contorted binding motifs, compelling the otherwise linear polymer chain to pack space inefficiently.

Generally, there are three types of microporous polymers: (i) hypercross-linked polymers, (ii) polymers of intrinsic microporosity (PIM), and (iii) COFs. All of these three types have a specific feature that provides the necessary stability. The underlying principles of these three types of microporous polymers are summarized in the following paragraphs.

The so-called hypercross-linked polymers can be obtained by excessive cross-linking of a swollen polymer (gels). Davankov et al. presented the archetypical example, microporous polystyrene via Friedel–Crafts hypercross-linking in the 1970s, and considerable progress has been made since then [30–32]. However, the method remains restricted to polymers, which can be easily post-modified into a very stiff network, which provides sufficiently high stability against pore closure.

The term polymers of intrinsic microporosity (PIM) was introduced by Budd *et al.* [33]. PIMs are microporous polymers whose microporosity is based on the inability of close chain packing in the solid state. Typically, rigid, kinked monomers (e.g., spiro compounds) are used. Polymerization of these monomers yield stiff, contorted chains or networks (depending on the functionality) that have a large and accessible free volume and are consequently considered to be microporous. Figure 7-1 summarizes the principle.

A large variety of polymer networks with intrinsic microporosity has been described, compared to only a few examples of soluble (noncross-linked) PIMs. The reason for this can most probably be found in the better stabilization of the microporosity in cross-linked polymers. It should be mentioned in short that PIMs still possess some flexibility, although they rely on very stiff architectures. This flexibility has some consequences on the porosity of polymers with different chemical compositions. For example it was shown that intermolecular interactions (e.g. hydrogen bonding) can yield pore closure [35].

In summary, the field of microporous polymers has considerably advanced during the last years and a large variety of microporous polymers (e.g. polyimides [34–38], polybenzodioxanes [33,39–43], polyorganosilanes [44,45], polyaryl-carbinols [46], and various conjugated microporous polymers (CMPs) [47] has been described. Among these, the semiconducting (conjugated) polymers are discussed in more detail in the next section. Finally, COFs are a third class of microporous polymer networks. A distinct feature of this polymer class, which was introduced by Côté *et al.* [48], is the presence of crystalline order on nano- or molecular scale. This can be achieved by thermodynamic control of the polymerization. The drawback of this prerequisite is that only a limited number covalent bond-forming reactions and consequently structures are available, as the bond formation has to be reversible. Up to now, COFs have been synthesized

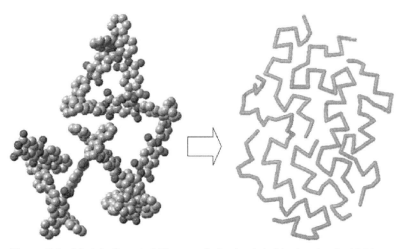

Figure 7-1 Model of a spirobifluorene-derived polyimide showing its highly contorted structure together with a cartoon representation of the space-inefficient packing of the polymer chains, yielding a polymer with intrinsic microporosity [19,34].

mainly on the basis of the condensation of boronic acids with each other or with aromatic alcohols [48–52]. Recently, it was shown that the ionothermal formation of triazines from aromatic dinitriles at elevated temperatures can also lead to COFs [53]. Another possibility is the use of the reversible Schiff-base chemistry to generate crystalline, imine-based frameworks [54]. Further details regarding COFs can be found in Section 7.3.3.

7.2.3   Analysis of Porosity in Porous Polymers

The measurement of gas adsorption/desorption isotherms is one of the most frequently used methods for the analysis of porous materials, as it provides access to a number of important parameters such as the specific surface area (S), the pore volume (V_p), pore size distribution (PSD), and porosity (ϕ) [55,56].

In a typical gas physisorption experiment, one measures the amount of gas adsorbed on the surface, depending on the relative pressure p/p_0, where p_0 is the saturation pressure of the gas at a given temperature. Depending on the used gas and temperature, one can observe monolayer formation, multilayer formation, and pore filling by condensation. The most widespread method is the analysis of nitrogen sorption isotherms measured at 77.36 K, the boiling point of nitrogen. This method allows the probing of micro- and mesoporosity, but has some drawbacks in the analysis of very small micropores, as discussed later.

From typical nitrogen sorption isotherms, it is possible to extract the specific surface area by applying the Brunauer–Emmett–Teller (BET) model. The BET model assumes the formation of multilayers by the adsorbate, which is more realistic than the assumption of monolayers. The BET model works reasonably well for powders and meso- and macroporous materials, but care has to be taken when analyzing microporous materials. This is due to the overlap of mono- or multilayer formation with pore filling in such small pores. Therefore, it is common to discuss apparent BET surface areas when discussing microporous materials. From the total uptake of nitrogen at $p/p_0 = 0.99$, one can estimate the total pore volume and calculate the porosity ϕ.

Gas adsorption also makes it possible to analyze the PSD of a given sample. A number of methods and models are available for the analysis. One of the most common methods for the evaluation of mesopore size distribution is the Barrett–Joyner–Halenda (BJH) approach. The BJH model relies on the well-known Kelvin equation, which describes the relation between the vapor pressure change and the radius of a curved liquid–gas interface. The method is, however, based on the assumption of cylindrical pores and might therefore underestimate pore sizes in the case of noncylindrical mesopores. A recent development in advanced analysis of PSDs is the so-called non-local density functional theory (NLDFT) [57–61]. This method relies on the fitting of the experimental isotherm with model isotherms and was shown to be a successful method for the description of various mesoporous materials. A major advantage of NLDFT methods is the fact that it can also describe microporosity, which is out of range for the BJH approach, as the Kelvin equation does not hold true for micropores. Furthermore, it was shown that the BJH method could underestimate the correct pore sizes by up to 100% [58].

However, care has to be taken when calculating PSDs of microporous soft matter. One often observes a low pressure hysteresis, that is, the adsorption and desorption branches do not close even at low relative pressures ($p/p_0 < 0.1$). This phenomenon is frequently attributed to structural changes in the course of gas adsorption [35,58]. If such changes occur, it is obvious that any calculation of PSDs is affected by these dynamic effects. Therefore, independent methods (e.g., small angle X-ray scattering (SAXS), positronium annihilation lifetime spectroscopy (PALS), ^{129}Xe-NMR) should be additionally employed if the knowledge of the precise micropore size is desired [38,62,63].

It should be noted that there are other useful analyte gases besides nitrogen. It was shown that hydrogen could be used as a probe molecule for the detection of ultramicropores that are not accessible by nitrogen [35,38,64]. The use of carbon dioxide sorption at ambient temperatures also has advantages in the analysis of ultramicroporous materials [65]. We have not discussed these methods in detail, but have highlighted the fact that nitrogen sorption alone is sometimes not sufficient to decide the presence or absence of microporosity.

7.3 POROUS π-CONJUGATED POLYMERS

7.3.1 From 3D Semiconducting Molecules to Porous Polymer Networks

Since the fundamental work on (semi)conducting polymers from Shirakawa, Heeger, and McDiarmid [66–68], credited with the Noble Prize in Chemistry in 2000, research on organic semiconductors based on π-conjugated systems has increased rapidly, especially because of the fascinating prospects for several applications. Displays or solar cells of low weight that are soft and flexible are just one example of applications that could be envisaged using these materials. Meanwhile, the first prototypes of OFETs, organic light emitting devices (OLEDs), and bulk heterojunction solar cells based on conjugated polymers have been assembled, and some devices have already entered the market. For an overview of the task of organic electronics, several reviews are available [7,69–73].

Depending on the desired function of the organic electronic device, there are certain requirements for the π-conjugated system. While for an OLED preferably high electroluminescence efficiency is needed, in an organic solar cell optimal absorption properties of the solar irradiation and a high absorption coefficient are required. These properties cannot be achieved exclusively by choosing the right chemical composition of the organic semiconductor, as molecular interactions and especially orientation has to be specifically controlled [74].

Semiconducting polymers are generally of a linear, that is, one dimensional structure which is significantly different from inorganic semiconductors, such as silicon which has a 3D, thus spatially isotropic, structure. For most applications, it is favorable when electron or charge transfer occurs in all three dimensions, and thus, such a transfer also has to take place in linear organic semiconductors between different molecular chains. This low dimensionality results in anisotropy of the optical and charge-transport properties. Material concepts yielding control over material organization and molecular orientation are therefore indispensable for the fabrication of devices. Several research groups were thus striving for organic devices with higher dimensionalities, finally yielding materials that could be described as "organic silicon" [74,75]. To increase the dimensionality of organic semiconductors, macromolecules that extend into two or even three dimensions have to be designed. The thus achieved higher spatial isotropy has consequences for electronic properties and was often associated with better solubility and processability. Owing to their extended π-conjugated structures, organic semiconductors feature relatively stiff molecular backbones, and hence they tend to crystallize, which decreases the solubility of the compounds. Star-shaped, dendritic, or hyperbranched polymers have, in contrast, a much lower tendency for crystallization, as they cannot organize into ordered, densely packed structures. This is favorable for their solubility and processing.

Interestingly, the structure-directing motifs enabling such 2D- or 3D-conjugated architectures are similar to those that are used to prepare conjugated microporous polymer (CMP) networks. Indeed, benzyl, triazine, triphenylamine (TPA), tetraphenylmethane, or spirobifluorene motifs have been used to synthesize large, fully conjugated, star-shaped molecules, which were subsequently applied to organic, optoelectronic devices, from OLEDs to solar cells [74–84]. Note that for CMPs comparable structure-directing motifs were used (Fig. 7-2).

Some of the structures of star-shaped or hyperbranched conjugated molecules are already quite close to the architectures that have later been reported to exhibit microporosity and high surface area. Indeed, often just a slight change in the reaction parameters and polymerization conditions yields materials ranging from dendritic or hyperbranched polymers to cross-linked microporous polymer networks. One example is the preparation of hyperbranched polythiophenes and polymer networks reported by Richter et al. [85] and Schmidt et al. [86] respectively. Starting with the same monomer, 1,3,5-(thienyl)benzene, polymerization at room temperature in chloroform with an excess of $FeCl_3$ results in a hyperbranched product, while a change in solvent and reaction temperature to acetonitrile and $-40°C$ yields a microporous network. For both protocols, precipitation of the polymer was observed after a short time, but while the precipitate could be redissolved in chlorobenzene in the first case, an insoluble product was found in the latter. This shows that the step from a hyperbranched, soluble polymer to a microporous, insoluble polymer network can indeed be rather small.

It can be speculated that acetonitrile is a better solvent for oligothiophene and therefore the first-formed oligomers are kept longer dissolved in solution and thus can form a more extended network that exhibits microporosity. A broad distribution of molecular weights (which cannot be determined for the insoluble products) from such an approach has been shown for the hyperbranched products. After reprecipitation from methanol, different fractions could be isolated by dissolution in trichloromethane and chlorobenzene, yielding products with molecular weights (determined using size exclusion chromatography)

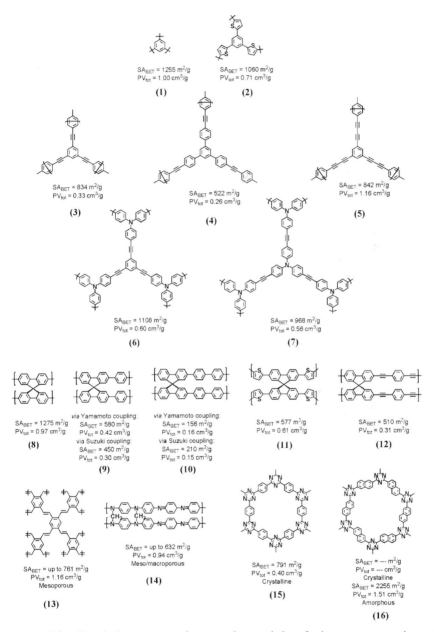

Figure 7-2   Chemical structures and porous characteristics of microporous π-conjugated polymer networks.

of 2220 and 6120 g/mol, respectively. Besides the necessity to bring the polymerization to high conversion or polymerization degrees, another important point for the generation of microporous networks can be deduced from these works: monomers with smaller angles between the functional groups (i.e., ortho instead of meta position) led to hyperbranched polythiophenes with, on average, smaller molecular weights and better solubility. These monomers are consequently not suitable for the formation of extended microporous networks.

Smaller angles between the bridging groups of the tectons have the disadvantage of yielding generally smaller molecular weights of the polymers. When assembled into a network, such structure-directing compounds also yield smaller pore sizes, finally leading to pores so small that they cannot accommodate nitrogen or other gases and therefore no accessible surface area can be detected. This finding is supported by comparing two aromatic polyimides with different tectons in the chain. Polyimides of intrinsic microporosity were produced using, on one hand, a diaminospirobifluorene tecton [34] ($90°$ and angle between the functionalities) and, on the other, a 2,2′-diamino-1,1′-binaphthalene [38] ($\sim78°$ and additionally some rotational freedom). While the first polymer is microporous with apparent BET surface areas of 550 m^2/g, for the latter no surface area from micropores could be assessed using nitrogen sorption. However, significant amounts of hydrogen and carbon dioxide, that is, gases with smaller kinetic diameters could be adsorbed. It can thus be concluded that the choice of the tecton has significant impact on the microporosity of the network.

7.3.2 Conjugated Microporous Polymers

Figure 7-2 shows some examples of meso- or microporous, conjugated polymer networks, together with the respective values for surface area and pore volume.

The first preparation of a microporous conjugated polymer was described by Cooper and coworkers in 2007 presenting poly(aryleneethynylene) networks (**3,4**) [87]. The networks are formed using the Sonogashira–Hagihara cross-coupling of ethynylene and iodine-substituted phenylenrings. The first surface areas reported were 834 m^2/g but could be later raised to values above 1000 m^2/g and the polymer networks were almost exclusively microporous [88].

This approach was highly suitable for the generation of networks with different functionalities. Microporous poly(phenylene butadiynylene) networks with BET surface area up to 842 m^2/g were obtained by the palladium-catalyzed homocoupling of 1,3,5-triethynylbenzene and 1,4-diethynylbenzene (**5**) [89]. Thus, a series of nitrogen-containing microporous polymers (NCMP) (**6,7**) based on poly(triethynylphenyl)amine (poly TPA) has been synthesized, which showed surface areas up to 1100 m^2/g [90].

The TPA group, which was introduced into this polymer network, opens up new vistas for using such networks for optoelectronic applications, as organic polymer semiconductors derived from TPA derivatives have been widely investigated as hole transporting materials [70,75]. Also, amorphous microporous poly(aryleneethynylene) networks using tetrahedral carbon- and silicon-centred monomers could be prepared with surface areas reaching 1213 m^2/g [91].

Other types of coupling reactions and tectons have also been used to create microporous conjugated polymer networks. The Suzuki-coupling of 2,2',7,7'-tetrabromospirobifluorene with benzenediboronic acids yielded networks with a surface area of 450 m^2/g (**9**) [92]. Yamamoto couplings of 2,2',7,7'-tetrabromospirobifluorene (**8**) and 1,3,5-tribromobenzene (**1**) were used for homocoupling reactions toward CMPs [93]. Nickel-catalyzed C–C coupling of bromine-functionalized aromatic compounds yielded polymer networks with surface areas of up to 1275 m^2/g. Such networks show an intensive and tunable photoluminescence. Changing the size or the linear linkers between the spirobifluorene tectons (9,10,12) not only changes the porous characteristics but also the optical properties of the networks [92,93].

Gilch-coupling reactions have been used to prepare poly(phenylenevinylene) (PPV) networks from 1,2,4,5-tetrakis(bromomethyl)benzene using potassium *tert*-butoxide in anhydrous tetrahydrofuran (THF). Surface areas of 761 m^2/g were reported (**13**) [94]. While polymers prepared via the scaffolding approach almost always yield microporous materials, here the nitrogen sorption isotherms show a pronounced hysteresis, which doubtlessly proves the mesoporous character of the sample, as seen in an average pore diameter of 5.6 nm for this network. At first glance, this is rather surprising as the molecular architecture of this polymer network does not seem to differ much from others reported; thus, in this case, small pores, that is, microporosity, would be expected. The authors of this work therefore suggested that the pore structure in these materials is determined more by the details of phase separation [95] than by the structure of the molecular framework. Although mesoporosity introduced via phase separation of a polymer network is hard to control and predict, this approach gives rise to further interesting applications of porous conjugated polymers.

Besides their conjugated nature, none of these polymers contains further chemical functionalities. Poly(thiophene) networks prepared by oxidative coupling of 2,2',7,7'-tetrathienylspirobifluorene (**11**) and 1,3,5-tristhienylbenzene (**2**) polymer networks showed surface areas of up to 1060 m^2/g and additionally contain large amounts of thio-functionalities (for example, S-content 26.4 wt% for the 1,3,5-tristhienylbenzene-derived network) [86]. These functionalities are part of the organic bridges of the framework, and hence accessible for compounds entering the porous system; in other words, these networks are deemed suitable as catalyst support for catalytically active adsorbates. This was exemplified by impregnation of the polythiophene network with a palladium-salt solution. Metal loadings of up to 15 wt% were achieved. A reduction in the palladium ions yielded very small, monodisperse, palladium clusters, with an average diameter of 1.5 nm homogeneously dispersed in the polymer matrix.

The design principle of the first described conjugated polymer networks, based on polyaryleneethynylenes, has also been used for the first systematic study toward the introduction of functional groups into porous organic networks. Dawson *et al.* prepared a variety of CMP networks using Sonogashira–Hagihara cross-coupling of 1,3,5-triethynylbenzene with a number of functionalized dibromobenzenes [96]. The properties of the networks could be controlled by monomer selection. For example, the dye sorption behavior of the networks was shown to be controlled by varying the hydrophobicity of the pore walls using different functional dibromocompounds. Indeed, the prospects of microporous polymers was largely expanded by this approach, as it allows the preparation of high-surface-area networks with properties that can be tailored for applications such as catalysis and separations.

Microporous conjugated polymers have also been produced by hypercross-linking. Commercial polyaniline was swollen in an organic solvent and hypercross-linked with either diiodoalkanes or paraformaldehyde [97]. The resulting polymers (14) exhibited permanent porosities and surface areas up to 632 m^2/g. The nitrogen sorption isotherms suggest a hybrid meso/macroporous structure for this material. Hypercross-linked polyanilines show very high enthalpies of adsorption of up to 9.3 kJ/mol (for polyaniline cross-linked with paraformaldehyde) and thus a very high affinity for hydrogen. This makes this class of polymers certainly interesting for application as hydrogen storage materials.

All the above-described polymer networks have two main points in common. First, all networks prepared by this approach are amorphous. Secondly, by increasing the length of the linker between the tectons, the overall surface area decreases.

The first finding, i.e. the amorphous nature of the networks, may surprise initially, since on paper, all the building blocks shown in Fig. 7-2 could be in principle organized into crystalline frameworks, due to their defined angles and rigid aromatic structures. Nevertheless, all these approaches were carried out under reaction conditions that enabled kinetically controlled structures. During the reactions, stable C–C bonds are formed, yielding irreversible accretion of the polymer network. Once a critical number of bonds are created, the polymer networks precipitate from solution. These reaction processes thus yield kinetically controlled, yet crystalline, structures but not necessarily the thermodynamically most stable products. Indeed, crystalline porous structures are always prepared under reversible thus thermodynamically controlled conditions. The most prominent examples are zeolites formed under basic and hydrothermal conditions. Reversible formation of the Si–O bond favors the formation of the thermodynamically (meta)stable structure. When no hydrothermal conditions are applied, amorphous silica are obtained as can be seen for sol–gel approaches. The other examples are MOFs, which are based on coordination bonds, which are likewise formed and broken reversibly under the used reaction conditions.

Intriguingly, the formation of an amorphous structure can be advantageous as porosity and surface areas can be increased compared to the crystalline analog. This is especially obvious for networks that would crystallize into extended, two-dimensional sheets or layers. Such layers are naturally close packed with layer-to-layer distances on the order of magnitude known from graphite (~0.32 nm). In case of eclipsed packing (A-A-A orientation of the layers), the holes in the layers can pile up to accessible cylindrical channels and thus porosity would be observed. In contrast, for a stacking in a staggered manner (e.g., A-B-A-B), the holes in the layers are obstructed by the tectons or linkers from adjacent layers. In the latter case, no porosity could be expected for these materials (and in a later example, we see that indeed such an architecture and a thermodynamically favored structure are possible). Changing such structures from the crystalline to the amorphous form can open up additional pore channels and connections between the pores, increasing the overall surface area.

The amorphous, more open structure of the CMPs gives rise to the second finding: with increasing length of the linkers, the surface area and porosity of the networks decreases. Indeed, cross-coupling reactions of one structure-directing tecton with linkers of different lengths have been achieved for several microporous, π-conjugated polymers, and in all the cases, increasing linker size (e.g., from benzene to a diphenyl linker) was accompanied by a decrease in porosity and surface area. Quite the opposite is expected at first glance. A longer linker is expected to increase the pore size of the networks, yielding an increased porosity. Exactly this observation is made for crystalline structures, especially for MOFs, where the exchange of a larger linker between the metal centers (the tectons in this case) yields bigger unit cells, larger pores, and higher porosity. As a defined tuning of porosity is possible this way, the process is called "reticular synthesis" [98]. But why is the opposite true for an amorphous system? Two main reasons might be responsible: first, many organic linkers show higher flexibility and rotational freedom with increasing length. While in crystalline networks, no deviation from the given angles of the crystal structure is possible, an amorphous network can easily use this higher flexibility to avoid the formation of pores (always note that "nature abhors the vacuum", i.e., any possibility to avoid the existence of small, empty cavities is used). Second, interpenetration of the networks can easily occur. Intercalation of a crystalline network occurs when the size of the cavity exceeds that of the biggest structure-directing part. Then, the two crystalline networks are intercalated in a defined manner. In amorphous networks, such interpenetration could occur easily, as again only certain angles of the network are fixed. Thus, longer linkers easily penetrate small voids in the network. However, as mentioned before, the analysis of the pore structure of an amorphous polymer network is quite difficult, and no explanation can be confirmed with certainty.

7.3.3 π-Conjugated Covalent Organic Frameworks

Rational synthesis of extended arrays of organic matter in bulk, in solution, in crystals, and in thin films has always been a paramount goal of chemistry. The classical synthetic tools to obtain long-range regularity are, however, limited to noncovalent interactions [99,100], in contrast to covalent polymerization reactions, which usually yield structurally more random products [101,102]. As mentioned before, the most challenging hurdle in the synthesis of extended, yet precisely defined, two- and three-dimensional structures based on covalent chemistry is the requirement that the reaction linking individual organic constituents should be reversible, allowing the scaffold to arrange to the thermodynamic, well-ordered product rather than the kinetic, amorphous structure [103,104]. Although the synthesis of crystalline inorganic materials largely relies on dynamic polymerization processes, only a few examples that make use of dynamic polymerizations for the generation of crystalline organic frameworks are known [48,53].

Hence, a combination of porosity and regularity in organic, covalently bonded materials requires not only the design of molecular building blocks that allow for growth into a nonperturbed, regular geometry (i.e., two or three latent bond-forming sites spanning an appropriate angle) but also a condensation mechanism that progresses under reversible, thermodynamic, self-optimizing conditions. Ideally, a single precursor molecule fulfils all these criteria, making structure-directing agents obsolete and discounting undesired chain-termination events (if any) known from A-B type of polymerization reactions. A recent work on the crystalline covalent organic frameworks (COFs) by Yaghi and Côté et al. highlighted two kinds of condensation reactions, yielding planar, six-membered boroxine rings (B_3O_3) and five-membered BO_2C_2 rings, which are formed in a dynamic and reversible manner and thus can be utilized as covalent linkers between organic units to generate 2D and 3D COFs [48,50].

The bor-oxygen linkages in these COFs intrinsically avoid the formation of an extended π-conjugated system. It should be noted that the boroxine molecule does not have a significant aromatic character or pronounced B–O π-bonding, which have been tested using UV photoelectron spectroscopy [105]. Nevertheless, larger π-conjugated molecules can certainly be connected using this binding motif and thus assembled into useful structures. This was shown by Jiang et al., who introduced larger organic spacers into covalent frameworks. Pyrene groups were introduced into COFs by either a condensation reaction of 2,3,6,7,10,11-hexahydroxytriphenylene (HHTP) and pyrene-2,7-diboronic acid (TP-COF) [106] or via self-condensation of pyrenediboronic acid (PPy-COF, Fig. 7-3) under solvothermal conditions [107]. Both networks align in a perfectly eclipsed manner, yielding a porous COF (PPy-COF) with surface areas of 868 and 932 m^2/g, respectively. For TP-COF, the channels formed were so large that a mesoporous material with pore sizes of 3.2 nm could be produced. Because of the eclipsed alignment of the sheets, PPy-COF shows a fluorescence shift as a consequence of the formation of excimers. Furthermore, it was proved that the materials show semiconductor characteristics. For example, PPy-COF shows effective photoconduction accompanied with quick response to light irradiation. Therefore, this work can indeed be seen as an important step from the synthesis of new covalent frameworks toward applications, in this case for organic, optoelectronic, and photovoltaic materials.

An extension of the conjugated system can be envisaged when boroxine rings (B_3O_3) are replaced by isoelectronic triazine rings (C_3N_3). In contrast to the boroxine ring, triazines are aromatic compounds, although their resonance energy is much lower compared to that of benzene, which is, for example, expressed in a much lower activity toward electrophilic substitution. Still, triazines have been part of several semiconducting molecules and polymers [77,108,109]. Triazines can be formed by simple cyclotrimerization and polymerization of aromatic nitriles. The cyclotrimerization reaction of cyanamide and dicyandiamide to melamine—a tri-amino-substituted triazine—is known since the days of Liebig [110]. It requires a high reaction temperature of 250°C and above, but it is highly dynamic and reversible at these conditions. Since the monomeric constituents of the framework are not preoriented in any way to yield an extended crystalline

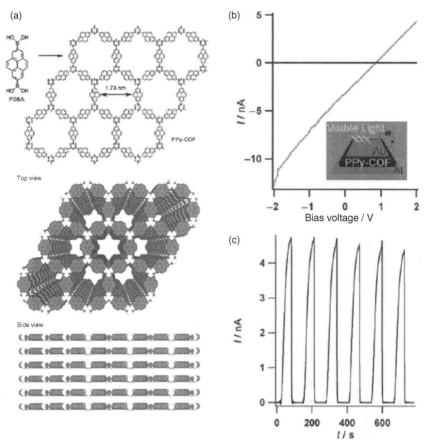

Figure 7-3   (a) Chemical and space-filling structure of PPy-COF. (b) I–V profile of PPy-COF sandwiched between sandwich type Al–Au electrodes (black curve: without light irradiation, red curve: with light irradiation). (c) Photocurrent during switching the light on and off [107].

network, not only should the condensation reaction of the monomers be reversible but also the mobility of monomers and their aggregates be ensured. Therefore, the above-described COF synthesis based on boron–oxygen always made use of suitable solvents to dissolve the monomers and oligomers during the reaction. Under the reaction conditions used for the trimerization reaction, however, standard solvents and even most ionic liquids are rendered unusable. Kanatzidis and coworkers have previously noted the limited choice of appropriate media for synthetic applications at intermediate temperatures (i.e., 150–350°C) and successfully applied alkali-metal polysulfide melts in the synthesis of low-dimensional ternary chalcogenides [111]. The cyclotrimerization of aromatic nitriles was hence performed under ionothermal conditions [112], that is, in molten zinc chloride, which acts both as a solvent and catalyst at temperatures between 400 and 600°C. Optimized reaction conditions yield apparent surface areas in the range of 1000 m^2/g for the periodic covalent triazine-based framework (CTF-1) network (see structure **15** in Fig. 7-2). In this framework, an eclipsed stacking of the sheet was assumed, opening up hexagonally arranged pore channels from the holes in the layers. As mentioned before, such an ordered structure does not yield the highest possible surface area for such arrangements, as just the holes and not the plane sides of the layers can accommodate gas molecules when they are densely packed. Indeed, it was calculated by O'Keefe, Yaghi, and coworkers that the surface area of a single infinite graphite sheet would reach 2965 m^2/g (calculating the Connolly surface areas of both sides). If the graphite sheet is additionally perforated, this would further add towards accessible surface area. By dividing this sheet into more and more fragments, higher surface areas to an upper limit for isolated, six-membered rings with values of 7745 m^2/g were calculated, although linking these fragments into a real material would not lead to complete realization of these theoretically limiting values [113].

Indeed, the formation of amorphous networks using triazine chemistry yielded materials with superior surface areas. When the para dicyanobenzne (pDCB) monomer was reacted with 5 molar equivalents of zinc chloride and 600°C reaction temperature [114], surface areas increased to 3270 m^2/g. At these conditions the chemical composition of the networks changes, though. Evidence of nitrogen-depletion following the high-temperature (HT) treatment and radii on the mesoscale of these newly created holes both suggest gradual in-plane triazine-cleavage and thus the creation of mesoporous transport channels through which latent, yet initially inaccessible, microporosity can be addressed.

While the inorganic chemist is used to evaluate crystallinity with a myriad of sharp X-ray diffraction peaks and the polymer chemist considers one broad reflex a sign of order, the world of two- and three-dimensional polymer frameworks with regular porosity lies in-between the two. Peak broadening is in most cases a cause of small crystalline domains on the order of a few microns, which in turn is mostly determined by the solubility product of the salt melt toward the starting monomer, its oligomers, and all subsequent aggregates up to the product. If the reaction is terminated by premature precipitation, extended domains of the thermodynamic product are very unlikely to form. The two CTFs with the most pronounced order are based on pDCB and, recently, 2,6-dicyanonaphthalene (DCN) (see structure (**16**) in Fig. 7-2), and are fully π-conjugated, 2D, layered polymer frameworks with regular porosity. While their in-plane lattice parameters can be assigned to the expected calculated distances with relative ease, there is a degree of ambiguity regarding the stacking arrangement of the sheets. Two principle packing modes are obvious—a staggered (AB) structure commonly associated with graphite and an eclipsed (AA) structure. However, owing to the weak interplanar forces (mostly van der Waals forces, dipole–dipole interactions, and dipole-induced dipole interactions) and the limited extension of the crystalline domain, many other possibilities come to mind to relate from one sheet to the next, with only marginal enthalpic gains or losses—varying not only by translation but also potentially by torsion. In the present case, CTF-1 presumably assumes an eclipsed (AA) structure, in which the atoms of each layer are placed above their analogs in the neighboring layers, while CTF-2 does not exhibit any conventional long-distance ordering in the z-direction. On the contrary, it is more likely that domains with eclipsed ordering are interspersed with domains of staggered or arbitrary packing motifs or other defects such as screw-dislocation. However, note that, regardless of the packing motif, the pseudo-(002) peak of both frameworks indicates an interplanar stacking distance of ~3.4 Å, a value that is comparable in magnitude to other known aromatic, lamellar systems [115,116].

Through the use of heteroatom-containing building blocks, CTFs possess structurally inherent metal coordination sites and sites for chemical functionalization. The networks were indeed used as catalyst support for the coordination and thus stabilization of metal particles. Palladium particles have been supported onto a CTF network (Pd@CTF), and the catalytic performance was compared to the same metal particles supported on activated carbon (Pd@AC) [117]. For the oxidation of glycerol, the Pd@CTF catalyst showed a much higher stability compared to that of the Pd@AC catalyst, which was ascribed to the better coordination ability of the CTF surface, preventing agglomeration and leaching of the noble metal nanoparticles.

Using heterocyclic, aromatic dicyano compounds, the synthetic introduction of further functional groups, that is, heteroatoms into such networks, could be enabled [114]. Such heteroatoms could be used for enhanced metal binding. In a recent example, platinum ions were introduced into a CTF network composed of 2,5 dicyanopyridine [118]. The resulting framework nearly resembles a polymerized form of a platinum–bipyridinium catalyst (Periana's catalyst) [119], which was active for the selective catalysis of methane to methanol. The catalytic reaction was carried out at 200°C in concentrated sulfuric acid, and the solid catalysts were stable at least over six reaction cycles, showing the stability of CTF networks even under harsh conditions. No leaching of the platinum species has been detected, and the solid catalyst showed activities and selectivities well comparable to the homogeneous one.

7.3.4   Porous π-Conjugated Polymers Prepared via Hard or Soft Templating

Different templating strategies have been used to introduce porosity into π-conjugated semiconducting polymers. Templating, as introduced in Section 7.1, has the advantage that pore sizes can be increased largely over the microporous range (>2 nm). For the construction of organic bulk heterojunction solar cells, this could be of major importance, as the size of the electron- and hole-conducting phase should be in the range of ~10 nm to ensure efficient exciton separation.

Although such templating strategies have been extensively used for mesoporous polymers and especially mesoporous carbon [3,18,24,25,120–123], reports considering the generation of mesoporous semiconducting polymers are scarce, with one exemption, which is discussed in the following section.

However, there exist a few reports describing the synthesis of porous polythiophenes using hard or soft templating. Copper nanoparticles of ~30 nm were used as hard templates to obtain porous and insoluble polythiophene films [124].

The films obtained, however, featured low porosities (<20%). Furthermore, the size of the voids was found to be too big to provide benefits from the presence of a nanostructure. As the size of the voids is just a replica of the particle size, a reduction in the template size could overcome these problems.

As indicated in Section 7.1, block copolymer templating is a powerful way for the synthesis of nanoporous polymers. Consequently, this approach was also used for the generation of porous poly(3-alkylthiophene) (P3AT) films [125]. P3AT-*b*-polylactide block copolymers with varying block ratios were synthesized starting from P3AT macroinitiators. Thin films (~37 nm height) were generated by spin coating and thermal annealing. Removal of the polylactide block was achieved by treatment with aqueous sodium hydroxide solution. Morphology analysis by atomic force microscopy (AFM) confirmed the generation of nanoporous films with pits of ~35 nm diameter and depths of 10 nm.

A comparable strategy was presented recently by Sivanandan *et al.* [126]. Poly(3-hexylthiophene)-*g*-Poly(styrene) polymers were prepared using P3HT macroinitiators. The connection point (a trityl ether) between the P3HT backbone chain and the poly(styrene) (PS) chains is acid labile, and cleaving of the PS chains was achieved by acid treatment after film formation and annealing. Again, thin and nanoporous (~35 nm domain size) films could be prepared. So far, no studies of a backfilling of such morphologies are available, but it can be expected that this topic will be handled by a number of research groups in the near future.

Mesoporous silica/semiconducting polymer composites have been reported, which show a significant anisotropy of the optical properties of the polymers due to the spatial restrictions provided by the mesoporous silica [127–132]. Semiconducting polymers infiltrated into mesoporous titania have even been used as compounds in composite solar cells [133]. Nevertheless, all these approaches have not been used to create porous semiconducting polymers by removing the supporting inorganic phase.

On the other hand, templating approaches have often been used to create other nanostructures of semiconducting polymers; some striking examples are presented in the following paragraphs.

Li *et al.* [134,135] and Ikegame *et al.* [136] introduced an intriguing approach toward π-conjugated polymer/mesoporous silica composites and functional polymer nanofibres. Thiophene-containing surfactants were used and assembled (similar to the classic MCM-41 synthesis) into monomer/silica composites, with the monomer molecules aligned in the cylindrical mesochannel. Polymerization of the monomers then occurred under retention of the assembled structures, providing polymer silica nanocomposites. Removal of the silica template yielded conjugated molecular polythiophene wires (Fig. 7-4). Thus, the monomers acted first as an organic template for the generation of the mesoporous silica, which then served in a second step as a hard template for the polymerization toward polymer nanofibers.

Several semiconducting polymers, such as polypyrrole, poly(3-methylthiophene), and polyaniline were nanostructured using anodized aluminum oxide (AAO) templates. The AAO templates possess hexagonally

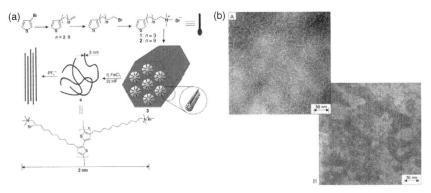

Figure 7-4   (a) Schematic presentation of the fabrication and organization of conjugated polythiophene molecular wires with mesoporous silica MCM-41. (b) Transmission electron microscopy (TEM) images of (A) the conjugated molecular wires from the polymerization of 2 and (B) the organized wires upon the addition of PF6− ions. *Source*: Reprinted with permission from Ref. [135]. Copyright 2003 Wiley-VCH.

ordered pore channels with tuneable diameters around 15–100 nm, which are assembled vertically with respect to an aluminum layer. Thus, replicating those structures can yield free-standing, rodlike nanostructures on an electrode, which is highly attractive for device fabrication. Monomers have been infiltrated into the channels, and an oxidative polymerization has been carried out either electrochemically or by using chemical oxidizing agents [137–141]. Also, more advanced architectures have been achieved, for example, segmented metal/polymer composite rods consisting of a gold and a polypyrrole block [142]. These composite rods were produced first by electrodeposition of gold into the porous alumina template followed by electrochemical polymerization of polypyrrole. The rods, consisting of hydrophilic hard and hydrophobic soft domains, assembled into mesoscopic aggregates, obeying the rules of the packing parameter, and different aggregate structures were observed depending on the length of the gold part in comparison to the polypyrrole part.

Porous semiconducting polymers with much larger pore diameters have been prepared using polymer lattices organized into colloidal crystals as templates. Here the voids of the colloidal crystal were filled with a suitable monomer, which was subsequently polymerized into the semiconducting polymer. Cassagneau and Caruso used an elegant method to produce inverse opals (as the replicas of colloidal crystals are called) from semiconducting polymers [143]. The colloidal crystal of polystyrene microparticles was assembled on a conducting indium tin oxide (ITO) glass and the uncovered surface passivated by coating this part with an epoxy resin. This limited the further reaction to the part where the colloidal crystal has been assembled. This assembly was then used as a working electrode, and pyrrole and thiophene monomers were electropolymerized into the voids of the opal. The PS template could be subsequently removed with THF. The resulting inverse opals were featuring highly organized pores in the size range of ∼1 μm; thus, they are regarded as macroporous materials, which would not be suitable for the above-mentioned applications. The larger pores allow linear, noncross-linked polymers to form the pore walls and maintain the porosity as much lower capillary forces and surface energies arise with larger pores. Colloidal crystals are well known for their property to reflect light of a certain wavelength, as their periodicity is on length scales comparable to that of visible light (therefore they are also called *photonic crystals*). The wing of a butterfly is a beautiful example of how nature is using this concept to create color. It is an interesting task to combine the optical properties derived from the porous structure with those derived from the chemical structure of the semiconducting polymer. Besides polypyrrole and polythiophene, other polymer compositions have also been templated with colloidal crystals, for example, polypyrrole, polyanilines, and polyphenylenevinylene [144–147].

7.4 POROUS GRAPHITIC CARBON NITRIDE SEMICONDUCTORS

7.4.1 Graphitic Carbon Nitride

Terminologically all binary compounds of carbon and nitrogen can be considered as carbon nitrides. Molecular carbon and nitrogen-containing compounds—analogous to carbon oxides, such as CO_2, CO, and other C_xO_y species—are known to every chemist. First attempts to prepare solid-state carbon nitride materials have been known since the early 1830s [110,148], and in 1922, Franklin described the formation of an amorphous carbon nitride material by thermolysis of mercury(II) thiocyanate [149], more commonly known from its use in pyrotechnics as the Pharaoh's snakes. The resulting materials, however, have eluded rigorous analysis and characterization for a long time mostly due to their insolubility in water and organic solvents. Interest in carbon nitrides—in particular its saturated, sp^3-hybridized, crystalline allotropes—was sparked in the 1980s by the work of Sung and Sung [150] and the theoretician Cohen [151], who speculated that the bulk modulus for a "tetrahedral compound ... formed between C and N" would be "significantly larger than for diamond." Although the shear modulus, G_0, rather than the bulk modulus, B_0, is a good indicator for hardness [152], the prospect of outstanding material properties, above all, extreme hardness and high thermal stability, has led to a large number of publications on carbon nitrides since the first theoretical paper was published, some of them claiming the successful formation of several of the initially predicted carbon nitride phases by chemical vapor deposition (CVD) and physical vapor deposition (PVD) techniques [153,154]. Reoccurring criticism from Fang [155], DeVries [156], Matsumoto [154], and Malkow [157], however, gives reason to believe that, to date, no satisfactory evidence has been presented that a crystalline solid of the proposed kind has been synthesized; moreover, that in some cases nonstoichiometric, amorphous solids were the dominant product.

Owing to the difficulties encountered in the synthesis of single-phase sp^3-hybridized carbon nitride phases on a preparative scale by physical methods, synthetic approaches shifted toward a chemical route in the 1990s. A large number of experiments were devised, and all of them utilized the condensation of suitable carbon- and nitrogen-containing molecular precursors to form graphitic carbon nitride materials (g-C_3N_4), which are regarded to be the most stable allotrope at ambient conditions [158]. Several C/N/H/X-containing compounds were identified, which were considered promising precursors for the chemical and bulk routes toward pure CN_x-phases such as s-triazines (in particular, their chlorides and fluorides) [159–161], s-heptazines [110,162,163], and thiocyanates [149,164]. However, note that bulk synthesis routes based on these precursors did not yield the desired, highly condensed, crystalline CN_x-phases either, but rather yielded incompletely condensed materials. Elemental microanalyses of these substances show hydrogen contents of up to 2 wt% resembling a stoichiometry closer to melon (poly[(8-amino-1,3,4,6,7,8,9,9b-heptaazaphenalen-2,5-diyl)imine]) (1.5 wt% hydrogen) [165]. Examples of the many methods used to synthesize C_3N_4 structures via high-pressure (HP), HT routes are slow thermal decomposition [166] and detonative synthesis [167] using tri-azido-s-triazine, $C_3N_3(N_3)_3$, which yielded a series of amorphous C/N/H/O-containing compounds: carbon nanotubes in the prior case and nanoparticles in the latter. It should be noted that the field of HP-HT synthesis was advanced by Riedel, Kroke, McMillan *et al.* very recently, showing that conversion of dicyandiamide under HP-HT conditions indeed yields a crystalline carbon nitride imide phase, $C_2N_2(NH)$ [168].

Kroke *et al.* proposed that g-C_3N_4 consists of sheets of highly ordered tri-s-triazine moieties connected through planarized, tertiary amino groups as depicted in Fig. 7-5a. The tri-s-triazine-based structure was postulated on the basis of the density-functional theory (DFT) calculations to be more stable at ambient conditions [169]. The condensation mechanism is essentially based on the molecular reaction of cyanamide and its successive condensation products, which is driven both by deamination and the formation of aromatic units (Fig. 7-5a) [170].

Incomplete condensation or polymerization in the bulk—as applied for this type of reaction—is widely acknowledged as a predominantly kinetic problem. Schnick *et al.* demonstrated that the bulk reaction does not proceed significantly past the polymeric form (melon) [162]. Employing electron diffraction and solid-state NMR spectroscopy, they have shown that the bulk condensation reaction of melem terminates at layers comprising infinite chains of "melem monomers." These chains are condensed via N–H bridges forming a closely packed, two-dimensional array [172]. Now, this principle problem is inherent to all of solid-state chemistry. If we want to achieve perfect condensation in a kinetically inhibited system, that is, systems in which dynamic and reversible chemistry yielding the thermodynamically preferred product as described above cannot be achieved, we require perfect prealignment to yield extended and covalent crystalline products. If the direction of growth and, hence, the alignment of our large, oligomeric precursors (melon) is statistically controlled, the number of terminal reactive groups with the right orientation toward one another will also be statistical, and the resulting product will be expected to be littered with unreacted termini. This problem of mono- and oligomer mobility can be overcome using an

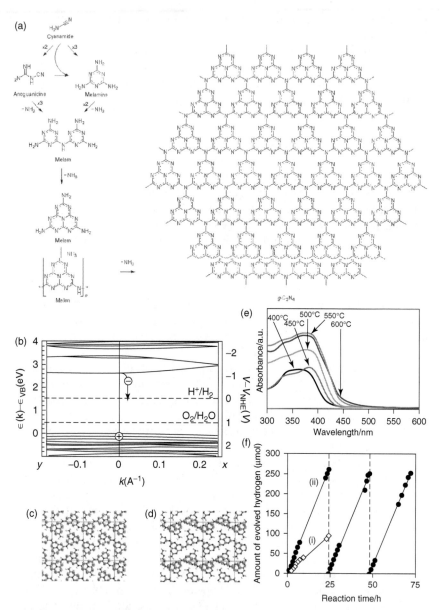

Figure 7-5   (a) Condensation reactions of cyanamide, yielding discrete oligomers, polymers, and extended networks. (b) Density-functional theory band structure for polymeric melon and the calculated spatial electron density coefficients for the valence band (c) of melon and for the conduction band (d), respectively. (e) Ultraviolet–visible diffuse reflectance spectrum of cyanamide heated to different temperatures. (f) A typical time course of H_2 production from water containing 10 vol% triethanolamine as an electron donor under visible light (of wavelength longer than 420 nm) by (i) unmodified g-C_3N_4 and (ii) 3.0 wt% Pt-deposited g-C_3N_4 photocatalyst. The reaction was continued for 72 h, with evacuation every 24 h (dashed line) [171].

appropriate solvent. The nature of the polycondensation reaction of s-heptazine derivatives, namely, the thermally induced reaction at temperatures of 400°C and above, however, discards all standard solvents, and is thus comparable to the problems facing the synthesis of the previously described, triazine-based CTFs, for which salt melts were employed. An eutectic mixture of lithium chloride and potassium chloride (45:55 wt%, T_m: 352°C) [173] has been used, as it shows good solvation of the small molecular precursors, and subsequent aggregates of higher molecular weight facilitate the condensation of the carbon nitride network. The product obtained is a highly crystalline, graphitic carbon nitride, which is deficient in hydrogen compared to the solvent-free preparations. PXRD analysis and high-resolution transmission electron microscopy show the presence of pronounced in-plane ordering and an interlayer distance of $d = 3.36$ Å. The organic crystals of graphitic carbon nitride are therefore based not on 0D or 1D constituents such as known molecular or short-chain polymeric crystals but on the packing motifs of extended 2D frameworks.

7.4.2 Porous Graphitic Carbon Nitrides by Hard Templating

The synthesis of graphitic carbon nitrides involves the heat treatment of precursors such as cyanamide, dicyandiamide, or melamine. At different temperatures, the carbon nitride species condense into fragments of ever increasing molecular weight via the melem unit ($C_6N_8(NH_2)_2$) up to the polymeric form melon and, in theory, finally to the two-dimensional graphitic structure. The idealized structure of graphitic carbon nitride features holes in the layers (Fig. 7-5a), comparable to the structures seen in COFs. However, it can be assumed that these holes are not accessible as the layers pack in a staggered manner. Moreover, the holes are much too small to accommodate guest molecules, and even gases such as nitrogen or hydrogen are excluded. Consequently, the materials show very low surface areas and porosities comparable to any other bulk material. However, as we will see in the next section, for their use as semiconductor in catalytic applications, a high and accessible surface area can be beneficial. Thus, strategies have been developed to introduce pores into graphitic carbon nitrides and consequently increase their surface area. As the thermal condensation is comparable to any other polymerization reaction, hard templating strategies should allow the introduction of pores. Another advantage of this system is that one of the precursors, cyanamide, is liquid at room temperature; thus, either suitable templates can be dispersed in cyanamide (endotemplates) or the cyanamide can be infiltrated into porous materials (exotemplate). Heat treatment of the resulting composite and removal of the template should yield graphitic carbon nitrides with structures replicating the nanostructure of the preceding template. Both approaches (using endo- or exotemplates) have been used to prepare mesoporous graphitic carbon nitrides (mpg-C_3N_4). In the first case, silica nanoparticles were dispersed in cyanamide, and in the second, cyanamide was infiltrated into mesoporous silica (SBA-15). In both cases, perfect replication of the silica templates was observed. In the former case, an mpg-C_3N_4 with spherical pores with a diameter of 12 nm was produced [174], while in the latter, mpg-C_3N_4 featuring a 2D hexagonal structure with cylindrical pores and a diameter of 4 nm was formed [175,176]. The surface areas of these materials reached up to 450 and 240 m^2/g, respectively.

7.4.3   Carbon Nitride-Based Semiconductor in Photocatalysis

In the context of sustainable energy sources, the conversion of sunlight into a form of direct or indirect energy or carrier—be it in the form of photovoltaics or the splitting of water into hydrogen gas—remains an urgent topic. The formal requirements for photocatalytic water splitting are that the band gap energy of the material should be in the range of the energy of visible light (i.e., 1.2–3.0 eV), that the position of the energy level of H^+ to H_2 reduction and the energy level of H_2O to O_2 oxidation should ideally be situated within the band gap, and that the valence and the conduction band are required to have a different character (i.e., location of the red-/ox-sites) to prevent immediate recombination of hydrogen and oxygen to water. In addition, it is desirable that the semiconductor is cheap in fabrication and ideally should not employ precious metals as cocatalysts—as is the case for most group VA and VIA semiconductors explored for this purpose so far [177–179].

In a recent work by Antonietti *et al.*, the thermal polycondensation reaction of the simple carbon- and nitrogen-containing organic monomers has been used to synthesize carbon nitride materials with various architectures [174,180], deliberately taking synthetic defects into account and utilizing these unreacted termini as sites for, for example, metal-free Friedel–Crafts catalysis [181], CO_2 activation [182], or—most interestingly—the photocatalytic cleavage of water [171].

It was found that the degree of conversion of the thermally induced polycondensation of cyanamide—which is driven toward completion at higher condensation temperatures through deamination—determines the location of the UV adsorption edge of the resulting carbon nitride material (cf. Figure 7-5e). DFT calculations suggest [183] that, while the product at 400°C is likely to resemble melem (*s*-heptazine) with a band gap of approximately 3.5 eV and the band gap of an ideal sheet of graphitic carbon nitride is calculated to be 2.1 eV, the product from the 550/600°C thermal condensation has a band gap of 2.7 eV (according to its UV/vis spectrum)—and hence is on the order of magnitude of the energy of visible light.

Figure 7-5c and d shows surface representations of the calculated spatial electron density coefficients for the valence band of chains of poly-melem (i.e., melon) and for the conduction band, respectively. Note that as a consequence the likely sites of H_2O to O_2 oxidation are the nitrogen atoms, while H^+ to H_2 reduction takes place on the carbon atoms. The corresponding reduction level of hydrogen and the oxidation level of oxygen are calculated *ab initio* and contrasted with the calculated levels of the valence and conduction band of melon in Fig. 7-5b [184,185]. In theory, the carbon nitride material obtained at 550/600°C also fulfils the criteria of spatially separate sites of reduction and oxidation, with the right energetic situation of the valence and conduction bands.

In practice, the carbon nitride material indeed achieves a steady evolution of hydrogen (cf. Fig. 7-5e,f), depending on the particle and surface morphologies as well as degree of condensation and amount of defects. Modification of the material with 3.0 wt% Pt facilitates the release and recombination of H^+ to H_2.

The process requires a sacrificial electron donor such as triethylamine, methanol, ethanol, or ethylenediaminetetraacetic acid, which effectively removes the nascent oxygen. On the other hand, by using sacrificial electron acceptors and a suitable co-catalyst (e.g. RuO_2) also water oxidation has been achieved with the carbon nitride [171]. Combining these two processes to finally obtain overall water-splitting into hydrogen and oxygen will be certainly one of the future targets for this polymeric photocatalyst. Nonetheless, this class of π-conjugated, polymeric carbon nitrides are comparatively cheap and easy to synthesize with a myriad of options for chemical and physical tuning, ranging from isoelectronic replacement of constituent atoms to the introduction of porosity through hard and soft templating.

Indeed, the carbon nitride-based photocatalyst has been furthermore optimized by introduction of a mesoporous structure. mpg-C_3N_4 has a similar band gap (2.7 eV) as that observed for the bulk materials and shows a remarkable improvement (factor 8) in the H_2 evolution activity [186].

7.5 CONCLUSION

Porous organic semiconductors can be prepared in various chemical compositions and structures on the nanoscale. Different methods introducing high porosities and high surface areas into these materials have been developed using, for example, scaffolding or templating approaches. Micro- and mesoporous conjugated polymers have been produced with surface areas reaching >1000 m^2/g. With their high and accessible surface areas, these materials have considerable prospects for optoelectronic or photocatalytic applications. It can be predicted that conjugated polymers will be filled with suitable compounds, having optical or electronic functions to from interpenetrating networks and bulk heterojunctions. However, before the first organic optoelectronic device will be assembled, several obstacles have to be overcome. First, analysis of the optoelectronic properties of porous conjugated networks so far is surprisingly scarce. Thus, up to now not much can be stated about important properties of these materials such as the electroluminescence efficiency, the absorption coefficient, the position of the highest occupied and lowest unoccupied molecular orbital (HOMO/LUMO), and so on. The current shortcomings in the analysis of the optoelectronic properties of such networks can be explained in the context of the second, more serious problem: so far, processing of conjugated polymer networks has simply not been possible. The reported synthetic methods for cross-linked conjugated polymer networks indeed always yield powders, which, due to the cross-linking of the chemical structures, can neither be melted nor dissolved and thus cannot be manufactured into films and electrodes. Overcoming this hurdle will be one of the most challenging topics for material chemists working in this field. Some possibilities present themselves: the first described PIMs just featured kinks in an otherwise linear polymer chain, that is, they were not cross-linked, but still microporous in the solid state. A comparable structure featuring π-conjugation in the chain could be easily processed from solution. Owing to their contorted structure, such polymers would even show much better solubility than their linear counterparts. Thus, the synthesis of such a PIM with a conjugated structure would certainly bring new advantages to the processing of optoelectronic devices. The intrinsic porous structure of the envisaged conjugated PIM would even allow intertwining with another (conjugated) polymer. The energy gain of closing the pores might even counteract the normally unfavorable free enthalpy of polymer mixing.

Furthermore, new polymerization techniques might bring further development. Electropolymerization of suitable tectons might directly yield the formation of microporous conjugated films, automatically attached to one electrode during the process of formation. Microporous polythiophenes could be produced via this method. Adding the second phase into the pore channels and placing the counter electrode on the final structure might be a very cheap, fast, and simple process to assemble a first organic optoelectronic device based on porous organic semiconductors.

Furthermore, it should be mentioned that the materials described in this chapter do not have to find their application in optoelectronic compounds alone. Indeed, first promising reports on the application of these materials for gas storage and catalysis open up a bright future for these types of polymers.

Considering the diminishing resources of fossil fuels, mankind has to find sustainable pathways toward new energy supplies. Direct water splitting using a photocatalyst with visible light is a highly attractive method to produce hydrogen, thus converting sunlight into storable and transportable energy. A polymeric photocatalyst such as graphitic carbon nitride based on cheap and available monomers—a polymer that sports a tuneable band gap and band positions that can be tuned by basic tools of the materials and polymer chemist—would certainly be an attractive model-system in the field of sustainable energy generation.

REFERENCES

1. Beck, J. S., Vartuli, J. C., Roth, W. J., Leonowicz, M. E., Kresge, C. T., Schmitt, K. D., et al. (1992). A new family of mesoporous molecular-sieves prepared with liquid-crystal templates. *J. Am. Chem. Soc.*, *114*, 10834–10843.
2. Zhao, D. Y., Feng, J. L., Huo, Q. S., Melosh, N., Fredrickson, G. H., Chmelka, B. F., et al. (1998). Triblock copolymer syntheses of mesoporous silica with periodic 50 to 300 angstrom pores. *Science*, *279*, 548–552.
3. Ryoo, R., Joo, S. H., Jun, S. (1999). Synthesis of highly ordered carbon molecular sieves via template–mediated structural transformation. *J. Phys. Chem. B*, *103*, 7743–7746.
4. Hu, X. D., Jenkins, S. E., Min, B. G., Polk, M. B., Kumar, S. (2003). Rigid-rod polymers: synthesis, processing, simulation, structure, and properties. *Macromol. Mater. Eng.*, *288*, 823–843.
5. Pinto, M. R., Schanze, K. S. (2002). Conjugated polyelectrolytes: synthesis and applications. *Synthesis*, *9*, 1293–1309.
6. Jiang, H., Taranekar, P., Reynolds, J. R., Schanze, K. S. (2009). Conjugated polyelectrolytes: synthesis, photophysics, and applications. *Angew. Chem. Int. Ed.*, *48*, 4300–4316.
7. Gunes, S., Neugebauer, H., Sariciftci, N. S. (2007). Conjugated polymer-based organic solar cells. *Chem. Rev.*, *107*, 1324–1338.

8. Hoppe, H., Sariciftci, N. S. (2004). Organic solar cells: An overview. *J. Mater. Res.*, *19*, 1924–1945.

9. Winder, C., Sariciftci, N. S. (2004). Low bandgap polymers for photon harvesting in bulk heterojunction solar cells. *J. Mater. Chem.*, *14*, 1077–1086.

10. Blom, P. W. M., Mihailetchi, V. D., Koster, L. J. A., Markov, D. E. (2007). Device physics of polymer: fullerene bulk heterojunction solar cells. *Adv. Mater.*, *19*, 1551–1566.

11. Yang, X., Loos, J. (2007). Toward high–performance polymer solar cells: the importance of morphology control. *Macromolecules*, *40*, 1353–1362.

12. Yang, X. N., Loos, J., Veenstra, S. C., Verhees, W. J. H., Wienk, M. M., Kroon, J. M., *et al.* (2005). Nanoscale morphology of high-performance polymer solar cells. *Nano Lett.*, *5*, 579–583.

13. Kim, Y., Choulis, S. A., Nelson, J., Bradley, D. D. C., Cook, S., Durrant, J. R. (2005). Device annealing effect in organic solar cells with blends of regioregular poly(3-hexylthiophene) and soluble fullerene. *Appl. Phys. Lett.*, *86*.

14. Berson, S., De Bettignies, R., Bailly, S., Guillerez, S. (2007). Poly(3-hexylthiophene) fibers for photovoltaic applications. *Adv. Funct. Mater.*, *17*, 1377–1384.

15. Nunzi, J. M. (2002). Organic photovoltaic materials and devices. *C. R. Phys.*, *3*, 523–542.

16. Po, R., Maggini, M., Camaioni, N. (2010). Polymer solar cells: recent approaches and achievements. *J. Phys. Chem. C*, *114*, 695–706.

17. Olson, D. A., Chen, L., Hillmyer, M. A. (2008). Templating nanoporous polymers with ordered block copolymers. *Chem. Mater.*, *20*, 869–890.

18. Thomas, A., Goettmann, F., Antonietti, M. (2008). Hard templates for soft materials: creating nanostructured organic materials. *Chem. Mater.*, *20*, 738–755.

19. Thomas, A., Kuhn, P., Weber, J., Titirici, M.-T., Antonietti, M. (2009). Porous polymers: enabling solutions for energy applications. *Macromol. Rapid Commun.*, *30*, 221–236.

20. Kosonen, H., Valkama, S., Nykanen, A., Toivanen, M., ten Brinke, G., Ruokolainen, J., *et al.* (2006). Functional porous structures based on the pyrolysis of cured templates of block copolymer and phenolic resin. *Adv. Mater.*, *18*, 201–205.

21. Meng, Y., Gu, D., Zhang, F. Q., Shi, Y. F., Cheng, L., Feng, D., *et al.* (2006). A family of highly ordered mesoporous polymer resin and carbon structures from organic–organic self–assembly. *Chem. Mater.*, *18*, 4447–4464.

22. Kailasam, K., Jun, Y. -S., Katekomol, P., Epping, J. D., Hong, W. H., Thomas, A. Mesoporous melamine resins by soft templating of block-co-polymer mesophases. *Chem. Mater.*, *22*, 428–434.

23. Goltner, C. G., Weissenberger, M. C. (1998). Mesoporous organic polymers obtained by "twostep nanocasting". *Acta Polym.*, *49*(12), 704–709.

24. Johnson, S. A., Ollivier, P. J., Mallouk, T. E. (1999). Ordered mesoporous polymers of tunable pore size from colloidal silica templates. *Science*, *283*, 963–965.

25. Weber, J., Antonietti, M., Thomas, A. (2007). Mesoporous poly(benzimidazole) networks via solvent mediated templating of hard spheres. *Macromolecules*, *40*, 1299–1304.

26. Weber, J., Bergström, L. (2009). Impact of cross-linking density and glassy chain dynamics on pore stability in mesoporous poly(styrene). *Macromolecules*, *42*, 8234–8240.

27. Schuth, F. (2003). Endo- and exotemplating to create high-surface-area inorganic materials. *Angew. Chem. Int. Ed.*, *42*, 3604–3622.

28. Hasegawa, J., Kanamori, K., Nakanishi, K., Hanada, T., Yamago, S. (2009). Pore formation in poly(divinylbenzene) networks derived from organotellurium-mediated living radical polymerization. *Macromolecules*, *42*, 1270–1277.

29. Hasegawa, J., Kazuyoshi, K., Kazuki, N., Teiichi, H., Shigeru, Y. (2009). Rigid crosslinked polyacrylamide monoliths with well-defined macropores synthesized by living polymerization. *Macromol. Rapid Commun.*, *30*, 986–990.

30. Tsyurupa, M. P., Davankov, V. A. (2002). Hypercrosslinked polymers: basic principle of preparing the new class of polymeric materials. *React. Funct. Polym.*, *53*, 193–203.

31. Tsyurupa, M. P., Davankov, V. A. (2006). Porous structure of hypercrosslinked polystyrene: state-of-the-art mini-review. *React. Funct. Polym.*, *66*, 768–779.

32. Wood, C. D., Tan, B., Trewin, A., Niu, H., Bradshaw, D., Rosseinsky, M. J., *et al.* (2007). Hydrogen storage in microporous hypercrosslinked organic polymer networks. *Chem. Mater.*, *19*, 2034–2048.

33. Budd, P. M., Ghanem, B. S., Makhseed, S., McKeown, N. B., Msayib, K. J., Tattershall, C. E. (2004). Polymers of intrinsic microporosity (PIMs): robust, solution–processable, organic nanoporous materials. *Chem. Commun.*, 230–231.
34. Weber, J., Su, O., Antonietti, M., Thomas, A. (2007). Exploring polymers of intrinsic microporosity-microporous, soluble polyamide and Polyimide. *Macromol. Rapid Commun.*, 28, 1871–1876.
35. Weber, J., Antonietti, M., Thomas, A. (2008). Microporous networks of high-performance polymers: elastic deformations and gas sorption properties. *Macromolecules*, 41, 2880–2885.
36. Farha, O. K., Spokoyny, A. M., Hauser, B. G., Bae, Y.-S., Brown, S. E., Snurr, R. Q., *et al.* (2009). Synthesis, properties, and gas separation studies of a robust diimide–based microporous organic polymer. *Chem. Mater.*, 21, 3033–3035.
37. Ghanem, B. S., McKeown, N. B., Budd, P. M., Al–Harbi, N. M., Fritsch, D., Heinrich, K., *et al.* (2009). Synthesis, characterization, and gas permeation properties of a novel group of polymers with intrinsic microporosity: PIM–polyimides. *Macromolecules*, 42, 7881–7888.
38. Ritter, N., Antonietti, M., Thomas, A., Senkovska, I., Kaskel, S., Weber, J. (2009). Binaphthalene-based, soluble polyimides: the limits of intrinsic microporosity. *Macromolecules*, 42, 8017–8020.
39. Budd, P. M., McKeown, N. B., Fritsch, D. (2005). Free volume and intrinsic microporosity in polymers. *J. Mater. Chem.*, 15, 1977–1986.
40. Du, N., Robertson, G. P., Pinnau, I., Guiver, M. D. (2009). Polymers of intrinsic microporosity derived from novel disulfone-based monomers. *Macromolecules*, 42, 6023–6030.
41. Du, N., Song, J., Robertson, G. P., Pinnau, I., Guiver, M. D. (2008). Linear high molecular weight ladder polymer via fast polycondensation of 5,5′,6,6′-tetrahydroxy-3,3,3′,3′-tetramethylspirobisindane with 1,4-dicyanotetrafluorobenzene. *Macromol. Rapid Commun.*, 29, 783–788.
42. McKeown, N. B., Budd, P. M. (2006). Polymers of intrinsic microporosity (PIMs): organic materials for membrane separations, heterogeneous catalysis and hydrogen storage. *Chem. Soc. Rev.*, 35, 675–683.
43. Staiger, C. L., Pas, S. J., Hill, A. J., Cornelius, C. J. (2008). Gas separation, free volume distribution, and physical aging of a highly microporous spirobisindane polymer. *Chem. Mater.*, 20, 2606–2608.
44. Rose, M., Bohlmann, W., Sabo, M., Kaskel, S. (2008). Element-organic frameworks with high permanent porosity. *Chem. Commun.*, 2462–2464.
45. Zhang, B., Wang, Z. (2009). Building ultramicropores within organic polymers based on a thermosetting cyanate ester resin. *Chem. Commun.*, 5027–5029.
46. Webster, O. W., Gentry, F. P., Farlee, R. D., Smart, B. E. (1992). Hypercrosslinked rigid-rod polymers. *Makromol. Chem. Macromol. Symp.*, 54–5, 477–482.
47. Cooper, A. I. (2009). Conjugated microporous polymers. *Adv. Mater.*, 21, 1291–1295.
48. Cote, A. P., Benin, A. I., Ockwig, N. W., O'Keeffe, M., Matzger, A. J., Yaghi, O. M. (2005). Porous, crystalline, covalent organic frameworks. *Science*, 310, 1166–1170.
49. Cote, A. P., El–Kaderi, H. M., Furukawa, H., Hunt, J. R., Yaghi, O. M. (2007). Reticular synthesis of microporous and mesoporous 2D covalent organic frameworks. *J. Am. Chem. Soc.*, 129, 12914–12915.
50. El-Kaderi, H. M., Hunt, J. R., Mendoza-Cortes, J. L., Cote, A. P., Taylor, R. E., O'Keeffe, M., *et al.* (2007). Designed synthesis of 3D covalent organic frameworks. *Science*, 316, 268–272.
51. Tilford, R. W., Gemmill, W. R., zur Loye, H. C., Lavigne, J. J. (2006). Facile synthesis of a highly crystalline, covalently linked porous boronate network. *Chem. Mater.*, 18, 5296–5301.
52. Tilford, R. W., Mugavero Iii, S. J., Pellechia, P. J., Lavigne, J. J. (2008). Tailoring microporosity in covalent organic frameworks. *Adv. Mater.*, 20, 2741–2746.
53. Kuhn, P., Antonietti, M., Thomas, A. (2008). Porous, covalent triazine-based frameworks prepared by ionothermal synthesis. *Angew. Chem. Int. Ed.*, 47, 3450–3453.
54. Uribe-Romo, F. J., Hunt, J. R., Furukawa, H., Klöck, C., O'Keeffe, M., Yaghi, O. M. (2009). A crystalline imine-linked 3-D porous covalent organic framework. *J. Am. Chem. Soc.*, 131, 4570–4571.

55. Sing, K. S. W., Everett, D. H., Haul, R. A. W., Moscou, L., Pierotti, R. A., Rouquerol, J., *et al.* (1985). Reporting physisorption data for gas solid systems with special reference to the determination of surface-area and porosity (Recommendations 1984). *Pure Appl. Chem.*, *57*, 603–619.

56. Rouquerol, F., Rouquerol, J., Sing, K. S. W. (2008). The experimental approach. In *Handbook of Porous Solids* (Eds.: Schuth, F., Sing, K. S. W., Weitkamp, J.), Wiley-VCH, Weinheim, pp. 236–275.

57. Ravikovitch, P. I., Neimark, A. V. (2001). Characterization of micro-and mesoporosity in SBA-15 materials from adsorption data by the NLDFT method. *J. Phys. Chem. B*, *105*, 6817–6823.

58. Ravikovitch, P. I., Neimark, A. V. (2002). Density functional theory of adsorption in spherical cavities and pore size characterization of templated nanoporous silicas with cubic and three-dimensional hexagonal structures. *Langmuir*, *18*, 1550–1560.

59. Tarazona, P. (1985). Free-energy density functional for hard spheres. *Phys. Rev. A*, *31*, 2672–2679.

60. Tarazona, P., Marconi, U. M. B., Evans, R. (1987). Phase equilibria of fluid interfaces and confined fluids—Non-local versus local density functionals. *Mol. Phys.*, *60*, 573–595.

61. Thommes, M., Smarsly, B., Groenewolt, M., Ravikovitch, P. I., Neimark, A. V. (2006). Adsorption hysteresis of nitrogen and argon in pore networks and characterization of novel micro- and mesoporous silicas. *Langmuir*, *22*, 756–764.

62. Jansen, J. C., Macchione, M., Tocci, E., De Lorenzo, L., Yampolskii, Y. P., Sanfirova, O., *et al.* (2009). Comparative study of different probing techniques for the analysis of the free volume distribution in amorphous glassy perfluoropolymers. *Macromolecules*, *42*, 7589–7604.

63. Yampolskii, Y. P. (2007). Methods for investigation of the free volume in polymers. *Russ. Chem. Rev.*, *76*, 59–78.

64. Germain, J., Svec, F., Fre'chet, J. M. J. (2008). Preparation of size-selective nanoporous polymer networks of aromatic rings: potential adsorbents for hydrogen storage. *Chem. Mater.*, *20*, 7069–7076.

65. Lozano-Castelló, D., Cazorla-Amorós, D., Linares-Solano, A. (2004). Usefulness of CO_2 adsorption at 273 K for the characterization of porous carbons. *Carbon*, *42*, 1233–1242.

66. Chiang, C. K., Fincher, C. R., Park, Y. W., Heeger, A. J., Shirakawa, H., Louis, E. J., *et al.* (1977). Electrical–conductivity in doped polyacetylene. *Phys. Rev. Lett.*, *39*, 1098–1101.

67. Shirakawa, H., Louis, E. J., Macdiarmid, A. G., Chiang, C. K., Heeger, A. J. (1977). Synthesis of electrically conducting organic polymers—halogen derivatives of polyacetylene, (CH)X. *J. Chem. Soc., Chem. Commun.*, 578–580.

68. Chiang, C. K., Druy, M. A., Gau, S. C., Heeger, A. J., Louis, E. J., Macdiarmid, A. G., *et al.* (1978). Synthesis of highly conducting films of derivatives of polyacetylene, (CH)X. *J. Am. Chem. Soc.*, *100*, 1013–1015.

69. Hide, F., DiazGarcia, M. A., Schwartz, B. J., Heeger, A. J. (1997). New developments in the photonic applications of conjugated polymers. *Acc. Chem. Res.*, *30*, 430–436.

70. Shirota, Y. (2000). Organic materials for electronic and optoelectronic devices. *J. Mater. Chem.*, *10*, 1–25.

71. Kelley, T. W., Baude, P. F., Gerlach, C., Ender, D. E., Muyres, D., Haase, M. A., *et al.* (2004). Recent progress in organic electronics: materials, devices, and processes. *Chem. Mater.*, *16*, 4413–4422.

72. Allard, S., Forster, M., Souharce, B., Thiem, H., Scherf, U. (2008). Organic semiconductors for solution–processable field–effect transistors (OFETs). *Angew. Chem. Int. Ed.*, *47*, 4070–4098.

73. Thompson, B. C., Frechet, J. M. J. (2008). Organic photovoltaics—polymer-fullerene composite solar cells. *Angew. Chem. Int. Ed.*, *47*, 58–77.

74. Roncali, J., Leriche, P., Cravino, A. (2007). From one- to three-dimensional organic semiconductors: in search of the organic silicon? *Adv. Mater.*, *19*, 2045–2060.

75. Thelakkat, M. (2002). Star-shaped, dendrimeric and polymeric triarylamines as photoconductors and hole transport materials for electro-optical applications. *Macromol. Mater. Eng.*, *287*, 442–461.

76. Ponomarenko, S. A., Kirchmeyer, S., Elschner, A., Huisman, B. H., Karbach, A., Drechsler, D. (2003). Star-shaped oligothiophenes for solution–processible organic field-effect transistors. *Adv. Funct. Mater.*, *13*, 591–596.

77. Meier, H., Holst, H. C., Oehlhof, A. (2003). Star-shaped compounds having 1,3,5-triazine cores. *Eur. J. Org. Chem.*, 4173–4180.

78. Cremer, J., Bauerle, P. (2006). Star-shaped perylene-oligothiophene-triphenylamine hybrid systems for photovoltaic applications. *J. Mater. Chem.*, *16*, 874–884.

79. Johansson, N., Salbeck, J., Bauer, J., Weissortel, F., Broms, P., Andersson, A., et al. (1998). Solid-state amplified spontaneous emission in some spiro–type molecules: a new concept for the design of solid-state lasing molecules. *Adv. Mater.*, *10*, 1136–1141.

80. Saragi, T. P. I., Spehr, T., Siebert, A., Fuhrmann-Lieker, T., Salbeck, J. (2007). Spiro compounds for organic optoelectronics. *Chem. Rev.*, *107*, 1011–1065.

81. Steuber, F., Staudigel, J., Stossel, M., Simmerer, J., Winnacker, A., Spreitzer, H., et al. (2000). White light emission from organic LEDs utilizing spiro compounds with high–temperature stability. *Adv. Mater.*, *12*, 130–133.

82. Leriche, P., Piron, F., Ripaud, E., Frere, P., Allain, M., Roncali, J. (2009). Star–shaped triazine–thiophene conjugated systems. *Tetrahedron Lett.*, *50*, 5673–5676.

83. Salbeck, J., Yu, N., Bauer, J., Weissortel, F., Bestgen, H. (1997). Low molecular organic glasses for blue electroluminescence. *Synth. Met.*, *91*, 209–215.

84. Inada, H., Shirota, Y. (1993). 1,3,5-Tris[4-(Diphenylamino)Phenyl]Benzene and its methyl-substituted derivatives as a novel class of amorphous molecular materials. *J. Mater. Chem.*, *3*, 319–320.

85. Richter, T. V., Link, S., Hanselmann, R., Ludwigs, S. (2009). Design of soluble hyperbranched polythiophenes with tailor-made optoelectronic properties. *Macromol. Rapid Commun.*, *30*, 1323–1327.

86. Schmidt, J., Weber, J., Epping, J. D., Antonietti, M., Thomas, A. (2009). Microporous conjugated poly(thienylene arylene) networks. *Adv. Mater.*, *21*, 702–705.

87. Jiang, J. X., Su, F., Trewin, A., Wood, C. D., Campbell, N. L., Niu, H., et al. (2007). Conjugated microporous poly(aryleneethynylene) networks. *Angew. Chem. Int. Ed.*, *46*, 8574–8578.

88. Jiang, J. X., Su, F., Trewin, A., Wood, C. D., Niu, H., Jones, J. T. A., et al. (2008). Synthetic control of the pore dimension and surface area in conjugated microporous polymer and copolymer networks. *J. Am. Chem. Soc.*, *130*, 7710–7720.

89. Jiang, J. X., Su, F., Niu, H., Wood, C. D., Campbell, N. L., Khimyak, Y. Z., et al. (2008). Conjugated microporous poly(phenylene butadiynylene)s. *Chem. Commun.*, 486–488.

90. Jiang, J. X., Trewin, A., Su, F. B., Wood, C. D., Niu, H. J., Jones, J. T. A., et al. (2009). Microporous Poly(tri(4-ethynylphenyl)amine) Networks: Synthesis, Properties, and Atomistic Simulation. *Macromolecules*, *42*, 2658–2666.

91. Stockel, E., Wu, X. F., Trewin, A., Wood, C. D., Clowes, R., Campbell, N. L., et al. (2009). High surface area amorphous microporous poly(aryleneethynylene) networks using tetrahedral carbon- and silicon-centred monomers. *Chem. Commun.*, 212–214.

92. Weber, J., Thomas, A. (2008). Toward stable interfaces in conjugated polymers: microporous poly(p–phenylene) and poly(phenyleneethynylene) based on a spirobifluorene building block. *J. Am. Chem. Soc.*, *130*, 6334–6335.

93. Schmidt, J., Werner, M., Thomas, A. (2009). Conjugated microporous polymer networks via yamamoto polymerization. *Macromolecules*, *42*, 4426–4429.

94. Dawson, R., Su, F. B., Niu, H. J., Wood, C. D., Jones, J. T. A., Khimyak, Y. Z., et al. (2008). Mesoporous poly(phenylenevinylene) networks. *Macromolecules*, *41*, 1591–1593.

95. Sherrington, D. C. (1998). Preparation, structure and morphology of polymer supports. *Chem. Commun.*, 2275–2286.

96. Dawson, R., Laybourn, A., Clowes, R., Khimyak, Y. Z., Adams, D. J., Cooper, A. I. (2009). Functionalized conjugated microporous polymers. *Macromolecules*, *4*, 8809–8816.

97. Germain, J., Frechet, J. M. J., Svec, F. (2007). Hypercrosslinked polyanilines with nanoporous structure and high surface area: potential adsorbents for hydrogen storage. *J. Mater. Chem.*, *17*, 4989–4997.

98. Yaghi, O. M., O'Keeffe, M., Ockwig, N. W., Chae, H. K., Eddaoudi, M., Kim, J. (2003). Reticular synthesis and the design of new materials. *Nature*, *423*, 705–714.

99. Blake, A. J., Champness, N. R., Hubberstey, P., Li, W. S., Withersby, M. A., Schroder, M. (1999). Inorganic crystal engineering using self–assembly of tailored building–blocks. *Coord. Chem. Rev.*, *183*, 117–138.

100. Decher, G. (1997). Fuzzy nanoassemblies: Toward layered polymeric multicomposites. *Science*, *277*, 1232–1237.

101. Kim, Y. H. (1998). Hyperbranchecs polymers 10 years after. *J. Polym. Sci., Part A: Polym. Chem.*, *36*, 1685–1698.

102. Voit, B. (2000). New developments in hyperbranched polymers. *J. Polym. Sci., Part A: Polym. Chem.*, *38*, 2505–2525.

103. Sakamoto, J., van Heijst, J., Lukin, O., Schluter, A. D. (2009). Two-dimensional polymers: just a dream of synthetic chemists? *Angew. Chem. Int. Ed.*, *48*, 1030–1069.

104. Yaghi, O. M., Li, H. L., Davis, C., Richardson, D., Groy, T. L. (1998). Synthetic strategies, structure patterns, and emerging properties in the chemistry of modular porous solids. *Acc. Chem. Res.*, *31*, 474–484.

105. Novak, I., Kovac, B. (2007). On the aromaticity of trimethylboroxine: a photoelectron spectroscopic study. *Chem. Phys. Lett.*, *440*, 70–72.

106. Wan, S., Guo, J., Kim, J., Ihee, H., Jiang, D. L. (2008). A belt-shaped, blue luminescent, and semiconducting covalent organic framework. *Angew. Chem. Int. Ed.*, *47*, 8826–8830.

107. Wan, S., Guo, J., Kim, J., Ihee, H., Jiang, D. L. (2009). A photoconductive covalent organic framework: self-condensed arene cubes composed of eclipsed 2D polypyrene sheets for photocurrent generation. *Angew. Chem. Int. Ed.*, *48*, 5439–5442.

108. Fink, R., Frenz, C., Thelakkat, M., Schmidt, H. W. (1997). Synthesis and characterization of aromatic poly(1,3,5-triazine-ether)s for electroluminescent devices. *Macromolecules*, *30*, 8177–8181.

109. Wen, G. A., Xin, Y., Zhu, X. R., Zeng, W. J., Zhu, R., Feng, J. C., *et al.* (2007). Hyperbranched triazine-containing polyfluorenes: efficient blue emitters for polymer light-emitting diodes (PLEDs). *Polymer*, *48*, 1824–1829.

110. Liebig, J. V. (1834). Ueber einige stickstoff-verbindungen. *Ann. Pharm.*, *10*, 1–47.

111. Kanatzidis, M. G., Park, Y. (1990). Molten-salt synthesis of low-dimensional ternary chalcogenides—novel structure types in the K/HG/S, K/HG/SE system. *Chem. Mater.*, *2*, 99–101.

112. Cooper, E. R., Andrews, C. D., Wheatley, P. S., Webb, P. B., Wormald, P., Morris, R. E. (2004). Ionic liquids and eutectic mixtures as solvent and template in synthesis of zeolite analogues. *Nature*, *430*, 1012–1016.

113. Chae, H. K., Siberio-Perez, D. Y., Kim, J., Go, Y., Eddaoudi, M., Matzger, A. J., *et al.* (2004). A route to high surface area, porosity and inclusion of large molecules in crystals. *Nature*, *427*, 523–527.

114. Kuhn, P., Thomas, A., Antonietti, M. (2009). Toward tailorable porous organic polymer networks: a high-temperature dynamic polymerization scheme based on aromatic nitriles. *Macromolecules*, *42*, 319–326.

115. Bojdys, M. J., Muller, J. O., Antonietti, M., Thomas, A. (2008). Ionothermal synthesis of crystalline, condensed, graphitic carbon nitride. *Chem.—Eur. J.*, *14*, 8177–8182.

116. Ruland, W., Smarsly, B. (2002). X-ray scattering of non–graphitic carbon: an improved method of evaluation. *J. Appl. Crystallogr.*, *35*, 624–633.

117. Chan-Thaw, C. E., Villa, A., Katekomol, P., Su, D. S., Thomas, A., Prati, L. (2010). Covalent Triazine Framework as Catalytic Support for Liquid Phase Reaction. *Nano Lett.*, *10*, 537–541.

118. Palkovits, R., Antonietti, M., Kuhn, P., Thomas, A., Schuth, F. (2009). Solid catalysts for the selective low-temperature oxidation of methane to methanol. *Angew. Chem. Int. Ed.*, *48*, 6909–6912.

119. Periana, R. A., Taube, D. J., Gamble, S., Taube, H., Satoh, T., Fujii, H. (1998). Platinum catalysts for the high-yield oxidation of methane to a methanol derivative. *Science*, *280*, 560–564.

120. Tiemann, M. (2008). Repeated templating. *Chem. Mater.*, *20*, 961–971.

121. Jun, S., Joo, S. H., Ryoo, R., Kruk, M., Jaroniec, M., Liu, Z., *et al.* (2000). Synthesis of new, nanoporous carbon with hexagonally ordered mesostructure. *J. Am. Chem. Soc.*, *122*, 10712–10713.

122. Ryoo, R., Joo, S. H., Kruk, M., Jaroniec, M. (2001). Ordered mesoporous carbons. *Adv. Mater.*, *13*, 677–681.

123. Stein, A., Wang, Z. Y., Fierke, M. A. (2009). Functionalization of porous carbon materials with designed pore architecture. *Adv. Mater.*, *21*, 265–293.

124. Andreasen, J. W., Jorgensen, M., Krebs, F. C. (2007). A route to stable nanostructures in conjugated polymers. *Macromolecules*, *40*, 7758–7762.

125. Boudouris, B. W., Frisbie, C. D., Hillmyer, M. A. (2008). Nanoporous poly(3-alkylthiophene) thin films generated from block copolymer templates. *Macromolecules*, *41*, 67–75.

126. Sivanandan, K., Chatterjee, T., Treat, N., Kramer, E. J., Hawker, C. J. High surface area poly(3-hexylthiophenes) thin films from cleavable graft copolymers. *Macromolecules*, *43*, 233–241.

127. Cardin, D. J. (2002). Encapsulated conducting polymers. *Adv. Mater.*, *14*, 553–563.

128. Moller, K., Bein, T. (1998). Inclusion chemistry in periodic mesoporous hosts. *Chem. Mater.*, *10*, 2950–2963.

129. Schwartz, B. J. (2003). Conjugated polymers as molecular materials: how chain conformation and film morphology influence energy transfer and interchain interactions. *Annu. Rev. Phys. Chem.*, *54*, 141–172.

130. Scott, B. J., Wirnsberger, G., Stucky, G. D. (2001). Mesoporous and mesostructured materials for optical applications. *Chem. Mater.*, *13*, 3140–3150.

131. Nguyen, T. Q., Wu, J. J., Doan, V., Schwartz, B. J., Tolbert, S. H. (2000). Control of energy transfer in oriented conjugated polymer-mesoporous silica composites. *Science*, *288*, 652–656.

132. Wu, J. J., Gross, A. F., Tolbert, S. H. (1999). Host–guest chemistry using an oriented mesoporous host: alignment and isolation of a semiconducting polymer in the nanopores of are ordered silica matrix. *J. Phys. Chem. B*, *103*, 2374–2384.

133. Coakley, K. M., Liu, Y. X., McGehee, M. D., Frindell, K. L., Stucky, G. D. (2003). Infiltrating semiconducting polymers into self-assembled mesoporous titania films for photovoltaic applications. *Adv. Funct. Mater.*, *13*, 301–306.

134. Li, G. T., Bhosale, S., Bhosale, S., Li, F. T., Zhang, Y. H., Guo, R. R., *et al.* (2004). Template synthesis of functionalized polystyrene in ordered silicate channels. *Chem. Commun.*, 1760–1761.

135. Li, G. T., Bhosale, S., Wang, T. Y., Zhang, Y., Zhu, H. S., Fuhrhop, K. H. (2003). Gram-scale synthesis of submicrometer-long polythiophene wires in mesoporous silica matrices. *Angew. Chem. Int. Ed.*, *42*, 3818–3821.

136. Ikegame, M., Tajima, K., Aida, T. (2003). Template synthesis of polypyrrole nanofibers insulated within one-dimensional silicate channels: hexagonal versus lamellar for recombination of polarons into bipolarons. *Angew. Chem. Int. Ed.*, *42*, 2154–2157.

137. Jerome, C., Demoustier-Champagne, S., Legras, R., Jerome, R. (2000). Electrochemical synthesis of conjugated polymer wires and nanotubules. *Chem.—Eur. J.*, *6*, 3089–3093.

138. Martin, C. R. (1991). Template synthesis of polymeric and metal microtubules. *Advanced Materials*, *3*, 457–459.

139. Martin, C. R. (1995). Template synthesis of electronically conductive polymer nanostructures. *Acc. Chem. Res.*, *28*, 61–68.

140. Martin, C. R., Parthasarathy, R., Menon, V. (1993). Template synthesis of electronically conductive polymers—a new route for achieving higher electronic conductivities. *Synth. Met.*, *55*, 1165–1170.

141. Parthasarathy, R. V., Martin, C. R. (1994). Template-synthesized polyaniline microtubules. *Chem. Mater.*, *6*, 1627–1632.

142. Park, S., Lim, J. H., Chung, S. W., Mirkin, C. A. (2004). Self-assembly of mesoscopic metal-polymer amphiphiles. *Science*, *303*, 348–351.

143. Cassagneau, T., Caruso, F. (2002). Semiconducting polymer inverse opals prepared by electropolymerization. *Adv. Mater.*, *14*, 34–38.

144. Luo, X. L., Killard, A. J., Smyth, M. R. (2007). Nanocomposite and nanoporous polyaniline conducting polymers exhibit enhanced catalysis of nitrite reduction. *Chem.—Eur. J.*, *13*, 2138–2143.

145. Tian, S. J., Wang, J. J., Jonas, U., Knoll, W. (2005). Inverse opals of polyaniline and its copolymers prepared by electrochemical techniques. *Chem. Mater.*, *17*, 5726–5730.

146. Sumida, T., Wada, Y., Kitamura, T., Yanagida, S. (2000). Electrochemical preparation of macroporous polypyrrole films with regular arrays of interconnected spherical voids. *Chem. Commun.*, 1613–1614.

147. Deutsch, M., Vlasov, Y. A., Norris, D. J. (2000). Conjugated-polymer photonic crystals. *Adv. Mater.*, *12*, 1176–1180.

148. Gmelin, L. (1835). Ueber einige Verbindungen des Melon's. *Ann. Pharm.*, *15*, 252–258.

149. Franklin, E. C. (1922). The ammono carbonic acids. *J. Am. Chem. Soc.*, *44*, 486–509.

150. Sung, C. M., Sung, M. (1996). Carbon nitride and other speculative superhard materials. *Mater. Chem. Phys.*, *43*, 1–18.

151. Cohen, M. L. (1985). Calculation of bulk moduli of diamond and zinc-blende solids. *Phys. Rev. B: Condens. Matter*, *32*, 7988–7991.

152. Teter, D. M. (1998). Computational alchemy: the search for new superhard materials. *MRS Bull.*, *23*, 22–27.

153. Malkow, T. (2000). Critical observations in the research of carbon nitride. *Mater. Sci. Eng. A-Struct. Mater. Prop. Microstruct. Process.*, *292*, 112–124.

154. Matsumoto, S., Xie, E. Q., Izumi, F. (1999). On the validity of the formation of crystalline carbon nitrides, C_3N_4. *Diamond Relat. Mater.*, *8*, 1175–1182.

155. Fang, P. H. (1995). On the beta–C_3N_4 search. *J. Mater. Sci. Lett.*, *14*, 536–538.

156. DeVries, R. C. (1997). Inventory on innovative research: the case of C_3N_4. *Mater. Res. Innovat.*, *1*, 161–162.

157. Malkow, T. (2000). Critical observations in the research of carbon nitride. *Mater. Sci. Eng., A*, *292*, 112–124.

158. Kroke, E., Schwarz, M. (2004). Novel group 14 nitrides. *Coord. Chem. Rev.*, *248*, 493–532.

159. Kawaguchi, M., Nozaki, K. (1995). Synthesis, structure, and characteristics of the new host material $[(C_3N_3)(2)(NH)(3)](N)$. *Chem. Mater.*, *7*, 257–264.

160. McMurran, J., Kouvetakis, J., Nesting, D. C. (1998). Synthesis of molecular precursors to carbon-nitrogen-phosphorus polymeric systems. *Chem. Mater.*, *10*, 590–593.

161. Montigaud, H., Tanguy, B., Demazeau, G., Courjault, S., Birot, M., Dunogues, J. (1997). Graphitic form of C_3N_4 through the solvothermal route. *C. R. Acad. Sci. Paris, Ser. IIb*, *325*, 229–234.

162. Komatsu, T. (2001). Attempted chemical synthesis of graphite–like carbon nitride. *J. Mater. Chem.*, *11*, 799–801.

163. Komatsu, T. (2001). The first synthesis and characterization of cyameluric high polymers. *Macromol. Chem. Phys.*, *202*, 19–25.

164. Shtrempler, G. I., Murzubraimov, B., Rysmendeev, K. (1982). Thermal-stability of complex-compounds based on thiosemicarbazide. *Zh. Neorg. Khim.*, *27*, 789–792.

165. Pauling, L., Sturdivant, J. H. (1937). The structure of cyameluric acid, hydromelonic a cid, and related substances. *Proc. Natl. Acad. Sci. U.S.A.*, *23*, 615–620.

166. Gillan, E. G. (2000). Synthesis of nitrogen–rich carbon nitride networks from an energetic molecular azide precursor. *Chem. Mater.*, *12*, 3906–3912.

167. Kroke, E., Schwarz, M., Buschmann, V., Miehe, G., Fuess, H., Riedel, R. (1999). Nanotubes formed by detonation of C/N precursors. *Adv. Mater.*, *11*, 158–161.

168. Horvath–Bordon, E., Riedel, R., McMillan, P. F., Kroll, P., Miehe, G., van Aken, P. A., et al. (2007). High–pressure synthesis of crystalline carbon nitride imide, $C_2N_2(NH)$. *Angew. Chem. Int. Ed.*, *46*, 1476–1480.

169. Kroke, E., Schwarz, M., Horath-Bordon, E., Kroll, P., Noll, B., Norman, A. D. (2002). Tri-s-triazine derivatives. Part I. From trichloro-tri-s-triazine to graphitic C_3N_4 structures. *New J. Chem.*, *26*, 508–512.

170. Jurgens, B., Irran, E., Schneider, J., Schnick, W. (2000). Trimerization of NaC_2N_3 to $Na_3C_6N_9$ in the solid: Ab initio crystal structure determination of two polymorphs of NaC_2N_3 and of $Na_3C_6N_9$ from X–ray powder diffractometry. *Inorg. Chem.*, *39*, 665–670.

171. Wang, X. C., Maeda, K., Thomas, A., Takanabe, K., Xin, G., Carlsson, J. M., et al. (2009). A metal-free polymeric photocatalyst for hydrogen production from water under visible light. *Nat. Mater.*, *8*, 76–80.

172. Lotsch, B. V., Doblinger, M., Sehnert, J., Seyfarth, L., Senker, J., Oeckler, O., et al. (2007). Unmasking melon by a complementary approach employing electron diffraction, solid-state NMR spectroscopy, and theoretical calculations-structural characterization of a carbon nitride polymer. *Chem.—Eur. J.*, *13*, 4969–4980.

173. Solomons, C., Goodkin, J., Gardner, H. J., Janz, G. J. (1958). Heat of fusion, entropy of fusion and cryoscopic constant of the LICL-KCL eutectic mixture. *J. Phys. Chem.*, *62*, 248–250.

174. Goettmann, F., Fischer, A., Antonietti, M., Thomas, A. (2006). Chemical synthesis of mesoporous carbon nitrides using hard templates and their use as a metal–free catalyst for Friedel-Crafts reaction of benzene. *Angew. Chem. Int. Ed.*, *45*, 4467–4471.

175. Jun, Y. S., Hong, W. H., Antonietti, M., Thomas, A. (2009). Mesoporous, 2D hexagonal carbon nitride and titanium nitride/carbon composites. *Adv. Mater.*, *21*, 4270–xxxx

176. Chen, X. F., Jun, Y. S., Takanabe, K., Maeda, K., Domen, K., Fu, X. Z., et al. (2009). Ordered mesoporous SBA–15 type graphitic carbon nitride: a semiconductor host structure for photocatalytic hydrogen evolution with visible light. *Chem. Mater.*, *21*, 4093–4095.

177. Lee, Y., Terashima, H., Shimodaira, Y., Teramura, K., Hara, M., Kobayashi, H., et al. (2007). Zinc germanium oxynitride as a photocatalyst for overall water splitting under visible light. *J. Phys. Chem. C*, *111*, 1042–1048.

178. Maeda, K., Takata, T., Hara, M., Saito, N., Inoue, Y., Kobayashi, H., et al. (2005). GaN: ZnO solid solution as a photocatalyst for visible-light-driven overall water splitting. *J. Am. Chem. Soc.*, *127*, 8286–8287.

179. Maeda, K., Teramura, K., Lu, D. L., Takata, T., Saito, N., Inoue, Y., et al. (2006). Photocatalyst releasing hydrogen from water—Enhancing catalytic performance holds promise for hydrogen production by water splitting in sunlight. *Nature*, *440*, 295–295.

180. Groenewolt, M., Antonietti, M. (2005). Synthesis of g-C_3N_4 nanoparticles in mesoporous silica host matrices. *Advanced Materials*, *17*, 1789–1792.

181. Goettmann, F., Fischer, A., Antonietti, M., Thomas, A. (2006). Metal-free catalysis of sustainable Friedel-Crafts reactions: direct activation of benzene by carbon nitrides to avoid the use of metal chlorides and halogenated compounds. *Chem. Commun.*, 4530–4532.

182. Goettmann, F., Thomas, A., Antonietti, M. (2007). Metal-free activation CO_2 by mesoporous graphitic carbon nitride. *Angew. Chem. Int. Ed.*, *46*, 2717–2720.

183. Clark, S. J., Segall, M. D., Pickard, C. J., Hasnip, P. J., Probert, M. J., Refson, K., et al. (2005). First principles methods using CASTEP. *Z. Kristallogr.*, *220*, 567–570.

184. Tissandier, M. D., Cowen, K. A., Feng, W. Y., Gundlach, E., Cohen, M. H., Earhart, A. D., et al. (1998). The proton's absolute aqueous enthalpy and Gibbs free energy of solvation from cluster-ion solvation data. *J. Phys. Chem. A*, *102*, 7787–7794.

185. Weast, R. C., Astle, M. J., Beyer, W. H. (1983). *Handbook of Chemistry and Physics* (64th ed.), 1983–1984 CRC Press Inc., Boca Raton, Florida.

186. Wang, X. C., Maeda, K., Chen, X. F., Takanabe, K., Domen, K., Hou, Y. D., et al. (2009). Polymer semiconductors for artificial photosynthesis: hydrogen evolution by mesoporous graphitic carbon nitride with visible light. *J. Am. Chem. Soc.*, *131*, 1680–1681.

CHAPTER 8

TWO-DIMENSIONAL SEMICONDUCTIVE π-ELECTRONIC FRAMEWORKS

Donglin Jiang, Xuesong Ding, and Jia Guo

Department of Materials Molecular Science, National Institutes of Natural Sciences, Okazaki, Japan

8.1 INTRODUCTION

Conjugated polymers are indispensable in molecular electronics and optoelectronics. Their capabilities in emission, photoinduced energy and electron transfer, and charge carrier transportation play vital roles in determining device performances. Various conjugated polymers have been developed after the discovery of electric conductivity of polyacetylene films [1–3]. The conjugated polymers thus far synthesized are mainly based on rigid conjugated backbone structures, that is, one-dimensional (1D) conjugated polymers. Their functions, such as exciton migration, emission, photoinduced energy and electron transfer, and redox activity, and semiconducting properties have been intensively investigated both at the single-molecular level and in the bulky state. The conjugated 1D polymers, on chemical modifications of side chains or introduction of functional segments upon block copolymerization, enable the controlled assembly of conjugated chain alignment to achieve various supramolecular structures. On the other hand, three-dimensional (3D) conjugated polymers such as hyperbranched and dendritic conjugated polymers have been shown to have advantages in constructing novel polymeric architectures.

Very recently, a new class of 3D conjugated polymers, that is, conjugated micro- and mesoporous polymers (CMPs) have been developed [4–15]. One of the significant structural characteristics of CMPs is their 3D cross-linked porous framework. Owing to their inherent porous nature with high surface areas, CMPs have emerged as a new media for gas adsorption. The pores are even accessible to ionic species, allowing the synthesis of metal nanoclusters spatially confined in the pores of CMPs. These hybrid systems serve as heterogeneous catalysts

Supramolecular Soft Matter: Applications in Materials and Organic Electronics, First Edition.
Edited by Takashi Nakanishi.
© 2011 John Wiley & Sons, Inc. Published 2011 by John Wiley & Sons, Inc.

for organic transformations. In sharp contrast, by integration of catalytic sites as building blocks to the skeleton of CMPs, a built-in heterogeneous catalyst has been developed [14]. This built-in catalyst is unique in that it bears a highly dense catalytic skeleton, has large surface areas, possesses well-controlled micro- and mesopores, and is chemically robust. Therefore, unlike other solid catalysts such as supported catalysts or metal-organic frameworks (MOFs), CMP-based solid catalysts are reusable, are highly active with excellent selectivity, and allow large-scale synthesis with a large Turnover number (TON) value. On the other hand, the entrapment of energy-accepting chromophores into the pores of energy-donating CMPs allows the construction, via noncovalent interaction, of the light-harvesting antennae, which is cooperative in triggering a directional, speedy, and efficient energy flow [15]. As demonstrated by these examples, CMPs allow the designing of the skeleton as well as the fine control of pore parameters, making them a new platform for novel polymers and advanced materials [4].

In contrast to the above-mentioned 1D and 3D conjugated polymers, two-dimensional (2D) polymers are very limited in both kind and number. Supramolecular approaches to 2D polymeric architectures via elaborate side chain modifications of 1D conjugated polymers have led to the fabrication of 2D anisotropic monolayers by Self-assembly monolayer (SAM) [16], Langmuir-Blodgett (LB) [17], and layer-by-layer (LBL) [18] techniques. However, owing to the noncovalent characteristics, they are easily deassembled into nonordered structures. Clearly, a covalent 2D polymer would assure a more robust polymer network with permanent order in its building blocks. The synthesis of such covalent 2D conjugated polymers has been seriously pursued, since they are expected to open entirely new vistas for future science and technology [19–21]. Recent attempts at synthesis have succeeded in the surface polymerization of special metal substrates to form single-layer 2D polymers under ultrahigh vacuum at high temperatures [22–29].

A covalent 2D polymer would resemble an sp^2-carbon-based graphene sheet, but its structure would have a different molecular skeleton formed by orderly linkage of building blocks to constitute a flat organic sheet. Nanocarbons are intriguing molecules and motifs with a discrete size and 2D conformation [28] but beyond the scope of this chapter. Using topological approaches and polycondensation schemes, 2D covalent organic frameworks (2D COFs) have been synthesized in solution phase [30]. 2D COFs are not single atomically thick sheets but consist of 2D sheets layered as a $\pi-\pi$ stack. Because the pore size and shape are determined by the structure of the monomers, 2D COFs represent a class of porous materials with tailor-made components and compositions. These 2D COFs serve as porous materials for gas adsorption and storage. On the other hand, on incorporation of π-electronic building blocks, 2D COF allows the design and construction of π-electronic polymers with luminescent, semiconducting, and photoconductive properties.

In this chapter, we review the advancement in 2D polymers, including 2D polymers on metal substrates and layered 2D polymers (2D COFs), with an emphasis on their design, synthesis, and unique functions.

8.2 TWO-DIMENSIONAL POLYMERS ON METAL SURFACES

Metal surfaces have been widely utilized for the assembly of organic molecules and the polymerization of 1D polymers. The specific interaction between organic molecules and the metal substrate allows the deposition of only one molecular layer of topologically designed monomers and thus the preparation of 2D porous polymers. For this purpose, condensation reactions have been developed for the preparation of such 2D structures, while multiple functionalized monomers are required to fulfill a 2D extension of the polymer latex.

COF-1

The self-condensation (trimerization) of benzene diboronic acid (BDBA) takes place on an Au(111) surface under ultrahigh vacuum, and the resulting 2D honeycomb network **COF-1** can be visualized by STM (scanning tunneling microscope) [22]. However, under ultrahigh-vacuum conditions at elevated temperature, such as 370–460 K, the by-product, that is, water molecules, is immediately removed, so that the condensation reaction is irreversible and loses the self-healing ability; this eventually leads to the structural defects. The coverage of the surface by the condensed polymer is low, at about <1%, when the temperature of metal surface is 300–500 K, but can be increased to near entire monolayer by raising the reaction temperature to 700 K. STM reveals that the networks of polygons contain not only a hexagonal structure but also pentagonal, heptagonal, and even octagonal rings. These defects are formed because of the deformation of hexagonal rings and incomplete ring closure reactions. Annealing of the network at high temperatures is not effective to remove these defects because the networks are covalently connected. In comparison with the self-condensed BDBA network, cocondensation of BDBA with hexahydroxytriphenylene

(HTTP) gives network **COF-5** on Au(111) surface with a relatively small number of defects. This suggests that a bimolecular reaction is much superior to trimolecular reaction in the preparation of a 2D polymer network.

A 2D conjugated network can be prepared using the chemical reaction of tetrakis(bromophenyl)porphyrin **1** evaporated on Au(111) surface [23]. On heating to 330°C, the dissociation of C−Br bonds generates phenyl radicals, while the focal porphyrin units remain stable. Under ultrahigh-vacuum conditions, these radical species on the Au(111) substrate couple with each other to form a square-lattice 2D polyporphyrin network **2**. On the other hand, when *trans*-dibromo-substituted tetrakis-phenylporphyrin **3** is used as the monomer, the heat-promoted reaction affords a 1D ladder-type polymer structure **4**. The domain of the 2D porphyrin network is of nanometer scale and has only a low coverage of the Au(111) surface.

3   **4**

In relation to the above approach, deposition of 1,3,5-tris(4-boromophenyl)benzene **5** on Cu(111) and Ag(110) substrates followed by homolysis at room temperature leads to the generation of 2D polyphenylene [26]. The low activation temperature is related to the interaction of the monomer with the metal substrate. The radical coupling reaction of the activated monomer on the substrates on heating afford a 2D network of polyphenylene **6** with pentagonal, hexagonal, and heptagonal pores. Annealing experiments at different temperatures confirm the formation of 2D covalent networks. This approach has been further extended for the synthesis of porous graphene. A specifically designed monomer, that is, iodo-substituted cyclohexa-*m*-phenylene (CHP) **7**, when deposited on the Ag(111) surface, forms hexagonal porous polyphenylene network **8** [27]. The C–I bond of CHP while adsorbed on the Ag(111) surface can be homolyzed at room temperature to the corresponding radical species, which on annealing above 570 K, react with each other to afford the polyphenylene honeycomb networks with a single-atom-wide pore and subnanometer periodicity.

The synthesis of 2D polymers on metal surfaces requires the deposition of monomers under ultrahigh vacuum and at high temperatures, while the self-assembly behavior of the monomers on the metal substrates determines the network. Moreover, in these cases, not the polymer itself, but the metal–polymer interface dominates the properties. Owing to the difficulty in the control of the molecular alignment on the metal surface, the continuity of the single-layer polymer domains is difficult to control. Therefore, the synthesis of a large-area monolayer on a substrate with a perfect structure is of further interest and remains a challenge. The controlled self-assembly of organic molecules on metal surface may provide a way to a well-defined 2D architecture with designated size and shape.

5   **6**   **7**   **8**

8.3 TWO-DIMENSIONAL POLYMERS WITH COVALENT ORGANIC FRAMEWORKS

In addition to the methodology of 2D polymer networks on metal substrates, a solution process to the synthesis of 2D polymer has been developed using a solvothermal condensation protocol. Owing to the crystalline nature of the polymer, the 2D polymers stack via π–π interaction to form a well-defined layer structure, where the channels can be utilized for further application, for example, gas adsorption. Different from the single layer on metal surface, 2D polymers thus prepared allow the investigation of their bulk properties.

The first examples of 2D polymers with COFs (**COF-1** and **COF-5**) are prepared by the reversible self-condensation reactions of 1,4-benzene diboronic acid (BDBA) and its cocondensation with 2,3,6,7,10,11-hexahydroxyl triphenylene (HHTP) in solvothermal conditions [30]. One of the significant structural features of COFs is that COFs consist of light elements (H, B, C, and O) and are linked with strong covalent bonds; therefore, they are light but strong porous materials. By using different building blocks, COFs can be predesigned to form 2D or 3D structures [31–40]. On the other hand, by integrating π-electronic components into the frameworks, COFs have been developed for designing new-type optoelectronic polymers [41,42], which are in sharp contrast to traditional 1D and 3D polymers in the terms of shape, morphology, and properties. We review the concept of molecular design and the synthesis of 2D COFs, emphasizing on the control of porous structure for gas storage applications and the incorporation of functional components for novel optoelectronic applications.

8.3.1 Topological Design of 2D Porous Materials

Porous materials represent an important subject from a scientific point of view to study phenomena related to surfaces and confined spaces and have significant practical technological applications [43–45]. The development of methodologies to precisely control pore size and distribution to achieve large surface areas is a central theme in the exploration of porous materials. Inorganic porous frameworks, also called zeolites, have attracted significant recent attention for their controllable porosity and systematically tailored chemical environment within the pores ranging from micro- (<2 nm) to mesoscales (>2 nm). Microporous zeolites, which are crystalline silicates or aluminosilicates, are typically synthesized under hydrothermal conditions from reactive gels in alkaline or acidic media at elevated temperatures [46–49]. On the other hand, mesoporous silicates are typical inorganic materials obtained using amphiphilic organic surfactants or block polymers as templates to form inorganic/organic mesostructured composites. On removal of the templates, the frameworks remain to give crystalline porous domains. These porous materials have shown high capability to absorb gases and small molecules in the pore and to serve as carriers to load catalysts and various functional small molecules and polymers. By integrating organic modules or chiral centers into the templates, zeolites with organic components or helical pore structures have been developed [50–53].

In contrast to zeolites, porous materials with organic components in the frameworks can be synthesized using coordination chemistry of metal ions [54–72]. Thus, the formation of MOFs is driven by noncovalent coordination bonds [73–75]. The diversity of metal ions and the availability of organic ligands allow for the construction of various crystalline frameworks with different structures and topologies. The pore size can be tuned systematically from micro- to mesometer scales by using organic ligands of different lengths, while the topology of the framework is retained [76]. MOFs with large surface areas are very useful for gas storage and separation [77,78]. MOFs have also been explored for their ability to act as catalysts for organic transformation [79] and polymerization [80]; to show chirality [81], conductivity [82], luminescence [83], and magnetism [84], and to trigger spin transition [85] and nonlinear optical phenomena [86].

It is well established that supramolecular self-assembly of template molecules directs the formation of the structure of zeolite frameworks. Thus, pore parameters are highly dependent on the noncovalent interactions and the size of preorganized aggregates. This fact makes it difficult to precisely predesign and/or control the pore parameters at the single-molecule level. For example, the same template would yield different zeolites when different self-assembly conditions are employed. Therefore, the pore topology is hardly correlated with the geometry or shape of the template molecules. In sharp contrast, due to the discrete angles and distances required for orbital overlap between metal ions and organic ligands, MOFs are unique in that they allow for topological design of porous structures. The discovery of MOFs has thus open up new possibilities for designable molecular porous materials. However, one drawback of MOFs is that their structures are based on coordination bonds, which are much weaker than covalent bonds and make the structure fragile.

From the molecular design point of view, one of the significant characters of 2D COFs is the topological construction of the framework. This is possible because the geometries of monomers direct the growth pattern of the 2D skeleton and the sizes of monomers determine the porosity and pore size. Thus, 2D COFs enable an elaborate design of "programmable" porous macromolecules with desired components and pore parameters. By using planar monomers, 2D COFs with a stacked sheet structure will yield one-dimensional open channels. For example, self-condensation of the C3 monomer or cocondensation with the C2 monomer leads to 2D COFs with a hexagonal pore structure. On the other hand, cocondensation of the C4 and the C2 monomers form 2D COFs with tetragonal pores.

Thus far, several synthetic approaches have been established for constructing 2D COFs. Namely, self-condensation and cocondensation of boronic acid derivatives [30–34], trimerization of cyano groups [35], and condensation of amino and aldehyde groups [37] are representative methods for the synthesis of COFs with boronic ester, triazine, and imine linkages, respectively.

8.3.2 Control of Pore Size and Parameters

2D COFs have high flexibility for designing and tuning pore parameters, which opens an avenue to predesignable porous functional materials. This character is of sharp contrast to other porous materials, including inorganic zeolites and MOFs.

The boronic acid condensation reaction is reversible, allowing for structure self-healing and leading to COFs with few structural defects. Boronic acid derivatives undergo self-condensation to form a six-membered boroxine unit. Self-condensation of BDBA in a mixture of dioxane/mesitylene results in **COF-1** with a staggered structure. **COF-1** possesses a 7-Å-diameter pore with a Brunauer–Emmett–Teller (BET) surface area of 711 m^2/g [30]. On the other hand, a large monomer, for example, pyrene-2,7-diboronic acid (PDBA) under otherwise identical conditions yields **PPy-COF** with a large pore size of 1.88 nm. The large pyrene building blocks prefer a superimposed π–π interaction. As a result, **PPy-COF** has an eclipsed stack structure [42]. By using a large building block such as triphenylene derivative as a comonomer, the pore size can be further enlarged. For example, the cocondensation of BDBA with HHTP leads to **COF-5** with a remarkable BET surface area of 1590 m^2/g and uniform mesopores of 2.7 nm in diameter [30]. The cocondensation of HHTP and PDBA yields **TP-COF** bearing much larger pores of 3.14 nm width [41].

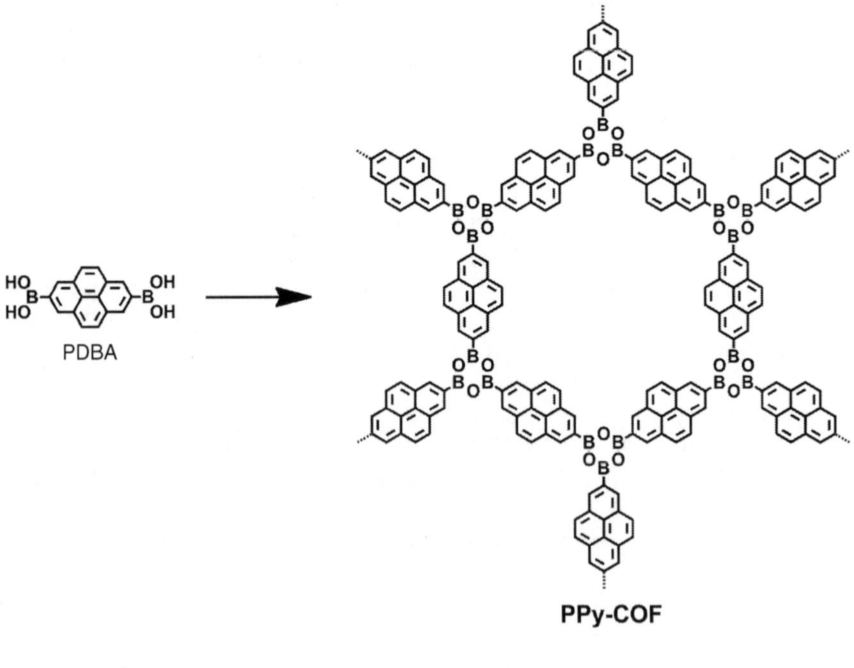

PDBA

PPy-COF

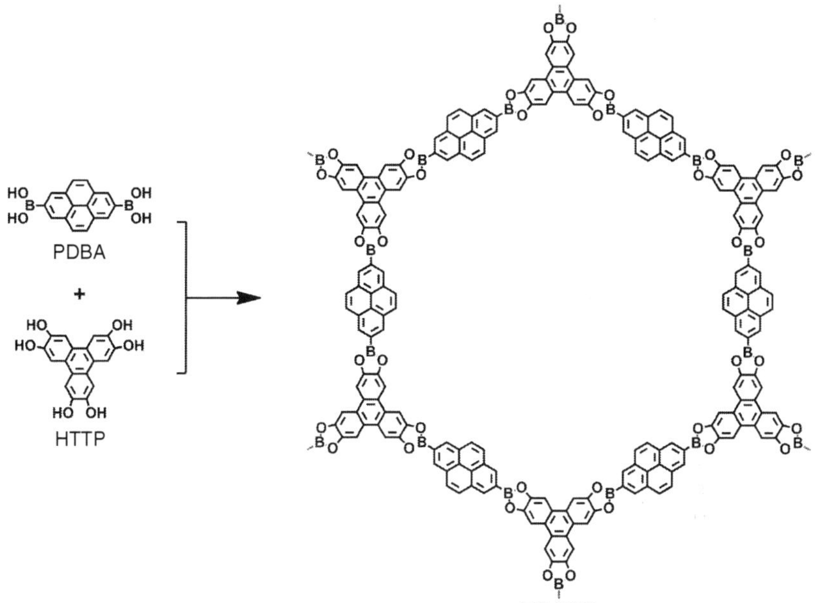

PDBA

+

HTTP

TP-COF

In contrast to the above solvothermal conditions, a reflux condition for the high throughput synthesis of **COF-18 Å** from monomers **9** and **10** has been developed with a yield as high as 96% [32]. On the other hand, microwave reaction leads to the efficient synthesis of **COF-5** with a BET surface area of 2019 m^2/g, which is much larger than that synthesized by the solvothermal method [34]. Another class of 2D porous COFs with regular porosity and large surface area is based on triazine materials, which can be synthesized by the dynamic trimerization reaction of dicyano compounds under ionothermal conditions [35]. Trimerization of 1,4-cyano benzene **11** in molten zinc chloride at 400°C yields a covalent triazine-based framework **CTF** with robust chemical and thermal stability.

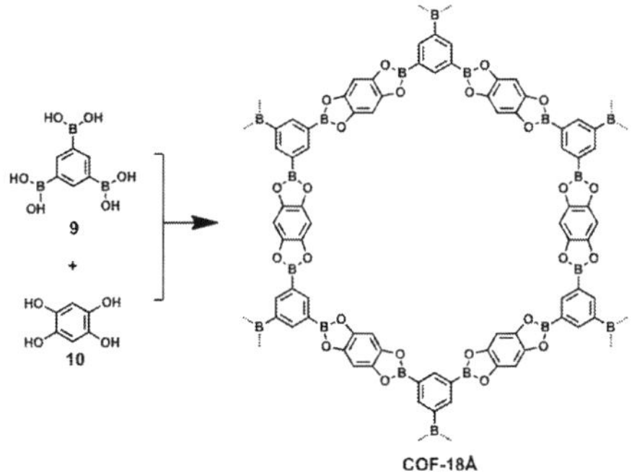

9

+

10

COF-18Å

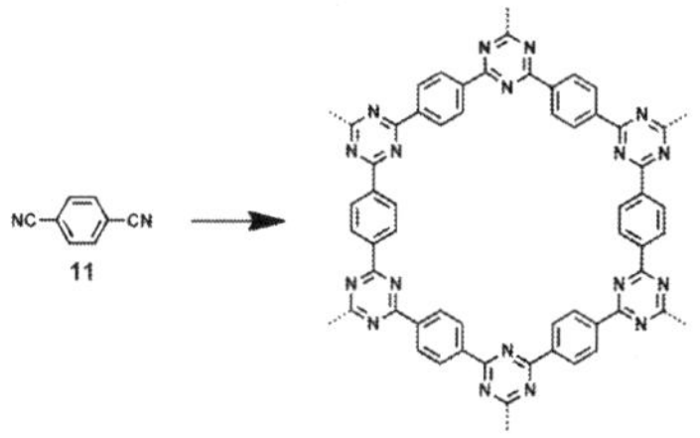

CTF

A unique method to tune the pore size is to incorporate sterically bulky groups in the pores based on the same 2D COF framework using alkyl chain-appended monomer **12** and comonomer **9** [33]. The alkyl-functionalized porous 2D COFs (**COF-18 Å, COF-16 Å, COF-14 Å, COF-11 Å**) have pore diameters tunable from 1 to 2 nm, depending on the length of the alkyl chain attached to the skeleton. The longer the alkyl chain, the smaller the pore size. The alkyl chain also affects the gas sorption capacity of the resulting COFs. The introduction of alkyl chains to the pore interior causes a decrease in the nitrogen uptake but an increase in hydrogen adsorption. The pendent alkyl chains not only fill the pore, thereby reducing the size and the surface area of the pores, but also create "pore corner cavities," which block the access of nitrogen molecules while accommodating the smaller hydrogen species. Furthermore, the alkyl chains also provide a surface for hydrogen sorption, which plays a role in enhancing hydrogen storage capacity of these materials.

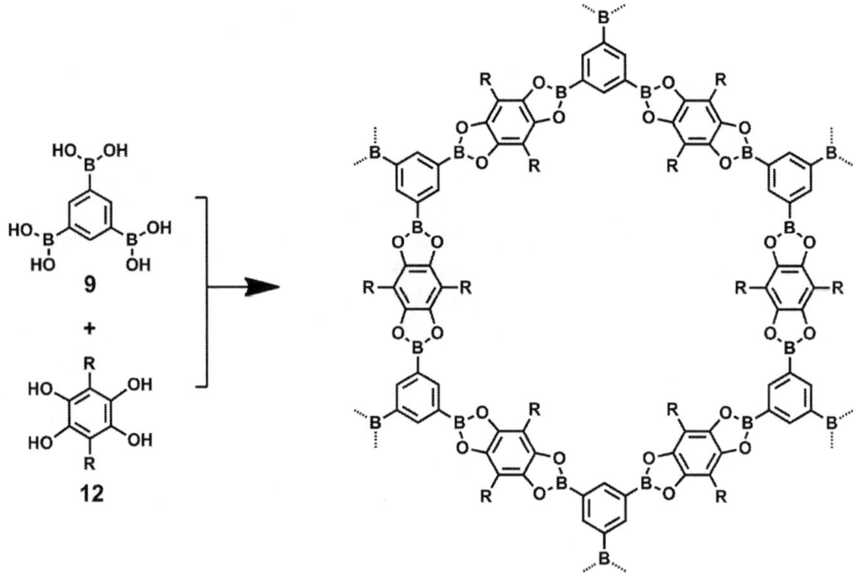

R = H: **COF-18Å**
R = CH$_3$: **COF-16Å**
R = CH$_2$CH$_3$: **COF-14Å**
R = CH$_2$CH$_2$CH$_3$: **COF-11Å**

8.3.3   Two-Dimensional COFs for Hydrogen Storage

Hydrogen has exceptional prospects as a future energy resource because of its high chemical abundance, high energy density, and environmentally friendly "green" characteristics. However, the wide use of hydrogen as an energy source has been limited. One major bottleneck is its safe and efficient storage. The target values for hydrogen storage set by the US Department of Energy (DOE) for 2010 (storage of 6 wt% and 45 kg H_2 per m^3) and 2015 (storage of 9 wt% and 81 kg H_2 per m^3) [87,88] appear to be clearly out of reach. Besides compression and cryogenic techniques for the storage of hydrogen, adsorption methods, either chemical or physical, have been proposed, where the materials are key issues to be resolved. However, thus far, physisorption has hardly reached the above targets, although some materials with high surface areas are indeed very promising for increased hydrogen storage capacities. Thus, if organic materials are envisioned as potential carriers for hydrogen storage, maximizing their advantages through precise pore design and/or integration of open metal sites or heteroatom into such materials is necessary.

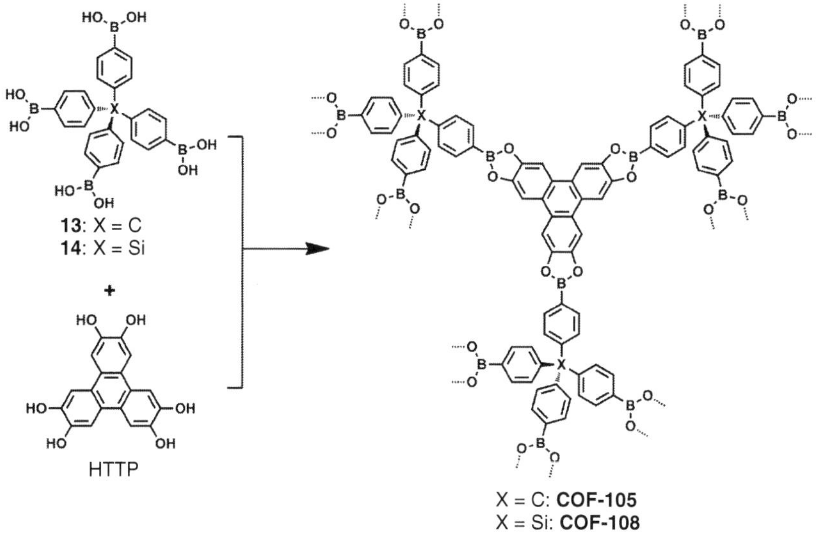

13: X = C
14: X = Si

HTTP

X = C: **COF-105**
X = Si: **COF-108**

Recent molecular dynamics simulations of hydrogen uptake by COFs have revealed that their hydrogen storage capacities are comparable with those of MOFs. Intriguingly, the calculations predict that lithium-doped **COF-105** and **COF-108**, which have been synthesized from the condensation of HTTP with **13** or **14**, respectively, have capacities for hydrogen adsorption as high as 6.84 and 6.73 wt% at 298 K and 100 bar [38]. In this sense, COFs are the most promising candidates for hydrogen storage. Systematical studies on the uptake behavior and capacities of several COFs for hydrogen storage show that COFs are one of the best hydrogen adsorbents. For example, the saturation uptake for **COF-102** reaches 72.4 mg/g, which is the highest among COFs and is comparable to the values for MOF-177 (75 mg/g) and MOF-5 (76 mg/g) [39]. These results show

COF-102

that COFs with well-defined pore structures are attractive media for hydrogen storage.

8.3.4 Two-Dimensional COFs for Other Gas Uptake

2D COFs containing boronate ester building blocks are studied for the adsorption of ammonia gas, because the Lewis acidic boron atoms may have a chance to interact with Lewis base such as ammonia. **COF-10** condensed from HTTP and **15** shows the highest uptake capacity of 15 mol/kg at 298 K and 1 bar [40], which is much higher than that of the state-of-the-art porous materials. Of interest, the ammonia adsorbed can be removed from the pores by heating at 200°C under vacuum. Repeated adsorption–desorption cycles cause the shift in the interlayer packing, which results in the decrease of surface area to N_2. However, the total uptake capacity of ammonia does not change much as a result of strong Lewis acid–base interaction and this allows the cycling of COFs for several times. These properties are interesting from the application point of view, for example, in the transportation of ammonia by storing ammonia in COFs. For this purpose, **COF-10** can be pressed to form a binder-free tablet at 2000 psi. The nitrogen adsorption surface area is slightly smaller than that of pristine **COF-10** crystalline powders as a result of mechanical disruption of wall structure. Nevertheless, the adsorption–desorption capacity of ammonia shows only a small decrease in comparison to that of **COF-10** crystalline powder. These results demonstrate that 2D COFs with a functionalized structure are promising carriers for the storage and transportation of specific gas, upon the interaction of the gas with the open walls of 2D COFs.

COF-10

8.3.5   Semiconducting Blue Luminescence 2D Polymers

Most studies on COFs to date have focused on the development of synthetic methodologies with the aim of controlling the pore size and enlarging the surface area. Recently, novel π-electronic COFs have been synthesized by using pyrene and triphenylene derivatives as building blocks. These 2D COFs display unique light-harvesting functions and photoconductive and semiconducting properties, which are highly correlated with their 2D sheet structure and eclipsed stack alignment.

The cocondensation of HTTP and PDBA under solvothermal conditions yields **TP-COF** with triphenylene and pyrene alternately linked in a hexagonal skeleton [41]. **TP-COF** adopts a belt shape, with length extending to several micrometers and a uniform width and thickness of ~300 and 100 nm, respectively (Fig. 8-1). No other shapes, such as spherical, fiberlike, rodlike, or irregular, were observed, suggesting high phase purity. High resolution tunneling electron microscopy (HRTEM) revealed that the two-dimensional polymer sheets were aligned parallelly, with an interval of 3.4 Å, which is reasonable for $\pi-\pi$ stacking. Powder X-ray diffraction (PXRD) measurement together with PXRD pattern simulation indicated that **TP-COF** has an eclipsed stack structure, wherein the triphenylene and pyrene units are perfectly superimposed on corresponding units of the neighboring sheet.

Nitrogen sorption isotherm measurement revealed a BET surface area of 868 m^2/g with a pore diameter of 3.14 nm. **TP-COF** shows a highly blue luminescence with the emission band centered at 474 nm on excitation of the pyrene units. It is of interest that when the triphenylene unit is selectively excited, **TP-COF** emits again at 474 nm, while the fluorescence from the triphenylene unit itself is hardly observable. This result suggests that energy transfer from triphenylene to pyrene takes place within the 2D **TP-COF** framework. Quantitative evaluation revealed that the energy transfer quantum yield is $\sim$60%. Therefore, **TP-COF** harvests a wide range of photons from the ultraviolet to the visible regions and converts them to blue luminescence. Owing to the eclipsed stack structure of π-electronic components in the framework, **TP-COF** is semiconducting, is capable of hole transportation, and allows on–off switching of electric current.

8.3.6 Photoconductive 2D Polymers

Self-condensation of PDBA under otherwise identical conditions to the above yields **PPy-COF** [42], which consists of an eclipsed alignment of two-dimensional sheet with single π-electronic component pyrene units on the edges of the hexagonal framework. **PPy-COF** assumes a cubic shape with length, width, and thickness in the micrometer range. Fluorescence microscopy displays that all the cubes are highly blue in luminescence. Investigations using fluorescence spectroscopy show that on excitation of the pyrene units at 414 nm, **PPy-COF** emits blue luminescence centered at 484 nm as a result of excimer formation of pyrene units. Fluorescence anisotropy measurements revealed that the fluorescence of **PPy-COF** is highly depolarized to give an anisotropy value of only 0.0001, which is much smaller than that of **TP-COF**

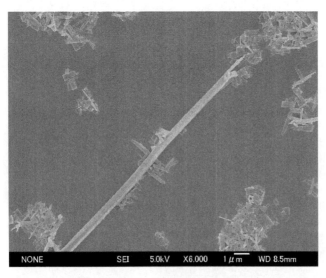

Figure 8-1 Field-emission scanning electron microscopy of TP-COF.

(0.017). This result suggests that excitons are not localized but can migrate over the **PPy-COF** framework. Compared to **TP-COF**, the single-component pyrene-based **PPy-COF** can facilitate exciton migration. Unlike in **TP-COF**, which contains two alternately linked components with different energy gaps, excitons in the single-component **PPy-COF** can flow not only over the sheet plane but also across the stacked layers as well.

PXRD measurements reveal strong diffraction peaks and allow the assignment of the molecular alignment of the 2D polymers. HRTEM displays that the 2D polymers extend flat and straight. These sheets stack face-to-face via π-stacking to form a layered structure. Molecular dynamics simulation studies confirm that the 2D sheets are most likely stacked in an eclipsed mode to give 2D COF with open 1D channels. Indeed, nitrogen sorption isotherm measurements display typical microporous profiles, while the pore size is about 1.9 nm, very close to the theoretical simulated value (1.88 nm).

Owing to the ultimate stacking of the large π-conjugated components, **PPy-COF** is a semiconductor, displaying an almost linear I–V profile. Doping with iodine results in an increment in the electric current, which suggests that **PPy-COF** is a p-type semiconductor. It is to be noted that **PPy-COF** facilitates charge carrier transportation and is photoconductive. On irradiation with visible light, **PPy-COF** generates a prominent photocurrent with a quick response to light irradiation. The photocurrent switch can be repeated many times with an on–off ratio as large as 8.0×10^4. In contrast, **TP-COF**, which consists of a network of cocondensed triphenylene and pyrene units, shows a low photocurrent with a one-fourth on–off ratio of that of **PPy-COF**. The single component together with the eclipsed alignment plays an important role in facilitating exciton migration and carrier transportation in the π-electronic COFs. These features are highly correlated with the structural characteristics of 2D COF and clearly demonstrate that 2D COFs provide a new platform for the design and construction of novel π-electronic polymers.

8.4   CONCLUSIONS

In summary, unlike the conventional 1D and 3D polymers, 2D polymers require different design principles and synthetic methodology. Particularly, from the viewpoint of materials science, 2D polymers with COFs represent a new class of porous crystalline materials. These porous 2D macromolecules are significantly different from traditional 1D and 3D polymers in the aspects of synthesis methodologies used, shape and morphology, and functionalities of the resulting polymers. 2D COFs are unique in that they allow for elaborate control of both the framework and pore parameters. In particular, the combined features of strong covalent bonds, light elements, and well-defined pore structures offer a new way to synthesize designable porous materials for gas sorption. In fact, several 2D COFs have already demonstrated their extremely large capacities for hydrogen storage and uptake of other gases. Incorporation of π-electronic components in the frameworks allows for the construction of π-electronic and photofunctional

COFs. The development of appropriate synthetic approaches and new monomers remains a challenge for expanding the scope and diversity of COFs. One of the important issues to be challenged is the synthesis and function of freestanding single sheets of these 2D polymers. Another interesting aspect is the precise control of the size and shape of the 2D polymers. The synthetic challenges together with the unique functions of these 2D polymers are crucial and will hold the key to the advancement of the chemistry and physics of 2D polymers.

REFERENCES

1. Shirakawa, H. (2001). The discovery of polyacetylene film: the dawning of an era of conducting polymers. *Angew. Chem. Int. Ed.*, *40*, 2575–2580.
2. Heeger, A. J. (2001). Semiconducting and metallic polymers: the fourth generation of polymeric materials. *Angew. Chem. Int. Ed.*, *40*, 2591–2611.
3. MacDiarmid, A. G. (2001). "Synthetic metals": a novel role for organic polymers. *Angew. Chem. Int. Ed.*, *40*, 2581–2590.
4. Jiang, J., Su, F., Trewin, A., Wood, C. D., Niu, H., Jones, J. T. A., Khimyak, Y. Z., Cooper, A. I. (2008). Synthetic control of the pore dimension and surface area in conjugated microporous polymer and copolymer networks. *J. Am. Chem. Soc.*, *130*, 7710–7720.
5. Cooper, A. I. (2009). Conjugated microporous polymers. *Adv. Mater.*, *21*, 1291–1295.
6. Wood, C. D., Tan, B., Trewin, A., Niu, H., Bradshaw, D., Rosseinsky, M. J., Khimyak, Y. Z., Campbell, N. L., Kirk, R., Stockel, E., Cooper, A. I. (2007). Hydrogen storage in microporous hypercrosslinked organic polymer networks. *Chem. Mater.*, *19*, 2034–2048.
7. Dawson, R., Su, F. B., Niu, H. J., Wood, C. D., Jones, J. T. A., Khimyak, Y. Z., Cooper, A. I. (2008). Mesoporous poly(phenylenevinylene) networks. *Macromolecules*, *41*, 1591–1593.
8. Germain, J., Svec, F., Fréchet, J. M. J. (2008). Preparation of size-selective nanoporous polymer networks of aromatic rings: potential adsorbents for hydrogen storage. *Chem. Mater.*, *20*, 7069–7076.
9. Jiang, J. X., Su, F., Trewin, A., Wood, C. D., Campbell, N. L., Niu, H., Dickinson, C., Ganin, A. Y., Rosseinsky, M. J., Khimyak, Y. Z., Cooper, A. I. (2007). Conjugated microporous poly(aryleneethynylene) networks. *Angew. Chem. Int. Ed.*, *46*, 8574–8578.
10. Jiang, J. X., Su, F., Trewin, A., Wood, C. D., Niu, H., Jones, J. T. A., Khimyak, Y. Z., Cooper, A. I. (2008). Synthetic control of the pore dimension and surface area in conjugated microporous polymer and copolymer networks. *J. Am. Chem. Soc.*, *130*, 7710–7720.
11. Jiang, J. X., Su, F., Niu, H., Wood, C. D., Campbell, N. L., Khimyak, Y. Z., Cooper, A. I. (2008). Conjugated microporous poly(phenylene butadiynylene)s. *Chem. Commun.*, 486–488.
12. Kuhn, P., Forget, A., Su, D., Thomas, A., Antonietti, M. (2008). From microporous regular frameworks to mesoporous materials with ultrahigh surface area: dynamic reorganization of porous polymer networks. *J. Am. Chem. Soc.*, *130*, 13333–13337.
13. Schwab, M. G., Fassbender, B., Spiess, H. W., Thomas, A., Feng, X., Müllen, K. (2009). Catalyst-free preparation of melamine-based microporous polymer networks through Schiff base chemistry. *J. Am. Chem. Soc.*, *131*, 7216–7217.
14. Chen, L., Yang, Y., Jiang, D. (2010). CMPs as scaffolds for constructing porous catalytic frameworks: a built-in heterogeneous catalyst with high activity and selectivity based on nanoporous metalloporphyrin polymers. *J. Am. Chem. Soc.*, *132*, 9138–9143.
15. Chen, L., Honsho, Y., Seki, S., Jiang, D. (2010). Light-harvesting conjugated microporous polymers: rapid and highly efficient flow of light energy with a porous polyphenylene framework as antenna. *J. Am. Chem. Soc.*, *132*, 6742–6748.
16. Kim, J., Swager, T. M. (2001). Control of conformational and interpolymer effects in conjugated polymers. *Nature*, *411*, 1030–1034.

17. Reitzel, N., Greve, D. R., Kjaer, K., Howes, P. B., Jayaraman, M., Savoy, S., McCullough, R. D., McDevitt, J. T., Bjørnholm, T. (2000). Self-assembly of conjugated polymers at the air/water interface. Structure and properties of Langmuir and Langmuir-Blodgett films of amphiphilic regioregular polythiophenes. *J. Am. Chem. Soc.*, *122*, 5788–5800.

18. Zotti, G., Vercelli, B., Berlin, A. (2008). Monolayers and multilayers of conjugated polymers as nanosized electronic components. *Acc. Chem. Res.*, *41*, 1098–1109.

19. Bartels, L. (2010). Tailoring molecular layers at metal surfaces. *Nat. Chem.*, *2*, 87–95.

20. Perepichka, D. F., Rosei, F. (2009). Chemistry: extending polymer conjugation into the second dimension. *Science*, *323*, 216–217.

21. Gourdon, A. (2008). On-surface covalent coupling in ultrahigh vacuum. *Angew. Chem. Int. Ed.*, *47*, 6950–6953.

22. Zwaneveld, N. A., Pawlak, R., Abel, M., Catalin, D., Gigmes, D., Bertin, D., Porte, L. (2008). Organized formation of 2D extended covalent organic frameworks at surfaces. *J. Am. Chem. Soc.*, *130*, 6678–6679.

23. Grill, L., Dyer, M., Lafferentz, L., Persson, M., Peters, M. V., Hecht, S. (2007). Nano-architectures by covalent assembly of molecular building blocks. *Nat. Nanotechnol.*, *2*, 687–691.

24. Weigelt, S., Busse, C., Bombis, C., Knudsen, M. M., Gothelf, K. V., Lægsgaard, E., Besenbacher, F., Trolle, R. Linderoth, T. R. (2008). Surface synthesis of 2D branched polymer nanostructures. *Angew. Chem. Int. Ed.*, *47*, 4406–4410.

25. Matena, M., Riehm, T., Stöhr, M., Jung, T. A., Gade, L. H. (2008). Transforming surface coordination polymers into covalent surface polymers: linked polycondensed aromatics through oligomerization of n-heterocyclic carbene intermediates. *Angew. Chem. Int. Ed.*, *47*, 2414–2417.

26. Gutzler, R., Walch, H., Eder, G., Kloft, S., Heckl, W. M., Lackinger, M. (2009). Surface mediated synthesis of 2D covalent organic frameworks: 1,3,5-tris(4-bromophenyl)benzene on graphite(001), Cu(111), and Ag(110). *Chem. Commun.*, 4456–4458.

27. Bieri, M., Treier, M., Cai, J., Aït-Mansour, K., Ruffieux, P., Gröning, O., Gröning, P., Kastler, M., Rieger, R., Feng, X., Müllen, K., Fasel, R. (2009). Porous graphenes: two-dimensional polymer synthesis with atomic precision. *Chem. Commun.*, 6919–6921.

28. Sakamoto, J., van Heijs, J., Lukin, O., Schlüter, A. D. (2009). Two-dimensional polymers: just a dream of synthetic chemists? *Angew. Chem. Int. Ed.*, *48*, 1030–1069.

29. Li, Y., Zhou, Z., Shen, P., Shen, Z. (2010). Two-dimensional polyphenylene: experimentally available porous graphene as a hydrogen purification membrane. *Chem. Commun.*, *46*, 3672–3674.

30. Côté, A. P., Benin, A. I., Ockwig, N. W., O'Keeffe, M., Matzger, A. J., Yaghi, O. M. (2005). Porous, crystalline, covalent organic frameworks. *Science*, *310*, 1166–1170.

31. Hunt, J. R., Doonan, C. J., LeVangie, J. D., Côté, A. P., Yaghi, O. M. (2008). Reticular synthesis of covalent organic borosilicate frameworks. *J. Am. Chem. Soc.*, *130*, 11872 11873.

32. Tilford, R. W., Gemmill, W. R., zur Loye, H. C., Lavigne, J. J. (2006). Facile synthesis of a highly crystalline, covalently linked porous boronate network. *Chem. Mater.*, *18*, 5296–5301.

33. Tilford, R. W., Mugavero, S. J., Pellechia, P. J., Lavigne, J. J. (2008). Tailoring microporosity in covalent organic frameworks. *Adv. Mater.*, *20*, 2741–2746.

34. Campbell, N. L., Clowes, R., Ritchie, L. K., Cooper, A. I. (2009). Rapid microwave synthesis and purification of porous covalent organic frameworks. *Chem. Mater.*, *21*, 204–206.

35. Kuhn, P., Antonietti, M., Thomas, A. (2008). Porous, covalent triazine-based frameworks prepared by ionothermal synthesis. *Angew. Chem. Int. Ed.*, *47*, 3450–3453.

36. El-Kaderi, H. M., Hunt, J. R., Mendoza-Cortés, J. L., Côté, A. P., Taylor, R. E., O'Keeffe, M., Yaghi, O. M. (2007). Designed synthesis of 3D covalent organic frameworks. *Science*, *316*, 268–272.

37. Uribe–Romo, F. J., Hunt, J. R., Furukawa, H., Klock, C., O'Keeffe, M., Yaghi, O. M. (2009). A crystalline imine-linked 3-D porous covalent organic framework. *J. Am. Chem. Soc.*, *131*, 4570–4571.

38. Cao, D., Lan, J., Wang, W., Smit, B. (2009). Lithium-doped 3D covalent organic frameworks: high-capacity hydrogen storage materials. *Angew. Chem. Int. Ed.*, *48*, 4730–4733.

39. Furukawa, H., Yaghi, O. M. (2009). Storage of hydrogen, methane, and carbon dioxide in highly porous covalent organic frameworks for clean energy applications. *J. Am. Chem. Soc.*, *131*, 8875–8883.

40. Doonan, C. J., Tranchemontagne, D. J., Glover, G. T., Hunt, J. R., Yaghi, O. M. (2010). Exceptional ammonia uptake by a covalent organic framework. *Nat. Chem.*, *2*, 235–238.

41. Wan, S., Guo, J., Kim, J., Ihee, H., Jiang, D. L. (2008). A belt-shaped, blue luminescent, and semiconducting covalent organic framework. *Angew. Chem. Int. Ed.*, *47*, 8826–8830.

42. Wan, S., Guo, J., Kim, J., Ihee, H., Jiang, D. L. (2009). A photoconductive covalent organic framework: self-condensed arene cubes composed of eclipsed 2D polypyrene sheets for photocurrent generation. *Angew. Chem. Int. Ed.*, *48*, 5439–5442.

43. Davis, M. E. (2002). Ordered porous materials for emerging applications. *Nature*, *417*, 813–821.

44. Cundy, C. S., Cox, P. A. (2003). The hydrothermal synthesis of zeolites: history and development from the earliest days to the present time. *Chem. Rev.*, *103*, 663–701.

45. Wan, Y., Zhao, D. (2007). On the controllable soft-templating approach to mesoporous silicates. *Chem. Rev.*, *107*, 2821–2860.

46. Sakamoto, Y., Kaneda, M., Terasai, O., Zhao, D. Y., Kim, J. M., Stucky, G. D., Shim, H. J., Ryoo, R. (2000). Direct imaging of the pores and cages of three-dimensional mesoporous materials. *Nature*, *408*, 449–453.

47. Che, S., Garcia-Bennett, A. E., Yokoi, T., Sakamoto, K., Kunieda, H., Terasaki, O. Tatsumi, T. (2003). A novel anionic surfactant templating route for synthesizing mesoporous silica with unique structure. *Nat. Mater.*, *2*, 801–805.

48. Shen, S. D., Garcia-Bennett, A. E., Liu, Z., Lu, Q. Y., Shi, Y. F., Yan, Y., Yu, C. Z., Liu, W. C., Cai, Y., Terasaki, O., Zhao, D. Y. (2005). Three-dimensional low symmetry mesoporous silica structures templated from tetra-headgroup rigid bolaform quaternary ammonium surfactant. *J. Am. Chem. Soc.*, *127*, 6780–6787.

49. Tan, B., Dozier, A., Lehmler, H. J., Knutson, B. L., Rankin, S. E. (2004). Elongated silica nanoparticles with a mesh phase mesopore structure by fluorosurfactant templating. *Langmuir*, *20*, 6981–6984.

50. Zhao, D. Y., Feng, J. L., Huo, Q. S., Melosh, N., Fredrickson, G. H., Chmelka, B. F., Stucky, G. D. (1998). Triblock copolymer syntheses of mesoporous silica with periodic 50 to 300 angstrom pores. *Science*, *279*, 548–552.

51. Zhao, D. Y., Huo, Q. S., Feng, J. L., Chmelka, B. F., Stucky, G. D. (1998). Nonionic triblock and star diblock copolymer and oligomeric surfactant syntheses of highly ordered, hydrothermally stable, mesoporous silica structures. *J. Am. Chem. Soc.*, *120*, 6024–6036.

52. Che, S., Liu, Z., Ohsuna, T., Sakamoto, K., Terasaki, O., Tatsumi, T. (2004). Synthesis and characterization of chiral mesoporous silica. *Nature*, *429*, 281–284.

53. Sun, J. L., Bonneau, C., Cantin, A., Corma, A., Diaz-Cabanas, M. J., Moliner, M., Zhang, D. L., Li, M. R., Zou, X. D. (2009). The ITQ-37 mesoporous chiral zeolite. *Nature*, *458*, 1154–1157.

54. Long, J. R., Yaghi, O. M. (2009). The pervasive chemistry of metal-organic frameworks. *Chem. Soc. Rev.*, *38*, 1213–1214.

55. O'Keeffe, M. (2009). Design of MOFs and intellectual content in reticular chemistry: a personal view. *Chem. Soc. Rev.*, *38*, 1215–1217.

56. Spokoyny, A. M., Kim, D., Sumrein, A., Mirkin, C. A. (2009). Infinite coordination polymer nano- and microparticle structures. *Chem. Soc. Rev.*, *38*, 1218–1227.

57. Uemura, T., Yanai, N., Kitagawa, S. (2009). Polymerization reactions in porous coordination polymers. *Chem. Soc. Rev.*, *38*, 1228–1236.

58. Düren, T., Bae, Y. S., Snurr, R. Q. (2009). Using molecular simulation to characterise metal-organic frameworks for adsorption applications. *Chem. Soc. Rev.*, *38*, 1237–1247.

59. Ma, L., Abney, C., Lin, W. (2009). Enantioselective catalysis with homochiral metal-organic frameworks. *Chem. Soc. Rev.*, *38*, 1248–1256.

60. Tranchemontagne, D. J., Mendoza-Cortés, J. L., O Keeffe, M., Yaghi, O. M. (2009). Secondary building units, nets and bonding in the chemistry of metal-organic frameworks. *Chem. Soc. Rev.*, *38*, 1257–1283.

61. Czaja, A. U., Trukhan, N., Müller, U. (2009). Industrial applications of metal-organic frameworks. *Chem. Soc. Rev.*, *38*, 1284–1293.

62. Murray, L. J., Dincă, M., Long, J. R. (2009). Hydrogen storage in metal-organic frameworks. *Chem. Soc. Rev.*, *38*, 1294–1314.

63. Wang, Z., Cohen, S. M. (2009). Postsynthetic modification of metal-organic frameworks. *Chem. Soc. Rev.*, *38*, 1315–1329.
64. Allendorf, M. D., Bauer, C. A., Bhakta, R. K., Houk, R. J. T. (2009). Luminescent metal-organic frameworks. *Chem. Soc. Rev.*, *38*, 1330–1352.
65. Kurmoo, M. (2009). Magnetic metal-organic frameworks. *Chem. Soc. Rev.*, *38*, 1353–1379.
66. Férey, G., Serre, C. (2009). Large breathing effects in three-dimensional porous hybrid matter: facts, analyses, rules and consequences. *Chem. Soc. Rev.*, *38*, 1380–1399.
67. Perry, J. J. IV, Perman, J. A., Zaworotko, M. J. (2009). Design and synthesis of metal-organic frameworks using metal-organic polyhedra as supermolecular building blocks. *Chem. Soc. Rev.*, *38*, 1400–1417.
68. Zacher, D., Shekhah, O., Wöll, C., Fischer, R. A. (2009). Thin films of metal-organic frameworks. *Chem. Soc. Rev.*, *38*, 1418–1429.
69. Shimizu, G. K. H., Vaidhyanathan, R., Taylor, J. M. (2009). Phosphonate and sulfonate metal organic frameworks. *Chem. Soc. Rev.*, *38*, 1430–1449.
70. Lee, J. Y., Farha, O. K., Roberts, J., Scheidt, K. A., Nguyen, S. T., Hupp, J. T. (2009). Metal-organic framework materials as catalysts. *Chem. Soc. Rev.*, *38*, 1450–1459.
71. Han, S. S., Mendoza-Cortés, J. L., Goddard, W. A. III. (2009). Recent advances on simulation and theory of hydrogen storage in metal-organic frameworks and covalent organic frameworks. *Chem. Soc. Rev.*, *38*, 1460–1476.
72. Li, J. R., Kuppler, R. J., Zhou, H. C. (2009). Selective gas adsorption and separation in metal-organic frameworks. *Chem. Soc. Rev.*, *38*, 1477–1504.
73. Asefa, T., MacLachlan, M. J., Coombs, N., Ozin, G. A. (1999). Periodic mesoporous organosilicas with organic groups inside the channel walls. *Nature*, *402*, 867–871.
74. Eddaoudi, M., Kim, J., Rosi, N., Vodak, D., Wachter, J., O'Keeffe, M., Yaghi, O. M. (2002). Systematic design of pore size and functionality in isoreticular MOFs and their application in methane storage. *Science*, *295*, 469–472.
75. James, S. L. (2003). Metal-organic frameworks. *Chem. Soc. Rev.*, *32*, 276–288.
76. Yaghi, O. M., O'Keeffe, M., Ockwig, N. W., Chae, H. K., Eddaoudi, M., Kim, J. (2003). Reticular synthesis and the design of new materials. *Nature*, *423*, 705–714.
77. Rowsell, J. L. C., Milward, A. R., Park, K. S., Yaghi, O. M. (2004). Hydrogen sorption in functionalized metal-organic frameworks. *J. Am. Chem. Soc.*, *126*, 5666–5667.
78. Chen, B. L., Liang, C. D., Yang, J., Contreras, D. S., Clancy, Y. L., Lobkovsky, E. B., Yaghi, O. M., Dai, S. (2006). A microporous metal-organic framework for gas-chromatographic separation of alkanes. *Angew. Chem. Int. Ed.*, *45*, 1390–1393.
79. Heitbaum, M., Glorius, F., Escher, I. (2006). Asymmetric heterogeneous catalysis. *Angew. Chem. Int. Ed.*, *45*, 4732–4762.
80. Kitagawa, S., Kitaura, R., Noro, S. (2004). Functional porous coordination polymers. *Angew. Chem. Int. Ed.*, *43*, 2334–2375.
81. Kepert, C. J., Prior, T. J., Rosseinsky, M. J. (2000). A versatile family of interconvertible microporous chiral molecular frameworks: the first example of ligand control of network chirality. *J. Am. Chem. Soc.*, *122*, 5158–5168.
82. Sadakiyo, M., Yamada, T., Kitagawa, H. (2009). Rational designs for highly proton-conductive metal-organic frameworks. *J. Am. Chem. Soc.*, *131*, 9906–9907.
83. Chandler, B. D., Cramb, D. T., Shimizu, G. K. H. (2006). Microporous metal-organic frameworks formed in a stepwise manner from luminescent building blocks. *J. Am. Chem. Soc.*, *128*, 10403–10412.
84. Zeng, M. H., Wang, B., Wang, X. Y., Zhang, W. X., Chen, X. M., Gao, S. (2006). Chiral magnetic metal-organic frameworks of dimetal subunits: magnetism tuning by mixed-metal compositions of the solid solutions. *Inorg. Chem.*, *45*, 7069–7076.
85. Agusti, G., Munoz, M. C., Gaspar, A. B., Real, J. A. (2009). Spin-crossover behavior in cyanide-bridged iron(II)-copper(I) bimetallic 1-3D metal-organic frameworks. *Inorg. Chem.*, *48*, 3371–3381.
86. Zhang, L. J., Yu, J. H., Xu, J. Q., Lu, J., Bie, H. Y., Zhang, X. (2005). Synthesis, structural characterization and third-order non-linear optical property of new three-dimensional metal-organic

framework K center dot Na center dot[M$_2$^{II}(μ_6-btc) (μ_2-ox)(H$_2$O)$_2$]·2H$_2$O (M=Ni, Co). *Inorg. Chem. Commun.*, *8*, 638–642.

87. Schlapbach, L., Zuttel, A. (2001). Hydrogen-storage materials for mobile applications. *Nature*, *414*, 353–358.

88. Rowsell, J. L. C., Yaghi, O. M. (2005). Strategies for hydrogen storage in metal-organic frameworks. *Angew. Chem. Int. Ed.*, *44*, 4670–4679.

CHAPTER *9*

POLYMER-FRIENDLY METAL–ORGANIC FRAMEWORKS

Takashi Uemura

Department of Synthetic Chemistry and Biological Chemistry,
Kyoto University, Kyoto, Japan

9.1 INTRODUCTION

One of the most important subjects in polymer science is the development of controlled polymerizations that allow the formation of well-defined structures and topologies useful for desired purposes. In fact, precision polymer synthesis to control polymer primary structures, such as stereo- and regioregularity, copolymer compositions, monomer sequence, and molecular weight and its distribution, is currently an important topic for enhancing and improving the properties of the polymeric materials. In addition, control of the secondary and higher ordered polymer structures by tuning the alignment and arrangement of polymer chains is of considerable interest for the construction of well-organized architectures, which lead to anisotropic, directional, as well as combined properties. In order to attain this chemistry, template polymerizations by using nanosized spaces has allowed spatial controls of primary structures of polymers and even two- and three-dimensional multilevel structuring [1]. A variety of nanoporous materials, such as organic hosts, zeolites, clays, and mesoporous materials, have played important roles in regulating the polymer structures by effective through-space interactions between the pore walls and the monomers [2].

A new type of porous material, metal–organic frameworks (MOFs), prepared by self-assembling processes from metal ions and organic ligands, have been extensively studied because of their wide applications in gas storage, molecular recognition and separation, and catalysis, utilizing their advantageous pore characteristics [3]. Recently, much effort has been devoted to developing characteristic features of MOFs that differ from those of conventional microporous

Supramolecular Soft Matter: Applications in Materials and Organic Electronics, First Edition.
Edited by Takashi Nakanishi.

materials, such as zeolites and activated carbons: (i) highly regular channel structures; (ii) controllable channel size approximating molecular dimensions; and (iii) designable surface potentials and functionality. These features must be of key importance for the creation of unique nanosize reaction fields based on the MOF materials [4]. As a successful application of MOFs, utilization of their regulated and tunable nanochannels for a field of polymerization has recently been reported [5]. The designable nanochannels of MOFs can allow multilevel regulations of polymerization to control not only the primary structures (stereoregularity, molecular weight, reaction sites, etc.) but also higher ordered structures (alignment and arrangement) of the resulting polymers. In addition, preparation of host–guest nanocomposites based on MOFs and polymers is also of interest for nanomaterial synthesis, host–guest synergistic properties, and nanosize-dependent properties.

In this chapter, I have discussed the recent progress and future perspectives of various polymerization systems utilizing the MOF nanochannels for controlling primary and secondary polymer structures as well as for obtaining specific properties based on low-dimensional assemblies of polymers and functional polymer-based nanohybrids.

9.2 CHARACTERISTIC FEATURES OF MOFs

9.2.1 Regularity

Usually, MOFs have regular nanosized pores that are suitable to confine guest molecules even when the intermolecular interaction between MOFs and guest molecules is governed only by the dispersion force, the so-called van der Waals force (Fig. 9-1). The regular channels based on MOFs, with their sharp distribution of pore size due to their high crystallinity, provide advantages for controlled polymerization. A variety of selective polymerizations (stereo, regio, and monomer selectivities) will proceed in such highly regular pores. By tuning the regularity of the potential field of the channels, we can create well-ordered monomer assemblies inside the MOF channels toward the target polymerization.

9.2.2 Pore Size and Shape

The reactivity of guest monomers in confined nanochannels is strongly dependent on monomer mobility and arrangement. Generally, the pore sizes of MOFs range from 0.4 to 2.0 nm by changing the combination of metal ions and organic ligands (Fig. 9-1). In addition, a variety of the pore shapes of MOFs, such as triangle, rectangle, and hexagon, has been prepared by the possible combinations of the components. From the viewpoint of channel dimensionality, not only one-dimensional but also layered or 3-D intersecting channel structures can be prepared in the MOF materials. By tuning the pore size and shape of MOFs, we can control the mobility, arrangement, and density of adsorbed molecules in the pores, which is advantageous for controlled polymerizations.

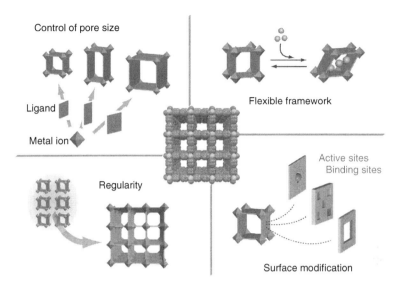

Figure 9-1 Characteristics of MOFs.

9.2.3   Pore Surface Functionality

All MOF compounds have metal ions that usually act as structure-directing components. If one can prepare MOFs containing unsaturated metal sites on the pore surfaces (Fig. 9-1) [6], they can be utilized for enhancing a variety of polymerization reactions (oxidative polymerization, Lewis acid-catalyzed polymerization, coordinative polymerization, etc.). Similarly, strong interactive organic sites can be introduced in MOF structures (Fig. 9-1) [7], which are also responsible for catalytic polymerizations. Introduction of relatively weak organic functional groups, such as carbonyl, hydroxyl, nitro, amide, and so on, into the nanochannels of MOFs would affect the monomer alignment, which may lead to stereo- and regioselective polymerizations. Of considerable interest is the use of the chiral channels [8]. In these pores, asymmetric polymerizations such as asymmetric selective polymerization of racemic monomers as well as asymmetric polymerization of prochiral monomers will proceed to give helical polymer conformations.

9.2.4   Flexibility

Dynamic structural transformation based on flexible frameworks is one of the most interesting phenomena and presumably characteristic of MOFs (Fig. 9-1), which cannot be attained by conventional microporous materials such as zeolites and activated carbons [9]. From the viewpoint of inclusion polymerization, utilization of such flexible frameworks based on MOFs would be a key principle for highly selective recognition, alignment, and reaction of the accommodated monomers, which is similar to the induced fit theory illustrated by enzymes in biological systems.

9.3   POLYMER SYNTHESIS IN ONE-DIMENSIONAL CHANNELS OF MOFs

MOFs often form regular and tunable one-dimensional channels in the structures. Considering the linear nature of organic polymer chains, it is suitable to use the one-dimenional MOF channels for controlling the primary structures (molecular weight, stereoregularity, regioselectivity, copolymer sequence, etc.) of polymers.

9.3.1   Radical Polymerization of Vinyl Monomers

There is a need for efficient methods for taming radical polymerizations into controlled ones for further development of functional polymer materials by tuning primary polymer structures. However, in the conventional process, the highly reactive free radical species induce the uncontrolled rapid chain growth without stereoselectivity and undergo inevitable termination via radical–radical coupling and disproportion. Thus, the development of precision radical polymerization systems for controlling molecular weight, streoregularity (tacticity), reaction position, copolymer sequence, and so on is currently an important topic in this area [10]. In order to control the primary structures of vinyl polymers, radical polymerization of various vinyl monomers has been recently performed in one-dimensional nanochannels of MOFs. For example, polymerization of styrene was examined in $[Cu_2(bdc)_2(ted)]_n$ (**1**; bdc = 1,4-benzenedicarboxylate, ted = triethylenediamine) and $[Cu(pzdc)_2(bpy)]_n$ (**2**; pzdc = pyrazine-2,3-dicarboxylate, bpy = 4,4′-bipyridine), which have regular and continuous one-dimensional nanochannels with cross-sections of 7.5 × 7.5 Å^2 and 8.2 × 6.0 Å^2, respectively [11]. In the case of polymerization using **1**, quantitative recovery of the accommodated polystyrene (PSt) from the **1**-PSt composites could be performed by decomposition of the host framework in 0.1 N NaOH after the polymerization of the styrene monomer inside the channels. From the gel permeation chromatography (GPC) measurement, the number-average molecular weight (M_n) and the polydispersity (M_w/M_n) of the recovered polymers were about 55,000 and 1.6, respectively. Interestingly, M_w/M_n of the PSt obtained in the channels of **1** was smaller than that of the corresponding bulk-synthesized polymer. This is because the propagating radical of PSt in this system was remarkably stabilized owing to effective entrapment in the nanochannel. Electron spin resonance (ESR) spectrum for propagating radicals of PSt in diamagnetic zinc analog of **1** ($[Zn_2(bdc)_2ted]_n$;**Zn1**) showed intense signals, and the maximum radical concentration of PSt in **Zn1** reached 2.6 kg^{-1}, which is much higher than that detected in conventional solution polymerizations (10^{-4}–10^{-5} mmol/kg). Note that the ESR signal of PSt in **Zn1** did not disappear over three weeks, even at 70°C. These results clearly indicate stabilization of the growing radical, caused by efficient suppression of termination reaction in the nanochannel. In high contrast to the system of **1**, only a trace amount of polymer was produced in the nanochannel of **2**, whose pore size is comparable to that of **1** [11]. The solid-state NMR (nuclear magnetic resonance) measurement has been used for explaining this different polymerizability of the encapsulated monomers. The shapes of the lines of the

^{2}H NMR spectra for styrene-d_8 adsorbed in **1** revealed that the guest styrene has high mobility with fast rotation in the nanochannels. However, the spectrum for styrene-d_8 in **2** showed completely solid-like behavior, even at the polymerization temperature, indicating that the mobility of styrene was highly restricted. Therefore, this restricted arrangement of the monomer in the nanochannel resulted in the poor reactivity.

In the MOF system of **1**, the nanochannel structure can be systematically controlled by changing the bridging dicarboxylate ligands to provide [Cu$_2$(L)$_2$ted]$_n$ (L = biphenyl-4,4'-dicarboxylate, pore size = 10.8 × 10.8 Å^2; L = 1,4-naphthalenedicarboxylate, pore size = 5.7 × 5.7 Å^2; L = 9,10-anthracenedicarboxylate, pore size = 4.8 × 4.3 Å^2) [12]. Interestingly, this series of MOF pores could be utilized for controlling the stereoregularity (tacticity) of vinyl polymers, such as PSt, poly(methyl methacrylate) (PMMA), and poly(vinyl acetate) (PVAc), which led to an increase in isotacticity of the resulting polymers [13]. In fact, the tacticity of PMMA strongly depends on the pore size of the MOFs; eventually, an increase of 9% in isotacticity was achieved in the narrow channel (5.7 × 5.7 Å^2) compared with that obtained in the bulk polymerization. In case of the polymerization of vinyl acetate, the ratio of isotactic units of the resultant PVAc structure certainly increased, and this is the first example of an increase in the isotactic units in the radical polymerization of vinyl acetate. These stereospecific polymerizations in MOFs can be explained by through-space interaction between the pore walls and the monomers. Because an isotactic unit requires a narrower conformational diameter (thickness) than the corresponding syndiotactic unit, the polymerizations in nanochannels of MOFs would result in an increase in the less stereobulky isotactic moiety.

Efficient methods for controlling or suppressing cross-linking structures in polymerization reactions are required for further development of functional polymer materials. Usually, radical polymerization of divinyl benzenes (DVBs) in bulk or solutions leads to the unavoidable formation of hyperbranched network polymers, because the reactivity of the two vinyl moieties in DVBs is equivalent. In contrast, radical polymerization of DVBs in one-dimensional channels of **1** and **Zn1** has produced linear polymers, leaving one pendant vinyl group in each benzene ring, which was confirmed by infrared (IR), UV–vis, and ^{1}H NMR spectra (Fig. 9-2) [14]. Because of the linearly extended formations, the polymers obtained were highly soluble in various organic solvents, such as tetrahydrofuran (THF), CHCl$_3$, dimethylformamide (DMF), dimethyl sulfoxide (DMSO), and so on, which is in great contrast to the insoluble cross-lined polymers obtained from the usual radical polymerization of DVBs in solution and bulk. Thus, it is of considerable interest that the one-dimensional nanochannels of MOFs could successfully direct the linear polymerization of multivinyl monomers because of effective entrapments of the reactive propagating radical mediators in the narrow nanochannels. In this polymerization system, unique effects of pore size and flexibility were observed (Fig. 9-2) [14]. Although polymerization of 1,3-disubstituted DVB (*m*-DVB; molecular dimension = 7.2 × 7.2 Å) could produce a linear polymer in the channel of **1** with a pore size of 7.5 × 7.5 Å^2, polymerization of

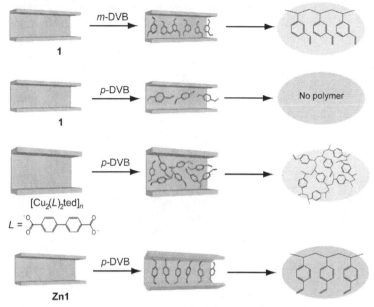

Figure 9-2 Effects of pore size and flexibility of MOFs on polymerization of DVB monomers.

1,4-disubstituted DVB (p-DVB; molecular dimension $= 8.5 \times 4.4$ Å) did not give any polymeric materials in **1**. Polymerization of p-DVB in larger open channels (pore size $= 10.8 \times 10.8$ Å^2) of [Cu$_2$(biphenyl-4,4′-dicarboxylate)$_2$ted]$_n$ gave an insoluble polymeric product, suggesting that semi-branching cross-linked structures were formed in this large channel. To overcome this problem, **Zn1** (pore size $= 7.5 \times 7.5$ Å^2) with a flexible framework on inclusion of specific guest molecules was employed. Introduction of p-DVB in the channels of **Zn1** induces a lattice expansion compared with the original host; however, the corresponding copper compound **1** did not show such structural changes. In addition, the adsorbed amount of p-DVB in **Zn1** was much larger than that in **1**. Consequently, average distances between p-DVB in the channels (along the c-axes) of **Zn1** and **1** could be estimated to be 4.5 and 10.6 Å, respectively, suggesting a polymerizable closed packing structure of p-DVB in the channel of **Zn1**. The host flexibility of **Zn1** is strongly associated with adsorption of the monomer by expanding the pore structure suited for adjacent arrangement of the monomer, resulting in the successful polymerization of p-DVB in the channels of **Zn1** (Fig. 9-2).

Owing to the nonconjugated nature of the propagating radical of PVAc, unfavorable chain transfer and termination reactions are particularly observed in the radical polymerization process; therefore, the resultant PVAc usually has many branching structures. It is noteworthy that branching formations of PVAc were effectively suppressed during polymerization in the nanochannels of **1**, which is confirmed by multiangle laser light scattering coupled with GPC [13]. This fact clearly suggests that the constrained chain growth in the narrow one-dimensional nanochannel of MOF can produce a less branching linear structure of PVAc.

Not only homopolymerization but also copolymerization of vinyl monomers in MOFs has been studied. Radical copolymerizations of vinyl monomers, such as styrene, methyl methacrylate, and vinyl acetate, in the nanochannels of **1** showed that compositions of styrene in the copolymers are lower than those obtained from the corresponding free radical systems [15]. This change of monomer reactivity in the nanochannels could be explained by the molecular size of the monomers. Because the molecular size of styrene (6.8 × 4.4 Å) is larger than that of methyl methacrylate (5.9 × 4.1 Å) and vinyl acetate (5.5 × 4.0 Å), the interaction potential between the styrene and the pore wall of **1** seems larger, decreasing the relative diffusion speed of styrene in the channel. Thus, these results suggest a possibility that copolymerization utilizing MOF nanochannels could have a significant influence on the composition and sequence of the copolymer structures, by efficient through-space inductions.

9.3.2   Catalytic Polymerization of Substituted Acetylenes

Stereocontrolled synthesis of poly(substituted acetylene)s is important for fine tuning of their characteristic properties, such as conjugation length, conductivity, suprastructures, and processability [16]. The nanochannels of **2** could induce a spontaneous and stereoselective polymerization of substituted acetylenes [17]. In this polymerization, the basic oxygen atoms of carboxylate ligands located at the pore wall of **2** strongly interact with the substituted acetylenes to produce reactive acetylide species that subsequently initiate anionic polymerization in the nanochannel. Compared with a controlled experiment using sodium benzoate as a discrete model catalyst, this polymerization system exhibited drastic acceleration of the polymerization. For example, the reaction of methyl propiolate with the model catalyst at room temperature gave only a trace amount of product even after reaction for one month. Increase in the reaction temperature to 70°C was also ineffective for the polymerization, whereby only unfavorable cyclic byproducts (trisubstituted benzenes) and a cis-geometric polymer were obtained in very low yields. However, the reaction of the acetylene monomers with **2** for 12 h at room temperature successfully provided the corresponding polymeric products. Interestingly, the narrow nanochannel structure could successfully direct the polymerization with trans-geometric addition, which contrasts with the result obtained by using the model catalyst. The experiments with various combinations of acetylene monomers and host MOFs showed that appropriate channel size, as well as the basic carboxylate moiety, is important for this spontaneous polymerization.

9.4   POLYMER SYNTHESIS IN HIGHER DIMENSIONAL CHANNELS OF MOFs

Control of polymer secondary structures is of considerable importance because many polymer properties can be improved and anisotropic functions would be obtained by appropriate arrangements of polymer chains. Nanospaces of MOFs can be used as templates to attain the control of polymer arrangements.

9.4.1 Preparation of Two-Dimensional Layered Polymers

Intercalative polymerization within crystalline-layered host materials is a useful method to restrict resulting polymers in the regular two-dimensional spaces [18]. If the host matrices can be removed without disturbing the polymer assembly, layered polymer objects, directed by the host structure, would be transferrably obtained. Such methodology will allow us to create novel polymer materials with structural order on the molecular level and with highly controlled morphologies and properties. Recently, oxidative polymerization of pyrrole has been performed within a redox-active two-dimensional layered framework, $[Ni(dmen)_2]_2[Fe^{III}(CN)_6](PhBSO_3)$ (**3**; dmen = 1,1-dimethylethylenediamine; $PhBSO_3^- = p$-phenylbenzenesulfonate), where the $[Fe^{III}(CN)_6]$ units oxidized and polymerized pyrrole monomers in the host framework [19]. The resulting polypyrrole (PPy) was intercalated between the 2-D sheets of the host, and a layer-by-layer-type crystalline nanocomposite could be synthesized. In this reaction, pyrrole monomers are intercalated into the nanoslits and are oxidized by the Fe^{III} ions in the host layers. As a consequence of the host–guest redox reaction, the Py monomers are converted to PPy, and the Fe^{III} ions are reduced to Fe^{II} ions, accompanied by the release of the pillar counter anions for charge compensation (Fig. 9-3a). Although MOFs with redox-active metal ions might be capable of providing controlled reactions for oxidized/reduced products in the resulting composites, only a few reports on the use of MOFs as such significant redox reaction fields have been reported [20] because, in many cases, the host frameworks are decomposed during the redox reaction.

In this work, removal of the host framework of **3** was performed in EDTA (ethylenediaminetetraacetic acid) solution, which allowed isolation of the intercalated PPy as an insoluble black precipitate. The controlled morphogenesis of the isolated polymer objects was observed in SEM (scanning electron microscope) measurements (Fig. 9-3b). In contrast to the granular morphology of the bulk PPy prepared by oxidation of Py with $K_3[Fe^{III}(CN)_6]$ in water, the morphology of the isolated PPy was found to be platy. XRPD (X-ray powder diffraction) measurements indicated that the bulk PPy was an amorphous compound; however, the XRPD profile of the isolated PPy showed the accumulation of aromatic planes of PPy. In addition, the PPy microsheet object obtained showed anisotropic conductivity; the conductivity parallel to the sheet was 20 times higher than that along the direction perpendicular to the sheet. Therefore, in this system, the crystalline MOF template successfully directed the well-structured and 2-D oriented organization of PPy assembly at the molecular level.

9.4.2 Preparation of Three-Dimensional Porous Polymers

Templating techniques for generating porous solids have been extensively studied in materials science. In particular, template synthesis of porous organic polymers

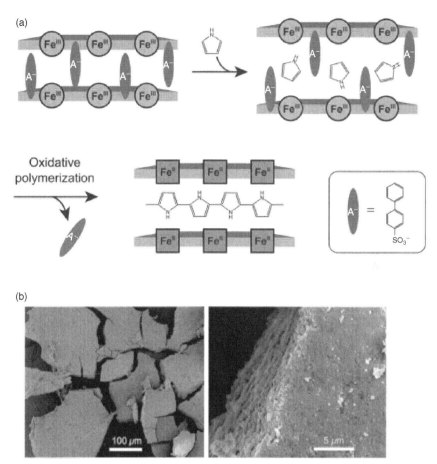

Figure 9-3   (a) Schematic illustration of the intercalation of pyrrole monomer and the oxidative polymerization of pyrrole by Fe(III) ions in the host layers. (b) SEM images of polypyrrole isolated from the layered MOF host.

has emerged as a promising subject because of the great potential of porous polymers in chromatography, sensing, and membranes [21]. Very recently, a feasible method for providing a porous structure of PPy has been reported by using three-dimensional channels of $[Cu_3btc_2]_n$ (**4**; btc = benzene-1,3,5-tricarboxylate) as a template [22]. In this work, the accessible Cu(II) sites on the pore surface catalyzed the oxidative polymerization of Py in the framework of **4** and finally provided a black powder composite. The accommodated polymer was isolated from the composite by complete dissolution of the host framework in various basic solutions. The porosity of the resulting polymer materials was examined by nitrogen adsorption measurements at 77 K. Although PPy prepared in the bulk condition did not adsorb the gas, PPy materials prepared in the MOF clearly showed an adsorption property, and the amount of adsorption gradually increased as the weaker bases were employed for the isolation of PPy. This phenomenon was explained by the affinity of PPy chains for strong bases. The PPy chains were probably restructured to pack the space efficiently during treatment with strong basic solutions. In this work, PPy isolated using ammonia showed the maximum adsorption, and the amount of adsorption was almost independent of the ammonia concentration in the isolation process. Unfortunately, the porosity of the PPy materials were still low; however, the pore-size distribution of the PPy isolated with ammonia showed a peak at 1.3 nm, where this pore size might reflect the thickness of the pore walls of **4** (ca. 1.2 nm). Interestingly, the porous structure of PPy is very stable and not affected by heating at 120°C or treatment with some common solvents. A detailed study on the porous properties showed a specific interaction between the porous PPy and oxygen molecule because of their electron sufficient and deficient natures, respectively. A high affinity of the porous PPy for hydrophilic molecules, such as water and MeOH, was also observed, which is ascribable to the effect of the polar amine groups in the PPy structure.

Porous carbon has also been prepared by using MOF channels as a template [23]. In this methodology, $[Zn_4O(bdc)_3]_n$ (**MOF-5**) was heated at 150°C for 48 h under an atmosphere of furfuryl alcohol vapor, when the furfuryl alcohol was polymerized in the pores of **MOF-5**. Carbonization of the polymer/**MOF-5** composite was then performed at 1000°C for 8 h with an Ar flow. Although the Zn species were not contaminated with the sample obtained by carbonizing at 1000°C, sharp XRD diffractions due to ZnO appeared in a sample obtained by carbonizing at 800°C. The authors explained that the observed ZnO was formed from the decomposition of the **MOF-5** framework between 425 and 525°C as observed in the thermogravimetric measurement. At temperatures higher than 800°C, ZnO was reduced during the carbonization process, and subsequently, Zn metal (boiling point 908°C) vaporized away along with the Ar flow, leaving the carbon species in the resulting sample. Thus, the carbonization temperature is critical for the structural evolution of the resultant carbon, because the samples prepared at low temperatures have low surface areas. However, the resulting nanoporous carbons exhibited a high hydrogen storage capacity as well as an excellent electrochemical property as an electrode material for electrochemical double-layered capacitor.

9.5   POLYMER–MOF COMPOSITES

Polymers confined in the nanosized spaces show distinctly different properties from those in the bulk state because of the formation of specific molecular assemblies and conformations. In particular, polymer inclusion in crystalline microporous hosts (pore size <2 nm) with ordered and well-defined nanochannel structures is attracting much attention. This is largely because this approach can prevent the entanglement of polymer chains and provide extended chains in restricted spaces, in contrast to amorphous bulk polymer systems and polymers in solution. In this field, microporous organic hosts such as ureas, perhydrotriphenylene, cyclotriphosphazenes, cholic acids, and cyclodextrins play dominant roles

[24]. However, these organic hosts exploit only hydrogen bonding or weak van der Waals interactions and generally produce narrow channels (ca. 4–6 AA) that often prevent the incorporation of polymer chains with bulky side chains, such as PSt and PMMA.

Inclusion of thick vinyl polymers can be attained by polymerization of the corresponding monomers accommodated in the nanochannels of MOFs, which can lead to a fundamental study on the properties of the confined single polymer chains. The usefulness of this method was demonstrated by the recent study of the conformation and dynamics of single PSt chains confined in **Zn1** [25]. Differential scanning calorimetry of bulk PSt showed a T_g at 105°C; in contrast, the PSt confined in the channels of **Zn1** did not show such a transition. The suppression of the T_g suggests that the chain assembly of PSt in the host–guest adduct was considerably different from that in the bulk state. From the N_2 adsorption measurements, an estimated density of the PSt residing in the channels was 0.55 g/cm^3, which is considerably lower than that of the bulk PSt (1.04–1.12 g/cm^3). This interesting result suggests that the polymer chains experience a larger space in the nanochannels than in the bulk. Because of the single chain nature, the PSt accommodated in regular 1-D nanochannels showed unique molecular dynamics (MD). Solid-state ^{2}H NMR measurement clearly showed that the bulk PSt provides a wide motional distribution that originates from the heterogeneous local environment of the phenyl rings; in contrast, PSt in the channel of **Zn1** represents a quasi-single-type phenyl flipping motion, indicating homogeneous side-chain mobility in the nanochannel (Fig. 9-4a). In addition, the activation energy for the phenyl flip of the PSt chain in **Zn1** was significantly lower ($E_a = 8.8$ kJ/mol) than that of bulk PSt because of the lower steric hindrance of the phenyl rings of the single PSt chain in the nanochannel.

The chain conformations of PSt oligomers in the 1-D channel of **Zn1**, simulated using the MD method, agreed with the unique molecular motion observed in the NMR measurements. In the MD simulation, the introduction of a higher molecular weight PSt into the nanochannel led to a highly extended linear conformation of the single PSt chain, which contrasts significantly with the random chain-entangled structures that essentially occur in the bulk state (Fig. 9-4b). These results indicate the usefulness of MOF nanovessels for providing significant insights into polymer-confinement effects, as well as inherent properties of single polymer chains, because MOFs can supply tailor-made nanopores with a variety of pore sizes and environments.

Wrapping of MOFs with organic polymers is another aspect of the formation of composite materials. This chemistry mainly focuses on the preparation of MOF nanoparticles for applications to various functional nanomaterials [26]. In particular, use of nanosized MOFs in biology is very interesting, because many MOFs can be dissolved rapidly by the addition of solvents or ligands that competitively bind to the metal ions. This characteristic of MOFs, coupled with size control and polymer wrapping, makes them attractive for drug delivery, as well as for biological label applications [27].

(a)

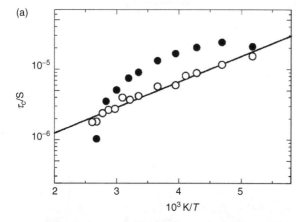

(b)

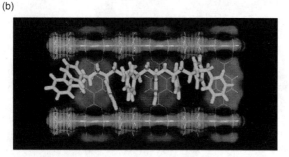

Figure 9-4 (a) Mean correlation times of bulk PSt (solid circle) and PSt in **Zn1** (open circle) as a function of temperature in ^{2}H NMR measurements. (b) Typical MD structures of PSt nonamer encapsulated in **Zn1**.

9.6   SUMMARY

Polymerization of monomers encapsulated in the nanochannels of functional MOFs has provided significant opportunities for precision controlled polymerizations. Utilization of the regular nanochannel structures of MOFs for polymerization can lead to the control of not only primary structures of the resulting polymers but also polymer arrangement and alignment (Fig. 9-5). Owing to the capability of designing nanochannels of MOFs (regular channel structures, controllable channel size and surface functionality, and flexible frameworks), systematic studies of inclusion polymerizations in microporous channels based on MOFs can be performed. We believe that this polymerization system can provide a new aspect of controlled polymerizations and is fundamentally important for the understanding of the role of pore size, geometry, and flexibility in attaining tailor-made polymerizations to obtain preferred polymer structures.

With respect to applications and technology, further research can be also directed to the quest of polymer properties and functions in the nanochannels of MOFs, as well as construction of polymer-MOF nanohybrids, so as to enhance the knowledge on applications to nanosize molecular-based devices. The regulated and tailor-made pores of MOFs are of key importance for achieving unique confinement of polymer materials. In the designable nanochannels of MOFs, we

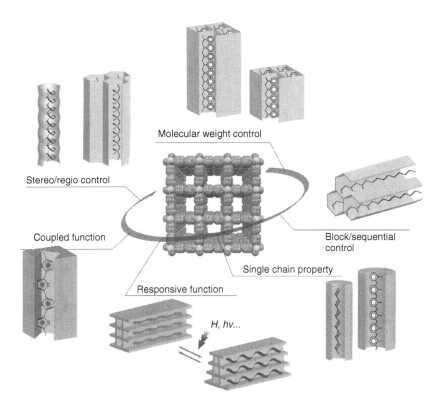

Figure 9-5   Polymer science that can be performed in nanochannels of MOFs.

can control polymer assemblies from the viewpoints of (i) the number of polymer chains; (ii) the environment of the polymers; and (iii) chain orientations. This concept will lead to quantitative analysis of low-dimensional properties of the polymer assemblies (single chain, double chain, triple chain, etc.) confined in specific porous environments and orientations and to the preparation of a new class of functional materials based on MOF–polymer hybridization (Fig. 9-5). Because of the infinite combinations of MOFs and polymers including inorganic polymers, such as metal oxides and sulfides, study on the fabrication of MOF–polymer nanocomposites (polymer-friendly MOFs), which is still in its infancy, will find a new class different from conventional composite materials and can contribute to the development of the field of nanomaterials. We will witness the beginning of what science develops among the research areas of polymer, inorganic, coordination, and materials chemistry.

REFERENCES

1. (a) Paleos, C. M., *Polymerization in Organized Media*. Gordon & Breach, New York, 1992.
   (b) Sada, K., Takeuchi, M., Fujita, N., Numata, M., Shinkai, S. (2007). Post-polymerization of preorganized assemblies for creating shapecontrolled functional materials. *Chem. Soc. Rev.*, *36*,

415–435; (c) Tajima, K., Aida, T. (2000). Controlled polymerization with constrained geometries. *Chem. Commun.*, 2399–2412.

2. (a) Miyata, M. (1996). Inclusion polymerization. *Comprehensive Supramolecular Chemistry*, Vol. *10*. Pergamon, Oxford, pp. 557–582; (b) Farina, M., Di Silvestro, G., Sozzani, P. (1996). Perhydrotriphenylene: a D_3 symmetric host. *Comprehensive Supramolecular Chemistry*, Vol. *10*. Pergamon, Oxford, pp. 371–398.

3. (a) Kitagawa, S., Kitaura, R., Noro, S.-I. (2004). Functional porous coordination polymers. *Angew. Chem. Int. Ed.*, *43*, 2334–2375; (b) Ockwig, N. W., Delgado-Friedrichs, O., O'Keefe, M., Yaghi, O. M. (2005). Reticular chemistry: occurrence and taxonomy of nets and grammar for the design of frameworks. *Acc. Chem. Res.*, *38*, 176–182; (c) Férey, G., Mellot-Draznieks, C., Serre, C., Millange, F. (2005). Crystallized frameworks with giant pores: are there limits to the possible? *Acc. Chem. Res.*, *38*, 217–225.

4. (a) Lee, J. Y., Farha, O. K., Roberts, J., Scheidt, K. A., Nguyen, S.-B. T., Hupp, J. T. (2009). Metal-organic framework materials as catalysts. *Chem. Soc. Rev.*, *38*, 1450–1459; (b) Wang, Z., Chen, G., Ding, K. (2009). Self-supported catalysts. *Chem. Rev.*, *109*, 322–359.

5. (a) Uemura, T., Horike, S., Kitagawa, S. (2006). Polymerization in coordination nanospaces. *Chem. Asian J.*, *1*, 36–44; (b) Uemura, T., Yanai, N., Kitagawa, S. (2009). Polymerization reactions in porous coordination polymers. *Chem. Soc. Rev.*, *38*, 1228–1236.

6. Higuchi, M., Horike, S., Kitagawa, S. (2007). Spatial and surface design of porous coordination polymers. *Supramol. Chem.*, *19*, 75–78.

7. (a) Seo, J. S., Whang, D., Lee, H., Jun, S. I., Oh, J., Jeon, Y. J., Kim, K. (2000). A homochiral metal-organic porous material for enantioselectiveseparation and catalysis. *Nature*, *404*, 982–986; (b) Hasegawa, S., Horike, S., Matsuda, R., Furukawa, S., Mochizuki, K., Kinoshita, Y., Kitagawa, S. (2007). Three-dimensional porous coordination polymer functionalized with amide groups based on tridentate ligand: selective sorption and catalysis. *J. Am. Chem. Soc.*, *129*, 2607–2614.

8. Ma, L., Abney, C., Lin, W. (2009). Enantioselective catalysis with homochiral metal-organic frameworks. *Chem. Soc. Rev.*, *38*, 1248–1256.

9. (a) Férey, G., Serre, C., (2009). Large breathing effects in three-dimensional porous hybrid matter: facts, analyses, rules and consequences. *Chem. Soc. Rev.*, *38*, 1380–1399; (b) Horike, S., Shimomura, S., Kitagawa, S. (2009). Soft porous crystals. *Nat. Chem.*, *1*, 695–704.

10. (a) Matyjaszewski, K., Davis, T. P. (2002). *Handbook of Radical Polymerization*. Wiley-Interscience, Hoboken, NJ; (b) Jagur-Grodzinski, J. (2006). *Living and Controlled Polymerization: Synthesis, Characterization and Properties of the Respective Polymers and Copolymers*. Nova Science, New York.

11. Uemura, T., Kiatagwa, K., Horike, S., Kawamura, T., Kitagawa, S., Mizuno, M., Endo, K. (2005). Radical polymerisation of styrene in porous coordination polymers. *Chem. Commun.*, 5968–5970.

12. Seki, K., Mori, W. (2002). Syntheses and characterization of microporous coordination polymers with open frameworks. *J. Phys. Chem. B*, *106*, 1380–1385.

13. Uemura, T., Ono, Y., Kitagawa, K., Kitagawa, S. (2008). Radical polymerization of vinyl monomers in porous coordination polymers: nanochannel size effects on reactivity, molecular weight, and stereostructure. *Macromolecules*, *41*, 87–94.

14. Uemura, T., Hiramatsu, D., Kubota, Y., Takata, M., Kitagawa, S. (2007). Topotactic linear radical polymerization of divinylbenzenes in porous coordination polymers. *Angew. Chem. Int. Ed.*, *46*, 4987–4990.

15. Uemura, T., Ono, Y., Kitagawa, S. (2008). Radical copolymerizations of vinyl monomers in a porous coordination polymer. *Chem. Lett.*, *37*, 616–617.

16. Lam, J. W. Y., Tang, B. Z. (2005). Functional polyacetylenes. *Acc. Chem. Res.* *38*, 745–754.

17. Uemura, T., Kitaura, R., Ohta, Y., Nagaoka, M., Kitagawa, S. (2006). Nanochannel-promoted polymerization of substituted acetylenes in porous coordination polymers. *Angew. Chem. Int. Ed.*, *45*, 4112–4116.

18. Cardin, D. J. (2002). Encapsulated conducting polymers. *Adv. Mater.*, *14*, 553–563.

19. Yanai, N., Uemura, T., Ohba, M., Kadowaki, Y., Maesato, M., Takenaka, M., Nishitsuji, S., Hasegawa, H., Kitagawa, S. (2008). Fabrication of two-dimensional polymer arrays: template synthesis of polypyrrole between redox-active coordination nanoslits. *Angew. Chem. Int. Ed.*, *47*, 9883–9886.

20. Moon, R., Kim, J. H., Suh, M. P. (2005). Redox-active porous metal-organic framework producing silver nanoparticles from AgI ions at room temperature. *Angew. Chem. Int. Ed.*, *44*, 1261–1265.

21. Thomas, A., Goettmann, F., Antonietti, M. (2008). Hard templates for soft materials: creating nanostructured organic materials. *Chem. Mater.*, *20*, 738–755.

22. Uemura, T., Kadowaki, Y., Yanai, N., Kitagawa, S. (2009). Template synthesis of porous polypyrrole in 3D coordination nanochannels. *Chem. Mater.*, *21*, 4096–4098.

23. Liu, B., Shioyama, H., Akita, T., Xu, Q. (2008). Metal-organic framework as a template for porous carbon synthesis. *J. Am. Chem. Soc.*, *130*, 5390.

24. (a) Liu, J., Mirau, P. A., Tonelli, A. E. (2002). Chain conformation and dynamics of crystalline polymers as observed in their inclusion compounds by solid-state NMR. *Prog. Polym. Sci.*, *27*, 357–401; (b) Sozzani, P., Bracco, S., Comotti, A., Simonutti, R. (2005). Motional phase disorder of polymer chains as crystallized to hexagonal lattices. *Adv. Polym. Sci.*, *181*, 153–177.

25. Uemura, T., Horike, S., Kitagawa, K., Mizuno, M., Endo, K., Bracco, S., Comotti, A., Sozzani, P., Nagaoka, M., Kitagawa, S. (2008). Conformation and molecular dynamics of single polystyrene chain confined in coordination nanospace. *J. Am. Chem. Soc.*, *130*, 6781–6788.

26. Spokoyny, A. M., Kim, D., Sumrein, A., Mirkin, C. A. (2009). Infinite coordination polymer nano- and microparticle structures. *Chem. Soc. Rev.*, *38*, 1218–1227.

27. (a) Rieter, W. J., Taylor, K. M., Lin, W. (2007). Surface modification and functionalization of nanoscale metal-organic frameworks for controlled release and luminescence sensing. *J. Am. Chem. Soc.*, *129*, 9852–9853; (b) Horcajada, P., Chalati, T., Serre, C., Gillet, B., Sebrie, C., Baati, T., Eubank, J. F., Heurtaux, D., Clayette, P., Kreuz, C., Chang, J.-S., Hwang, Y. K., Marsaud, V., Bories, P.-N., Cynober, L., Gil, S., Férey, G., Couvreur, P., Gref, R. (2010). Porous metal-organic-framework nanoscale carries as a potential platform for drug delivery and imaging. *Nat. Mater.*, *9*, 172–178.

RECENT TRENDS OF ORGANIC RADICAL MATERIALS

CHAPTER 10

MULTIDIMENSIONAL SUPRAMOLECULAR ORGANIZATIONS BASED ON POLYCHLOROTRIPHENYL-METHYL RADICALS

V. Mugnaini, M. Mas-Torrent, I. Ratera, C. Rovira, and J. Veciana

Insitut de Ciència de Materials de Barcelona (ICMAB-CSIC) and Centro de Investigación Biomédica en Red en Bioingeniería, Biomateriales y Nanomedicina (CIBER-BBN), Campus Universitari de la Universitat Autònoma de Barcelona, Bellaterra, Spain

10.1 INTRODUCTION

10.1.1 Organic Radicals: What Are They?

Organic free radicals are molecular species containing an unpaired electron that singly occupies an outer molecular orbital. This peculiar open-shell electronic configuration makes these species extremely reactive, since they can easily take part in reactions, such as dimerization or recombination, that lead to the loss of the open-shell character. Nevertheless, it is possible to have persistent organic radicals by screening the paramagnetic center with bulky substituents, hence protecting it from undesired reactions. Another way to produce persistent radicals consists in increasing the delocalization of the unpaired electron over a large part of an atomic skeleton. Examples of families of radicals showing persistency are the nitroxides and α-nitronyl nitroxides [1], where the unpaired electrons are delocalized between the nitrogen and the oxygen atoms; the verdazyl radicals [2], where the unpaired electrons are instead delocalized onto the heterocycle containing the nitrogen atoms; and the polychlorinated trityl radicals [3], where the central methyl sp^2 carbon is the spin-bearing atom being sterically protected.

Supramolecular Soft Matter: Applications in Materials and Organic Electronics, First Edition.
Edited by Takashi Nakanishi.
© 2011 John Wiley & Sons, Inc. Published 2011 by John Wiley & Sons, Inc.

These three families of persistent radicals find applications in many different fields, such as biochemistry [4], catalysis [5], and molecular magnetism [6]. In the last application, the stable radicals act as magnetic synthons, whose assembly, based on different types of interactions through which the magnetism can propagate, leads either to completely organic molecular magnetic materials or to metal–organic frameworks.

10.1.2 Polychlorotriphenylmethyl Radicals: Molecular Structure and Properties

We herein use the term polychlorotriphenylmethyl (PTM) radicals to refer to a wide family of open-shell molecules with a triphenylmethyl skeleton polysubstituted with chlorine atoms [3]. In these perchlorinated triphenylmethyl derivatives, the central carbon, where most of the spin density resides, is encapsulated by three pairs of chlorine atoms, whose sterical hindrance is responsible for the high persistency of the radical character (Fig. 10-1a).

From a structural point of view, these six chlorine atoms in *ortho* impose a torsion between the mean planes of the three polychlorinated aromatic rings and generates the typical propeller-shaped conformation found in this family of radicals that confers them their stereochemical characteristics [7].

As far as the properties are concerned, in addition to their inherent magnetic moment, these radicals are fluorescent in the red region of the spectrum and present a rich electrochemical behavior, as they can be easily and reversibly reduced to the anionic form or even oxidized to the cationic species. Moreover, the high thermal and chemical stability of PTMs allows, by means of demanding synthetical routes, the removal of the chlorine atoms in the *meta* or *para* position and the insertion of a bunch of different substituents, making it possible to prepare a large variety of multifunctional molecular materials.

It has been reported that red fluorescent monosubstituted perchlorinated radical derivatives designed to be amorphous glasses are good candidates to be incorporated in electroluminescent devices [8]. When the perchlorinated radicals are monosubstituted with a specific fluorophore [9], the compound obtained acts as a fluorescent and electron paramagnetic resonance probe for the detection of superoxide radical anion, paving the way towards the use of such compounds in biological systems under oxidative stress-mediated conditions.

The low reduction potential of perchlorinated radicals makes them good acceptor groups in donor–spacer–acceptor systems. For this reason, they have been used as building blocks together with electron donor moieties to obtain new valence tautomeric systems [10,11], as well as push–pull compounds with nonlinear optical (NLO) properties [12]. Other systems formed by two PTM moieties linked through different diamagnetic conjugated spacers have also been used to obtain mixed valence systems that present intramolecular electron transfer processes through the connecting bridge [13–15]. Interestingly, the electroactive character of PTMs has also been exploited to prepare molecular redox switches in solution [16] and on solid supports [17,18] since the magnetic and optical

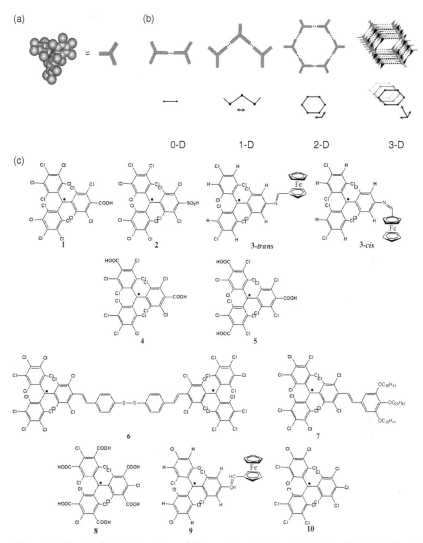

Figure 10-1   (a) Crystal structure of the perchlorotriphenylmethyl radical. (b) Strategies used to control the dimensionality of the supramolecular networks. From left to right: monotopic ligand, 0-D dimer; ditopic ligand, 1-D chains; tritopic ligand, 2-D planes; polytopic ligands, 3-D networks. (c) Molecular structures of the PTM radicals whose supramolecular networks are described in this chapter.

(*i.e.,* absorbance and emission) properties are simultaneously modified by the generation of the PTM anionic species.

The particular structural and electronic characteristics of PTM radicals have also permitted obtaining organic open-shell dendrimers with high-spin ground states and low-spin excited states, inaccessible even at room temperature

[19,20]. More recently, on the search for new molecular spintronic systems, PTM radicals have also been grafted on surfaces exhibiting interesting electronic transport properties [21,22].

By substituting chlorine atoms with functional groups with specific supramolecular binding capabilities, it has been possible to use PTM derivatives as robust multifunctional (paramagnetic, electroactive, and fluorescent) synthons for the preparation of purely organic or metal–organic networks. To achieve multidimensional supramolecular networks, different types of noncovalent interactions have been used, such as hydrogen bonds (HBs) [23], coordination bonds [24], or electrostatic interactions [11]. Interestingly, the multidimensional supramolecular networks obtained preserve the magnetic and electroactive properties typical of the PTM single molecule and hence are multifunctional materials. Moreover, since the molecular bulkiness and rigidity of PTM radicals prevent close packing of the molecular units, they are expected to give highly porous networks. In fact, porous and magnetic frameworks showing solvent selectivity and the shrinking–breathing phenomenon have been successfully prepared [25].

10.1.3 Strategies Towards Multidimensionality

With the aim to prepare multifunctional multidimensional networks, much effort has been devoted to the design and synthesis of appropriately substituted perchlorinated triphenylmethyl derivatives that could efficiently give the desired supramolecular interactions. Once the most appropriate substituent has been chosen, the general strategy to obtain a desired dimensionality has been based on the number (from 1 to 6) and position of the substituents (*para* or *meta*), that is, on the connectivity of the organic ligand, as illustrated schematically in Fig. 10-1b. However, as discussed in detail throughout this chapter, the final dimensionality cannot be always predicted since it can also depend on additional factors such as the presence of solvent molecules that can take part in the primary structure or on the topicity of the involved metal ion in the case of hybrid frameworks. Remarkably, the variable—but always large[1] number of chlorine atoms present in the PTM radical and its derivatives allows a large number of short Cl$\cdots$Cl contacts that play a fundamental role in the hierarchical aggregation of the single molecules [26] or of the dimeric networks, from 1-D chains or 2-D layers into 3-D networks and can be considered as an intrinsic characteristic in the supramolecular organization of these perchlorinated derivatives.

In the following paragraphs, we have given a brief overview of the different multidimensional type of networks so far obtained with the substituted PTM radical derivatives shown in Fig. 10-1c, in crystalline phase, in solution, and on surfaces. We have described the supramolecular organizations in order from low to high dimension. The criteria chosen for the classification of the supramolecular networks into 0-D, 1-D, 2-D, and 3-D are based on the dimensionality dictated by the driving noncovalent interactions different from Cl$\cdots$Cl short contacts rather than on the connectivity and topicity of the PTM synthon.

---

[1] The fully perchlorinated PTM derivative has 18 Cl atoms, while the derivative with less chlorine atoms synthesized up to now has 8 chlorine atoms.

10.2 ZERO DIMENSIONAL (0-D) SUPRAMOLECULAR ORGANIZATIONS

The PTM-based supramolecular organizations with 0-D connectivity are dimers (Fig. 10-1b) formed by the monotopic PTM derivatives **1**, **2**, and **3** (Fig. 10-1c). These dimers can be either purely organic, where HBs act as a driving force between the two interacting PTM units [27–31], or metal–organic, where one metal ion acts as connector between two PTM moieties [28,32,33].

10.2.1 Hydrogen-Bonded Dimers

Since PTMs are spin-bearing moieties, their supramolecular organization has been investigated in order to prepare materials with a specific magnetic behavior. In this frame, hydrogen bonding interactions have been chosen to form supramolecular networks of paramagnetic tectons since they not only exert high structural control but also favor the magnetic exchange interactions between the bound organic radical molecules without the need for connectors between the spin-bearing moieties [6].

In order to obtain hydrogen-bonded 0-D dimers with ferromagnetic interactions, Maspoch *et al.* [27] investigated the assembly of the monotopic *para*-mono carboxy PTM derivative **1**. The propagation of the magnetic interactions can occur only through unidirectional HBs that link the two radicals; therefore, ferromagnetic behavior could be observed only when using crystallization conditions that lead to the formation of dimers with no interference of solvent molecules. If the solvent molecules prevented the formation of radical dimers because of the crystallization conditions, only antiferromagnetic interactions, due to the Cl···Cl short contacts of the secondary structure, were observed (Fig. 10-2a).

10.2.2 Metal–Organic Complexes

The carboxylate group is known not only as a good HB donor or acceptor but also for its capability to coordinate metal ions. Therefore, the monotopic PTM carboxylate ligand **1** has also been used to prepare 0-D metal–organic complexes [32,33], where two monotopic ligands dimerize coordinating one metal ion whose coordination sphere is then filled by solvent molecules or auxiliary ligands, such as pyridine. By preparing 0-D complexes with PTM ligands, it has also been possible to demonstrate the active role of this trigonal open-shell molecule in the propagation of magnetic interactions through coordination bonds (Fig. 10-2b).

Particularly, by using radical **1** and paramagnetic Cu(II) ions as inorganic metal units with an appropriate stoichiometry as well as crystallization conditions, PTM dimers, in which the copper takes a paddle-wheel geometry, were prepared. The magnetic properties of the resulting coordination complex follow a butterfly spin-frustrated model with an exotic magnetic behavior [32a]. This study paved the way toward the investigation of the effect of the nature of inorganic connectors (copper, cobalt [33], or europium [34]) on the supramolecular organization and its magnetic interaction.

To further explore the coordinating capability and magnetic relay properties of monotopic PTM derivatives, Ribas *et al.* [28] carried out the substitution of

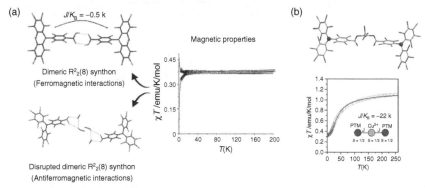

Figure 10-2 (a) Temperature dependence of the magnetic susceptibility, χ, of polycrystalline samples of α ($\triangle$) and β (O) phases of radical **1**, which are plotted as χT versus T. Solid lines are the best fit of the experimental data. *Source*: Figure reproduced from Ref. [23], Copyright (2007), with permission from Elsevier (b) Structure and temperature dependence of the magnetic susceptibility of the metal–organic dimer formed by two molecules of derivative **1** and copper metal ions. Inset, linear trimer arrangement of the metal ion and organic radicals, in which an exchange coupling constant Cu(II)-**1** of $J/K_B = -22$ K ($H = -2 J S_1 S_2$) was determined. *Source*: Figure reproduced from Ref. [24], Copyright (2005), with permission from Elsevier.

the carboxylic functional group by a sulfonic group, as in radical **2**. The coordination mode of the PTM derivative **2** with an appropriate copper square planar cation turned out to be similar to one of the geometries obtained with the mono-carboxylic derivative **1**. Nevertheless, in this case, the interactions were weakly antiferromagnetic, thus underlying the importance not only of the supramolecular motif but also of the functional group involved in the propagation of magnetic interactions.

10.2.3 Hydrogen-Bonded Dimers with Tuneable Properties

Despite the enormous interest in having switchable magnetic devices, hydrogen-bonded supramolecular magnetic materials whose properties may be systematically tuned and/or controlled by external stimuli are limited to one example [29–31]. This device is made with radical **3**, a ferrocene-PTM radical dyad with a conjugated bridge. Such a conjugated bridge was designed to fulfill two main requirements: (i) experience a reversible photoinduced structural change, such as a *trans/cis*-photoinduced isomerization and (ii) promote the formation of supramolecular species through noncovalent interactions such as hydrogen bonding. Among photosensitive bridges whose structural changes may be controlled by light irradiation, imino bridges are expected not only to exhibit a *trans/cis*-photoisomerization but also to enhance the presence of HBs. The relaxation from a photoinduced geometrical change in these imino compounds is extremely rapid, and this feature may offer advantages for some light-driven devices. In solution, the *trans* isomer of **3** (Fig. 10-1c) exists as a monomeric species, while the *cis* isomer dimerizes in solution, forming a thermodynamically stabilized hydrogen-bonded diradical species, $(cis)_2$. In both compounds, the donor character of the ferrocene unit and the acceptor character of the radical unit ensure a reasonable spin delocalization over the bridge so that, once the supramolecular species is formed, a magnetic exchange between both species can exist. The radical *trans* interconverts on irradiation at 415 nm into the dimeric species $(cis)_2$, in which relatively strong antiferromagnetic interactions are developed. This means that two doublet species can be converted into one dimeric species with a singlet ground state by inducing *trans* to *cis* photoisomerization. This photoinduced self-assembly process represents an interesting example of a switchable photomagnetic system based on a supramolecular phenomenon. This concept may be extended to other novel compounds bearing other organic and inorganic magnetic units, providing valuable access to this new class of supramolecular photomagnetic materials with which interesting supramolecular devices can be achieved. The translation into photofunctional magnets will open new possibilities for future applications of molecular compounds such as photorecording media or optical switches.

10.3   ONE DIMENSIONAL (1-D) SUPRAMOLECULAR ORGANIZATIONS

The strategy to increase the final dimensionality of the supramolecular assembly of PTM derivatives relies on increasing the topicity and connectivity of the organic ligand, as depicted schematically in Fig. 10-1b. While 0-D supramolecular organizations have been prepared using monotopic PTM derivatives, to prepare 1-D networks employing noncovalent IIB interactions or coordination bonds as driving force, ditopic ligands, such as **4**, were predicted to be appropriate synthons. Nevertheless, the purely organic supramolecular structures given by the ditopic ligand **4** gave unexpected 2-D networks due to the formation of an extended net of HBs, as described in Section 10.4, rather than the expected 1-D chain-like structure.

1-D supramolecular organizations could be instead obtained by using derivative **4** and metal ions, such as copper and cobalt [35]. For example, in the case of copper, the ditopic synthon **4** is able to coordinate two different copper ions in a monodentate fashion, whose coordination sphere is completed with ancillary pyridine ligands. This repetitive unit propagates generating infinite corrugated chains—where consecutive PTM molecules present opposite *plus* and *minus* helicities—with antiferromagnetic ordering. The repetitive unit of these chains is structurally/geometrically similar to that found in the coordination complexes formed by the monotopic ligand **2**, but the 2-D connectivity of radical **4** and its geometry are responsible for the zig-zag conformation of these metal–organic chains, where the copper ions act only as linker.

One-dimensional metal–organic networks have also been obtained, unexpectedly, by using as organic synthon the asymmetric and multitopic radical **8**, the hexa *meta*-carboxy PTM derivative (Fig. 10-1c). This molecule, because of its 3-D connectivity, was designed to form 3-D supramolecular networks when coordinating polytopic inorganic synthons, such as europium. Instead, it was serendipitously found that the presence of solvent molecules in its coordination sphere led to the formation of polymeric chains [34].

10.4 TWO-DIMENSIONAL (2-D) SUPRAMOLECULAR ORGANIZATIONS

Several 2-D purely organic [36,37] or metal–organic [25] layered structures have been obtained using multitopic ligands, such as radicals **4** [36] or **5** [25,37], or by growing a molecular layer confined on a surface [17,18,38].

10.4.1 Purely Organic Open Frameworks

As introduced in the previous section, the ditopic ligand **4** was expected to give 1-D purely organic chains, but this was not the case since the formation of hydrogen-bonded motifs in the supramolecular organization of **4** resulted in a purely organic magnetic and robust nanoporous 2-D lattice [36]. The primary structure obtained by crystallization of the dicarboxylic radical **4** in the absence of solvents susceptible to give HBs consists of a repetitive unit of a hexameric [$R_6^6(24)$] motif of hydrogen-bonded radicals through one carboxylic group. The second carboxylic group of each radical, instead, acts as a connecting element between the hexameric units by the formation of two complementary HBs between two carboxylic groups (as in the case of **1** under appropriate crystallization conditions). In this way, each hexamer is linked with six more identical units in a hexagonal topology, extending the infinite HB network along the *ab* plane and forming two-dimensional hydrogen-bonded sheets along this plane (Fig. 10-3a, left).

The tritopic ligand **5** is also responsible for the formation of 2-D hydrogen-bonded layers (Fig. 10-3a, right) [37]. Also in this case, as it happens for **4**, the repeating unit consists of an atypical hexameric [$R_6^6(24)$] hydrogen-bonded motif formed by six molecules of **5** with alternating *plus* and *minus* helicities in their three-bladed propeller-like substructures. In this motif, each radical is hydrogen bonded to two neighboring radicals through one carboxylic group. However, since every radical unit contains three carboxylic groups, each molecule of **5** participates in the construction of three identical hexameric units that propagate along the *ab* plane. Interestingly, because of the supramolecular assembly of the PTM units, weak ferromagnetic interactions develop in this network, and accordingly with its 3-D character, a bulk magnetic order is observed at low temperatures.

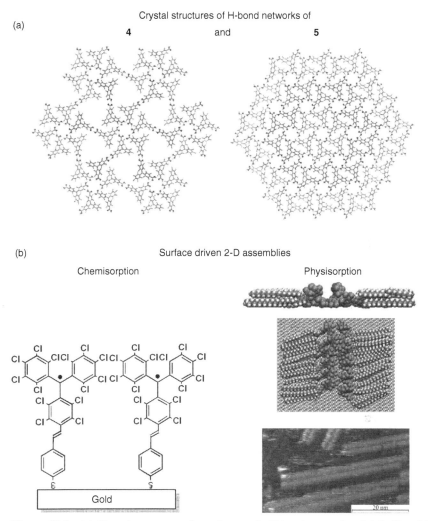

Figure 10-3 (a) Crystal structures of purely organic H-bond networks of **4** (left) and **5** (right). *Source*: Figure reprinted from Ref. [23], Copyright (2007), with permission from Elsevier. (b) Left: Example of chemisorption: PTM derivative **6** SAM on gold. Right: Example of physisorption: PTM derivative **7** on HOPG. Top and center: side and top view of the assembly of PTM derivative **7**. Bottom: STM image at the liquid–solid interface. *Source*: Figure adapted with permission from Ref. [38]. Copyright 2009 American Chemical Society.

When describing 1-D networks, we mentioned the unexpected behavior of the hexatopic derivative **8** designed to give high dimensional networks and instead forming, in the absence of solvent molecules, polymeric H-bonded chains [34]. In the presence of solvent molecules, instead, different 2-D hydrogen-bonded networks can be prepared as a result of the active role of the solvent molecules tetrahydrofuran or diethylic ether [39]) in HB formation.

10.4.2 Metal–Organic Radical Open Frameworks

The good capabilities of carboxylic PTM derivatives as coordinating synthons and magnetic relays, demonstrated with the monotopic compound **1**, were further studied with the tritopic derivative **5** [25,40] since having three coordinating moieties rather than one, it was expected to give higher dimensional assemblies. Derivative **5** was used as an organic synthon in the preparation of metal–organic frameworks with different metal ions. Particularly, when using copper ions, 2-D planes with a honeycomb structure were obtained, where each **5**$^{3-}$ molecule coordinates three copper ions and each copper ion coordinates two neighboring **5**$^{3-}$. These 2-D planes show magnetic ordering at low temperatures [25].

10.4.3 Surface-Driven 2-D Assemblies

Another example of 2-D supramolecular assemblies is given by the surface-confined molecular organizations of PTM derivatives [22], deposited on solid surfaces by means of two approaches, physisorption [38,41,42] and chemisorption [17,18,21]. In these cases, the pattern substitution of the PTM radicals will play a completely different role than when the organization is formed in the bulk solid state. Thus, for chemically bonding a molecular monolayer to the surface, it is necessary to design a derivative bearing only one functional group able to interact with the surface. The lateral ordering within the monolayer will then be determined by lateral noncovalent interactions, mainly Cl···Cl interactions in the case of the PTM radicals. Following this approach, Self-Assembled Monolayers (SAMs) of PTM derivatives have successfully been anchored on SiO$_2$-based as well as on Au substrates [17,18,21]. For this purpose, two different approaches were followed: (i) direct anchoring of the PTM radical on the surface and (ii) growth of a coupling SAM, which then interacts with the PTM derivative either by the formation of a covalent bond or via electrostatic interactions in a two-step approach. For the direct anchoring strategy [18], a conjugated PTM diradical (**6**), which incorporates a disulfide binding group to be anchored on gold, was synthesized. On the other hand, the two-step grafting procedure was achieved by first functionalizing either SiO$_2$ substrate with a SAM of *N*-[3-(trimethoxysilyl)-propyl]ethylenediamine [17] or an Au substrate with 4-aminothiophenol [21]. In both cases, the second step consisted of immersing the substrates in a solution of the chlorocarbonyl radical derivative **1**, resulting in the formation of a covalent amide bond between the PTM derivative and the first coupling monolayer. The preparation of the PTM radical SAM solely *via* interlayer electrostatic interactions was similarly accomplished, but here, the amino group of the first monolayer was protonated by rinsing the sample with 4-morpholineethanesulfonic acid monohydrate buffer (pH 5.6) and subsequently immersing the substrate in a solution of **1** [17].

All the PTM-functionalized surfaces were fully characterized, demonstrating that the optical, electroactive, and magnetic properties distinctive of PTM derivatives were preserved, resulting in multifunctional surfaces. In addition, these systems are also very attractive since they work as chemical or electrochemical redox switches with optical and magnetic responses or as molecular wires. An interesting observation revealed that the PTM radical is more conductive than the hydrogenated analog because of the large differences in their electronic structure [21].

Surface PTM organizations can also be realized by physisorption of the molecules [38,41,42]. In this process, the formation of the molecular assemblies is driven solely by weak intermolecular or molecular-surface interactions. The requirement for following this approach relies on the functionalization of the molecules with appropriate chemical groups so that they can interact with the surface. Very recently, the self-assembly of a PTM derivative bearing long alkyl chains (7) on highly oriented pyrolytic graphite (HOPG) was investigated (Fig. 10-3b, right) [38]. This derivative was designed taking into account that the conformation of the PTM moiety was propeller-like and that previous works showed that in similar bulky systems, such as C_{60}, it is necessary to have more than one long alkyl chain to ensure the adsorption of the molecules on graphite [43]. Overall, the scanning tunneling microscopic (STM) imaging of the assemblies formed at the liquid–solid interface combined with molecular modeling pointed to the fact that the alkyl-substituted PTM derivative hierarchically self-assembles as follows (Fig. 10-3b): (i) PTM molecules interact head-to-head, giving rise to dimeric structures sustained by favorable intermolecular interactions through Cl atoms and phenyl groups; (ii) the dimers stack maximizing the van der Waals interactions among adjacent alkyl chains, forming double rows that show a two-leg ladder topology; (iii) the alkyl chains drive the orientation of the assembled structures on the surfaces via CH–π interactions, maximizing the commensurability between the alkyl chains and the graphite lattice. In such PTM assemblies, the long alkyl chains have hence two main effects; first, they help to obtain well-ordered spin-containing two-leg ladder with a specific space between them and second, they act as diamagnetic barriers between neighboring radical units. Remarkably, atomic force microscopic (AFM) studies confirmed that this tendency to form double rows composed of the PTM magnetic heads surrounded by the alkyl chains is also maintained after the complete evaporation of the solvent. Undoubtedly, the fabrication of surface-confined nanoarchitectures of organic radicals represents an important step forward in the field of molecular electronics and molecular magnetism, for which important physical properties are foreseen.

In addition, preliminary results have also shown that the deposition of radical **5** by Ultra-High Vacuum (UHV) evaporation on Au(111) surfaces lead to the formation of an extended HB network, pointing out that promising PTM supramolecular assemblies can be obtained not only by processing them from solution but also by employing vacuum deposition techniques [42].

10.5 THREE-DIMENSIONAL (3-D) SUPRAMOLECULAR ORGANIZATIONS

10.5.1 The Role of Chlorine–Chlorine Short Contacts in the 3-D Packing of PTM Radicals

The presence of chlorine atoms on the triphenylmethyl skeleton favors the 3-D packing of PTM radical derivatives. Before describing the role of chlorine–chlorine interactions on the 3-D hierarchical assemblies based on PTM derivatives, whose primary structures have been already introduced, we will refer to the pioneering crystallographic studies lead by Veciana and coworkers, which, in the late 1980s, reported 3-D PTM networks obtained both in the absence [26] and in the presence [44] of solvent molecules. In the latter case, it was showed that the PTM radical **10** acts as a radical host compound [44] because of its two remarkable clathratogenic features: (i) molecular bulkiness, which limited conformational flexibility, and (ii) high molecular symmetry, showing a propeller-like conformation (D_3 symmetry) with a high enantiomerization barrier for the reversal of propeller helicity, due to the congestion of the three pairs of *ortho*-chlorine atoms. Particularly, radical **10** was crystallized in the presence of many different solvent molecules, such as benzene, fluorobenzene, chlorobenzene, bromobenzene, toluene, 1,4-dioxane, tetrahydrofurane, cyclohexane, and cyclohexene, with a host–guest ratio of 1:1 for most of them. In the case of benzene as the guest molecule, the packing scheme shows the C_6H_6 molecules located in continuous channels, which are formed by the packing of six neighbor host radical molecules through their coplanar phenyl groups (apparently a $\pi-\pi$-type interaction), with a structural arrangement different from the one observed for most families of hosts with trigonal symmetry. Interestingly, these clathrate compounds are stable towards vacuum drying at room temperature, but on heating there is a release of solvent molecules, occurring with different kinetics and at different temperatures according to the solvent included in the clathrand. In the absence of solvent molecules [26], instead, $Cl \cdots \pi$ interactions drive the 3-D close packing with no formation of channels for guest inclusion. On the other hand, in the presence of guest molecules, such as benzene, the leading structural interactions are $\pi-\pi$ and $Cl \cdots Cl$ short contacts. Solvent molecules are trapped inside the crystals and somehow mediate the separation between PTM radicals.

The role of supramolecular electrostatic interactions has also been demonstrated for the PTM-based valence tautomeric donor–acceptor (D–A) compound **9**, with a D group (ferrocene, Fc) linked by an ethylenic spacer to an A group (a perchlorotriphenylmethyl radical derivative) [45]. Interestingly, in the crystalline phase, Mössbauer experiments show the coexistence of both neutral (FcPTM) and zwitterionic (Fc^+PTM^-) forms at room conditions, while only the neutral form is found at 4 K [11,46]. This molecule offers the first experimental demonstration of bistability induced by 3-D supramolecular electrostatic interactions in crystals of valence tautomeric donor–acceptor molecules [11,47].

The 0-D, 1-D, and 2-D networks so far described and classified on the basis of the driving noncovalent interactions hierarchically assemble in 3-D as solids, also because of the leading role played by chlorine–chlorine short contacts. Therefore, in these multidimensional networks, the primary structure is based on noncovalent weak interactions such as hydrogen bonding or coordination bonds and the secondary structure on Cl···Cl short contacts.

For example, in the case of the *para*-mono carboxy PTM derivative **1**, the 0-D purely organic dimers are strengthened by chlorine–chlorine interactions, giving rise to the crystalline 3-D organization (Fig. 10-4a) [32].

In the case of radical compound **4** [36], the purely organic 2-D hydrogen-bonded layers (Section 10.4) pack together in a shifted manner through several weak Cl···Cl contacts, leading to an organic open-framework structure with 1-D channels formed by narrowed polar windows and larger hydrophobic cavities. Therefore, the trigonal symmetry and the substitution of two of the chlorine in *para* of the three phenyl rings provided a template for getting a 3-D framework with 1-D channels held together by HBs through the two carboxylic groups.

Also, in the case of the tricarboxy PTM derivative **5** [37], the hydrogen-bonded 2-D planes (Section 10.4) assemble through several Cl···Cl contacts (12 per molecule) between neighboring layers that confer rigidity to the resulting 3-D framework, whose secondary structure consists of the stacking of different layers with an ABAB alternation along the *c* axis. Surprisingly, the stacking of layers along the *c* axis generates a three-dimensional structure that has tubular nanochannels surrounded by a second set of small pores. The significant steric congestion caused by the large number of bulky chlorine atoms can be considered the main reason for obtaining this noncatenated crystal packing and a robust porous structure. The location of the carboxylic groups at the inner walls of the largest channels confers these pores a highly polar and hydrophilic environment.

(a)   (b)

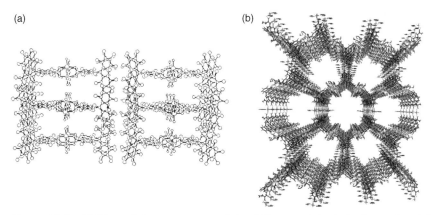

Figure 10-4   (a) View of shortest contacts giving 3-D network in the hydrogen-bonded dimers of radical **1** with copper paddle wheel. *Source*: Figure reproduced from Ref. [32b] with permission from the Royal Society of Chemistry, http://dx.doi.org/10.1039/b316458f. (b) Stacking of 2-D metal–organic honeycomb layers formed by derivative **5** and copper. *Source*: Figure reproduced from Ref. [24], Copyright (2005), with permission from Elsevier.

This arrangement may account for the lack of guest solvent molecules used for crystallization (*n*-hexane and/or CH_2Cl_2) within the nanochannels.

In the case of the metal–organic framework of the tricarboxy radical **5** and copper ions (Section 10.4) [24,25], the 2-D honeycomb layers stack together by means of weak $\pi \cdots \pi$ and van der Waals interactions, resulting in a 3-D open-framework structure with nanochannels. Interestingly, in this structure, the void volume amounts to 65% of the total unit cell volume, thus making it highly porous (Fig. 10-4b). From the magnetic point of view, the lack of compensation of the $S = 1/2$ spin of PTM and the Cu(II) units in ratio 2:3 is responsible for the antiferromagnetic interactions measured within the 2-D layers. These interactions occur only in the presence of the open-shell ligand **5** that is able to interact with the three coupled copper ions extending the magnetic interactions. Moreover, at low temperatures, interplane magnetic interactions probably originated by dipolar interactions lead to a bulk magnetic ordering in the 3-D assembly. Thus, this molecular material can be considered as a ferrimagnet with an overall three-dimensional magnetic ordering at low temperatures ($T_c \sim 2K$), which shows the behavior of a soft magnet.

10.5.2 Truly 3-D Frameworks Based on PTM Radicals

To obtain truly 3-D networks based on PTM derivatives linked by noncovalent interactions different from $Cl \cdots Cl$ short contacts, the strategy employed was the use of polytopic radical ligands, such as radicals **5** and **8** in coordinated hybrid structures. Thus, metal–organic frameworks based on PTM derivatives **5** and **8** were prepared by coordination to metal ions with low coordinances combined with additional auxiliary linkers (such as 4,4′-bipyridine(bpy) [48,49]) to increase the dimensionality or to high coordinances metal ions [34,50], such as the lanthanide ones, without the need of such auxiliary linkers.

In the case of transition metal ions with low coordinances, such as copper, 3-D coordinating supramolecular networks have been obtained by including the copper ion in a macrocyclic complex that acts as a 2-D directional bonding supramolecular entity. The supramolecular assembly of three macrocyclic dicopper complexes that clip two tricarboxylate radical linkers **5** gives paramagnetic 3-D trigonal prismatic metal–organic cages of the A_3B_2-type (where A is the copper complex and B is the tricarboxylate ligand) [48]. This was the first example of a molecular cage—obtained in a one-pot reaction—with an open-shell electronic structure, where the copper complex represents the walls of the cage and the PTM radicals the upper and lower caps.

When using Co(II) as the metal ion with radical **5** along with 4,4′- bipyridine as the auxiliary ligand, the resulting 3-D network showed relevant structural characteristics. In this 3-D network, **5** acts as a trigonal 3-connecting spacer unit, and each Co(II) unit, as a 5-connecting center since all three coordinated water molecules form HBs with two COO(H) groups of different radicals of **5**. The supramolecular arrangement of both the 3- and 5-connecting units creates (6,3) hexagonal planes in which each hexagon is defined by three PTM radicals **5** and three octahedral Co(II) centers. Furthermore, the 4,4′-bpy ligands interconnect the

Co(II) centers of neighboring layers, leading to a supramolecular 3-D structure that shows an unprecedented paramagnetic noninterpenetrated (3,5)-connected network with helical nanopores. Other geometrically interesting frameworks were obtained by using ionic salts instead of neutral molecules [33].

Another successful approach to prepare open and magnetic 3-D networks consisted in the use of a hexatopic coordinating ligand with an open-shell electronic structure as the hexa-*meta*-carboxy PTM **8** and transition metal ions with additional auxiliary ditopic ligands [49].

In the case of radical **8**, Cu(II) and 4,4'-bipyridine metal–organic supramolecular network exhibit a 3-D organization that can be ascribed to the octahedral geometry and the hexatopic nature of the PTM ligand. This supramolecular network is described as an interpenetrated and slightly distorted cubic framework where the metallic Cu(II) nodes are linked through 4,4'-bipyridine connecting entities (Fig. 10-5A). From this connectivity, the resulting network can be considered as a 3-D net constructed from two interpenetrating connected primitive cubic metal–radical frameworks [49]. This structure exhibits ferromagnetic interactions at low temperatures, which can be ascribed to the *meta*-location of the carboxylate groups on the PTM skeleton. At very low temperatures, an antiferromagnetic ordering is observed, which is directly related to the connection of the metal–radical cubic frameworks by the 4,4'-bipyridine linkers. Such antiferromagnetic couplers limit the propagation of ferromagnetic interactions and afford an overall 3-D antiferromagnetic ordering at very low temperatures.

Both the ability of lanthanide ions to provide 3-D structures and the ability of the radical **5** to transmit ferromagnetic interactions were combined to design novel porous 3-D metal–organic frameworks with magnetic ordering. Thus, PTM ligand **5** was used as an organic synthon in coordination with lanthanide metal ions (such as terbium) that show high coordination number and connectivity [50]. As a result, the first example of a 3-D open framework built using an organic radical was reported, where the open-shell tritopic ligand acts as a magnetic relay and allows the transmission of magnetic interactions. This open framework shows a complex T topology (Fig. 10-5B) with very large channels and, interestingly, has a guest-induced reversible crystal to amorphous transformation. Moreover, this 3-D open and porous metal–organic framework shows ferromagnetic metal–radical interactions at low temperatures.

10.5.3   Three-Dimensional Assemblies on Surfaces Based on PTM Radicals

Finally, we would like to present the possibility of building 3-D structures on solid surfaces, which are generally considered as a support for the 2-D confined assembly of different types of organic molecules.

One of the strategies used to achieve the third dimension onto solid surfaces is the step-by-step procedure [51] that allows the sequential deposition of inorganic and organic material onto carboxy-ended alkane thiol-passivated gold surfaces. While the lateral 2-D organization is substantially led by the van der

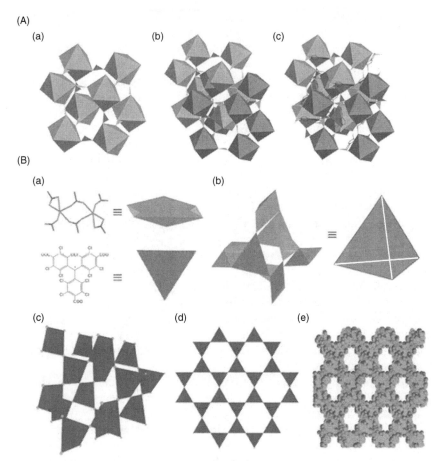

Figure 10-5   (A) Crystal structure of ligand **8**, copper, and 4,4′ bipyridine. (a) A metal–radical cube. (b) Two interpenetrated cubes. (c) Illustration of the connectivity between two interpenetrated cubes. Radicals **8** are represented as octahedrals, Cu(II) ions as tri-connected units, and 4,4′-bipyridine as black ligands. *Source*: Figure reproduced with permission from the Royal Society of Chemistry from Ref. [49] (http://dx.doi.org/10.1039/b713705b). (B) Crystal structure of [Tb(**5**)(DMF)$_3$]. (a) Representation of the structural motifs of the dinuclear Tb(III) subunits (six connecting distorted octahedral units) and the radical **5** (three connecting trigonal units). (b) Tetrahedral secondary building unit created from the linkage of Tb(III) dimers (vertices) with radicals **5** (faces). (c) View of the connectivity of framework [Tb(**5**)(DMF)$_3$], showing the typical lattice complex T network. (d) View of the framework along the *b* axis. (e) Space-filling representation showing the large channels in this 3-D structure. *Source*: Figure reproduced with permission from the Royal Society of Chemistry from Ref. [50] (http://dx.doi.org/10.1039/b802196a). (*See insert for color representation of the figure.*)

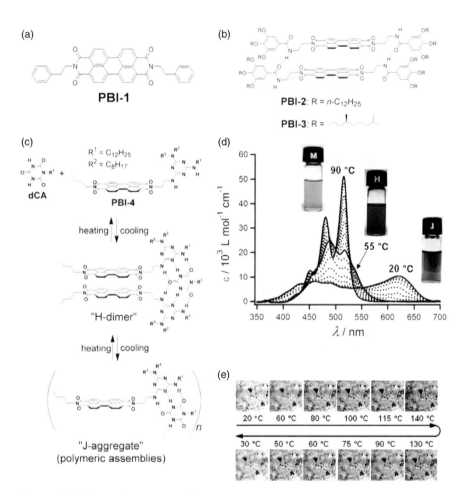

(a)

PBI-1

(b)

PBI-2: R = *n*-C₁₂H₂₅

PBI-3: R =

(c)

R¹ = C₁₂H₂₅
R² = C₈H₁₇

dCA

PBI-4

heating ⇅ cooling

"H-dimer"

heating ⇅ cooling

"J-aggregate"
(polymeric assemblies)

(d)

90 °C

55 °C

20 °C

M   H   J

ε / 10³ L mol⁻¹ cm⁻¹

λ / nm

(e)

20 °C   60 °C   80 °C   100 °C   115 °C   140 °C

30 °C   50 °C   60 °C   75 °C   90 °C   130 °C

Figure 5-1   See caption on page 80.

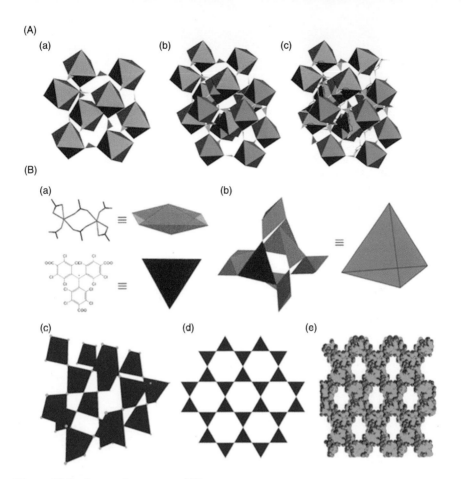

Figure 10-5 See caption on page 208.

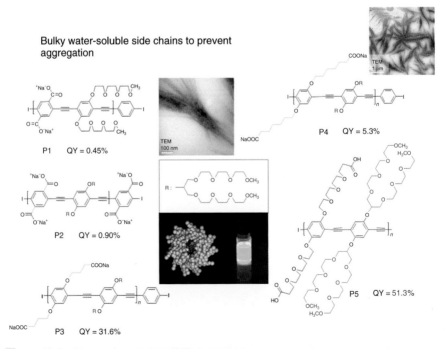

Bulky water-soluble side chains to prevent aggregation

P1   QY = 0.45%

P2   QY = 0.90%

P3   QY = 31.6%

P4   QY = 5.3%

P5   QY = 51.3%

ure 12.2   See caption on page 260

Figure 16-6 See caption on page 341.

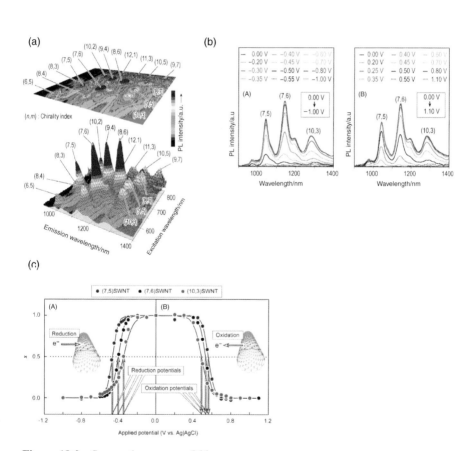

Figure 18-2 See caption on page 366.

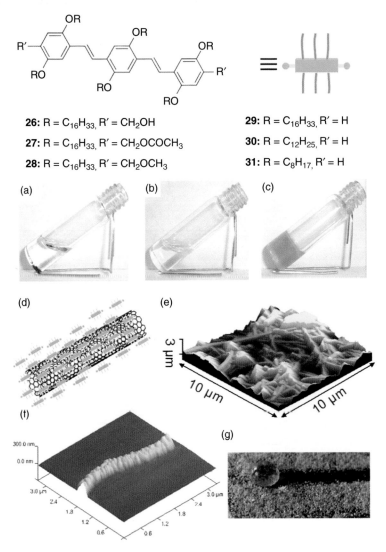

26: R = C$_{16}$H$_{33}$, R′ = CH$_2$OH

27: R = C$_{16}$H$_{33}$, R′ = CH$_2$OCOCH$_3$

28: R = C$_{16}$H$_{33}$, R′ = CH$_2$OCH$_3$

29: R = C$_{16}$H$_{33}$, R′ = H

30: R = C$_{12}$H$_{25}$, R′ = H

31: R = C$_8$H$_{17}$, R′ = H

Figure 19-2 See caption on page 390.

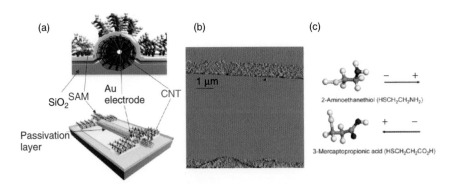

Figure 20-1 See caption on page 410.

Waals interactions between neighboring alkane thiol molecules onto which inorganic ions (such as copper) coordinate, 3-D growth is obtained by the subsequent coordination of the metal ions with a trigonal molecule, such as the radical **1** [52]. The result would be the 3-D growth of a metal–organic coordination complex from solution. In this specific case, coordination bonds are the interactions leading to the 3-D assembly, not the short chlorine–chlorine interactions. The possibility to graft metal–organic complexes onto surfaces may pave the way toward the growing of 3-D open networks based on multitopic ligands (such as PTM derivative **5**), where the chlorine–chlorine short contacts or other types of noncovalent weak interactions should further stabilize the 3-D supramolecular framework.

Three-dimensional PTM organizations on surfaces have been observed with the physisorbed PTM radicals such as radical **7** bearing the long alkyl chains, as described in the previous section. As mentioned before, in the STM images of the liquid–solid interface at low concentrations, a monolayer of oriented PTM molecules with interacting head-to-head subunits forming rows was observed. Such molecular rows were oriented in three different directions following the symmetry of graphite. However, in more concentrated solutions, the STM images surprisingly revealed that the double PTM rows lied over each other, forming a multilayer structure driven by the molecule–molecule interactions. Additionally, it is important to notice that the double rows from the top layer were still oriented with the angle of 120° with respect to the layer below, which suggests that the symmetry of the HOPG substrate influences the 3-D organization, making it possible to transmit the organizational information vertically from the substrate to the topmost molecular layers. PTMs bearing long alkyl moieties constitute very peculiar systems since they have two clearly distinct regions that interact very differently between themselves but in a cooperative manner (interactions between the PTM heads and between the aliphatic chains). This type of molecular building block highly resembles the one reported by Nakanishi and colleagues, in which C_{60} was also functionalized with long alkyl chains [43]. In that case, the fullerene derivatives formed bilayer structures that self-assembled forming complex 3-D architectures [53]. Preliminary results also demonstrate that by playing with different solvents, it is possible to obtain novel spin-containing 3-D supramolecular objects with these PTM derivatives [54].

10.6 CONCLUSIONS

In the past 20 years, PTM radical derivatives have been successfully used as paramagnetic, fluorescent, and electroactive building blocks for the preparation of multifunctional and multidimensional materials.

As it has been shown throughout this chapter, crystal engineering applied to these radicals has generally allowed the control of their structure and dimensionality, leading to the preparation of multidimensional supramolecular networks, based either on HBs or on coordination chemistry and eventually reinforced by weak chlorine–chlorine interactions.

Interestingly, the open-shell electronic configuration of the PTM radical derivatives can be considered of crucial importance for the transmission of magnetic interactions within these purely organic or hybrid molecular networks. Moreover, the rigid structure of PTM radical derivatives makes them synthons of choice for the preparation of porous frameworks showing magnetic behavior at low temperatures. Additionally, their fluorescence and electroactivity has allowed combining in one molecular network many different properties, making them ligands of choice for the preparation of multifunctional supramolecular networks.

REFERENCES

1. (a) Osiecki, J. H., Ullman, E. F. (1968). Studies of free radicals. I. α-Nitronyl nitroxides, a new class of stable radicals. *J. Am. Chem. Soc.*, *90*, 1078–1079; (b) Cirujeda, J., Ochando, L. E., Amigo, J. M., Rovira, C., Rius, J., Veciana, J. (1995). Structure determination from powder X-ray diffraction data of a hydrogen-bonded molecular solid with competing ferromagnetic and antiferromagnetic interactions: the 2-(3,4-dihydroxyphenyl)-α-nitronyl nitroxide radical. *Angew. Chem. Int. Ed.*, *34*, 55–57.
2. (a) Hicks, R. G. (2004). Adventures in stable radical chemistry. *Can. J. Chem.*, *82*, 1119–1127; (b) Hicks, R. G., Lemaire, M. T., Öhrström, L., Richardson, J. F., Thompson, L. K., Xu, Z. (2001). Strong supramolecular-based magnetic exchange in π-stacked radicals. Structure and magnetism of a hydrogen-bonded verdazyl radical: hydroquinone molecular solid. *J. Am. Chem. Soc.*, *123*, 7154–7159.
3. Armet, O., Veciana, J., Rovira, C., Riera, J., Castañer, J., Molins, E., Rius, J., Miravitlles, C., Olivella, S., Brichfeus, J. (1987). Inert carbon free radicals. 8. Polychlorotriphenylmethyl radicals. Synthesis, structure, and spin-density distribution. *J. Phys. Chem.*, *91*, 5608–5616.
4. (a) Hubbel, W. L., Gross, A., Langen, R., Lietzow, M. A. (1998). Recent advances in site-directed spin labelling of proteins. *Curr. Opin. Struct. Biol.*, *8*, 649–656; (b) Fanucci, G. E., Cafiso, D. S. (2006). Recent advances and applications of site-directed spin labeling. *Curr. Opin. Struct. Biol.*, *16*, 644–653.
5. (a) Stubbe, J., van der Donk, W. A. (1998). Protein radicals in enzyme catalysis. *Chem. Rev.*, 98, 705–762; (b) Amorati, R., Lucarini, M., Mugnaini, V., Pedulli, G. F., Minisci, F., Recupero, F., Fontana, F., Astolfi, P., Greci, L. (2003). Hydroxylamines as oxidation catalysts: thermochemical and kinetic studies. *J. Org. Chem.*, *68*, 1747–1754.
6. (a) Gatteschi, D. (1994). Molecular magnetism: a basis for new materials. *Adv. Mater.*, *6*, 635–645; (b) Train, C., Norel, L., Baumgarten, M. (2009). Organic radicals, a promising route towards original molecule-based magnetic materials. *Coord. Chem. Rev.*, *253*, 2342–2351; (c) Roques, N., Mugnaini, V., Veciana, J. (2010). Magnetic and Porous Molecule-Based Materials. *Top. Curr. Chem.*, *293*, 207–258, and references therein.
7. Sedó, J., Ventosa, N., Molins, M. A., Pons, M., Rovira, C., Veciana, J (2001). Stereoisomerism of molecular multipropellers. 2. Dynamic stereochemistry of bis- and tris-triaryl systems. *J. Org. Chem.*, *66*, 1579–1589.
8. Castellanos, S., Velasco, D., López-Calahorra, F., Brillas, E., Julia, L. (2008). Taking advantage of the radical character of tris(2,4,6-trichlorophenyl)methyl to synthesize new paramagnetic glassy molecular materials. *J. Org. Chem.*, *73*, 3759–3767.

9. Wang, J., Dang, V., Zhao, W., Lu, D., Rivera, B. K., Villamena, F. A., Wang, P. G., Kuppusamy, P. (2010). Perchlorotrityl radical-fluorophore conjugates as dual fluorescence and EPR probes for superoxide radical anion. *Bioorg. Med. Chem.*, *18*, 922–929.

10. Ratera, I., Ruiz-Molina, D., Renz, F., Ensling, J., Wurst, K., Rovira, C., Gütlich, P., Veciana, J. (2003). A new valence tautomerism example in an electroactive ferrocene substituted triphenylmethyl radical. *J. Am. Chem. Soc.*, *125*, 1462–1463.

11. D'Avino, G., Grisanti, L., Guasch, J., Ratera, I., Veciana. J., Painelli, A. (2008). Bistability in Fc-PTM crystals: the role of intermolecular electrostatic interactions. *J. Am. Chem. Soc.*, *130*, 12064–12072.

12. (a) Ratera, I., Marcen, S., Montant, S., Ruiz-Molina, D., Rovira, C., Veciana, J., Létard, J. F., Freysz, E. (2002). Nonlinear optical properties of polychlorotriphenylmethyl radicals: towards the design of 'super-octupolar' molecules. *Chem. Phys. Lett.*, *363*, 245–251; (b) Heckmann, A., Lambert, C., Goebel M., Wortmann, R. (2004). Synthesis and photophysics of a neutral organic mixed-valence compound. *Angew. Chem. Int. Ed.*, *43*, 5851–5856.

13. (a) Bonvoisin, J., Launay, J.-P., Rovira, C., Veciana, J. (1994). Purely organic mixed-valence molecules with nanometric dimensions showing long-range electron transfer. Synthesis, and optical and EPR studies of a radical anion derived from a bis (triarylmethyl)diradical. *Angew. Chem. Int. Ed.*, *33*, 2106–2109; (b) Sedó, J., Ruiz, D., Vidal-Gancedo, J., Rovira, C., Bonvoisin, J., Launay, J.-P., Veciana, J. (1996). Intramolecular electron transfer phenomena in purely organic mixed-valence high-spin ions: a triplet anion case. *Adv. Mater.*, *8*, 748–752.

14. (a) Lloveras, V., Vidal-Gancedo, J., Ruiz-Molina, D., Figueira-Duarte, T. M., Nierengarten, J.-F., Veciana, J., Rovira C. (2006). Influence of bridge topology and torsion on the intramolecular electron transfer. *Faraday Discuss.*, *131*, 291–305. (b) Lloveras, V., Gancedo-Vidal, J., Figueira-Duarte, T. M., Nierengarten, J-F., Novoa, J. J., Mota, F., Ventosa, N., Rovira, C., Veciana, J. (2011). Tunneling versus hopping in mixed-valence oligo-*p*-phenylenevinylene polychlorinated bis(triphenylmethyl) radical anions. *J. Am. Chem. Soc.*, *133*, 5818–5833.

15. Chopin, S., Cousseau, J., Levillain, E., Rovira, C., Veciana, J., Sandanayaka, A. S. D., Araki, Y., Ito, O. (2006). [60]Fullerene-perchlorotriphenylmethide anion triads. Synthesis and study of photoinduced intramolecular electron transfer processes. *J. Mater. Chem.*, *16*, 112–121.

16. Sporer, C., Ratera, I., Ruiz-Molina, D., Zhao, Y., Vidal-Gancedo, J., Wurst, K., Jaitner, P., Clays, K., Persoons, A., Rovira, C., Veciana, J. (2004). A molecular multiproperty switching array based on the redox behaviour of a ferrocenyl polychlorotriphenylmethyl radical. *Angew. Chem. Int. Ed.*, *43*, 5266–5268.

17. Crivillers, N., Mas-Torrent, M., Perruchas, S., Roques, N., Vidal-Gancedo, J., Veciana, J., Rovira, C., Basabe-Desmonts, L., Ravoo, B. J., Crego-Calama, M., Reinhoudt, D. N. (2007). Self-assembled monolayers of a multifunctional organic radical. *Angew. Chem. Int. Ed.*, *46*, 2215–2219.

18. (a) Crivillers, N., Mas-Torrent, M., Vidal-Gancedo, J., Veciana, J., Rovira, C. (2008). Self-assembled monolayers of electroactive polychlorotriphenylmethyl radicals on Au(111). *J. Am. Chem. Soc.*, *130*, 5499–5506. (b) Simao, C., Mas-Torrent, M., Crivillers, N., Lloveras, V., Artes, J. M., Gorostiza, P., Veciana, J., Rovira, C. (2011). A robust molecular platform for non-volatile memory devices with optical and magnetic responses. *Nature Chem.*, *3*, 359–364.

19. (a) Veciana, J., Rovira, C., Crespo, M. I., Armet, O., Domingo, V. M., Palacio, F. (1991). Stable polyradicals with high-spin ground states. 1. Synthesis, separation, and magnetic characterization of the stereoisomers of 2,4,5,6-tetrachloro $\alpha,\alpha,\alpha',\alpha'$-tetrakis(pentachlorophenyl)-m-xylylene biradical. *J. Am. Chem. Soc.*, *113*, 2552–2561; (b) Veciana, J., Rovira, C., Ventosa, N., Crespo, M. I., Palacio, F. (1993). Stable polyradicals with high-spin ground states. 2. Synthesis and characterization of a complete series of polyradicals derived from 2,4,6-trichloro, $\alpha,\alpha,\alpha',\alpha',\alpha'',\alpha''$ hexakis (pentachlorophenyl)mesitylene with $S = 1/2$, 1, and 3/2 ground states. *J. Am. Chem. Soc.*, *115*, 57–64.

20. Ruiz-Molina, D., Veciana, J., Palacio, F., Rovira, C. (1997). Drawbacks arising from the high steric congestion in the synthesis of new dendritic polyalkylaromatic polyradicals. *J. Org. Chem.*, *62*, 9009–9017.

21. (a) Crivillers, N., Munuera, C., Mas-Torrent, M., Simao, C., Bromley, S. T., Ocal, C., Veciana, J., Rovira, C. (2009). Dramatic influence of the electronic structure on the conductivity through

open- and closed-shell molecules. *Adv. Mater.*, *21*, 1177–1181. (b) Crivillers, N., Paradinas, M., Mas-Torrent, M., Bromley, S. T., Rovira, C., Ocal, C., Veciana, J. (2011). Negative differential resistance (NDR) in similar molecules with distinct redox behaviour. *Chem. Commun.*, *47*, 4664–4666.

22. Mas-Torrent, M., Crivillers, N., Mugnaini, V., Ratera, I., Rovira, C., Veciana, J. (2009). Organic radicals on surfaces: towards molecular spintronics. *J. Mater. Chem.*, *19*, 1691–1695.

23. Roques, N., Maspoch, D., Datcu, A., Wurst, K., Ruiz-Molina, D., Rovira, C., Veciana, J. (2007). Self-assembly of carboxylic substituted PTM radicals: from weak ferromagnetic interactions to robust porous magnets. *Polyhedron*, *26*, 1934–1948.

24. Maspoch, D., Domingo, N., Ruiz-Molina, D., Wurst, K., Tejada, J., Rovira, C., Veciana, J. (2005). Carboxylic-substituted polychlorotriphenylmethyl radicals, new organic building-blocks to design nanoporous magnetic molecular materials. *C. R. Chim.*, *8*, 1213–1225.

25. Maspoch, D., Ruiz-Molina, D., Wurst, K., Domingo, N., Cavallini, M., Biscarini, F., Tejada, J., Rovira, C., Veciana, J. (2003). A nanoporous molecular magnet with reversible solvent-induced mechanical and magnetic properties. *Nat. Mater.*, *2*, 190–195.

26. Rius, J., Miravitlles, C., Molins, E., Crespo, M., Veciana, J. (1990). Crystal structure of the organic free radical perchlotriphenylmethyl from powder X-ray diffraction data. Comparison with its clathrate structures. *Mol. Cryst. Liq. Cryst.*, *187*, 155–163.

27. Maspoch, D., Catala, L., Gerbier, P., Ruiz-Molina, D., Vidal-Gancedo, J., Wurst, K., Rovira, C., Veciana, J. (2002). Radical para-benzoic acid derivatives: transmission of ferromagnetic interactions through hydrogen bonds at long distances. *Chem.—Eur. J.*, *8*, 3635–3645.

28. Ribas, X., Maspoch, D., Wurst, K., Veciana, J., Rovira, C. (2006). Coordination capabilities of a novel organic polychlorotriphenylmethyl monosulfonate radical. *Inorg. Chem.*, *45*, 5383–5392.

29. Ratera, I., Ruiz-Molina, D., Vidal-Gancedo, J., Wurst, K., Daro, N., Letard, J.-F., Rovira, C., Veciana, J. (2001). A new photomagnetic molecular system based on photoinduced self-assembly of radicals. *Angew. Chem. Int. Ed.*, *40*, 919–922.

30. Ratera, I., Ruiz-Molina, D., Vidal-Gancedo, J., Rovira, C., Veciana, J. (2001). EPR study of the *trans* and *cis* isomers of a ferrocenyl Schiff-based polychlorotriphenylmethyl radical. *Polyhedron*, *20*, 1643–1646.

31. Ratera, I., Ruiz-Molina, D., Vidal-Gancedo, J. Novoa, J. J., Wurst, K., Letard, J.-F., Rovira, C., Veciana, J. (2004). Supramolecular photomagnetic materials: photoinduced dimerization of ferrocene-based polychlorotriphenylmethyl radicals. *Chem.—Eur. J.*, *10*, 603–616.

32. (a) Maspoch, D., Ruiz-Molina, D., Wurst, K., Rovira, C., Veciana, J. (2002). A very bulky carboxylic perchlorotriphenylmethyl radical as a novel ligand for transition metal complexes. A new spin frustrated metal system. *Chem. Commun.*, 2958–2959; (b) Maspoch, D., Ruiz-Molina, D., Wurst, K., Vidal-Gancedo, J., Rovira, C., Veciana, J. (2004). Synthesis, structural and magnetic properties of a series of copper (II) complexes containing a monocarboxylated perchlorotriphenylmethyl radical as a coordinating open-shell ligand. *Dalton Trans.*, 1073–1082.

33. Maspoch, D., Domingo, N., Ruiz-Molina, D., Wurst, K., Hernández, J. M., Lloret, F., Tejada, J., Rovira, C., Veciana, J. (2007). First-row transition-metal complexes based on a carboxylate polychlorotriphenylmethyl radical: trends in metal-radical exchange interactions. *Inorg. Chem.*, *46*, 1627–1633.

34. Roques, N., Perruchas, S., Maspoch, D., Datcu, A., Wurst, K., Sutter, J.-P., Rovira, C., Veciana, J. (2007). Europium (III) complexes derived from carboxylic-substituted polychlorotriphenylmethyl radicals. *Inorg. Chim. Acta*, *360*, 3861–3869.

35. Roques, N., Domingo, N., Maspoch, D., Wurst, K., Rovira, C., Tejada, J., Ruiz-Molina, D., Veciana, J. (2010). Metal-radical chains based on polychlorotriphenylmethyl radicals: synthesis, structure and magnetic properties. *Inorg. Chem.*, *49*, 3482–3488.

36. Maspoch, D., Domingo, N., Ruiz-Molina, D., Wurst, K., Tejada, J., Rovira, C., Veciana, J. (2004). A robust nanocontainer based on a pure organic free radical. *J. Am. Chem. Soc.*, *126*, 730–731.

37. Maspoch, D., Domingo, N., Ruiz-Molina, D., Wurst, K., Vaughan, G., Tejada, J., Rovira, C., Veciana, J. (2004). A robust purely organic nanoporous magnet. *Angew. Chem. Int. Ed.*, *43*, 1828–1832.

38. Crivillers, N., Furukawa, S., Minoia, A., ver Heyen, A., Mas-Torrent, M., Sporer, C., Linares, M., Volodin, A., Van Haesendonck, C., van der Auweraer, M., Lazzaroni, R., De Feyter, S., Veciana, J.,

Rovira, C. (2009). Two-leg molecular ladders formed by hierarchical self-assembly of an organic radical. *J. Am. Chem. Soc.*, *131*, 6246–6252.

39. Roques, N., Maspoch, D., Domingo, N., Ruiz-Molina, D., Wurst, K., Tejada, J., Rovira, C., Veciana, J. (2005). Hydrogen-bonded self-assemblies in a polychlorotriphenylmethyl radical derivative substituted with six meta-carboxylic acid groups. *Chem. Commun.*, 4801–4803.

40. (a) Maspoch, D., Ruiz-Molina, D., Wurst, K., Rovira, C., Veciana, J. (2004). A new $(6^3) \cdot (6^9 8^1)$ non-interpenetrated paramagnetic network with helical nanochannels based on a tricarboxylic perchlorotriphenylmethyl radical. *Chem. Commun.*, 1164–1165; (b) Maspoch, D., Ruiz-Molina, D., Wurst, K., Vaughan, G., Domingo, N., Tejada, J., Rovira, C., Veciana, J. (2004). Open-shell channel-like salts formed by the supramolecular assembly of a tricarboxylated perchlorotriphenylmethyl radical and a $[Co(bpy)_3]^{2+}$ cation. *CrystEngComm*, *6*, 573–578.

41. Mugnaini, V., Fabrizioli, M., Ratera, I., Mannini, M., Caneschi, A., Gatteschi, D., Manassen, Y., Veciana, J. (2009). Towards the detection of single polychlorotriphenylmethyl radical derivatives by means of Electron Spin Noise STM. *Solid State Sci.*, *11*, 956–960.

42. Grillo, F., Mugnaini, V., Oliveros, M., Francis, S. M., Choi, D. J., Rastei, M. V., Limot, L., Cepek, C., Pedio, M., Bromley, S. T., Richardson, N. V., Bucher, J. P., Veciana, J. *submitted*.

43. Nakanishi, T., Miyashita, N., Michinobu, T., Wakayama, Y., Tsuruoka, T., Ariga, K., Kurth, D. G. (2006). Perfectly straight nanowires of fullerenes bearing long alkyl chains on graphite. *J. Am. Chem. Soc.*, *128*, 6328–6329.

44. Veciana, J., Carilla, J., Miravitlles, C., Molins, E. (1987). Free radicals as clathrate hosts: crystal and molecular structure of 1:1 perchlorotriphenylmethyl radical-benzene. *J. Chem. Soc., Chem. Commun.*, 812–814.

45. Ratera, I., Sporer, C., Ruiz-Molina, D., Ventosa, N., Baggerman, J., Brouwer, A. M., Rovira, C., Veciana, J. (2007). Solvent tuning from normal to inverted Marcus region of intramolecular electron transfer in ferrocene-based organic radicals. *J. Am. Chem. Soc.*, *129*, 6117–6129.

46. Grisanti, L., D'Avino, G., Painelli, A., Guasch, J., Ratera, I., Veciana, J. (2009). Essential state models for solvatochromism in donor-acceptor molecules: the role of the bridge. *J. Phys. Chem. B*, *113*, 4718–4725.

47. D'Avino, G., Grisanti, L., Painelli, A., Guasch, J., Ratera, I., Veciana, J. (2009). Cooperativity from electrostatic interactions: understanding bistability in molecular crystals. *CrystEngComm*, *11*, 2040–2047.

48. Company, A., Roques, N., Guell, M., Mugnaini, V., Gomez, L., Imaz, I., Datcu, A., Solà, M., Luis, J. M., Veciana, J., Ribas, X., Costas, M. (2008). Nanosized trigonal prismatic and antiprismatic Cu^{II} coordination cages based on tricarboxylate linkers. *Dalton Trans.*, 1679–1682.

49. Roques, N., Maspoch, D., Luis, F., Camòn, A., Wurst, K., Datcu, A., Rovira, C., Ruiz-Molina, D., Veciana J. (2008). A hexacarboxylic open-shell building block: synthesis, structure and magnetism of a three-dimensional metal–radical framework. *J. Mater. Chem.*, *18*, 98–108.

50. Roques, N., Maspoch, D., Imaz, I., Datcu, A., Sutter, J.-P., Rovira, C., Veciana, J. (2008). A three-dimensional lanthanide-organic radical open-framework. *Chem. Commun.*, 3160–3162.

51. Shekhah, O., Wang, H., Kowarik, S., Schreiber, F., Paulus, M., Tolan, M., Sternemann, C., Evers, F., Zacher, D., Fischer, R. A., Wöll, C. (2007). Step-by-step route for the synthesis of metal-organic frameworks. *J. Am. Chem. Soc.*, *129*, 15118–15119.

52. Shekhah, O., Roques, N., Mugnaini, V., Munuera, C., Ocal, C., Veciana, J., Wöll, C. (2008). Grafting of monocarboxylic substituted polychlorotriphenylmethyl radicals onto a COOH-functionalized self-assembled monolayer through copper (II) metal ions. *Langmuir*, *24*, 6640–6648.

53. (a) Nakanishi, T., Schmitt, W., Michinobu, T., Kurth, D. G., Ariga, K. (2005). Hierarchical supramolecular fullerene architectures with controlled dimensionality. *Chem. Commun.*, 5982–5984; (b) Nakanishi, T., Ariga, K., Michinobu, T., Yoshida, K., Takahashi, H., Teranishi, T., Möhwald, H., Kurth, D. G. (2007). Flower-shaped supramolecular assemblies: hierarchical organization of a fullerene bearing long aliphatic chains. *Small*, *3*, 2019–2023.

54. Vera, F. *et al.*, in preparation.

CHAPTER 11

PHOTOSWITCHING PROPERTY OF DIARYLETHENES IN MOLECULAR MAGNETISM AND ELECTRONICS

Kenji Matsuda and Kenji Higashiguchi

Department of Synthetic Chemistry and Biological Chemistry,
Kyoto University, Kyoto, Japan

11.1 INTRODUCTION

"Molecular electronics" is a rapidly expanding field in nanoscience and nanotechnology. Molecular electronics can be defined narrowly as the study of electrical and electronic processes by accessing the individual molecules with electrodes and exploiting the molecular structure to control the flow of electrical signals through them [1].

Organic molecules have an uncountable variety of geometrical and electronic structures and also have a wide variety of functions. In particular, π-conjugated compounds show interesting functions owing to the delocalized π-electrons. Conductive polymers, organic light-emitting diodes, and photochromic compounds are good examples. Therefore, organic compounds are expected to play an important role in the field of molecular electronics.

Control of molecular conductance by controlling molecular structure is a key issue in molecular electronics [2,3]. The relationship between the structure and the conductance is being unveiled. Meanwhile, photochromic compounds change their geometrical and electronic structures along with their color change. The change in the structures induces a change in the interaction between the functional units that are located at each end of the photochromic molecules.

In this chapter, photoswitching of the magnetic exchange interaction between two unpaired electrons that are located at each end of the π-conjugated

Supramolecular Soft Matter: Applications in Materials and Organic Electronics, First Edition.
Edited by Takashi Nakanishi.
© 2011 John Wiley & Sons, Inc. Published 2011 by John Wiley & Sons, Inc.

photochromic diarylethene is overviewed first. Second, photoswitching of the conductance of the network prepared from the photochromic diarylethene molecules and noble metal nanoparticles is described. The relationship between structural change and switching effect is also discussed.

11.2 MOLECULAR MAGNETISM

Magnetism generally originates from the electron spin of an unpaired electron in transition metal. When there is no interaction between individual spins, magnetism is paramagnetic. The exchange interaction makes the electron spins align parallel or antiparallel. When the exchange interaction is ferromagnetic ($J > 0$), the alignment is parallel, and when the exchange interaction is antiferromagnetic ($J < 0$), the alignment is antiparallel. When such an alignment is achieved throughout the material, the electron spins have a long-range order and exhibit a cooperative phenomenon, such as ferromagnetism or ferrimagnetism.

Not only the electron spins of the transition metals but also the spins of the organic radicals can participate in the construction of magnetic materials. "Molecular Magnetism" is the field of science that deals with magnetic materials that are designed and synthesized from several kinds of metallic ions, organic radicals, and bridging units [4,5]. In this field, one of the ultimate goals was to make ferromagnets only from organic components. In 1991, Kinoshita et al. reported the first organic ferromagnet 2-(p-nitrophenyl)nitronyl nitroxide (4,4,5,5-tetramethyl-2-(4-nitrophenyl)imidazoline-1-oxyl-3-oxide) **1** [6]. However, this compound utilizes an intermolecular interaction that depends on the crystal structure. The magnetic property is controlled by the crystal structure, which is difficult to predict.

The exchange interaction can operate through chemical bonds within the molecule even more effectively. Dicarbene **2** is a classic example of this intramolecular interaction; it has a quintet ground state due to the ferromagnetic exchange interaction between two carbene centers [7,8]. The intramolecular interaction through the chemical bond depends on the molecular structure; therefore, the interaction can be regulated by molecular design. Therefore, by combining the organic radical and the functional organic molecule, highly integrated multifunctional magnetic materials will be constructed.

1   2

11.3   INTRAMOLECULAR MAGNETIC INTERACTION

When two unpaired electrons are placed in proximity, the exchange interaction operates between the two electrons. The exchange interaction results in the separation of the energy of the singlet state and the triplet state. In the singlet state, the two spins align antiparallel, while the two spins align parallel in the triplet state. This means that the exchange interaction itself is the magnetic interaction. From the Heisenberg Hamiltonian equation (Eq. 11.1), the singlet–triplet energy gap ΔE_{S-T} can be expressed by the exchange interaction J as in Eq. 11.2. When J is positive, the ground state is triplet; when J is negative, the ground state is singlet.

$$\hat{H} = -2JS_1 \cdot S_2 \tag{11.1}$$

$$\Delta E_{S-T} = 2J \tag{11.2}$$

When an unpaired electron is placed at each end of a π-conjugated system, the two spins of the unpaired electrons interact magnetically through the π-system effectively. Therefore, the π-conjugated system can be regarded as a "spin coupler" [9]. The magnitude of the interaction becomes weaker exponentially with an increase in the spacer length.

   The classification of biradicals should be noted here. If a biradical has no resonant closed-shell structure, the biradical is classified as a nonKekulé biradical. The nonKekulé biradical can be further classified into disjoint and nondisjoint biradicals [10]. Trimethylenemethane (TMM) **3** is a typical example of a nondisjoint biradical. The two singly occupied molecular orbitals (SOMOs) of **3** overlap in space. On the other hand, tetramethyleneethane (TME) **4** is a disjoint biradical. The two SOMOs of **4** do not overlap in space. If a biradical has a resonant closed-shell structure, the closed-shell structure is more stable and the molecule therefore exists as a normal closed-shell Kekulé molecule. 1,3-Butadiene **5′** is an example of a normal Kekulé molecule; it has a resonant biradical structure, 2-butene-1,4-diyl **5**. In this case, the closed-shell structure **5′** is the ground state, so the ground electronic state has no unpaired electrons. In other words, the magnetic interaction between the two spins in structure **5** is strongly antiferromagnetic.

11.4 PHOTOCHROMIC SPIN COUPLER

Photochromism is a reversible phototransformation of a chemical species between two forms having different absorption spectra [11]. Photochromic compounds reversibly change not only the absorption spectra but also their geometrical and electronic structures. The geometrical and electronic structural changes induce some changes in physical properties, such as fluorescence, refractive index, polarizability, and electric conductivity. When the photochromic compounds are used as "spin couplers," the magnetic interaction can be controlled by photoirradiation. Not only our group but also others [12–16] have recently been interested in the photocontrol of magnetism of organic radical-based magnetic materials. Diarylethenes with heterocyclic aryl groups are well known as thermally irreversible, highly sensitive, and fatigue-resistant photochromic compounds [17]. The photochromic reaction is based on a reversible transformation between the open-ring isomer with hexatriene structure and the closed-ring isomer with cyclohexadiene structure according to the Woodward–Hoffmann rule. While the open-ring isomer **6a** is colorless in most cases, the closed-ring isomer **6b** is yellow, red, or blue, depending on the molecular structure. The difference in color is due to the differences in the geometrical and electronic structures. In the open-ring isomer, free rotation is possible between the ethene moiety and the aryl group. Therefore, the open-ring isomer is nonplanar, and the π-electrons are localized in the two aryl groups. On the other hand, the closed-ring isomer has a planar polyene structure and the π-electrons are delocalized throughout the molecule. These geometrical and electronic structural differences resulted in some differences in the physical properties. For example, the closed-ring isomer has a high polarizability because the closed-ring isomer has more delocalized π-electrons [18]. Not only the change in polarizability but also the switching in fluorescence [19] and electronic conduction [20] has been developed.

6a   6b

11.5 DIARYLETHENE AS A PHOTOSWITCH

There is a characteristic feature in the electronic structural changes of diarylethenes. The following equation shows the open-ring isomer **7a** and closed-ring isomer **7b** of the radical-substituted diarylethenes with simplified structures. While there is no resonant closed-shell structure for **7a**, **7b′** exists as the resonant quinoid-type, closed-shell structure for **7b**. **7a** is a nonKekulé biradical, and **7b** is a normal Kekulé molecule. In other words, **7a** has two unpaired electrons, while **7b** has no unpaired electrons. The calculated shapes of two SOMOs of **7a** are separated in the molecule and there is no overlap. This configuration is a typical disjoint biradical, in which the intramolecular radical–radical interaction is weak. In the open-ring isomer, the bond alternation is discontinued at the 3-position of the thiophene rings. This is the origin of the disjoint nature of the electronic configuration of **7a**. **7a** is a disjoint nonKekulé biradial and corresponds to TME **4** in the former example. However, the closed-ring isomer **7b′** is a normal Kekulé molecule. In this case, the ground electronic state has no unpaired electrons. In this singlet ground state, the magnetic interaction is strongly antiferromagnetic. **7b′** corresponds to butadiene **5′** in the former example.

The electronic structural change of radical-substituted diarylethenes accompanying the photoisomerization is the transformation of a disjoint nonKekulé structure to a closed-shell Kekulé structure. One may infer from the above consideration that the interaction between spins in the open-ring isomer of diarylethene is weak, while significant antiferromagnetic interaction takes place in the closed-ring isomer. In other words, the open-ring isomer is the "OFF" state and the closed-ring isomer is the "ON" state.

11.6 PHOTOSWITCHING OF MAGNETIC INTERACTION

Photochromic biradical **8a** was designed and synthesized by choosing 1,2-bis(2-methyl-1-benzothiophen-3-yl)perfluorocyclopentene as a photochromic spin coupler and nitronyl nitroxides as a spin source [21,22] (Fig. 11-1a).

In solution, **8a** showed ideal photochromic behavior by irradiation with UV and visible light. Although the radical moiety absorbed in the region from 550 to 700 nm, this did not prevent the photochromic reaction. Almost 100% photochemical conversions were observed in both the cyclization from the open-ring isomer **8a** to the closed-ring isomer **8b** and the cycloreversion from **8b** to **8a**. For practical use of photochromic devices, high conversion is one of the most important characteristics.

Magnetic susceptibilities of **8a** and **8b** were measured on a superconducting quantum interference device (SQUID) susceptometer in microcrystalline form. $\chi T - T$ plots are shown in Fig. 11-1b. The data were analyzed in terms of a modified singlet–triplet two-spin model (the Bleaney–Bowers type), in which

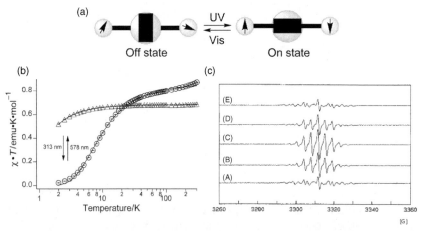

Figure 11-1 (a) Schematic representation of photoswitching of intramolecular magnetic interaction. (b) Temperature dependence of the magnetic susceptibility of **8a** (△) and **8b** (O) ($\chi T - T$ plot). (c) ESR spectral change of **9a** ($n = 1$) along with photochromism (benzene solution, 1.1×10^{-4} M): (A) initial; (B) irradiation with 366-nm light for 1 min; (C) 4 min; (D) irradiation with > 520-nm light for 20 min; and (E) 50 min. *Source:* Adopted from Refs [21] and [23] © American Chemical Society.

two spins ($S = 1/2$) couple antiferromagnetically within a biradical molecule by exchange interaction J [24]. The best-fit parameters obtained by means of a least squares method were $2J/k_B = -2.2 \pm 0.04$ K for **8a** and $2J/k_B = -11.6 \pm 0.4$ K for **8b**. Although the interaction between the two spins in the open-ring isomer **8a** was weak, the spins of **8b** had remarkable antiferromagnetic interaction.

The open-ring isomer **8a** had a twisted molecular structure and a disjoint electronic configuration. On the other hand, the closed-ring isomer **8b** had a planar molecular structure and a nondisjoint electronic configuration. The photoinduced change in magnetism agreed well with the prediction that the open-ring isomer has an "OFF" state and the closed-ring isomer has an "ON" state.

Although the switching of exchange interaction was detected by the susceptibility measurement of biradical **8**, both open- and closed-ring isomers **8a** and **8b** had 9-line electron spin resonance (ESR) spectra because the exchange interaction between the two radicals was much stronger than the hyperfine coupling constant in both isomers. To detect the change in the exchange interaction by ESR spectroscopy, the value of the interaction should be comparable to the hyperfine coupling constant. Therefore, biradicals **9** ($n = 1, 2$), in which p-phenylene spacers are introduced to control the strength of the exchange interaction, were designed and synthesized [23,25]. As described earlier, when two nitronyl nitroxides are magnetically coupled via an exchange interaction, the biradical gives a 9-line ESR spectrum. If the exchange interaction is weaker than the hyperfine coupling, the two nitroxide radicals are magnetically independent and give a 5-line spectrum. In intermediate situations, the spectrum becomes complex [26,27]. Diarylethenes **9a** ($n = 1, 2$) also underwent reversible photochromic reactions by alternative irradiation with UV and visible light. The changes in the ESR spectra accompanying the photochromic reaction were examined for diarylethenes **9a** ($n = 1, 2$). Figure 11-1c shows the ESR spectra at different stages of the photochromic reaction of **9a** ($n = 1$). The ESR spectrum of **9a** ($n = 1$) showed a complex 15 line spectrum. This suggests that the two spins of nitronyl nitroxide radicals were coupled by an exchange interaction that is comparable to the hyperfine coupling constant. Upon irradiation with 366-nm light, the spectrum converted completely to a 9-line spectrum, corresponding to the closed-ring isomer **9b** ($n = 1$). The 9-line spectrum indicates that the exchange interaction between the two spins in **9b** ($n = 1$) is much stronger than the hyperfine coupling constant. The ESR spectral change was also observed for **9a** ($n = 2$). The open-ring isomer **9a** ($n = 2$) had a 5-line spectrum, while the closed-ring isomer **9b** ($n = 2$) had a distorted 9-line spectrum. A simulation of the ESR spectra was performed to estimate the value of the exchange interaction. The exchange interaction decreases with increase in the π-conjugated chain length, as shown in Table 11-1. The exchange interaction change in **9a** ($n = 1, 2$) upon photoirradiation was more than 30-fold. This result shows a very large switching effect of diarylethenes and suggests the superiority of diarylethenes as molecular switching units. Although the absolute value of the exchange interaction is small, the information of the spins can be clearly transmitted through the closed-ring isomer, and the switching can be detected by ESR spectroscopy.

TABLE 11-1   Magnetic Interaction between Two Nitronyl Nitroxides Connected by Diarylethenes

	Open-Ring Isomer		Closed-Ring Isomer	
	ESR Line Shape	$|2J/k_B|/K$	ESR Line Shape	$|2J/k_B|/K$
8	9 lines	2.2	9 lines	11.6
9 $(n = 1)$	15 lines	$\begin{cases} 1.2 \times 10^{-3} \\ < 3 \times 10^{-4} \end{cases}$	9 lines	> 0.04
9 $(n = 2)$	5 lines	$< 3 \times 10^{-4}$	Distorted 9 lines	0.010
10 $(n = 1)$	13 lines	$\begin{cases} 5.6 \times 10^{-3} \\ < 3 \times 10^{-4} \end{cases}$	9 lines	> 0.04
10 $(n = 2)$	5 lines	$< 3 \times 10^{-4}$	9 lines	> 0.04

Oligothiophenes are good candidates for conductive molecular wires. The thiophene-2,5-diyl moiety has been used as a molecular wire unit for energy and electron transfer and can serve as a stronger magnetic coupler than p-phenylene [28,29]. Therefore, diarylethenes **10a** $(n = 1, 2)$ having one nitronyl nitroxide radical at each end of a molecule containing oligothiophene spacers was synthesized and their photo- and magnetochemical properties were studied.

Photochromic reactions and an ESR spectral change were also observed for **10a** $(n = 1, 2)$. Table 11-1 lists the exchange interaction between the two diarylethene-bridged nitronyl nitroxide radicals. For all five biradicals, the closed-ring isomers have stronger interactions than the open-ring isomers. The exchange interactions through oligothiophene spacers were stronger than the corresponding biradicals with oligophenylene spacers. The efficient π-conjugation in thiophene spacers resulted in strong exchange interactions between the two nitronyl nitroxide radicals. In the case of bithiophene spacers, the exchange interaction difference between open- and closed-ring isomers was estimated to be more than 150-fold.

11.7 REVERSED PHOTOSWITCHING USING BIS(2-THIENYL)ETHENE

Photoswitching using bis(2-thienyl)ethene **11**, in which the 2-thienyl group is connected to the perfluorocyclopentene ring, was performed [30]. In the case of regular bis(3-thienyl)ethenes, the bond alternation is discontinued in the open-ring isomer, but in the closed-ring isomer, the π-electron is delocalized throughout the molecule. Therefore, the open-ring isomer is in the "OFF" state because of the disconnection of the π-system and the closed-ring isomer is in the "ON" state because of the delocalization of the π-conjugated system. The situation is reversed when thiophene rings are substituted to the ethene moiety at the 2-position. The bond alternation is continued throughout the molecule in the open-ring isomer, while in the closed-ring isomer two aryl rings are separated by the sp^3 carbon and sulfur atoms. The magnetic interaction between the two unpaired electrons at 5-positions in the closed-ring isomer is expected to become much weaker than that in the open-ring isomer. In other words, the open-ring isomer is in the "ON" state and the closed-ring isomer is in the "OFF" state.

The synthesized bis(2-thienyl)ethene **11a** did not undergo photochromism, but the precursor underwent photochromic reaction. The closed-ring isomer of the precursor was obtained and the closed-ring isomer was transformed to the desired molecule. The obtained closed-ring isomer **11b** was converted to the open-ring isomer **11a** by irradiation with visible light. The ESR spectrum of the closed-ring isomer **11b** was a 5-line spectrum, but the spectrum of the open-ring isomer **11a** was a distorted 9-line spectrum. This suggests that there are stronger magnetic interactions in the open-ring isomer than those in the closed-ring isomer. The switching direction is reversed by using bis(2-thienyl)ethene.

11a   Vis   11b

11.8 PHOTOSWITCHING USING ARRAY OF PHOTOCHROMIC MOLECULES

In Section 11.7, it was demonstrated that the exchange interaction between two nitronyl nitroxide radicals located at either end of a diarylethene was photoswitched reversibly by alternate irradiation with ultraviolet and visible light. The difference in the exchange interaction between the two switching states was more than 150-fold. ESR spectra can be used as a good tool for detecting small magnetic interaction changes in the molecular systems. In this section, photoswitching of intramolecular magnetic interaction using a diarylethene dimer is presented [31].

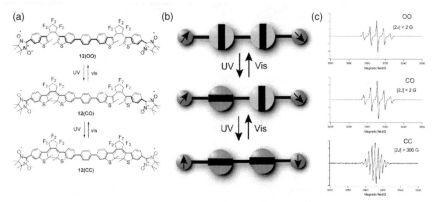

Figure 11-2 (a) Photochromic reaction and (b) schematic illustration of diarylethene **12**. (c) ESR spectra of **12(OO)**, **12(CO)**, and **12(CC)**. *Source*: Adopted from Ref. [31] © American Chemical Society.

When a diarylethene dimer is used as a switching unit, there are three kinds of photochromic states: open–open (OO), closed–open (CO), and closed–closed (CC). From the analogy of an electric circuit, one may infer that the dimer has two switching units in series. Diarylethene dimer **12**, which has 28 carbon atoms between two nitronyl nitroxide radicals, was synthesized (Fig. 11-2). When the two radicals are separated by the 28 conjugated carbon atoms, the 5-line and 9-line spectra were clearly distinguishable upon irradiation. A p-phenylene spacer was introduced so that the cyclization reaction could occur at both diarylethene moieties. Bond alternation is discontinued at the open-ring moieties of **12(OO)** and **12(CO)**. As a result, the spins at either end of **12(OO)** and **12(CO)** cannot interact with each other. On the other hand, the π-system of **12(CC)** is delocalized throughout the molecule, and the exchange interaction between the two radicals is expected to occur.

12(OO) underwent photochromic reaction by alternate irradiation with UV and visible light. Upon irradiation of the ethyl acetate solution of **12(OO)** with 313-nm light, an absorption at 560 nm appeared. This absorption grew and shifted and the system reached the photostationary state after 120 min. The color of the solution changed from pale blue to red–purple, and then to blue–purple. Such a red spectral shift suggests the formation of **12(CC)**. The isosbestic point was maintained at an initial stage of irradiation, but it later deviated. The blue–purple solution was bleached to pale blue by irradiation with 578-nm light. **12(CO)** and **12(CC)** were isolated from the blue–purple solution by high-performance liquid chromatography (HPLC). **12(CC)** has an absorption maximum at 576 nm, which is redshifted as much as 16 nm in comparison with its location in **12(CO)**. ESR spectra of isolated **12(OO)**, **12(CO)**, and **12(CC)** were measured in benzene at room temperature (Fig. 11-2c). The spectra of **12(OO)** and **12(CO)** are 5-line spectra, suggesting that the exchange interaction between the two nitronyl nitroxide radicals is much weaker than the hyperfine coupling constant ($|2J/k_{\rm B}| < 3 \times 10^{-3}$ K).

However, the spectrum of **12(CC)** has clear 9 lines, indicating that the exchange interaction between the two spins is much stronger than the hyperfine coupling constant ($|2J/k_B| > 0.04$ K). The result indicates that each diarylethene chromophore serves as a switching unit to control the magnetic interaction. The magnetic interaction between terminal nitronyl nitroxide radicals was controlled by the switching units in series.

11.9   SWITCHING ON ARYL GROUP

In the previous photoswitches, a radical unit is placed at each side of the diarylethene photoswitching unit and separated by an extended π-conjugated chain. When the π-conjugated chain length between the radical becomes longer, both photocyclization and cycloreversion reactivities are reduced. This is attributed to the reduced excitation density at the central diarylethene unit [32]. The excitation density localizes at the center of both sides of the π-conjugated aryl unit. To solve the problem, it is necessary to develop new switching systems, in which the excitation density at the switching unit is not strongly reduced. The proposed new switching molecule has its switching unit located in the middle of the π-conjugated chain.

We propose a new switching unit, in which two radicals are placed in the same aryl unit and the π-conjugated chain is extended from 2- and 5-positions of the thiophene ring in one aryl unit of the diarylethene [33]. The photochromic behavior of the newly developed diarylethene **13** is shown below. The photocyclization reaction of the diarylethene unit breaks the π-conjugation in the 2,5-bis(arylethynyl)-3-thienyl unit because of the change of the orbital hybridization from sp^2 to sp^3 at the 2-position of the thiophene ring. An imino nitroxide radical is introduced at each end of the 2,5-bis(arylethynyl)thiophene π-conjugated chain, and m-phenylene is chosen as a spacer between the radicals and the reactive center [34]. A methoxy group is introduced at the reactive carbon to reduce the cycloreversion quantum yield [35]. The magnetic interaction between the two radicals via the π-conjugated chain can be altered by the photocyclization. The open-ring isomer represents the "ON" state because the π-conjugated system is delocalized between two radicals, while the closed-ring isomer represents the "OFF" state because the π-conjugated system is disconnected at the 2-position of the thiophene ring.

Compound **13** underwent a photochromic reaction upon alternate irradiation with UV and visible light. ESR spectral change of the toluene solution containing compound **13** was followed by keeping the sample in the ESR cavity during irradiation with UV and visible light. The photoreaction was started from the isolated closed-ring isomer **13b**. The closed-ring isomer **13b** showed a 7-line spectrum, which is a spectrum of an isolated imino nitroxide ($|2J/k_B| < 3 \times 10^{-4}$ K). Upon irradiation with 578-nm light, the closed-ring isomer converted to the open-ring isomer. At the same time, the 13-line spectrum with the ratio of 1:2:5:6:10:10:13:10:10:6:5:2:1 appeared. Upon further irradiation, the spectrum was completely converted to the 13-line spectrum. The 13-line spectrum indicates that the exchange interaction takes place between the two radicals ($|2J/k_B| > 0.04$ K). Subsequent irradiation with 365-nm light regenerated a 7-line spectrum along with the regeneration of the closed-ring isomer **13b**. The difference in the exchange interaction between the open- and the closed-ring isomers was estimated to be larger than 150-fold. The difference in the exchange interaction is attributed to the change of the hybridization of the carbon atom from sp^2 to sp^3 atom at the reaction center.

11.10 NOBLE METAL NANOPARTICLE

Noble metal nanoparticles are currently attracting interest because they have unique physical properties [36]. The electromagnetic field of the incident light is enhanced by a localized surface plasmon and the nanoparticles have discrete charge states, which bring about a quantum mechanical effect. The physical properties are regulated by the size and shape of the nanoparticles. The size of the nanoparticles is several nanometers, which is comparable to the size of the molecules. Therefore, the network structure fabricated from the metal nanoparticles and organic molecules is attracting interest for the multifunctional integrated materials.

Among several interesting applications of the noble metal nanoparticles, photochemical applications have been examined considerably [37]. The photochemical reaction on the noble metal nanoparticles is not considered to be very efficient because the excited state on the noble metal surface is easily quenched by the surface plasmon resonance. For example, fluorescence from the molecule attached on the noble metal nanoparticles is completely quenched [38]. Therefore, for the realization of the photoswitching of the conductance of the network composed of photochromic molecules and metal nanoparticles, photochromic reactions on the noble metal nanoparticles have to be examined.

11.11 PHOTOREACTION ON METAL NANOPARTICLES

Photophysical properties of the metal nanoparticles are different from those of the bulk state. For example, plasmon absorption is clearly observed depending on the size of the particles. Gold nanoparticles have a surface plasmon resonance absorption band at around 520 nm [39], while silver nanoparticles have a surface plasmon resonance absorption at around 420 nm [40], which is shorter than that of

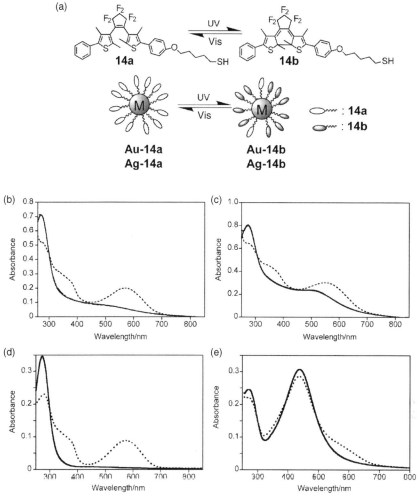

Figure 11-3 (a) Photochromic reaction of diarylethene **14** and **Au-** and **Ag-14**. Absorption spectra of (b) **Au-14a**(1:1), (c) **Au-14a**(3:1), (d) **Ag-14a**(1:1), and (e) **Ag-14a**(3:1) in ethyl acetate. Solid lines denote the open-ring isomer. Dotted lines denote the photostationary state under irradiation with 313-nm light. *Source*: Adopted from Ref. [44] © Chemical Society of Japan.

gold nanoparticles. Photochemical properties of photochromic dyes on the surface of metal nanoparticles should also be affected by the metal. Gold nanoparticles capped with azobenzene [41] and spiropyran [42] have been prepared and their photochromic reactivity examined.

The photochromic reaction of diarylethenes on the gold and silver nanoparticles is investigated [43,44] (Fig. 11-3). Diarylethene **14a** has a thiol unit, which forms a self-assembled monolayer on the metal nanoparticle core. Gold nanoparticles capped with thiol **14a** (**Au-14a**) were prepared by the Brust and Schiffrin protocol [45]. Silver nanoparticles (**Ag-14a**) were synthesized by Kim's method, which is a modification of the Brust's protocol [46]. Nanoparticles of different sizes were prepared by changing the molar ratio between diarylethene **14a** and the metal source.

Generally, metals are known to readily quench the electronic excited states of the molecule placed on their surfaces. Energy transfer and electron transfer to the metal surfaces are the main causes of the quenching [47,48]. These quenching processes are faster than fluorescence process; the time constant of energy transfer and electron transfer on the metal surfaces is normally around several picoseconds and fluorescence lifetime is normally around several nanoseconds. On the other hand, the time constant of the photocyclization reaction is normally around several picoseconds. Therefore, the cyclization reaction can compete with the quenching process.

The photochromic reaction of the gold and silver nanoparticles **Au-14** and **Ag-14** was investigated. The conversion was estimated as 74 and 64% for gold and silver nanoparticles respectively. The conversion gives information regarding the ratio of the cyclization and the cycloreversion quantum yields, and the lower conversion as compared to that of the free ligand indicates that the cyclization quantum yield is suppressed because of the metal. Although the conversion was suppressed by the metal, it was proved in this study that a photochromic reaction takes place on the surface of noble metal nanoparticles. Provided that quenching occurs by an energy-transfer mechanism, the cyclization reaction should be quenched more effectively than the cycloreversion reaction. The gold nanoparticles have a plasmon band at a longer wavelength than silver nanoparticles, which can explain the stronger quenching by the silver nanoparticles.

The optical properties of gold and silver nanoparticles capped with diarylethenes are summarized in Table 11-2. The plasmon absorption intensity decreases when the particle size becomes small. **Ag-14a**(1:1) did not show any distinctive plasmon absorption because the size of this nanoparticle was very small. The absorption maxima of the closed-ring isomer were neither strongly dependent on the metal element nor on the particle size.

TABLE 11-2   Properties of Gold and Silver Nanoparticles Capped with Diarylethene 14

Entry	TEM (Diameter, nm)	DLS (Diameter, nm)	Plasmon Absorption λ_{max} (nm)	Absorption Maxima of the Closed-Ring Isomer λ_{max} (nm)	IR Stretching (cm^{-1})
14	N/A	N/A	N/A	577	2864, 2935
Au-14a(1:1)	2.2 ± 0.3	N/A	520	578	2852, 2921
Au-14a(3:1)	3.2 ± 0.4	N/A	520	582	2856, 2926
Ag-14a(1:1)	1.2 ± 0.2	3.6	None	573	2855, 2926
Ag-14a(3:1)	6.7 ± 1.5	9.0	435	574	2857, 2928
Au-14b′	2.9 ± 0.7	N/A	520	582	N/A
Ag-14b′	11.4 ± 3.6	N/A	422	573	N/A

[a]The ratio in the entry indicates the ratio between the metal and ligand in preparation.

11.12 CONDUCTANCE PHOTOSWITCHING OF DIARYLETHENE–GOLD NANOPARTICLE NETWORK

The photoswitching of the conductance can be achieved along with intramolecular magnetic exchange interaction by using a diarylethene π-switching system carrying an electrode at each end. Although there are increasing numbers of reports dealing with the conductance of single molecules, studies on photoswitchable molecules are rare [49–51]. Besides, studies on the conductance of networks prepared with organic molecules and Au nanoparticles on interdigitated nanogapped electrodes are attracting interest because of the relatively easy preparation and the applicability to the small number of molecules [52,53].

The gold nanoparticles are bridged by diarylethene molecules with two thiol units and make a conducting path between the interdigitated nanogapped gold electrodes (Fig. 11-4a) [54,55]. Because the excited state of the organic molecule on noble metal surface is known to be easily quenched by the surface plasmon resonance, the photoswitching unit should be placed distant from the surface. On the other hand, considering that a conjugated molecule has much better conductance than a nonconjugated molecule, the sulfur atom should be attached directly to the π-conjugated system. The diarylethene–gold nanoparticle networks are characterized using transmission electron microscopy (TEM) and scanning electron microscopy (SEM). A TEM image of the nanoparticles network shows the existence of the extended network. An SEM image of the network on the interdigitated nanogapped gold electrode shows that the network bridges the electrodes.

Conductance is measured along with alternate irradiation with UV and visible light (Fig. 11-4b–e). For **Au-15a** and **Au-16a**, upon irradiation with UV light the conductance increased significantly and then on irradiation with visible light the conductance decreased. This implies that photoisomerization of the diarylethene unit brings about switching of the π-conjugated system as described above. The photocycloreversion reaction of the **Au-15a** nanoparticle network is very slow. Even after 56 h of irradiation with visible light, the conductance decreased by only 18%. On the contrary, the cycloreversion reaction of **Au-16a** completed in 8 h, and the system showed reversible photoswitching behavior. The

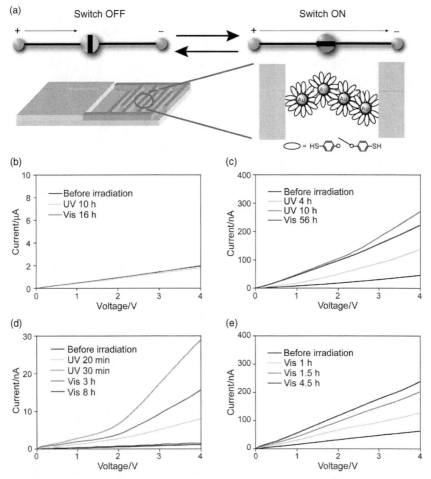

Figure 11-4   (a) A schematic drawing of a diarylethene–gold nanoparticle network. Changes in the *I – V* curves of diarylethene–gold nanoparticle networks: (b) **Au-18**; (c) **Au-15a**; (d) **Au-16a**; (e) **Au-17b**. Measurement was carried out at different stages of photoreaction. The open-ring isomer nanoparticle networks **Au-15a** and **Au-16a** were initially UV light irradiated and then visible light irradiated. For the closed-ring isomer nanoparticle network **Au-17b**, only visible irradiation was performed. Source: Adopted from Ref. [55] © American Chemical Society.

maximum ON/OFF ratio of the conductance was 25-fold for **Au-16a**. Because **16a** has high quantum yields of both cyclization and cycloreversion reactions, complete reversibility has been achieved. Nonphotoreactive **Au-18** did not show any photoinduced conductance change.

To confirm that the origin of the switching is the photochromism of the diarylethene unit, 2-thienyl-type diarylethene **17**, which has an opposite direction to the switching of the π-conjugated system, was investigated. Diarylethene **17** is considered to show opposite behavior to 3-thienyl-type diarylethene **15** and **16**. The **Au-17a** nanoparticle network did not undergo cyclization reaction by irradiation with UV light. However, the gold nanoparticle network **Au-17b**, which is prepared from the separated closed-ring isomer **17b** in the dark, showed the cycloreversion reaction upon irradiation with visible light. Perturbation by the gold surface is considered to suppress the reaction as a cycloreversion reaction of **Au-15b** and **Au-16b**. For the **Au-17b** nanoparticle network, upon irradiation with visible ($\lambda > 470$ nm) light, the conductance increased. The maximum ON/OFF ratio was 3.8-fold. The opposite behavior of **Au-17** demonstrates that the switching of the diarylethene plays an essential role in the control of the conductance.

11.13 CONDUCTANCE SWITCHING OF DIARYLETHENE–GOLD NANOPARTICLE NETWORK BY OXIDIZATION

The switching should proceed in both directions smoothly for the realization of practical molecular switching devices. The quenching of the photoexcited state by the metal surface and nanoparticles should be avoided. The electrochromic reaction of diarylethenes could be a strong candidate for an alternative external stimulus because isomerization is induced by electrochemical oxidation or chemical oxidation whose mechanism is different from that of photoisomerization [56]. Diarylethene linker **19** shows both photo- and electrochromic reactivities [57]. The cyclization reaction of the thiophene-substituted dithienylethene is strongly quenched when placed on gold nanoparticles and surfaces. Only one-way photoswitching properties have been reported for single-molecular devices. This might be overcome by an electrochromic oxidative cyclization [58].

The conductance of the **Au-19a** network hardly changed under irradiation with UV for 2 h. This result shows the strong quenching effect of the cyclization reaction of the thienyl-substituted dithienylethene by gold nanoparticles, which was not severe for phenyl-substituted dithienylethene, as suggested in other reports [59]. The conductive switching was then performed by the electrochromic reaction. The fabricated electrode was soaked in FeCl$_3$ in acetone for 1 min and moved to the vacuum chamber again. The conductance of the **Au-19a** network was increased by about fivefold. The increase of the conductance can be explained by the electrochromic reaction from "OFF" (the open-ring isomer) to "ON" (the closed-ring isomer). In order to verify the increase in conductance, the cycloreversion reaction was carried out by irradiation with visible ($\lambda > 510$ nm) light. The conductance of the network slowly decreased under irradiation with visible light. This switching behavior suggests that the closed-ring isomer in the network showed cycloreversion by photoexcitation.

11.14 CONCLUSIONS

Here we have reviewed the application of diarylethene as a molecular switch in the field of molecular magnetism and electronics. This inherently interdisciplinary field has developed by expanding its scope to organic, inorganic, physical chemistry, and solid state physics. In this chapter, first, several aspects of the intramolecular magnetic interaction have been overviewed, starting from some basic considerations of the interaction mechanism. The difference between the disjoint and nondisjoint biradicals was a key to the realization of the reversible photoswitching system. Second, the conductance photoswitching of the network prepared from diarylethene molecules and noble metal nanoparticles has been described. Photo- and electrochromism of the diarylethene ligand overcame the quenching effect of the photoexcited state on metal nanoparticles. These molecules will realize a new switching behavior in the molecular electronics field where organic π-conjugated compounds play an important role.

REFERENCES

1. Metzger, R. M. (2003). Unimolecular electrical rectifiers. *Chem. Rev.*, *103*, 3803–3834.
2. Joachim, C., Gimzewski, J. K., Aviram, A. (2000). Electronics using hybrid-molecular and mono-molecular devices. *Nature*, *408*, 541–548.
3. Moth-Poulsen, K., Bjørnholm, T. (2009). Molecular electronics with single molecules in solid-state devices. *Nat. Nanotechnol.*, *4*, 551–556.
4. Kahn, O. (1993). *Molecular Magnetism*. Wiley-VCH, New York.
5. Miller, J. S. (2006). Magnetically ordered molecule-based assemblies. *Dalton Trans.*, 2742–2749.
6. Kinoshita, M., Turek, P., Tamura, M., Nozawa, K., Shiomi, D., Nakazawa, U., Ishikawa, M., Takahashi, M., Awaga, K., Inabe, T., Maruyama, Y. (1991). An organic radical ferromagnet. *Chem. Lett.*, *20*, 1225–1228.
7. Itoh, K. (1967). Electron spin resonance of an aromatic hydrocarbon in its quintet ground state. *Chem. Phys. Lett.*, *1*, 235–238.
8. Wasserman, E., Murray, R. W., Yager, W. A., Trozzolo, A. M., Smolinski, G. (1967). Quintet ground states of *m*-dicarbene and *m*-dinitrene compounds. *J. Am. Chem. Soc.*, *89*, 5076–5078.
9. Iwamura, H., Koga, N. (1993). Studies of organic di-, oligo-, and polyradicals by means of their bulk magnetic properties. *Acc. Chem. Res.*, *26*, 346–351.

10. Borden, W. T., Davidson, E. R. (1977). Effects of electron repulsion in conjugated hydrocarbon diradicals. *J. Am. Chem. Soc.*, *99*, 4587–4594.

11. Dürr, H., Bouas-Laurent, H. (2003). *Photochromism: Molecules and Systems*. Elsevier, Amsterdam.

12. Tanaka, K., Toda, F. (2000). A novel photochromism of biindenylidene in crystal form. *J. Chem. Soc., Perkin Trans. 1*, 873–874.

13. Xu, L., Sugiyama, T., Huang, H., Song, Z., Meng, J., Matsuura, T. (2002). Photoinduced ground-state singlet biradical—novel insight into the photochromic compounds of biindenylidenediones. *Chem. Commun.*, 2328–2329.

14. Kurata, H., Takehara, Y., Kawase, T., Oda, M. (2003). Synthesis, structure, and properties of a dibenzo-*ortho*-terphenoquinone. The First *o*-terphenoquinone derivative. *Chem. Lett.*, *32*, 538–539.

15. Teki, Y., Miyamoto, S., Nakatsuji, M., Miura, Y. (2001). π-Topology and spin alignment utilizing the excited molecular field: observation of the excited high-spin quartet ($S = 3/2$) and quintet ($S = 2$) states on purely organic π-conjugated spin systems. *J. Am. Chem. Soc.*, *123*, 294–305.

16. Kaneko, T., Akutsu, H., Yamada, J., Nakatsuji, S. (2003). Photochromic radical compounds based on a naphthopyran system. *Org. Lett.*, *5*, 2127–2129.

17. Irie, M. (2000). Diarylethenes for memories and switches. *Chem. Rev.*, *100*, 1685–1716.

18. Kawai, T., Fukuda, N., Gröschl, D., Kobatake, S., Irie, M. (1999). Refractive index change of dithienylethene in bulk amorphous solid phase. *Jpn. J. Appl. Phys.*, *38*, L1194–L1196.

19. Tsivgoulis, G. M., Lehn, J.-M. (1995). Photonic molecular devices: reversibly photoswitchable fluorophores for nondestructive readout for optical memory. *Angew. Chem. Int. Ed. Engl.*, *34*, 1119–1122.

20. Kawai, T., Kunitake, T., Irie, M. (1999). Novel photochromic conducting polymer having diarylethene derivative in the main chain. *Chem. Lett.*, *28*, 905–906.

21. Matsuda, K., Irie, M. (2000). A diarylethene with two nitronyl nitroxides: photoswitching of intramolecular magnetic interaction. *J. Am. Chem. Soc.*, *122*, 7195–7201.

22. Matsuda, K., Irie, M. (2000). Photoswitching of intramolecular magnetic interaction: a diarylethene photochromic spin coupler. *Chem. Lett.*, *29*, 16–17.

23. Matsuda, K., Irie, M. (2000). Photoswitching of intramolecular magnetic interaction using a photochromic spin coupler: an ESR study. *J. Am. Chem. Soc.*, *122*, 8309–8310.

24. Bleaney, B., Bowers, K. D. (1952). Anomalous paramagnetism of copper acetate. *Proc. R. Soc. London, Ser. A*, *214*, 451–465.

25. Matsuda, K., Irie, M. (2001). Photochromism of diarylethenes with two nitronyl nitroxides: photo-switching of an intramolecular magnetic interaction. *Chem. Eur. J.*, *7*, 3466–3473.

26. Brière, R., Dupeyre, R.-M., Lemaire, H., Morat, C., Rassat, A., Rey, P. (1965). Nitroxydes XVII: biradicaux stables du type nitroxide. *Bull. Soc. Chim. Fr.*, *11*, 3290–3297.

27. Glarum, S. H., Marshall, J. H. (1967). Spin exchange in nitroxide biradicals. *J. Chem. Phys.*, *47*, 1374–1378.

28. Matsuda, K., Matsuo, M., Irie, M. (2001). Photoswitching of intramolecular magnetic interaction using photochromic diarylethene spin coupler: introduction of thiophene spacer. *Chem. Lett.*, *30*, 436–437.

29. Matsuda, K., Matsuo, M., Irie, M. (2001). Photoswitching of intramolecular magnetic interaction using diarylethene with oligothiophene π-conjugated chain. *J. Org. Chem.*, *66*, 8799–8803.

30. Matsuda, K., Matsuo, M., Mizoguti, S., Higashiguchi, K., Irie, M. (2002). Reversed photoswitching of intramolecular magnetic interaction using a photochromic bis(2-thienyl)ethene spin coupler. *J. Phys. Chem. B*, *106*, 11218–11225.

31. Matsuda, K., Irie, M. (2001). Photoswitching of intramolecular magnetic interaction using a diarylethene dimer. *J. Am. Chem. Soc.*, *123*, 9896–9897.

32. Bens, A. T., Frewert, D., Kodatis, K., Kryschi, C., Martin, H.-D., Trommsdorf, H. P. (1998). Coupling of chromophores: Carotenoids and photoactive diarylethenes–photoreactivity versus radiationless deactivation. *Eur. J. Org. Chem.*, 2333–2338.

33. Tanifuji, N., Irie, M., Matsuda, K. (2005). New photoswitching unit for magnetic interaction: diarylethene with 2,5-bis(arylethynyl)-3-thienyl group. *J. Am. Chem. Soc.*, *127*, 13344–13353.

34. Tanifuji, N., Matsuda, K., Irie, M. (2005). Effect of imino nitroxyl and nitronyl nitroxyl groups on the photochromic reactivity of diarylethenes. *Org. Lett.*, *7*, 3777–3780.

35. Shibata, K., Kobatake, S., Irie, M. (2001). Extraordinarily low cycloreversion quantum yields of photochromic diarylethenes with methoxy substituents. *Chem. Lett.*, *30*, 618–619.

36. Daniel, M.-C., Austruc, D. (2004). Gold nanoparticles: assembly, supramolecular chemistry, quantum-size-related properties, and applications toward biology, catalysis, and nanotechnology. *Chem. Rev.*, *104*, 293–346.

37. Thomas, K. G., Kamat, P. V. (2003). Chromophore-functionalized gold nanoparticles. *Acc. Chem. Res.*, *36*, 888–898.

38. Avouris, P., Persson, B. N. J. (1984). Excited states at metal surfaces and their non-radiative relaxation. *J. Phys. Chem.*, *88*, 837–848.

39. Hostetler, M. J., Wingate, J. E., Zhong, C.-J., Harris, J. E., Vachet, R. W., Clark, M. R., Londono, J. D., Green, S. J., Stokes, J. J., Wignall, G. D., Glish, G. L., Porter, M. D., Evans, N. D., Murray, R. W. (1998). Alkanethiolate gold cluster molecules with core diameters from 1.5 to 5.2 nm: core and monolayer properties as a function of core size. *Langmuir*, *14*, 17–30.

40. Manna, A., Imae, T., Aoi, K., Okada, M., Yogo, T. (2001). Synthesis of dendrimer-passivated noble metal nanoparticles in a polar medium: comparison of size between silver and gold particles. *Chem. Mater.*, *13*, 1674–1681.

41. Manna, A., Chen, P.-L., Akiyama, H., Wei, T.-X., Tamada, K., Knoll, W. (2003). Optimized photo-isomerization on gold nanoparticles capped by unsymmetrical azobenzene disulfides. *Chem. Mater.*, *15*, 20–28.

42. Ipe, I., Mahima, S., Thomas, K. G. (2003). Light-induced modulation of self-assembly on spiropyran-capped gold nanoparticles: a potential system for the controlled release of amino acid derivatives. *J. Am. Chem. Soc.*, *125*, 7174–7175.

43. Matsuda, K., Ikeda, M., Irie, M. (2004). Photochromism of diarylethene-capped gold nanoparticles. *Chem. Lett.*, *33*, 456–457.

44. Yamaguchi, H., Ikeda, M., Matsuda, K., Irie, M. (2006). Photochromism of diarylethenes on gold and silver nanoparticles. *Bull. Chem. Soc. Jpn.*, *79*, 1413–1419.

45. Brust, M., Walker, M., Bethell, D., Schiffrin, D. J., Whyman, R. (1994). Synthesis of thiol-derivatised gold nanoparticles in a two-phase liquid–liquid system. *J. Chem. Soc., Chem. Commun.*, 801–802.

46. Kang, S. Y., Kim, K. (1998). Comparative study of dodecanethiol-derivatized silver nanoparticles prepared in one-phase and two-phase systems. *Langmuir*, *14*, 226–230.

47. Dulkeith, E., Morteani, A. C., Niedereichholz, T., Klar, T. A., Feldmann, J., Levi, S. A., van Veggel, F. C. J. M., Reinhoudt, D. N., Möller, M., Gittins, D. I. (2002). Fluorescence quenching of dye molecules near gold nanoparticles: radiative and nonradiative effects. *Phys. Rev. Lett.*, *89*, 203002.

48. Barazzouk, S., Kamat, P. V., Hotchaudani, S. (2005). Photoinduced electron transfer between chlorophyll a and gold nanoparticles. *J. Phys. Chem. B*, *109*, 716–723.

49. Jong, J. J. D., Bowden, T. N., van Esch, J., Feringa, B. L., van Wees, B. J. (2003). One-way optoelectronic switching of photochromic molecules on gold. *Phys. Rev. Lett.*, *91*, 207402.

50. Taniguchi, M., Nojima, Y., Yokota, K., Terao, J., Sato, K., Kambe, N., Kawai, T. (2006). Self-organized interconnect method for molecular devices. *J. Am. Chem. Soc.*, *128*, 15062–15063.

51. Whalley, A. C., Steigerwald, M. L., Guo, X., Nuckolls, C. (2007). Reversible switching in molecular electronic devices. *J. Am. Chem. Soc.*, *129*, 12590–12591.

52. Ogawa, T., Kobayashi, K., Masuda, G., Takase, T., Maeda, S. (2001). Electronic conductive characteristics of devices fabricated with 1,10-decanedithiol and gold nanoparticles between 1-μm electrode gaps. *Thin Solid Films*, *393*, 374–378.

53. Bernard, L., Kamdzhilov, Y., Calame, M., van der Molen, S. J., Liao, J., Schönenberger, C. (2007). Spectroscopy of molecular junction networks obtained by place exchange in 2D nanoparticle arrays. *J. Phys. Chem. C*, *111*, 18445–18450.

54. Ikeda, M., Tanifuji, N., Yamaguchi, H., Irie, M., Matsuda, K. (2007). Photoswitching of conductance of diarylethene-Au nanoparticle network. *Chem. Commun.*, 1355–1357.

55. Matsuda, K., Yamaguchi, H., Sakano, T., Ikeda, M., Tanifuji, N., Irie, M. (2008). Conductance photoswitching of diarylethene-gold nanoparticle network induced by photochromic reaction. *J. Phys. Chem. C*, *112*, 17005–17010.

56. Koshido, T., Kawai, T., Yoshino, K. (1995). Optical and electrochemical properties of cis-1,2-dicyano-1,2-bis(2,4,5-trimethyl-3-thienyl)ethane. *J. Phys. Chem.*, *99*, 6110–6114.

57. Peters, A., Branda, N. R. (2003). Electrochemically induced ring-closing of photochromic 1,2-dithienylcyclopentenes. *Chem. Commun.*, 954–955.

58. Yamaguchi, H., Matsuda, K. (2009). Photo- and electrochromic switching of diarylethene–gold nanoparticle network on interdigitated electrodes. *Chem. Lett.*, *38*, 946–947.

59. Kudernac, T., van der Molen, S. J., van Wees, B. J., Feringa, B. L. (2006). Uni- and bi-directional light-induced switching of diarylethenes on gold nanoparticles. *Chem. Commun.*, 3597–3599.

ORGANOGELS AND POLYMER ASSEMBLY

CHAPTER 12

SELF-OSCILLATING POLYMER GELS

Ryo Yoshida

Department of Materials Engineering, The University of Tokyo, Tokyo, Japan

12.1 INTRODUCTION

"Polymer gels" is a research field of polymer science that has seen rapid progress during the past 20–30 years. A gel can be widely defined as a cross-linked polymer network, which gets swollen by absorbing large amounts of solvents such as water. A theoretical study of the characteristics of a gel had already started in the 1940s, and the principle of swelling by water absorption based on thermodynamics had been clarified by Flory. As an application of gel research, soft contact lenses were developed in the 1960s, and subsequently gels have been widely used in medical and pharmaceutical fields. Since a polymer that can absorb about 1000 times as much water as its own weight was developed in the United States in the 1970s, gels have been applied as super absorbent polymers in several industrial fields and are mainly applied in sanitary items, disposable diapers, and so on. Further, in 1978, it was discovered by Tanaka [1] that gels change volume reversibly and discontinuously in response to environmental changes, such as in solvent composition, temperature, pH, and so on (called "volume phase transition" phenomenon). With this discovery as a turning point, research on the use of gels as functional materials for artificial muscles, robot hands (actuator), stimuli-responsive drug delivery systems (DDSs), separation or purification, cell culture, biosensors, shape memory materials, and so on was accelerated [2–8].

Until now, fundamental and applied research, which includes many different fields such as elucidation of gelation mechanisms, analysis of physical properties and structure, functional control by molecular design, and so on, has been carried out. In particular, from the early 1990s, new functional gels, which include the following three functions in themselves, sensing an external signal (sensor function), judging it (processor function), and taking action (actuator function), have been developed by many researchers as "intelligent gels" or "smart gels."

Supramolecular Soft Matter: Applications in Materials and Organic Electronics, First Edition.
Edited by Takashi Nakanishi.
© 2011 John Wiley & Sons, Inc. Published 2011 by John Wiley & Sons, Inc.

Further, in recent years, the usefulness of gels has also been shown in the field of micromachines and nanotechnology. In addition, new synthetic methods have been attempted to give unique functions by molecular design in the nano-order scale, including supramolecular design, the design and construction of micro- or nanomaterials systems with the biomimetic functions of motion, mass transport, transformation and transmission of information, molecular recognition, and so on.

So far, many researchers have developed stimuli-responsive polymer gels that change volume abruptly in response to a change in their surroundings, such as solvent composition, temperature, pH, supply of electric field, and so on. Their ability to swell and deswell according to conditions makes them interesting candidates for use in new intelligent materials. In particular, their applications in biomedical fields have been extensively studied. One of the strategies of these applications is to develop biomimetic materials systems with stimuli-responding functions, that is, systems in which the materials can sense environmental changes and get activated. For these systems, the on–off switching of external stimuli is essential to instigate the action of the gel. Upon switching on, the gels provide only one unique action, either swelling or deswelling.

This stimuli-responding behavior is a temporary action toward an equilibrium state. In contrast, there are many physiological phenomena in our body that continue their own native cyclic changes. These phenomena exist over a wide range from the cell to the body level, namely, the cell cycle, cyclic reaction in glycolysis, pulsatile secretion of hormones, pulsatile potential of nerve cells, brain waves, heartbeat, peristaltic motion in the digestive tract, human biorhythms, and so on. If such self-oscillation could be achieved for gels, possibilities would emerge for new biomimetic intelligent materials that exhibit an autonomous rhythmical motion.

We succeeded in developing such a self-oscillating polymer and gels by incorporating an oscillating chemical reaction in the polymer network, that is, by constructing a built-in circuit of the energy-conversion cycle, producing mechanical oscillation within the polymer network itself. Under the coexistence of the reactants in a closed and constant-conditioned solution, the polymer undergoes spontaneous cyclic soluble–insoluble changes or swelling–deswelling changes (in the case of gel) without any on–off switching of external stimuli, differently from conventional stimuli-responsive gels. Since first reported in 1996 as a "self-oscillating gel" [9], we have been systematically studying the self-oscillating polymer and gel as well as their applications to biomimetic or smart materials. In this chapter, a new design concept for polymer gels that exhibits spontaneous and autonomous periodic swelling–deswelling changes under constant conditions without on–off switching of external stimuli is introduced.

12.2   DESIGN OF THE SELF-OSCILLATING GEL

In order to realize the autonomous polymer system by a tailor-made molecular design, we focused on the Belousov–Zhabotinsky (BZ) reaction [10,11], which is well known for exhibiting temporal and spatiotemporal oscillating phenomena. The BZ reaction is often analogically compared with the tricarboxylic acid (TCA) cycle (Krebs cycle), which is a key metabolic process taking place in the living body. The overall process of the BZ reaction is the oxidation of an organic substrate, such as malonic acid (MA) or citric acid, by an oxidizing agent (bromate ion) in the presence of a strong acid and a metal catalyst. In the course of the reaction, the catalyst undergoes spontaneous redox oscillation. When the solution is homogeneously stirred, the color of the solution periodically changes, like a neon sign, based on the redox changes in the metal catalyst. When the solution is placed as a thin film in stationary conditions, concentric or spiral wave patterns develop in the solution. The wave of the oxidized state propagating in the medium at a constant speed is called a "chemical wave." The significance of the BZ reaction has been recognized as a chemical model for understanding some aspects of biological phenomena, such as glycolytic oscillations or biorhythms, cardiac fibrillation, self-organization of amoeba cells, pattern formation on animal skin, visual pattern processing on retina, and so on.

We attempted to convert the chemical oscillation of the BZ reaction to the mechanical changes of gels and generate an autonomic swelling–deswelling oscillation under nonoscillatory outer conditions [9,12–14]. A copolymer gel, which consists of NIPAAm and ruthenium tris(2,2'-bipyridine) (Ru(bpy)$_3{}^{2+}$), was prepared. Ru(bpy)$_3{}^{2+}$, acting as a catalyst for the BZ reaction, is pendent to the polymer chains of NIPAAm (Fig. 12-1a). The poly(NIPAAm-co-Ru(bpy)$_3{}^{2+}$) gel has a phase-transition temperature because of the thermosensitive constituent NIPAAm. The oxidation of the Ru(bpy)$_3{}^{2+}$ moiety causes not only an increase in the swelling degree of the gel but also a rise in the transition temperature. These characteristics may be interpreted by considering an increase in the hydrophilicity of the polymer chains due to the oxidation of Ru(II) to Ru(III) in the Ru(bpy)$_3$ moiety. As a result, it is expected that the gel undergoes a cyclic swelling–deswelling alteration when the Ru(bpy)$_3$ moiety is periodically oxidized and reduced under constant temperature. When the gel is immersed in an aqueous solution containing the substrates of the BZ reaction (MA, acid, and oxidant) except for the catalyst, the substrates penetrate into the polymer network and the BZ reaction occurs in the gel. Consequently, periodical redox changes induced by the BZ reaction produce periodical swelling–deswelling changes in the gel (Figure 12-1a).

12.3 SELF-OSCILLATING BEHAVIORS OF THE GEL

12.3.1 Self-Oscillation of the Gel Smaller than the Chemical Wavelength

Figure 12-1b shows the observed oscillating behavior under a microscope for the miniature cubic poly(NIPAAm-co-Ru(bpy)$_3{}^{2+}$) gel (each of length of about 0.5 mm). In miniature gels, sufficiently smaller than the wavelength of the chemical wave (typically several mm), the redox change of the ruthenium catalyst occurs

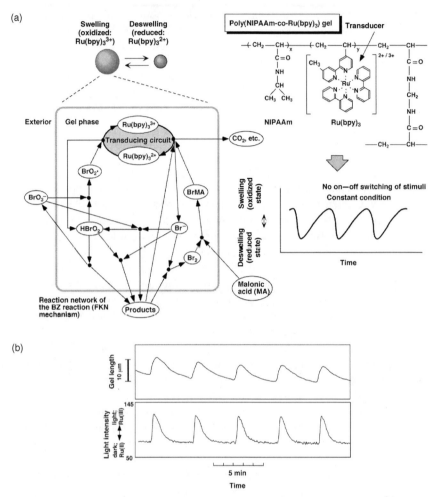

Figure 12-1 (a) Mechanism of self-oscillation of poly(NIPAAm-co-Ru(bpy)$_3$$^{2+}$) gel coupled with the Belousov–Zhabotinsky reaction. (b) Periodic redox changes in the miniature cubic poly(NIPAAm-co-Ru(bpy)$_3$$^{2+}$) gel (lower) and the swelling–deswelling oscillation (upper) at 20°C. Color changes in the gel accompanied by redox oscillations (orange: reduced state, light green: the oxidized state) were converted to eight-bit grayscale changes (dark: reduced, light: oxidized) by image processing. Transmitted light intensity is expressed as an eight-bit grayscale value. Outer solution: [MA] = 62.5 mM; [NaBrO$_3$] = 84 mM; [HNO$_3$] = 0.6 M.

homogeneously without pattern formation [15]. Owing to the redox oscillation of the immobilized Ru(bpy)$_3$$^{2+}$, the mechanical swelling–deswelling oscillation of the gel autonomously occurs with the same period as that for the redox oscillation. The volume change is isotropic, and the gel beats as a whole, like a heart muscle cell. The chemical and mechanical oscillations are synchronized without a phase difference (i.e., the gel exhibits swelling during the oxidized state and deswelling during the reduced state).

12.3.2 Control of Oscillating Behaviors

Typically, the oscillation period increases with a decrease in the initial concentration of substrates. Further, in general, the oscillation frequency (the reciprocal of the period) of the BZ reaction tends to increase as the temperature increases, in accordance with the Arrhenius equation. The swelling–deswelling amplitude of the gel increases with an increase in the oscillation period and amplitude of the redox changes. Therefore, the swelling–deswelling amplitude of the gel is controllable by changing the initial concentration of substrates as well as the temperature.

As an inherent behavior of the BZ reaction, the abrupt transition from the steady state (nonoscillating state) to the oscillating state occurs with a change in controlling parameters such as chemical composition, light, and so on. By utilizing these characteristics, reversible on–off regulation of self-beating triggered by addition and removal of MA was successfully achieved [16]. We showed the oscillating behavior of the gel when the stepwise change in [MA] was repeated between lower concentration (10 mM) in the steady state and higher concentration (25 mM) in the oscillating state. At [MA] = 10 mM, the redox oscillation does not occur and consequently the gel exhibits no swelling–deswellng changes. As soon as the concentration was increased to 25 mM, the gel started self-beating. The beating stopped again as soon as the concentration was decreased back to the initial value. In this manner, the reversible on–off regulation of self-beating triggered by MA was successfully achieved. Since there are some organic acids that can be the substrate for the BZ reaction (e.g., citric acid), the same regulation of beating is possible by using those organic acids instead of MA. Also, as the gel has thermosensitivity due to the NIPAAm component, the beating rhythm can be controlled by temperature [17].

12.3.3 Peristaltic Motion of Gels with Propagation of Chemical Wave

When the gel size is larger than the chemical wavelength, the chemical wave propagates in the gel by coupling with diffusion of intermediates. Then, the peristaltic motion of the gel is created. Figure 12-2a shows the cylindrical gel that is immersed in an aqueous solution containing the three reactants of the BZ reaction. The chemical waves propagate in the gel at a constant speed in the direction of the gel length [18]. Considering that the orange (Ru(II)) and green (Ru(III)) zones represent the shrunken and swollen parts, respectively, the locally swollen and shrunken parts move with the chemical wave, similar to the peristaltic motion of living worms. The tensile force of the cylindrical gel with oscillation was also measured [19].

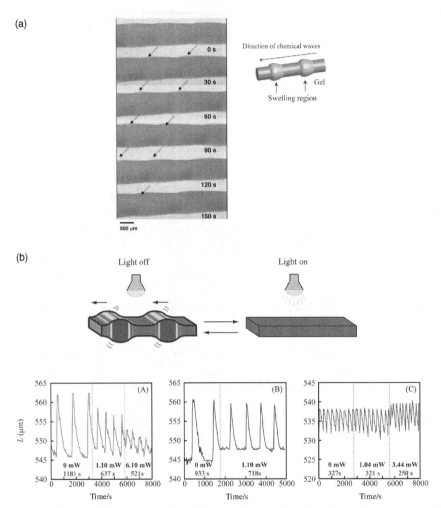

Figure 12-2 (a) Time course of peristaltic motion of the cylindrical self-oscillating gel in a solution of the BZ substrates at 18°C. The light and dark colors correspond to the oxidized and reduced states of the Ru moiety in the gel, respectively. (b) Irradiated light intensity and sodium bromate concentration tuning of self-sustaining peristaltic motion of the porous gel at 13°C. Outer solution: [MA] = 62.5 mM; [HNO₃] = 0.890 M; [NaBrO₃] = 42 mM (A), 60 mM (B), and 84 mM (C). The periods of the self-sustaining peristaltic motions for each condition are shown.

It is well known that the period of oscillation is affected by light illumination of the Ru(bpy)₃²⁺-catalyzed BZ reaction [20]. Therefore, we can intentionally make a pacemaker with a desired period (or wavelength) by local illumination of the laser beam to the gel, or we can change the period (or wavelength) by local illumination to a pacemaker, which already exists in the gel [21]. Chemical and optical controls of the wormlike peristaltic motion of the gel were demonstrated (Fig. 12-2b) [22–24].

12.4 DESIGN OF THE BIOMIMETIC MICRO-/NANOACTUATOR USING SELF-OSCILLATING POLYMER AND GEL

12.4.1 Self-Walking Gel

Further, we have successfully developed a novel biomimetic walking-gel actuator made of the self-oscillating gel [25]. To produce a directional movement of the gel, asymmetrical swelling–deswelling is desired. For this purpose, as a third component, hydrophilic 2-acrylamido-2-methylpropanesulfonic acid (AMPS) was copolymerized into the polymer to lubricate the gel and cause anisotropic contraction. During polymerization, the monomer solution faces two different surfaces of plates: a hydrophilic glass surface and a hydrophobic Teflon surface. Since the $Ru(bpy)_3^{2+}$ monomer is hydrophobic, it easily migrates to the Teflon surface side. As a result, a nonuniform distribution along the height is formed by the components, and the resulting gel has gradient distribution for the content of each component in the polymer network.

In order to convert the bending and stretching changes to one-directional motion, we employed a ratchet mechanism. A ratchet base with an asymmetrical surface structure was fabricated. On the ratchet base, the gel repeatedly bends and stretches autonomously, resulting in the forward motion of the gel, while sliding backward is prevented by the teeth of the ratchet. Figure 12-3a shows successive profiles of the "self-walking" motion of the gel, similar to a looper, in the BZ substrate solution under constant temperature. The walking velocity of the gel actuator is approximately 170 μm/min. Since the oscillating period and the propagating velocity of chemical wave change with the concentration of substrates in the outer solution, the walking velocity of the gel can be controlled. By using the gel with a gradient structure, the other type of actuator that generates a pendulum motion also starts working [26].

12.4.2 Mass Transport Surface Utilizing Peristaltic Motion of the Gel

Further, we attempted to transport an object by utilizing the peristaltic motion of poly(NIPAAm-*co*-Ru(bpy)$_3$-*co*-AMPS) gels. As a model object, a cylindrical poly(acrylamide) (PAAm) gel was put on the gel surface. It was observed that the PAAm gel was transported on the gel surface with the propagation of the chemical wave as it rolled [27] (Fig. 12-3b). We have proposed a model to describe the mass transport phenomena based on the Hertz contact theory and have also investigated the relation between transportability and peristaltic motion. The functional gel surface generating autonomous and periodic peristaltic motion has a potential for use in several new applications such as a conveyer to transport soft materials, a formation process for ordered structures of micro- and/or nanomaterials, a self-cleaning surface, and so on.

(a)

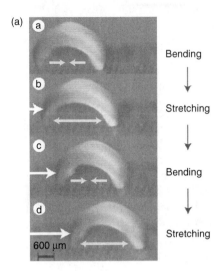

Bending

↓

Stretching

↓

Bending

↓

Stretching

(b)

Cylindrical PAAm gel

Self-oscillating gel sheet

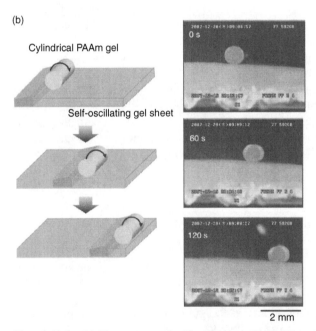

2 mm

Figure 12-3 (a) Time course of self-walking motion of the gel actuator. During stretching, the front edge can slide forward on the base, but the rear edge is prevented from sliding backward. Opposite to this, during bending, the front edge is prevented from sliding backward, while the rear edge can slide forward. This action is repeated, and as a result, the gel walks forward. (b) Schematic illustration of mass transport on the peristaltic surface (left) and observed transport of cylindrical PAAm gel on the self-oscillating gel sheet (right).

12.4.3 Microfabrication of the Self-Oscillating Gel for Microdevices

Recently, microfabrication technologies such as photolithography have also been attempted for preparation of microgels. Since any shape of the gel can be created by these methods, its application as a new manufacturing method for soft microactuator, microgel valve, gel display, and so on is expected. Microfabrication of the self-oscillating gel has also been attempted by photolithography for application in such microdevices [28,29].

In these devices, the possible application to DDS is as a self-oscillatory, drug-release microchip. If the microfabricated, self-oscillating gel can be used as a beating micropump to push and pull a diaphragm that separates the drug reservoir in the microchip, it starts releasing the drug periodically. Since self-beating of the gel occurs in a closed solution containing BZ reactants as an energy source, a complete stand-alone microchip without electric wiring and external apparatus is possible. Periodic release of the drug with preprogrammed periods under constant conditions can be achieved, and this would result in several advantages such as application to chronopharmacotherapy to release hormones synchronized with biorhythms, decreasing drug tolerance, and so on.

Microfabrication of the self-oscillating gel has also been attempted by lithography for application to the ciliary motion actuator (artificial cilia) [30]. The gel membrane with a micro-projection array on the surface was fabricated by using the X-ray lithography (LIGA) method. With the propagation of a chemical wave, the micro-projection array exhibits a dynamic rhythmic motion, similar to cilia. The actuator may also serve as a microconveyer.

12.4.4 Control of Chemical Wave Propagation in a Self-Oscillating Gel Array

A chemomechanical actuator utilizing a reaction-diffusion wave across the gap junction was constructed to obtain a novel microconveyer by a micropatterned, self-oscillating gel array [31]. By unidirectional propagation of the chemical wave, the BZ reaction was induced on gel arrays. When a triangle-shaped gel was used as an element of the array, the chemical wave propagated from the corner side of the triangle gel to the plane side of the other gel (C to P) across the gap junction, whereas it propagated from the plane side to the corner side (P to C) when the pentagonal gel array was used. Numerical analysis based on the theoretical model was carried out for understanding the mechanism of unidirectional propagation in triangle and pentagonal gel arrays. By fabricating different shapes of gel arrays, control of the direction is possible. The swelling and deswelling changes in the gels followed the unidirectional propagation of the chemical wave. Application of gels to novel microconveyers is expected.

12.4.5   Self-Oscillating Polymer Chains as a "Nano-Oscillator"

The periodic changes in linear and uncross-linked polymer chains can be easily observed as cyclic transparent and opaque changes for the polymer solution with color changes due to the redox oscillation of the catalyst [32]. Synchronized with the periodical changes between Ru(II) and Ru(III) states of the Ru(bpy)$_3^{2+}$ site, the polymer becomes hydrophobic and hydrophilic and exhibits cyclic soluble–insoluble changes (Fig. 12-4a).

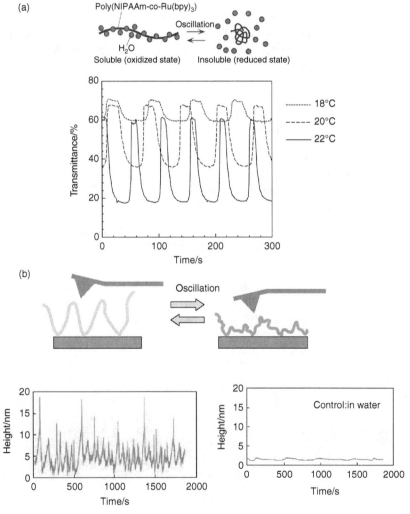

Figure 12-4   (a) Oscillating profiles of optical transmittance for poly(NIPAAm-co-Ru(bpy)$_3^{2+}$) solution at constant temperatures. (b) Self-oscillating behavior of immobilized polymer in the BZ substrate solution measured by AFM.

By grafting the polymers or arraying the gel beads on the surface of substrates, we have attempted to design self-oscillating surfaces as nano conveyers. The self-oscillating polymer was covalently immobilized on a glass surface and self-oscillation was directly observed at a molecular level by atomic force microscopy (AFM) [33]. The self-oscillating polymer with the N-succinimidyl group was immobilized on an aminosilane-coupled glass plate. While no oscillation was observed in pure water, nanoscale oscillation was observed in an aqueous solution containing the BZ substrates (Fig. 12-4b). The amplitude was ~10–15 nm, and the period was ~70 s, although some irregular behavior was observed because of no stirring. The amplitude was less than that in solution, as observed by DLS (23.9 and 59.6 nm). This smaller amplitude may be because the structure of the immobilized polymer was a loop-train tail: the moving regions were shorter than that of the soluble polymer, as illustrated in Fig. 12-4b. The amplitude and frequency were controlled by the concentration of the reactant, as observed in the solution. Here, nanoscale molecular self-oscillation was observed for the first time. The oscillation polymer chain may be used as a component of a nanoclock or a nanomachine.

12.4.6   Self-Flocculating/Dispersing Oscillation of Microgels

We prepared submicron-sized poly(NIPAAm-co-Ru(bpy)$_3$$^{2+}$) gel beads by surfactant-free aqueous precipitation polymerization and analyzed the oscillating behaviors [34–39]. Figure 12-5A shows the oscillation profiles of transmittance for microgel dispersions. At low temperatures (20–26.5°C) and on raising the temperature, the amplitude of the oscillation became larger. The increase in amplitude is due to increased deviation of the hydrodynamic diameter between the Ru(II) and Ru(III) states. Furthermore, a remarkable change was observed in the waveform between 26.5 and 27°C. Then the amplitude of the oscillations dramatically decreased at 27.5°C, and finally the periodic transmittance changes could no longer be observed at 28°C. The sudden change in oscillation waveform should be related to the difference in the colloidal stability between the Ru(II) and Ru(III) states. Here, the microgels should be flocculated because of the lack of electrostatic repulsion when the microgels are deswollen. The remarkable change in waveform was observed only at higher dispersion concentrations (greater than 0.225 wt%). The self-oscillating property makes microgels more attractive for future developments such as microgel assembly, optical and rheological applications [37,38], and so on.

12.4.7   Fabrication of Microgel Beads Monolayer

As discussed in the previous section, we have been interested in the construction of micro/nanoconveyers by grafting or arraying self-oscillating polymer or gel beads. For this purpose, a fabrication method for organized monolayers of microgel beads was investigated [40]. A 2D closely packed array of thermosensitive microgel beads was prepared by double template polymerization (Fig. 12-5B). First, a 2D colloidal crystal of silica beads with 10-μm diameter was obtained by

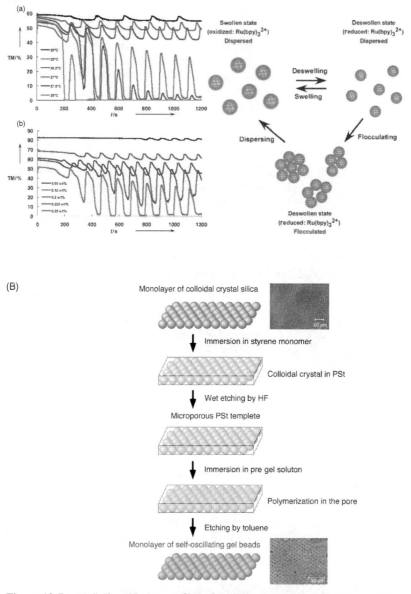

Figure 12-5 (a) Self-oscillating profiles of optical transmittance for microgel dispersions. The microgels were dispersed in aqueous solutions containing MA, NaBrO$_3$, and HNO$_3$. (A) Profiles measured at different temperatures. (B) Profiles measured at different microgel dispersion concentrations at 27°C. (b) Preparation of self-oscillating gel bead monolayer by two-step template polymerization.

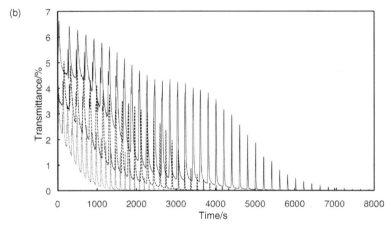

Figure 12-6   Chemical structure of poly(NIPAAm-co-Ru(bpy)$_3^{2+}$-co-AMPS -co-MAPTAC) (a) and the oscillating profiles of the optical transmittance for the polymer solution at 12°C when only MA (0.7 M: fine dotted line, 0.5 M: rough dotted line, 0.3 M: solid line) is added to the solution (b).

solvent evaporation. This monolayer of colloidal crystals served as the first template for preparation of macroporous polystyrene. The macroporous polystyrene trapping the crystalline order was used as a negative template for fabricating a gel bead array. By this double template polymerization method, functional surfaces using thermosensitive PNIPAAm gel beads were fabricated. It was observed that the topography of the surface changed with temperature. The fabrication method demonstrated here was so versatile that any kind of gel beads could be obtained. This method may be a key technology that could be used to create a new functional surface. The monolayer of self-oscillating microgel beads was fabricated by this method, and the chemical wave propagation was observed on the monolayer.

12.4.8 Attempts of Self-Oscillation under Physiological Conditions

However, in this self-oscillating polymer system, the operating conditions are limited to the nonphysiological environment where the strong acid and the oxidant coexist. For extending the application field to biomaterials, a more sophisticated molecular design is required to cause self-oscillation under physiological conditions. For this purpose, we constructed the integrated polymer system, where all of the BZ substrates other than the biorelated organic substrate were incorporated into the polymer chain [41]. We synthesized the quaternary copolymer, which includes both the pH-control and oxidant-supplying sites in the poly(NIPAAm-co-Ru(bpy)$_3$) chain at the same time. In the polymer, AMPS was incorporated as a pH-control site and methacrylamidopropyltrimethylammonium chloride (MAP-TAC) with a positively charged group was incorporated as a capture site for an anionic oxidizing agent (bromate ion). By using the polymer, self-oscillation under biological conditions where only the organic acid (MA) exists was achieved (Fig. 12-6).

REFERENCES

1. (a) Tanaka, T. (1978). Collapse of gels and the critical endpoint. *Phys. Rev. Lett.*, *40*, 820–823; (b) Tanaka, T. (1981). Gels. *Sci. Am.*, *244*, 124–136.
2. Yoshida, R. (2005). Design of functional polymer gels and their application to biomimetic materials. *Curr. Org. Chem.*, *9*, 1617–1641.
3. Yoshida, R., Sakai, K., Okano, T., Sakurai, Y. (1993). Modern hydrogel delivery systems. *Adv. Drug Delivery Rev.*, *11*, 85–108.
4. Okano, T., ed. (1998). *Biorelated Polymers and Gels: Controlled Release and Applications in Biomedical Engineering*. Academic Press, Boston.
5. Miyata, T. (2002). Stimuli-responsive polymer and gels. in *Supramolecular Design for Biological Applications* (Ed.: Yui, N.). CRC Press, Boca Raton, FL, pp. 191–225.
6. Osada, Y., Khokhlov, A. R., eds. (2002). *Polymer Gels and Networks*. Marcel Dekker, New York.
7. Yui, N., Mrsny, R. J., Park, K., eds. (2004). *Reflexive Polymers and Hydrogels—Understanding and Designing Fast Responsive Polymeric Systems*. CRC Press, Boca Raton, FL.
8. Peppas, N. A., Hilt, J. Z., Khademhosseini, A., Langer, R. (2006). Hydrogels in biology and medicine: from molecular principles to bionanotechnology. *Adv. Mater.*, *18*, 1345–1360.
9. Yoshida, R., Takahashi, T., Yamaguchi, T., Ichijo, H. (1996). Self-oscillating gel. *J. Am. Chem. Soc.*, *118*, 5134–5135.
10. Field, R. J., Burger, M., eds. (1985). *Oscillations and Traveling Waves in Chemical Systems*. John Wiley & Sons, Inc.. New York.
11. Epstein, I. R., Pojman, J. A. (1998). *An Introduction to Nonlinear Chemical Dynamics: Oscillations, Waves, Patterns, and Chaos*. Oxford University Press, New York.
12. Yoshida, R., Takahashi, T., Yamaguchi, T., Ichijo, H. (1997). Self-oscillating gels. *Adv. Mater.*, *9*, 175–178.
13. Yoshida, R. (2008). Self-oscillating polymer and gels as novel biomimetic materials. *Bull. Chem. Soc. Jpn.*, *81*, 676–688.
14. Yoshida, R., Sakai, T., Hara, Y., Maeda, S., Hashimoto, S., Suzuki, D., Murase, Y. (2009). Self-oscillating gel as novel biomimetic materials. *J. Controlled Release*, *140*, 186–193.
15. Yoshida, R., Tanaka, M., Onodera, S., Yamaguchi, T., Kokufuta, E. (2000). In-phase synchronization of chemical and mechanical oscillations in self-oscillating gels. *J. Phys. Chem. A*, *104*, 7549–7555.

16. Yoshida, R., Takei, K., Yamaguchi, T. (2003). Self-beating motion of gels and modulation of oscillation rhythm synchronized with organic acid. *Macromolecules*, *36*, 1759–1761.

17. Ito, Y., Nogawa, N., Yoshida, R. (2003). Temperature control of the Belousov-Zhabotinsky reaction using a thermo-responsive polymer. *Langmuir*, *19*, 9577–9579.

18. Maeda, S., Hara, Y., Yoshida, R., Hashimoto, S. (2008). Peristaltic motion of polymer gels. *Angew. Chem. Int. Ed.*, *47*, 6690–6693.

19. Sasaki, S., Koga, S., Yoshida, R., Yamaguchi, T. (2003). Mechanical oscillation coupled with the Belousov-Zhabotinsky reaction in gel. *Langmuir*, *19*, 5595–5600.

20. Amemiya, T., Ohmori, T., Yamaguchi, T. (2000). An oregonator-class model for photoinduced behavior in the Ru(bpy)$_3$$^{2+}$-catalyzed Belousov-Zhabotinsky reaction. *J. Phys. Chem. A*, *104*, 336–344.

21. Yoshida, R., Sakai, T., Tabata, O., Yamaguchi, T. (2002). Design of novel biomimetic polymer gels with self-oscillating function. *Sci. Technol. Adv. Mater.*, *3*, 95–102.

22. Takeoka, Y., Watanabe, M., Yoshida, R. (2003). Self-sustaining peristaltic motion on the surface of a porous gel. *J. Am. Chem. Soc.*, *125*, 13320–13321.

23. Shinohara, S., Seki, T., Sakai, T., Yoshida, R., Takeoka, Y. (2008). Chemical and optical control of peristaltic actuator based on self-oscillating porous gel. *Chem. Commun.*, 4735–4737.

24. Shinohara, S., Seki, T., Sakai, T., Yoshida, R., Takeoka, Y. (2008). Photoregulated wormlike motion of a gel. *Angew. Chem. Int. Ed.*, *47*, 9039–9043.

25. Maeda, S., Hara, Y., Sakai, T., Yoshida, R., Hashimoto, S. (2007). Self-walking gel. *Adv. Mater.*, *19*, 3480–3484.

26. Maeda, S., Hara, Y., Yoshida, R., Hashimoto, S. (2008). Control of dynamic motion of a gel actuator driven by the Belousov-Zhabotinsky reaction. *Macromol. Rapid Commun.*, *29*, 401–405.

27. Murase, Y., Maeda, S., Hashimoto, S., Yoshida, R. (2009). Design of a mass transport surface utilizing peristaltic motion of a self-oscillating gel. *Langmuir*, *25*, 483–489.

28. Yoshida, R., Omata, K., Yamaura, K., Ebata, M., Tanaka, M., Takai, M. (2006). Maskless microfabrication of thermosensitive gels using a microscope and application to a controlled release microchip. *Lab Chip*, *6*, 1384–1386.

29. Yoshida, R., Omata, K., Yamaura, K., Sakai, T., Hara, Y., Maeda, S., Hashimoto, S. (2006). Microfabrication of functional polymer gels and their application to novel biomimetic materials. *J. Photopolym. Sci. Technol.*, *19*, 441–444.

30. Tabata, O., Hirasawa, H., Aoki, S., Yoshida, R., Kokufuta, E. (2002). Ciliary motion actuator using self-oscillating gel. *Sens. Actuators, A*, *95*, 234–238.

31. Tateyama, S., Shibuta, Y., Yoshida, R. (2008). Direction control of chemical wave propagation in self-oscillating gel array. *J. Phys. Chem. B*, *112*, 1777–1782.

32. Yoshida, R., Sakai, T., Ito, S., Yamaguchi, T. (2002). Self-oscillation of polymer chains with rhythmical soluble-insoluble changes. *J. Am. Chem. Soc.*, *124*, 8095–8098.

33. Ito, Y., Hara, Y., Uetsuka, H., Hasuda, H., Onishi, H., Arakawa, H., Ikai, A., Yoshida, R. (2006). AFM observation of immobilized self-oscillating polymer. *J. Phys. Chem. B*, *110*, 5170–5173.

34. Suzuki, D., Sakai, T., Yoshida, R. (2008). Self-flocculating/self-dispersing oscillation of microgels. *Angew. Chem. Int. Ed.*, *47*, 917–920.

35. Suzuki, D., Yoshida, R. (2008). Temporal control of self-oscillation for microgels by cross-linking network structure. *Macromolecules*, *41*, 5830–5838.

36. Suzuki, D., Yoshida, R. (2008). Effect of initial substrate concentration of the Belousov-Zhabotinsky reaction on self-oscillation for microgel system. *J. Phys. Chem. B*, *112*, 12618–12624.

37. Suzuki, D., Taniguchi, H., Yoshida, R. (2009). Autonomously oscillating viscosity in microgel dispersions. *J. Am. Chem. Soc.*, *131*, 12058–12059.

38. Taniguchi, H., Suzuki, D., Yoshida, R. (2010). Characterization of autonomously oscillating viscosity induced by swelling/deswelling oscillation of the microgels. *J. Phys. Chem. B*, *114*, 2405–2410.

39. Sakai, T., Yoshida, R. (2004). Self-oscillating nanogel particles. *Langmuir*, *20*, 1036–1038.

40. Sakai, T., Takeoka, Y., Seki, T., Yoshida, R. (2007). Organized monolayer of thermosensitive microgel beads prepared by double-template polymerization. *Langmuir*, *23*, 8651–8654.

41. Hara, Y., Yoshida, R. (2008). Self-oscillating polymer fueled by organic acid. *J. Phys. Chem. B*, *112*, 8427–8429.

CHAPTER 13

SELF-ASSEMBLY OF CONJUGATED POLYMERS AND THEIR APPLICATION TO BIOSENSORS

David Bilby and Jinsang Kim

Department of Materials Science and Engineering, University of Michigan, Ann Arbor, MI,

13.1 INTRODUCTION

Conjugated polymers (CPs) have attracted much interest because of their tunable optical and electronic properties [1–4]. Applications ranging from microcircuitry to organic light emitting diodes, photovoltaics, and biosensors benefit from the electronic and optical properties of CPs as well as the lower cost production routes as compared to inorganic equivalents [5–7]. Generally, these devices rely on some form of morphology or structure that is on the order of a micrometer or smaller. Conventional methods used in the assembly of small-scale features in inorganics involve high vacuum and high temperature, which may be incompatible with or may degrade macromolecules. Therefore, polymers are designed to self-assemble, due to electrostatic, surface energy, or other intermolecular interactions, into the desired morphology.

More than just a means of achieving small-scale features, self-assembly leads to control over molecular conformation and therefore electronic and optical properties. The optical absorption, fluorescence emission, molecular orbital energy level, and intramolecular energy transfer characteristics of CPs depend on their inter- and intramolecular configurations. For instance, CP backbone twisting shifts the optical absorption and fluorescence emission to shorter wavelengths. Self-assembly is a means to control the spacial arrangement of CPs, including influences on backbone twisting, and can therefore be used to control these optical properties. As a result, many applications can benefit from the control over both small-scale features and optoelectronic properties that self-assembly affords.

Supramolecular Soft Matter: Applications in Materials and Organic Electronics, First Edition. Edited by Takashi Nakanishi.
© 2011 John Wiley & Sons, Inc. Published 2011 by John Wiley & Sons, Inc.

CP-based biosensors are a prototypical example of the self-assembly of smart polymers for an optical application because they utilize a wide variety of self-assembly mechanisms. They benefit from self-assembly-mediated control over the aforementioned optical properties to create sensory signals. Conjugated polyelectrolytes are CPs containing water-soluble side chains and are commonly used for biosensor development. Biosensors are devices that convert a change induced by recognition of a specific biological molecule into a (typically optical or electronic) signal that can be qualitatively or quantitatively measured. For example, a recognition event with a target biological molecule can trigger a change in conformation or molecular spacing of a sensory CP, which changes the optical absorption or emission of the polymer. In this example, the molecular conformation or spacing change takes place on the nanometer scale, and the optical absorption or emission signal that is generated is often facilitated by the self-assembled structure of the system. Different signaling mechanisms, varying from Förster resonance energy transfer (FRET) from well-defined CP films to a dye to complexation between a CP and a dye-labeled ss-DNA (single strand deoxyribonucleic acid), lead to a variety of self-assembly methods [8–11].

Biosensors are used in many applications including those related to health care, environmental monitoring, and food analysis [12]. These applications benefit from the recognition of several molecules or ions, including DNA (deoxyribonucleic acid), glucose, cholesterol, lactate, ethanol, polyphenols, potassium, and many others [12–20]. The detection of these biomolecules is important in the diagnosis of genetic disease, diabetes, and heart disease and in the identification of toxins in the environment and of the composition of foods [12,14–20]. Therefore, accurate biosensors, which are self-signaling, sensitive, and selective toward interesting molecules, are important to improving the quality of life. This goal is the motivating factor behind this discussion of self-assembly and its effects on the optical properties of CPs.

An effective biosensor must have good sensitivity and selectivity to small amounts of the correct analyte molecule. Also, a good biosensor is self-signaling so that additional labeling steps of the analyte molecule are not needed. These requirements bring about some design rules for effective biosensors. Biosensors must have good sensitivity, but owing to the low concentrations of some analyte molecules in biological systems, the sensor must have an amplification mechanism built in [10,13,21,22]. Signal amplification avoids the laborious processes that increase the amount of analyte material. This factor makes CPs ideal for application in biosensors; their extended π-orbital conjugation enables the transport and delocalization of a remote sensory signal along the conjugated backbone of the polymer, thus amplifying the signal. Furthermore, self-assembly is utilized to control intra- and intermolecular conformation, and when a positive recognition in a biosensor changes this conformation, the optical properties change on the macroscale, providing a noticeable signal. Good sensitivity also requires a high signal to noise ratio, that is, optical biosensors require a good photoluminescent quantum yield of conversion from an excitation wavelength to a signal wavelength so that a positive recognition event can be identified.

Even if a positive signal is triggered, if the signal is too weak and cannot be perceived then the device has failed.

Specificity toward a particular biological molecule and self-signaling mechanisms require clever molecular design that incorporates biomolecule-recognizing probe molecules and control over functional groups, dye molecules, and side chains that direct molecular self-assembly. Incorporating specificity toward a biomolecule is a fairly straightforward practice; a receptor molecule that has the specificity built into it is affixed to the sensor and provides selective recognition. Self-signaling, on the other hand, is more complex since it is dependent on the nature of the signaling mechanism. Since the signaling mechanism depends on inter- and intramolecular configuration and conformation, CP self-assembly is of utmost importance. Keeping these considerations in mind, the self-assembly of CPs for optical biosensors can be appreciated.

13.2 BACKGROUND

Biosensors rely on molecular design and self-assembly for some of their most important properties. The optical absorption, emission, and energy transfer characteristics of CPs are sensitive to molecular spacing, conformation, and composition. Also, since most biological molecules are soluble in aqueous environments, CPs for use in homogeneous solution-version biosensors must also conform to the constraints of this solvent. Finally, the molecule must maintain the functional groups necessary for biomolecule recognition and, more importantly, for self-assembly. This section outlines the influence of self-assembly on fluorescence, light absorption, and FRET and the molecular design background about self-assembly of CPs. These combined effects are necessary in understanding the design of CPs for biosensors and the importance of self-assembly in optoelectronic applications.

13.2.1 Influence of Self-Assembly on Optical Absorption and Emission of Conjugated Polymers

Self-assembly of CPs has a large influence on their optical absorption and fluorescence emission properties through the control of intramolecular conformation and intermolecular packing. A photon with energy greater than the electronic band gap of the CP can be absorbed and has multiple relaxation paths available, including thermal dissipation (phonon creation) and fluorescence (typically a Stokes shift and photon emission). In order to promote one path over another, careful molecular design is considered. For instance, Kim and Swager designed Langmuir monolayers of polyphenylene ethnylene (PPE) with control over molecular conformation and packing to show how these factors influence the optical properties [23]. The Langmuir films, or monolayers of CPs at the air–water interface, self-assembled differently because of varying side chain character [23]. Four types of monomers, one with two para-positioned bifurcated alkyl chains (hydrophobic), one with two para-positioned ethylene oxide chains (hydrophilic),

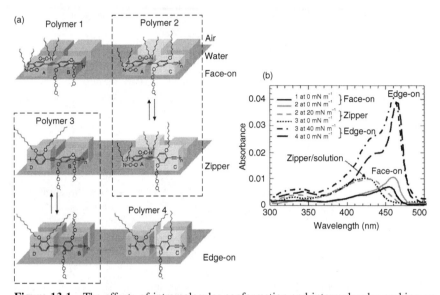

Figure 13-1 The effects of intramolecular conformation and intermolecular packing on the absorption of Langmuir films of PPEs. As the polymer backbone is twisted, the effective conjugation length decreases and there is a blueshift in absorption, as in the zipper conformation [23]. As the polymers form intermolecular $\pi - \pi$ interaction, as shown in the edge-on structure, the fluorescence quantum yield drops down significantly. (a) Illustration of polymer conformations and packing and (b) the absorption spectrum of the different conformations.

one with para-positioned alkoxy and ethylene oxide chains (amphiphilic), and one with two ortho-positioned alkoxy chains (hydrophobic), were designed to take particular orientations (standing or laying) at the air–water interface and were combined in pairs into polymers, as seen in Fig. 13-1 [23]. The fully hydrophobic monomers stood on edge (ortho-positioned chains; the phenyl ring perpendicular to the surface) or on their face (para-positioned chains), whereas the hydrophilic or amphiphilic monomers could take on either conformation depending on the imposed surface pressure [23]. This allowed the polymers to take on fully face-on, fully edge-on, or zipper (twisted backbone) conformation [23]. In addition, depending on the compression stress state of the films, reversible switching between zipper and face-on or edge-on conformations takes place in polymers with the hydrophilic or amphiphilic monomers [23].

The intramolecular conformation and intermolecular packing of CPs determine their optical properties. The PPE Langmuir films exhibit differing optical absorption and emission peaks and intensities depending on polymer packing and backbone twisting [23]. The absorption and emission peaks depend on the effective conjugation length of the polymers. Therefore, when the polymer is in solution or when the Langmuir film is in a zipper conformation, the effective conjugation length is short and there is a blueshift in the absorption and fluorescence

peaks [23]. Conversely, when the polymer is in a face-on or edge-on conformation, the conjugation length is extended and there is a redshift in the absorption
and fluorescence emission peaks [23]. Additionally, there is a further redshift in
both peaks, and the emission peak intensity is decreased in edge-on conformation
polymers due to $\pi - \pi$ stacking and self-quenching by aggregation [23]. These
effects convey the importance of self-assembly on the optical properties of CPs.

13.2.2   Molecular Structure, Self-Assembly, and Quantum Yield of Conjugated Polyelectrolytes

Luminescence quantum yield refers to the efficiency of converting absorbed photons of one wavelength to emitted photons of another wavelength. It is important
to maximize this parameter in sensors in which the absorbing CP directly (by
fluorescence or fluorescence quenching) or indirectly (by energy transfer) participates in generating the optical sensing response. A high quantum yield in
a biosensor allows for greater distinction between the background (reflection
and emission) signal of a nonrecognition event with an analyte molecule and a
positive, highly responsive signal. As discussed previously, the optical properties of CPs are highly dependent on their configuration and conformation, and
maximization of the quantum yield requires control over these parameters.

Self-assembly, through careful molecular design, is utilized to maximize the
quantum yield of CPs. Lee *et al*. have investigated the influence of aggregation,
side chain length, and proximity of ionic pendant groups on the fluorescence
quantum yield of PPE polyelectrolytes [24]. The conventional method to make
conjugated polyelectrolytes is by attaching water-soluble side chains such as a
charge group or ethylene oxide. However, the strong self-aggregation tendency of
CPs due to their rigid and hydrophobic backbone only allows them to be soluble
in water in extremely small amounts. The authors have shown that when the
conventionally designed conjugated polyelectrolyte, with alternating ionic and
ethylene oxide side chains (polymer P1 in Fig. 13-2), was put into water, it had
a low quantum yield of 0.45% [24]. In water, the polymer self-aggregated into
treelike, fractal fibers, as seen in Fig. 13-2, because of hydrophobic interactions
between the rigid polymer backbone chains [24]. Polymer aggregation reduces
the quantum yield by promoting a nonemissive exciton relaxation pathway. When
the aggregation is reduced by adding a better solvent (methanol) to the water,
by adding a charged surfactant molecule, or by using solely a good solvent
(chloroform), the quantum yield can be increased up to 45% [24]. Although
the quantum yield can be increased by these additives, they are not compatible
with analyte molecules in a biosensor, so aggregation effects have to be reduced
through other molecular design routes.

In order to reduce self-aggregation between the hydrophobic polymer
chains, Lee *et al*. replaced the ethylene oxide side chains with bifurcated ethylene
oxide chains, which are bulkier, shielding the backbone chains from each other
and preventing self-aggregation (polymer P2 in Fig. 13-2) [24]. Despite this
change and the significantly reduced self-aggregation, the polymer showed a low
quantum yield in water (0.9%), likely due to photoinduced charge transfer from

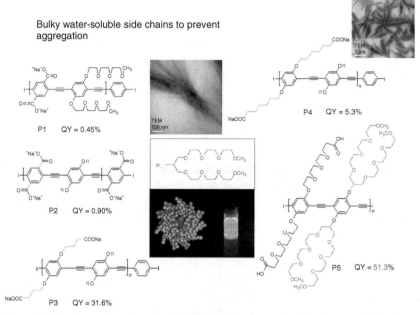

Figure 13-2   The series of water-soluble PPEs that were developed to maximize the luminescent quantum yield. When the polymer side chains are not bulky enough, as in P1, the CP self-aggregates (transmission electron microscopic (TEM) image in upper middle) in water, leading to fluorescence quenching. Also, even when bulky side groups are added, as in P2, adjacent charged groups participate in charge transfer and quench the fluorescence. Thus, alkyl spacers are added to distance the charge group from the CP backbone, as in P3 and P4. However, when the alkyl spacers are too long, they also induce self-aggregation, as in P4 and TEM image shown in upper right. Therefore, long, bulky, water-soluble side chains completely surround the polymer, as in P5, and distance it from its pendant charge groups, allowing for water solubility and a high luminescent quantum yield (lower middle) [24]. (*See insert for color representation of the figure.*)

the conjugated main chain to the directly attached charged side chain [24]. Lee *et al.* prepared PPEs with different spacers between the charged side groups and the fluorescent main chain to investigate the origin of the quenching [24]. By introducing a propyl spacer group between the conjugated main chain and the charged group, the quantum yield is increased to 31% (polymer P3 in Fig. 13-2) [24]. However, by increasing the length of the alkyl chain further to a hexyl group (polymer P4 in Fig. 13-2), because of its hydrophobicity, the polymer begins to show self-aggregation and a lowered quantum yield again [24]. Lastly, by switching the alkyl spacer to a water-soluble ethylene oxide spacer (polymer P5 in Fig. 13-2), the self-aggregation and charge transfer quenching mechanisms can be suppressed and the quantum yield raised to 51% [24].

This example clearly shows the influence of molecular design on self-assembly and the influence of self-assembly on quantum yield. When the CP is designed with compact or long hydrophobic side chains, its natural tendency to self-aggregate in water leads to closely packed self-assembly and fluorescence quantum yield quenching. Furthermore, when the CP design holds ionic side groups too close, intramolecular charge transfer quenches the quantum yield. Thus, highly luminescent conjugated polyelectrolytes can be achieved by careful design control over self-assembly between CP chains and between ionic side groups and polymer main chains.

13.2.3   Influence of Self-Assembly on Intramolecular Energy Transfer

Optically signaling biosensors rely on fluorescence for signal generation. Even when a polymer has a high fluorescence quantum yield, if it has a small Stokes shift, then the incident excitation light may blend into the slightly shifted signal light, and a positive signal may be lost in the noise. In order to circumvent this and further increase the signal to noise ratio, sensors are usually designed with a mechanism that increases the shift between the incident and fluoresced light. For instance, intramolecular hydrogen bonding between adjacent moieties in the conjugated backbone can induce a highly planar, lowered energy state, which causes a redshift in emission (similar to the self-assembly effects discussed earlier) [25]. Similarly, adjacent hydrogen bond acceptors and donors in a polymer backbone can undergo an excited-state intramolecular hydrogen transfer; a hydrogen transfer mechanism uses some of the absorbed energy, reducing that available for fluorescence, thus causing a redshift in the emission and leading to an increased Stokes shift of about 200 nm [26–31]. A third mechanism, most popular in the literature, to increase the shift between absorption and emission involves energy transfer from an excited molecule to an emissive molecule in a process known as FRET. This process is enabled by self-assembled close proximity between two chromophores.

Biosensors benefit greatly from FRET. It offers mechanisms for amplification, analyte–receptor binding recognition, and sensor noise reduction. FRET is the transfer of energy through space from a high-energy excited-state molecule to another molecule that has a lower band gap. When the energy accepting molecule relaxes, it emits a photon at a lower energy than that used to excite the donor, thus distinguishing a positive signal from background noise through a wavelength shift. Because FRET occurs between two molecules, self-assembly must be utilized to bring the molecules close and signal generation and bioreceptor binding functions are separated. For instance, in many biosensors, FRET occurs between a dye and a CP; the polymer serves as an absorbing antenna, and the dye serves as a highly emissive, wavelength-shifted chromophore. The polymer is typically attached to a receptor biomolecule, whereby it collects more energy than a small molecule could, creating an energy harvesting source, and the dye is typically attached to the analyte molecule, whereby its small size allows for the pair to be mobile. Since the donor and acceptor are bound to conjugate analyte–receptor pairs, a recognition event allows the donor and acceptor to be brought close enough for FRET to occur.

The efficiency of FRET depends on many factors; as mentioned above, one of those factors is the spacing between donor and acceptor molecules [32,33]. More specifically, the efficiency, E, of FRET can be described by

$$E = \frac{R_o^6}{R_o^6 + R_{DA}^6}; \quad R_o = 0.2108 \times \left(\frac{J_{DA}(\lambda) \cdot \kappa^2 \cdot \phi_D}{n^4} \right)^{\frac{1}{6}}$$

where R_o is the Förster radius in angstroms, R_{DA} is the distance between the donor and acceptor, κ is an orientation factor, n is the refractive index of the medium, ϕ_D is the quantum yield of the donor, and $J_{DA}(\lambda)$ is the spectral overlap integral [32]. Each of these factors can be controlled in a self-assembled system, especially intermolecular spacing. For this reason, self-assembly is critical in optoelectronic applications in which control over intermolecular energy transfer is desired.

One energy relaxation route that competes with FRET, photoinduced charge transfer, will take precedence if the donor–acceptor distance or energy level alignment is not optimized for FRET [33]. Efficient FRET requires that the donor and acceptor are in close proximity and that the donor energy levels lie outside of those of the acceptor (i.e., the HOMO (highest occupied molecular orbital) and LUMO (lowest unoccupied molecular orbital) of the donor are lower and higher, respectively, than those of the acceptor) [33,34]. Photoinduced charge transfer has a stronger distance dependence than FRET and, in contrast, works better if the energy levels are staggered (i.e., the HOMO and LUMO of the donor are higher than those of the acceptor) [33]. Because of this, efficient FRET requires careful molecular design and self-assembly to ensure that the energy level alignment is correct and that the donor and acceptor are close but not too close. Once again, this distance dependence emphasizes the importance of self-assembly on the intermolecular energy transfer characteristics of a molecular system.

Additionally, FRET requires that the emission spectrum of the donor and the absorption spectrum of the acceptor overlap well and that the molecules have a proper orientation relative to each other [32,33]. The orientation dependence of FRET can be very strong in some systems. Lee et al. utilized self-assembly to investigate the influence of the orientation factor on the FRET efficiency between two anisotropic, rod-shaped chromophores, a PPE and a CdTe nanowire [32]. A water-soluble PPE, with N-hydroxysulfosuccinimide-substituted carboxyl groups at the polymer chain ends, was bound to the nanowire surface in different orientations [32]. The spectral overlap of and distance between the donor and acceptor were correct for FRET [32]. Since the CdTe nanowires were much larger than the individual polymer chains, control over their surface binding orientation is related to their concentration. In low concentrations, both ends of the PPEs would bind to the amine stabilizers on the nanowires, creating a parallel configuration; in high concentrations, a perpendicular arrangement was fabricated [32]. When the chromophore dipoles are aligned (parallel conformation), excitation of the polymer leads to FRET emission enhancement by 4.2 times for the nanowire, while there was no FRET in the perpendicular conformation [32]. This example shows how orientation control of molecular self-assembly of CPs in a multichromophore system can be used to control FRET efficiency.

Aside from FRET and photoinduced charge transfer, close proximity conjugated molecules can participate in other interactions such as excimer creation. An excimer is a stabilized excited state; two molecules, one in its ground state and the other in an excited state, can share an exciton and stabilize it with a slight energy relaxation [35]. Emission from an excimer results in a redshifted light wavelength because of the energy relaxing stabilization [35,36]. Excimers can form when two planar, cofacial chromophores are in close proximity (3–4Å), such as in CP aggregates or in Langmuir films [35]. Therefore, self-assembly is again critical in controlling this intermolecular optical phenomenon.

These examples show the influence that self-assembly has on the optical absorption, fluorescence emission, and intermolecular energy transfer in CP systems. Self-assembly is used to control self-aggregation, intermolecular spacing, intermolecular conformation, and intramolecular conformation. These physical controls in turn influence the peak shifts in optical absorption and emission as well as the fluorescence intensities and intermolecular energy transfer characteristics. This control over optical properties makes self-assembly of CPs a versatile design tool for optical biosensors and other optoelectronic applications.

13.3   SELF-ASSEMBLY OF CONJUGATED POLYMERS

Many different routes exist for the self-assembly of CPs for optoelectronic applications. Typically, surface energy interactions, such as those between hydrophilic and hydrophobic groups, between oppositely charged groups, and between polar groups, are used to direct self-assembly. These interactions can lead to multilayer films, vesicle or micelle formation, monolayer formation, and directed thin film growth [13,36]. In the biosensor example, different analtye biomolecules or sensing mechanisms can take advantage of certain self-assembly constructs. For instance, CPs could serve as FRET donors that are activated or quenched upon the recognition of a specific molecule or ion, or a polydiacetylene liposome, which is stress sensitive, can render analyte-specific colorimetric detection [13]. However, because no particular assembly is favored for an analyte, many variations can be fabricated. Therefore, biosensor fabrication is only constrained by the ideal "lab-on-a-chip" case, where the sensory CP is fixed to a substrate in an array.

13.3.1   Langmuir–Blodgett

Langmuir–Blodgett (LB) films are monolayers of typically amphiphilic or hairy-rod-type polymeric molecules at the air–water interface [37,38]. Amphiphilic molecules self-assemble with their hydrophilic portions in contact with the water phase and their hydrophobic parts as far away as possible, creating an ordered assembly [36]. Since CPs are rather rigid and hydrophobic in nature, and thus tend to self-aggregate, it is very difficult to form a stable interface with water unless hydrophilic side chains or hairy, main-chain-surrounding side chains are included [36,38]. However, once a stable interface is formed, the monolayer can be compressed into a changing conformation [23,37,39]. Also, a stable mono-layer can be transferred to a substrate by dipping vertically or horizontally a hydrophilic, hydrophobic, or patterned surface through the monolayer. As an example of these concepts, highly regioregular polythiophene with alternating hydrophilic and hydrophobic side chains assembles into tightly packed π-stacked monolayers on water, with their faces perpendicular to the interface [39]. When these monolayers were compressed, they formed highly oriented nanowires [39]. Finally, the monolayers were transferred to patterned solid substrates to demonstrate that their ordering could be retained [39]. Alternatively, by choosing proper hydrophilic or hydrophobic side chains on a rodlike CP such as PPE, which have two attaching points, one can control the orientation of each unit of the polymer with respect to the water interface [23]. For instance, by choosing one monomer with two hydrophilic side chains and putting this next to a monomer with a hydrophilic and a hydrophobic side group, the polymer will take on a zipper-like conformation [23]. By careful choice of side groups and backbone structure, the conformation of the polymer in the self-assembled monolayer can be controlled. As discussed earlier, this control is important because the conformation of the polymer has an influence on the optical (absorption and emission) and electronic (conductivity) properties of the film.

13.3.2 Surface Interaction

Similar to interactions with the air–water interface, polymer–substrate interactions can also be used to direct self-assembly. The π-interaction between CPs can be a strong force that drives aggregation into a $\pi-\pi$ stacking formation, but a strong π-interaction between the polymer and the substrate can direct monolayer assembly [36]. Highly oriented pyrolytic graphite competes with interpolymer π-stacking to form a layer of CP with the π-plane and side chains of each moiety parallel to the substrate [40]. With regioregular poly(3-alkylthiophene), this self-assembly leads to regular hairpin structures that are tightly packed and parallel to the substrate [41]. This self-assembly method has also been demonstrated with PPE; the π-interactions between the polymer and the substrate direct assembly [40]. When the substrate is switched to a hydrophobic, non-π-interacting substrate such as mica or glass, the PPE forms π-stacked ribbons perpendicular to the substrate, leaving only the side chains in contact [42]. These surface energy interactions can help control the alignment of the π-planes of the polymer's backbone, leading to controlled conjugation lengths and therefore controlled optical and electronic properties associated with this.

13.3.3 Block Copolymers

When immiscible polymers are mixed, they phase separate; when immiscible polymer sections are combined in a block copolymer, their phase separation can be controlled. This phenomenon is well understood in flexible coil-type polymers, where choosing the relative volume fraction of the blocks and the correct solvents can selectively form micelle, cylinder, lamellae, or bicontinuous phase separation morphologies [43]. However, self-assembly of rod-like conjugated block copolymers is not as straightforward because CP blocks do not have a coiled structure, and control of the molecular weight of each block does not predictably change the curvature of the interface. Moreover, traditional synthetic methods to create block copolymers are incompatible with CPs [36]. Despite this, some rod-coil block copolymers have been synthesized by joining conjugated blocks and nonconjugated blocks that are prepared separately [44]. With such a polymer, poly(phenylquinoline)-*b*-polystyrene, Jenekhe and Chen were able to, using a solvent selective for the coil block, create hollow CP cores surrounded by polystyrene through block copolymer self-assembly, as seen in Fig. 13-3 [45]. On casting, the spheres assembled further into microporous films [45]. Additionally, they were able to form cylinders, lamellae, and vesicles by varying the solvent composition to partially favor both blocks in different proportions [44].

Although there is no thorough understanding of the phase separation properties of block copolymers containing conjugated units, interesting properties can be obtained by mixing functional polymer blocks with conjugated blocks. For

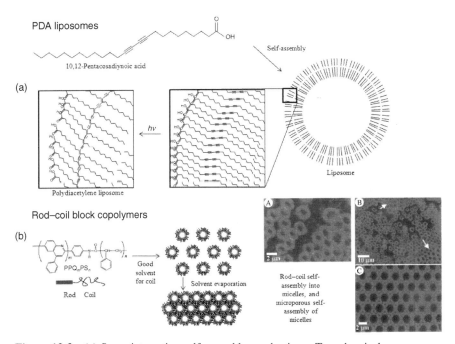

Figure 13-3   (a) Some interesting self-assembly mechanisms. Topochemical polymerization of polydiacetylene (PDA) involves the self-assembly of the amphiphilic diacetylene monomers into liposomes followed by UV polymerization. (b) Rod-coil block copolymers self-assemble into micelles based on solvent selectivity of each block [45]. The fluorescence image on the right shows the self-assembled micelles, with arrows indicating ordered microporous domains.

instance, poly(N-isopropylacrylamide) (PNIPAAM) exhibits a lower critical solution temperature (LCST) (32°C), above which it is no longer soluble in aqueous solution [46]. That is, at low temperatures, the polymer is swollen and solvated by water, and at high temperatures, it contracts into insoluble aggregates. The combination of PNIPAAM, a coil-like polymer, and polyfluorene (PF), a rodlike CP, in a block copolymer can take advantage of the optoelectronic and assembly characteristics afforded by each block. Wang *et al.* report such a block copolymer, with ionic side chains on the PF blocks to impart water solubility [46]. At 15°C, the block copolymer is soluble in water, and because of the conjugated unit, it is fluorescent. On increasing the temperature, the polymer assembles into micelles with PNIPAAM cores and PF shells because of the LCST of PNIPAAM [46]. The solubility of the PF block is relatively temperature independent; however, at high temperatures, the PNIPAAM blocks condense into aggregates and pull the PF blocks along into the assembly. More so, the assembly temperature can be tuned by choosing the hydrophobicity of the second block. Since PF is conjugated and the stiff backbone is relatively hydrophobic, its presence in the copolymer lowers the LCST to 27°C [46]. In other words, since the PF block is not quite soluble in water, it makes the PNIPAAM to come out of solution at a lower temperature. The temperature-induced micellization of the PF-*b*-PNIPAAM brings many PF chains into close proximity, leading to interesting optical properties. If incorporated into a biosensor, for instance, dual functionality could be achieved. An assembly-dependent quenching or FRET amplification effect could be utilized in a biosensor; small amounts of a dye or quencher bound to an analyte molecule would have unnoticeable effects on the fluorescence of the polymer until it was assembled into micelles. Alternatively, a hydrophobic or hydrophilic analyte could alter the LCST of the block copolymer upon a binding recognition effect, changing the assembly characteristics, which could be probed through FRET or fluorescence quenching.

While assembly can be designed on a case-by-case basis, there is no good understanding of general phase separation in CP-containing block copolymers. For certain rod–coil block copolymers, nanowire structures are observed to have assembled in thin films; this unusually high aspect ratio structure could have formed as a result of $\pi-\pi$ interactions between rod-shaped conjugated blocks [36,47]. Different secondary bonding mechanisms are at work in the assembly of this kind of polymer, so the predictability afforded by coil-coil-type polymers is lost. Furthermore, rod-rod-type block copolymers have been synthesized, but typically do not exhibit long range order in their phase separation, likely due to the rigid nature of the constituents [36,48,49]. In an example contrary to this, a rod–rod copolymer based on polythiophene with different aliphatic side chains in each block, poly(3-octylthiophene)-*b*-poly(3-butylthiophene), exhibits a long-range nanowire formation [50]. The homopolymers of each block exhibit spontaneous aggregation into nanowires in solution [50]. The block copolymer also assembles into quite crystalline nanowires (~13.5 nm by 250–1000 nm) in solution [50]. This shows that the $\pi-\pi$ interaction is stronger than side chain interactions in determining the microstructure on the nanometer scale [50]. While this is not a traditional block copolymer with chemically immiscible blocks, the different side chain lengths influence the equilibrium packing lengths of each block in a solid film. This leads to an interesting phase separation within the nanowires; crystalline domains of each block separate to form periodic (~25.7) lamellar, smectic C phases [50]. Therefore, side chain interactions seem to be the dominant factor for determining the assembly of the blocks. This interplay of different interactions needs to be well controlled to design a particular microstructure. In this case, since both blocks are based on a polythiophene backbone, the absorption and emission spectra of the block copolymer match quite well with either homopolymer [50]. Self-assembly into nanowires has important implications for conductivity. However, other interesting optical and electronic effects could be engineered if more was understood about controlling the secondary bonding interactions that govern self-assembly in conjugated block copolymers. Once block copolymer self-assembly is further developed for CPs, it will have many applications, including in organic solar cells, biosensors, and light emitting diodes, where phase separation, aggregation, and phase percolation are crucial for tuning the charge carrier mobility and absorption and emission characteristics of the polymer.

13.3.4 Liposomes and Polydiacetylene

CP liposomes can be fabricated by designing self-assembling diacetylene molecules with hydrophilic and hydrophobic groups. Liposomes are similar in structure to the lipid bilayer wall of living cells. First, the small amphiphilic molecules in chloroform are dried via evaporation; next, an aqueous solvent is added and the molecules assemble during tip-probe sonication [51–53]. The aqueous solvent forces the hydrophobic parts of the molecule to congregate, surrounded by the hydrophilic parts, into a spherical bilayer structure, as seen in Fig. 13-3. The advantage of such a process is that when the diacetylene units (two adjacent triple bonds) are placed in between the hydrophobic and hydrophilic parts of the molecule, as its center, they are held in close proximity after the assembly. These highly reactive groups can then be polymerized, upon 254-nm ultraviolet (UV) light irradiation, into polydiacetylene [13,51–53]. Polydiacetylene is an interesting CP for optoelectronic applications because it exhibits a blueshift in absorption under external stimuli (heat, stress, etc.) and becomes weakly fluorescent [13,51,53,55]. The polymerization of polydiacetylene involves the reaction of one molecule's triple bond with that of another, thus requiring close proximity of the monomers. Therefore, liposome self-assembly is a practical procedure for the ordering and synthesis of polydiacetylene.

Since the hydrophilic surface of the polydiacetylene liposome can be designed with functional groups, biorecognition molecules can be attached to give this assembly a sensing function. When an analyte binds to the receptor on a functionalized polydiacetylene liposome, it stresses the polymer backbone, producing the aforementioned mechanochromism signal. Because of this intrinsic chromic response, polydiacetylene is very versatile; much attention is focused on developing biosensors and self-assembly for polydiacetylene [54].

Aside from liposomes, polydiacetylene can be synthesized by other means of monomer-aligning self-assembly. For instance, two triethoxy silane molecules, bridged by a diacetylene molecule, self-assemble around surfactant molecules into mesoporous, hollow cylinders [56]. The assembly holds the diacetylenes close enough for UV polymerization and subsequent thermochromism [56]. Another monomer-aligning route involves electrospinning the diacetylene monomers within a poorly soluble matrix polymer, forcing them to self-assemble [57]. Next, the electrospun matrix-polymer mat is exposed to UV light, polymerizing the diacetylene monomers within, producing a colorimetric thermoresponsive composite [57]. This versatile material can be formed in any system that can align the monomers; thus, many other polydiacetylene self-assembly methods exist.

13.3.5 Electrostatic Assemblies

Electrostatic interactions are also used to assemble conjugated polyelectrolytes, CPs with charged side groups. A charged substrate is dipped alternately between solutions of positively and negatively charged molecules, at least one of which is a CP, in a layer-by-layer (LBL) deposition process [58]. Alternate layers are adsorbed with a rinse in between creating a layered structure with control over thickness, number of layers, and material properties (imbued from the constituents). In order for deposition to work correctly, one must optimize the charge interactions of the system; correct choice of solvent, ionic strength, pH, molecular morphology, substrate, and substrate functionalization are crucial to control the charge shielding and deposition interactions that will compete with and enhance film formation [58]. For instance, common charged side groups include SO_3^-, CO_2^-, $N(CH_3)_3^+$, NH_3^+, and PO_4^- [59]. Proper control over these factors can allow the deposition of films with hundreds of layers [60]. More interesting, perhaps, is that any charged molecule or ion can be deposited as a layer in the film or that any molecule codissolved with an ionic polymer can be trapped in the film during deposition [58,60,61]. Colloids, proteins, bioreceptors, DNA, and nanoparticles are just a few materials that can be added to the films [58,60,61]. These doped multilayer films can have controlled conductivity and optical properties (absorption, emission, etc.) and sensitivity to different analytes. Careful control over doping materials must be exercised to prevent interference with the deposition process through modification of pH, charge shielding, or other effects that may lead to film destabilization [58]. Nonetheless, this self-assembly method can be practical for biosensors and other optoelectronic applications based on CPs.

Electrostatic interactions are also used to assemble small molecules with conjugated polyelectrolytes. For instance, one of the most widely investigated systems is the binding, or complexation, between a cationic polymer and DNA, which is anionic [9–11]. This simple assembly involves charge–charge binding between the polymer and dual strand or ss-DNA. Once the DNA–polymer complex is formed, the average effective conjugation length of the polymer, or its proximity to a DNA-bound dye can influence fluorescence or FRET luminescence for use in biosensors [9–11]. For instance, a cationic polythiophene will form a duplex with an ss-DNA molecule [11]. This duplex is flexible but mainly has the polymer stretched out to contact the DNA molecule as much as possible. This conformation extends the effective conjugation length of the polymer, reducing its band gap and changing the solution color from yellow ($\lambda_{abs} = 397$ nm) when randomly coiled to red ($\lambda_{abs} = 527$ nm) when the duplex is formed [11]. As a biosensor, when the correct analyte DNA is added, hybridization occurs between DNA chains and a triplex is formed. This change causes a blueshift in the absorption ($\lambda_{abs} = 421$ nm) as the polymer is contorted around the double strand DNA complex [11]. This example shows how self-assembly between CPs and DNA influences the optical absorption of the CP.

Self-assembly competition between DNA, PNA (peptide nucleic acid, a charge neutral DNA analog) and a conjugated polyelectrolyte produces changes in intermolecular energy transfer character [10]. When this system is used as a biosensor, an analyte DNA strand is added to a solution containing a dye-conjugated, probe PNA and a cationic CP [10]. Since the PNA is charge neutral, before the DNA is added the distance between the dye and CP is too large for FRET to occur [10]. When the correct DNA strand is added, it hybridizes with the PNA and electrostatically attracts the CP [10]. Then FRET can occur between the CP and dye in the nearby DNA:PNA–dye complex [10]. The sensor's selectivity is provided both by the PNA sequence and by the self-assembly between the complex and polymer; if the analyte DNA is the incorrect sequence, it will still assemble with the polymer; however, it will not bring the PNA-dye molecule close, and no appreciable FRET will occur [10]. Thus, the electrostatic and hydrogen-bonding DNA/PNA hybridization self-assembly mechanisms influence intermolecular energy transfer in this system.

Similar sensors can be made utilizing electrostatic interactions with ribonucleic acid (RNA) instead of DNA; charged proteins interact predictably with RNA and can influence the luminescence characteristic of a conjugated polyelectrolyte:RNA complex [62]. With these intentions, it is important to consider molecular design because factors such as charge density and polymer backbone shape can influence the efficiency of DNA complexation. For example, conjugated polyelectrolytes with varying amounts of meta and para connections between monomers were synthesized to change the shape of the polymer [63]. Since the meta linkers create bends ($120°$) in the polymer backbone (as opposed to the straight para connections), the resulting polymer is convoluted. Although the bent backbone has a shorter conjugation length than a straight polymer (leading to a blueshifted absorption), it is also expected to be able to conform better to an analyte shape. The effect of this backbone change on the assembly and optical characteristics of a polyfluorene-phenylene was investigated through complexation with single strand (dye-hybridized) and dual strand DNA (with an intercalating dye) [63]. In both cases a greater meta-linker content, and therefore greater backbone convolution, led to better FRET efficiency with the dye that was involved. The bent polymer bound more closely to the ss-DNA (and dye) or had more favorable conformations for FRET energy transfer. The dual strand DNA was also bound closely, holding the intercalated dye in good proximity for FRET [63,64]. Therefore, proper design of conjugated polyelectrolytes can enhance the intermolecular energy transfer within a polymer:small molecule complex by influence on self-assembly strength.

Complexes between DNA and cationic CPs can be directed to self-assemble into highly ordered nanostructures. Dendrite-shaped fractal aggregates, which look similar on many micron or smaller length scales, spontaneously assemble when dual strand DNA or ss-DNA oligomers are put in solution with cationic CPs [64]. DNA is anionic and does not aggregate in solution because of electrostatic repulsion. However, a charge neutral cationic polymer:DNA complex will aggregate. These complexes diffuse on a random walk path, encounter each other periodically, and stick to each other in a fractal morphology, as seen in Fig 13-4. The length and width of the microstructure can be controlled through various mixing parameters [64]. For instance, since the assembly process is diffusion limited, increasing the diffusion length between particles (low concentration) influences the shape of the dendrites [64]. Also, aggregation between DNA:polymer complexes depends on charge density; increasing the charge density of one constituent over the other will increase repulsive forces between complexes, lowering collision speed [64]. Therefore, different fractal aggregates, with long or short and thick or thin dendrites, can be formed. This assembly affects the packing between CP chains and therefore the optoelectronic properties (absorption and fluorescence emission peaks and intensities, etc.) of the assemblies.

13.3.6 Surfactant-Directed Liquid Crystalline Assembly

Complexes between conjugated polyelectrolytes and small surfactant molecules can be used to direct liquid crystalline self-assembly. Yoon *et al.* combined anionic, regiorandom poly(3-thiopheneacteate) and cationic dialkyl chains of various lengths to create polythiophene complexes with electrostatically bound side chains [65]. Once the side chains were bound in place, the polymer assembles as a result of interactions between the side chains, leading to a closely packed, aligned-chain structure. More so, by varying the length of the side chains, the stability, intermolecular spacing, and therefore optical absorption and photoluminescence emission peaks of the assemblies could be controlled [65]. The assemblies exhibit both low- and high-angle diffraction peaks; the side chains assemble into regular structures forcing the molecular complexes to form layered smectic structures at a larger length scale [65]. Increasing the chain length intuitively increases the interplanar spacing on both assembly length scales. Higher molecular weight side chains also lead to stronger interactions and higher melting and decomposition temperatures [65]. Also, the longer chains induced a redshift in the maximum absorbance and emission peaks of the regiorandom films [65]. Interestingly, this effect is not seen in regiorandom poly(3-alkylthiophene) films for various alkyl chain lengths because the regiorandom structure is highly contorted [65]. Therefore, Yoon *et al.* have shown that the surfactant complexes afford control over the polymer backbone twisting that is not conventionally available; the strong

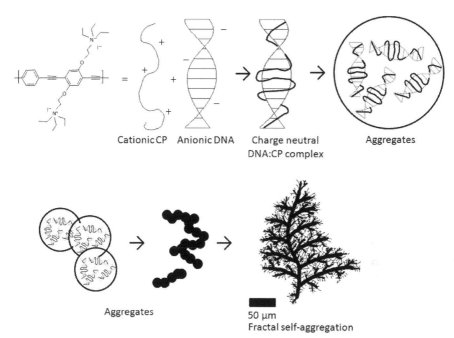

Cationic CP   Anionic DNA   Charge neutral   Aggregates
DNA:CP complex

Aggregates

50 μm
Fractal self-aggregation

Figure 13-4 The assembly between cationic CPs and DNA creates a charge–neutral complex. This complex diffuses in solution and self-assembles into fractal-like aggregates [63].

interaction between the cationic side chains and the polymer backbone offers another route for control of the absorption and emission peak wavelengths [65].

The aforementioned mechanisms of self-assembly are only a small fraction of those available to CPs and an even smaller sampling of self-assembly as a whole. Self-assembly is designed to take advantage of secondary bonding mechanisms, geometry, and thermodynamics to produce controlled morphologies. These assemblies, changes in packing, and relative orientations in macromolecules, in turn influence the optical, electronic, thermal, and many other properties of the bulk materials. CPs offer additional benefits in assemblies because of their electronic structure; however, they also present new challenges. Assembly of CPs will continue to be an interesting research field for many years because of its diversity, challenge, and benefits.

This background section has offered an elementary introduction to the optical phenomena and assembly mechanisms that are at play in the design of CPs and their assemblies for biosensors. Luminescent emission, aggregation, energy transfer, and self-assembly are all important considerations in the design of a CP-based biosensor. The aforementioned science is too narrow a window through which to view the operation of biosensors; numerous sensing-target molecules exist, and each of these may require a unique system with specific luminescence or assembly characteristics. However, the ideas that were presented provide a necessary foundation for basic understanding of the field.

13.4 APPLICATIONS OF SELF-ASSEMBLY OF CONJUGATED POLYMERS TO BIOSENSORS

The assembly and packing of CPs advantageously influences their optoelectronic properties for biosensors. Since there are molecules and ions whose presence, conformation, and configuration are interesting to identify in mixed materials systems, there are an equally large number of sensory systems available to perform this task. This section introduces some exemplary CP-based systems that utilize the self-assembly mechanisms, discussed previously, in biosensors.

13.4.1 DNA Sensors

DNA is an interesting biomolecule because its constitutional sequence determines genes; it can implicate particular diseases or body traits. Owing to selective hydrogen bonding between base pairs on the backbone of ss-DNA, a complementary DNA sequence is the perfect receptor molecule for selective binding or recognition of a particular DNA sequence. Therefore, many DNA biosensors utilize hybridization between complementary DNA strands as part of their selective signal-generating mechanisms.

Lee *et al.* have developed a series of DNA biosensors based on conjugated molecules. For instance, a PPE was synthesized to have both anionic propyloxy sulfonate and bifurcated ethylene oxide side chains, which, as described earlier, provide water solubility [14]. This polymer was also given carboxylic acid terminal groups; amine-functionalized probe DNA strands were covalently bound to the ends of the PPE by amine-carboxylate reactivity [14]. Both turn-on and FRET-type sensors were constructed. The latter, FRET-type sensor, had a simple probe DNA strand attached to the ends of the PPE. An analyte DNA was labeled with a hexachlorofluorescein dye molecule, whose absorption was well overlapped with the emission of PPE [14]. When the analyte DNA sequence was hybridized with the probe DNA, it formed the common DNA double helix complex and held the dye close to the PPE [14]. Then, when the PPE was selectively excited with 365 nm light, the dye fluoresced via FRET at about 550 nm [14]. This FRET-based fluorescence was much brighter than direct excitation of the dye because of the energy harvesting capability of the large CP chromophore. In the turn-on type sensor, the so-called molecular beacon-style DNA probe was constructed with self-recognition at its ends such that a hairpin-shaped complex would form, holding the tail of the DNA strand near its head [14]. The head of the looped probe DNA strand was bound to the highly fluorescent PPE, and a fluorescence quencher was bound to its tail [14]. Therefore, the PPE fluorescence was quenched because the quencher was held near. When an appropriate analyte DNA strand was introduced, it complexed with the middle-loop part of the probe DNA, opening up the hairpin and moving the head away from the tail [14]. This also removed the quencher from contact with the PPE and allowed it to fluoresce again [14]. This sensor type is self-signaling, eliminating the cost and time-consuming labeling of the analyte DNA. This is a prototypical example of a self-signaling (because the polymer is naturally fluorescent, and there is no labeling step involved), selective (due to the selectivity of the probe DNA strand), and sensitive (due to the drastic changes in fluorescence intensity) biosensor.

A solution-based sensor is nonideal for long-term stability and storage, thus solid-state films are desired for biosensors. Pun *et al*. have synthesized a PPE with ethylene oxide and carboxylic acid-capped hexyloxy side chains for application in DNA biosensors [8]. A chloroform solution (1 mg/mL) of this polymer was spread at the air–water interface for LB thin film assembly [8]. Owing to the alternating hydrophilic (ethylene oxide) and hydrophobic (alkyl) side chains, the polymer assembles into a face-on monolayer conformation, with the π-plane of the monomers lying parallel to the air–water interface [8]. The pressure area isotherms for the film confirm this conformation [8]. This monolayer film is transferred to a hydrophobic glass slide via the LB technique [8]. Finally, the single strand probe DNA molecules, with amine functional groups, are bound to the polymer film by utilizing the pendant carboxylic acid groups [8]. When the appropriate dye-tethered analyte DNA strand is introduced to the sensory CP film, hybridization between the DNA strands occurs, bringing the dye into close proximity with the film. Since the CP is designed to be a FRET donor and an energy harvesting unit, efficient FRET can occur between the dye and CP film, thus amplifying the emission of the dye and allowing higher sensitivity. The LB monolayer assembly of the sensory CP helps create highly sensitive (via the FRET mechanism) and selective solid-state DNA sensors.

Besides optoelectronic and assembly properties, the stability of a CP must be considered during molecular design; UV light and acidic conditions are typically used for solid-state synthesis of peptides or nucleotides; however, these conditions cause CP degradation [66]. To resolve this, Lee *et al*. have designed a polymer containing oxadiazole units for photo- and acid-stability and utilized it in a solid-state DNA microarray [22,66]. The stability of the polymer on acid or UV irradiation allows on-chip probe DNA synthesis, enabling a high throughput DNA microarray. The on-chip DNA synthesis involves the sequential feeding and deprotection of DNA monomers (5′-(4,4′-dimethoxytrityl)-protected nucleophosphoramidite) by a photogenerated acid at selective areas on the CP-coated substrate via UV illumination through a mask [22]. This method allows for the fabrication of various probe DNA sequences for a high throughput solid-state DNA microarray. On target DNA–probe DNA recognition (hybridization), effective FRET (from the CP layer to the target-DNA-tethered dye) amplifies the emission [22].

Lee *et al*. added the label-free feature to the signal amplifying CP-based DNA microarray by introducing either an intercalating dye or the molecular beacon concept [67,68]. An intercalating dye is a molecule that preferably binds to double helical DNA than to ss-DNA. Typically, in systems without FRET, the dye would have to be in high concentrations to be detectable for confirmation of a positive DNA match [67]. However, at high concentrations, the intercalating dye's selectivity toward double helical DNA is significantly reduced, leading to a false-positive. The FRET system allows for a small concentration of the dye, removing the possibility of a false-positive, to produce a detectable signal through energy transfer and amplification [67]. Alternatively, the molecular beacon concept, described in the beginning of this section, was applied to the DNA microarray, with either a quencher or a dye on a hairpin DNA probe for turn-on and turn-off self-signaling-type sensors [68]. This system has high selectivity because of the DNA probe, it has high sensitivity because of the FRET amplification mechanism, and it is self-signaling because the analyte DNA does not have to be labeled.

Many simple, solution-based DNA biosensors take advantage of the charge–charge interactions between anionic DNA and cationic conjugated polyelectrolytes. Refer to the section about electrostatic DNA–CP assemblies (Section 13.3.5) for specific examples. In addition to DNA interactions, these sensors usually utilize CP self-assembly for FRET or for control over CP conformation.

Many DNA sensor variations on the above themes exist. They take advantage of FRET to provide sensitive detection. They also take advantage of single strand probe DNA selectivity. Finally, many introduce interesting, shape-changing parts, which move quenchers or dye molecules away from or toward the CP, to imbue self-signaling attributes to the sensors. Self-assembly, usually driven by electrostatic or hydrogen bonding (within DNA complexes), remains key to the function of these sensors.

13.4.2 Ion Sensitive Field Effect Transistors

Another interesting biomolecule, acetylcholine, participates as a chemical transmitter in neurotransmission in humans [60]. This molecule is interesting for biosensing because a low concentration is associated with Alzheimer's disease [60]. Liu *et al.* have devised an ion sensitive field effect transistor, fabricated via LBL self-assembly, which is sensitive to acetylcholine [60]. The device was fabricated as follows: first, three alternating layers of cationic poly(dimethyldiallyl-ammonium chloride) (PDDA) and anionic poly(styrenesulfonate) (PSS) were deposited onto a substrate between two electrodes as precursor layers [60]. Next, the channel material, polyaniline (PANI), (in acidic solution to provide cationic characteristics) was alternately deposited along with PSS to form a five-layer conduction channel for the transistor on top of the precursor layer [60]. Then, the gate dielectric, SiO_2, was deposited alternately with PDDA [60]. Finally, the selective material was deposited; alternate layers of PDDA and PSS blended with acetylcholine esterase were coated on top of the dielectric [60].

While the preceding discussion about self-assembly of CPs focused on optical applications, many similar materials parameters, such as intermolecular proximity, aggregation, and effective conjugation length, influence the electronic character of CPs too. In this example, Liu *et al.* utilized the precise layered fabrication control that LBL deposition affords to create reproducible field effect transistor-based sensors. The device functions as a regular n-channel field effect transistor until acetylcholine is added in aqueous solution; the acetylcholine esterase catalyzes the hydrolysis of acetylcholine, producing acetic acid, lowering the pH, and affecting the transistor [60]. The lowered pH increases the number of hydrogen ions present; the hydrogen ions penetrate into the conducting PANI layer, increasing the number of charge carriers and thus the drain current [60].

The device gains selectivity through acetylcholine esterase. It is self-signaling because measurements of conductivity changes do not require a labeling process on the analyte. Finally, because conductivity in CPs is sensitive to defects (packing, planarity of molecules, and added ions), the sensor can detect acetylcholine in concentrations as low as 1 µM [60].

13.4.3 Polydiacetylene Liposome-Based Ion Sensors

The presence of both mercury and potassium are interesting to quantify in the body, for instance, mercury accumulation in the body leads to brain damage [51]. Lee *et al.* have created a series of polydiacetylene liposome-based ion sensors to detect these ions [13,51]. In the case of potassium ion detection, polydiacetylene liposomes are fabricated with some of the constituents having N-hydroxysuccinimide functionalization [13]. This functionalization was used to attach a guanine-rich ss-DNA molecule (aptamer) to the outside of the liposome as a probe molecule [13]. Then, the liposomes were arrayed onto amine-functionalized glass [13]. On introduction of potassium cations, the anionic DNA aptamer complexes with the ion to form an electrostatically bound G-quadruplex [13]. Interestingly, despite electrostatic interactions, the G-rich DNA aptamer is selective to potassium, even in the presence of other biologically relevant cations (such as sodium) [13]. Upon quadruplex formation, the bulky complexes physically interfere with each other, creating stress on the polydiacetylene to which they are bound [13]. This stress reduces the effective π-conjugation length of the polydiacetylene backbone, causing a blueshift in its absorption, making the microarray appear red, and also activating red fluorescence [13]. Selectivity is provided by the DNA aptamer forming a G-quadruplex exclusively with potassium, and sensitivity (down to 0.5 mM) and self-signaling are provided by the self-assembled conjugated polydiacetylene backbone [13].

A more practical liposome array sensor would have probe molecules attached after liposome microarraying, thus enabling the end user to control which analyte molecules would be targeted. To this end, Lee *et al.* have created epoxy functionalization on some of the constituents of a polydiacetylene liposome [51]. The epoxy functionalization was used both to secure the liposome to an amine-functionalized glass substrate and to bind the probe molecule to the liposome [51]. A thymine rich ss-DNA aptamer was chosen as the probe because it selectively forms a complex with mercury [51]. Similar to the potassium sensor, when the target molecule (mercury) is added, a bulky complexation takes place, putting stress on the polymer liposome's conjugated backbone [51]. This creates an absorption shift and fluorescence activation, which serve as self-signaling mechanisms in the sensor [51]. Also, selectivity is provided by the DNA probe, and sensitivity (down to 0.03 mM Hg^{2+}) is provided by the CP through the aforementioned signaling mechanisms [51].

There are obviously many other interesting biomolecules related to diseases, toxins, and biological processes. The small sampling of biomolecules discussed above was chosen to showcase the CP systems used to detect them. Although this list of systems is not complete, it shows how self-assembly is used to control molecular packing, aggregation, and molecular conformation of CPs. The sensitivity of CPs to backbone twisting, intermolecular dye proximity, and impurities such as ions or packing defects make them ideal aids for the transduction of biomolecule recognition events to optical or electronic signals in self-assembled systems. These polymers are designed to assemble, driven by secondary bonding interactions, into controlled structures. The wide variety of self-assembly and signaling mechanisms, along with the drive to detect many different biomolecules, will make research on the self-assembly of smart polymers for optoelectronic applications interesting for many years to come.

13.5 OUTLOOK AND FUTURE WORK

Macromolecules are prototypical smart materials because they can be designed to self-assemble into larger, ordered structures. This ability, combined with the tunable optical and electronic properties of CPs, makes them highly interesting and applicable smart materials. The preceding discussion outlines some of the design considerations that influence luminescence and self-assembly of these materials, as well as some applications of these considerations to biosensors. Only a small glimpse into the range of assembly interactions was provided; self-assembly is highly diverse, utilizing many secondary bonding interactions and influencing optical and electronic properties of materials. Also, the diversity of this field means that there is much to be discovered.

Self-assembly is directed through control over secondary bonding forces, and in the future more complex systems will be explored. In systems where multiple forces may act, competition between the forces can be used to direct assembly. Forces including $\pi-\pi$ bonding, hydrogen bonding, van der Waals bonding, halogen bonding, side chain entanglement, electrostatic bonding, and surface/interface energy minimization may act between molecules. Hierarchical competition between these forces could possibly be used to direct multistage assembly once better understanding of these interactions has been attained. The wide range of forces available for assembly and the wide range of molecular design to incorporate these forces will allow for tactile control over intra- and intermolecular conformation.

CPs will have more than just moiety-tunable optical properties; smart self-assembly will lead to conformation-controllable optical and electrical properties. Future research will lead to better understanding of the influence of molecular packing on conductivity and molecular energy levels. Self-assembly will enable control over the alignment and packing of molecules in three dimensions, thus controlling the three-dimensional conductivity and molecular energy levels. This will produce interesting photonic organic crystals and provide control over the absorption, emission, and energy transfer character of the materials.

Smart CPs will continue to command research interest. Control over the properties, through molecular design and self-assembly, make these materials practical for many optoelectronic applications. More so, stimuli-responsive materials, as embodied by biosensors, will allow for dynamic tuning of the properties. This effect, combined with the variety of assembly methods and ensuing control over microstructure and optoelectronic properties, makes CPs a highly versatile class of materials, which will be investigated for many years.

ACKNOWLEDGMENTS

The authors appreciate the financial support from the National Science Foundation (BES 0428010), the NSF CAREER Award (DMR 0644864), and the Center for Chemical Genomics-ThermoFisher grant for the work described in this chapter. J. K. acknowledges the financial support from the WCU (World Class University) program through the National Research Foundation of Korea funded by the Ministry of Education, Science, and Technology (R31-2008-000-10075-0) for the preparation of this chapter.

REFERENCES

1. Wu, P.-T., Kim, F., Champion, R. Jenekhe, S. (2008). Conjugated donor-acceptor copolymer semiconductors. Synthesis, optical properties, electrochemistry, and field-effect carrier mobility or pyridopyrazine-based copolymers. *Macromolecules*, *41*, 7021–7028.
2. Gurunathan, K., Murugan, A., Marimuthu, R., Mulik, U., Amalnerkar, D. (1999). Electrochemically synthesized conducting polymeric materials for applications towards technology in electronics, optoelectronics and energy storage devices. *Mater. Chem. Phys.*, *61*, 173–191.
3. Pron, A., Rannou, P. (2002). Processible conjugated polymers: from organic semiconductors to organic metals and superconductors. *Prog. Polym. Sci.*, *27*, 135–190.
4. Moliton, A., Hiorns, R. C. (2004). Review of electronic and optical properties of semiconducting π-conjugated polymers: applications in optoelectronics. *Polym. Int.*, *53*, 1397–1412.
5. Shinar, J., Shinar, R. (2008). Organic light-emitting devices (OLEDs) and OLED-based chemical and biological sensors: an overview. *J. Phys. D: Appl. Phys.*, *41*, 133001–133026.
6. Gunes, S., Neugebauer, H., Sariciftci, N. (2007). Conjugated polymer-based solar cells. *Chem. Rev.*, *107*, 1324–1338.
7. Teles, F., Fonseca, L. (2008). Applications of polymers for biomolecule immobilization in electrochemical biosensors. *Mater. Sci. Eng., C*, *28*, 1530–1543.
8. Pun, C.-C., Lee, K., Kim, H.-J., Kim, J. (2006). Signal amplifying conjugated polymer-based solid-state DNA sensors. *Macromolecules*, *39*, 7461–7463.
9. Liu, B., Bazan, G. (2004). Homogeneous fluorescence-based DNA detection with water-soluble conjugated polymers. *Chem. Mater.*, *16*, 4467–4476.
10. Gaylord, B., Heeger, A., Bazan, G. (2002). DNA detection using water-soluble conjugated polymers and peptide nucleic acid probes. *Proc. Natl. Acad. Sci. U.S.A.*, *99*, 10954–10957.
11. Ho, H.-A., Boissinot, M., Bergeron, M., Corbeil, G., Doré, K., Boudreau, D., Leclerc, M. (2002). Colorimetric and fluorometric detection of nucleic acids using cationic polythiophene derivatives. *Angew. Chem. Int. Ed.*, *41*, 1548–1551.
12. Gerard, M., Chaubey, A., Malhotra, B. (2002). Application of conducting polymers to biosensors. *Biosens. Bioelectron.*, *17*, 345–359.
13. Lee, J., Kim, H.-J., Kim, J. (2008). Polydiacetylene liposome arrays for selective potassium detection. *J. Am. Chem. Soc. Commun.*, *130*, 5010–5011.
14. Lee, K., Povlich, L., Kim, J. (2007). Label-free and self-signal amplifying molecular DNA sensors based on bioconjugated polyelectrolytes. *Adv. Funct. Mater.*, *17*, 2580–2587.

15. Singh, S., Chaubey, A., Malhotra, B. D. (2004). Amperometric cholesterol biosensor based on immobilized cholesterol esterase and cholesterol oxidase on conducting polypyrrole films. *Anal. Chim. Acta*, *502*, 229–234.

16. Pejcic, B., De Marco, R., Parkinson, G. (2006). The role of biosensors in the detection of emerging infectious diseases. *Analyst*, *131*, 1079–1090.

17. Baeumner, A. (2003). Biosensors for environmental pollutants and food contaminants. *Anal. Bioanal. Chem.*, *377*, 434–445.

18. Mello, L. D., Kubota, L. T. (2002). Review of the use of biosensors as analytical tools in the food and drink industries. *Food Chem.*, *77*, 237–256.

19. Iqbal, S., Mayo, M., Bruno, J., Bronk, B., Batt, C., Chambers, J. (2000). A review of molecular recognition technologies for detection of biological threat agents. *Biosens. Bioelectron.*, *15*, 549–578.

20. Arya, S., Datta, M., Malhotra, B. (2008). Recent advances in cholesterol biosensor. *Biosens. Bioelectron.*, *23*, 1083–1100.

21. Zhu, S. S., Carrol, P., Swager, T. M. (1996). Conducting polymetallorotaxanes: a supramolecular approach to transition metal ion sensors. *J. Am. Chem. Soc.*, *118*, 8713–8714.

22. Lee, K., Rouillard, J.-M., Pham, T., Gulari, E., Kim, J. (2007). Signal-amplifying conjugated polymer-DNA hybrid chips. *Angew. Chem.*, *46*, 4667–4670.

23. Kim, J., Swager, T. M. (2001). Control of conformational and interpolymer effects in conjugated polymers. *Nat. Lett.*, *411*, 1030–1034.

24. Lee, K. Yucel, T., Kim, H.-J., Pochan, D., Kim, J. Design principle of conjugated polyelectrolytes to make them water-soluble and highly emissive. Manuscript in progress, (manuscript).

25. Bolton, O., Kim, J. (2007). Design principles to tune the optical properties of 1,3,4-oxadiazole-containing molecules. *J. Mater. Chem.*, *17*, 1981–1988.

26. Lee, J.-K., Kim, H.-J., Kim, T., Lee, C.-H., Park, W., Kim, J., Lee, T. (2005). A new synthetic approach for polybenzoxazole and light-induced fluorescent patterning on its film. *Macromolecules*, *38*, 9427–9433.

27. Kim, H.-J., Lee, J., Kim, T.-H., Lee, T., Kim, J. (2008). Highly emissive self-assembled organic nanoparticles having dual color capacity for targeted immunofluorescence labeling. *Adv. Mater.*, *20*, 1117–1121.

28. Chang, D. W., Kim, S., Park, S. Y. (2000). Excited-state intramolecular proton transfer via a preexisting hydrogen bond in semirigid polyquinoline. *Macromolecules*, *33*, 7223–7225.

29. Kim, S., Chang, D. W., Park, S. Y. (2002). Excited-state intramolecular proton transfer and stimulated emission photoautomerizable polyquinoline film. *Macromolecules*, *35*, 6064–6066.

30. Seo, J., Kim, S., Gihm, S. H., Park, C. R., Park, S. Y. (2007). Highly fluorescent columnar liquid crystals with elliptical molecular shape: oblique molecular stacking and excited-state intramolecular proton-transfer fluorescence. *J. Mater. Chem.*, *17*, 5052–5057.

31. Park, S., Kim, S., Seo, J., Park, S. Y. (2008). Application of excited-state intramolecular proton transfer (ESIPT) principle to functional polymeric materials. *Macromol. Res.*, *16*, 385–395.

32. Lee, J., Kim, H.-J., T. Chen, Lee, K., Kim, K.-S., Glotzer, S., Kim, J., Kotov, N. (2009). Control of energy transfer to CdTe nanowires via conjugated polymer orientation. *J. Phys. Chem. C*, *113*, 109–116.

33. Bazan, G. (2007). Novel organic materials through control of multichromophore interactions. *J. Org. Chem.*, *72*, 8615–8635.

34. Pu, K.-Y., Liu, B. (2009). Optimizing the cationic conjugated polymer-sensitized fluorescent signal of dye labeled oligonucleotide for biosensor applications. *Biosens. Bioelectron.*, *24*, 1067–1073.

35. Jenekhe, S. A., Osaheni, J. (1994). Excimers and exciplexes of conjugated polymers. *Science*, *265*, 765–768.

36. Kim, J. (2002). Assemblies of conjugated polymers. Intermolecular and intramolecular effects on the photophysical properties of conjugated polymers. *Pure Appl. Chem.*, *74*, 2031–2044.

37. Kim, J., Levitsky, I., McQuade, T., Swager, T. W. (2002). Structural control in thin layers of poly(p-phenyleneethynylene)s: photophysical studies of Langmuir and Langmuir–Blodgett films. *J. Am. Chem. Soc.*, *124*, 7710–7718.

38. Wegner, G. (2003). Nanocomposites of hairy-rod macromolecules: concepts, constructs, and materials. *Macromol. Chem. Phys.*, *204*, 347–357.

39. Bjørnholm, T., Hassenkam, T., Greve, D., McCullough, R., Jayaraman, M., Savoy, S., Jones, C., McDevitt, J. (1999). Polythiophene nanowires. *Adv. Mater. Commun.*, *11*, 1218–1221.
40. Samori, P., Severin, N., Müllen, K., Rabe, J. (2000). Macromolecular fractionation of rod-like polymers at atomically flat solid-liquid interfaces. *Adv. Mater.*, *12*, 579–582.
41. Mena-Osteritz, E. (2002). Superstructures of self-organizing thiophenes. *Adv. Mater.*, *14*, 609–616.
42. Samori, P., Sikharulidze, I., Francke, V., Müllen, K., Rabe, J. (1999). Nanoribbons from conjugated macromolecules on amorphous substrates observed by SFM and TEM. *Nanotechnology*, *10*, 77–80.
43. Bates, F., Fredrickson, G. (1995). Block copolymers-designer soft materials. *Phys. Today*, *52*, 32–38.
44. Jenekhe, S. A., Chen, X. L. (1998). Self-assembled aggregates of rod-coil block copolymers and their solubilization and encapsulation of fullerenes. *Science*, *279*, 1903–1907.
45. Jenekhe, S. A., Chen, X. L. (1999). Self-assembly of ordered microporous materials from rod-coil block copolymers. *Science*, *283*, 372–375.
46. Wang, W., Wang, R., Zheng, C., Lu, S., Liu, T. (2009). Synthesis, characterization and self-assembly behavior in water as fluorescent sensors of cationic water-soluble conjugated polyfluorene-*b*-poly(N-isopropylacrylamide) diblock copolymers. *Polymer*, *50*, 1236–1245.
47. Leclérc, P., Parente, V., Brédas, J., François, B., Lazzaroni, R. (1998). Organized semiconducting nanostructures from conjugated block copolymer self-assembly. *Chem. Mater.*, *10*, 4010–4014.
48. Rubatat, L., Kong, X., Jenekhe, S. A., Ruokolainen, J., Hojeij, M., Mezzenga, R. (2008). Self-assembly of polypeptide/π-conjugated polymer/polypeptide triblock copolymers in rod-rod-rod and coil-rod-coil conformations. *Macromolecules*, *41*, 1846–1852.
49. Schmitt, C., Nothofer, H. G., Falcou, A., Scherf, U. (2001). Conjugated polyfluorene/polyaniline block copolymers. *Macromol. Rapid Commun.*, *22*, 624–628.
50. Wu, P.-T., Ren, G., Li, C., Mezzenga, R., Jenekhe, S. (2009). Crystalline diblock copolymers: synthesis, self-assembly, and microphase separation of poly(3-butylthiophene)-*b*-poly(3-octylthiophene). *Macromolecules*, *42*, 2317–2320.
51. Lee, J., Jun, H., Kim, J. (2009). Polydiacetylene-liposome microarrays for selective and sensitive mercury (II) detection. *Adv. Mater.*, *21*, 1–4.
52. Pan, J., Charych, D. (1997). Molecular recognition and colorimetric detection of cholera toxin by poly (diacetylene) liposomes incorporating G_{m1} ganglioside. *Langmuir*, *13*, 1365–1367.
53. Yoon, B., Lee, S., Kim, J. M. (2009). Recent conceptual and technological advances in polydiacetylene-based supramolecular chemosensors. *Chem. Soc. Rev.*, *38*, 1958–1968.
54. Lee, K., Povlich, L. K., Kim, J. (2010). Recent Advances in Fluorescent and Colorimetric Conjugated Polymer-based Biosensors. *Analyst*, *135*, 2179–2189.
55. Park, H. K., Chung, S., Park, H. G., Cho, J.-H., Kim, M., Chung, B. (2008). Mixed self-assembly of polydiacetylenes for highly specific and sensitive strip biosensors. *Biosens. Bioelectron.*, *24*, 480–484.
56. Peng, H., Tang, J., Yang, L., Pang, J., Ashbaugh, H., Brinker, C., Yang, Z., Lu, Y. (2006). Responsive periodic mesoporous polydiacetylene/silica nanocomposites. *J. Am. Chem. Soc.*, *128*, 5304–5305.
57. Chae, S. K., Park, H., Yoon, J., Lee, C. H., Ahn, D. J., Kim, J. M. (2007). Polydiacetylene supramolecules in electrospun microfibers: fabrication, micropatterning, and sensor applications. *Adv. Mater.*, *19*, 521–524.
58. Lange, U., Roznyatovskaya, N., Mirsky, V. (2008). Conducting polymers in chemical sensors and arrays. *Anal. Chim. Acta*, *614*, 1–26.
59. Jiang, H., Taranekar, P., Reynolds, J., Schanze, K. (2009). Conjugated polyelectrolytes: synthesis, photophysics, and applications. *Angew. Chem.*, *48*, 4300–4316.
60. Liu, Y., Erdman, A., Cui, T. (2007). Acetylcholine biosensors based on layer-by-layer self-assembled polymer/nanoparticle ion-sensitive field-effect transistors. *Sens. Actuators, A*, *136*, 540–545.
61. Kim, H.-J., Lee, K., Kumar, S., Kim, J. (2005). Dynamic sequential layer-by-layer deposition method for fast and region-selective multilayer thin film fabrication. *Langmuir*, *21*, 8532–8538.
62. Wang, S., Bazan, G. (2003). Optically amplified RNA-protein detection methods using light-harvesting conjugated polymers. *Adv. Mater.*, *15*, 1425–1428.
63. Liu, B., Wang, S., Bazan, G., Mikhailovsky, A. (2003). Shape-adaptable water-soluble conjugated polymers. *J. Am. Chem. Soc.*, *125*, 13306–13307.

64. Gan, H., Li, Y., Liu, H., Wang, S., Li, C., Yuan, M., Liu. X., Wang. C., Jiang, L., Zhu, D. (2007). Self-assembly of conjugated polymers and ds-oligonucleotides directed fractal-like aggregates. *Biomacromolecules*, *8*, 1723–1729.
65. Yoon, Y.-S., Park, K.-H., Lee, J.-C. (2009). Self-assembly behavior and optical properties of poly(3-thiopheneacetate)/dialkyldimethylammonium complexes. *Macromol. Chem. Phys.*, *210*, 1510–1518.
66. Lee, K., Kim, H.-J., Cho, J.-C., Kim, J. (2007). Chemically and photochemically stable conjugated poly (oxadiazole) derivatives: a comparison with polythiophenes and poly(*p*-phenyleneethynylenes). *Macromolecules*, *40*, 6457–6463.
67. Lee, K., Maisel, K., Rouillard, J.-M., Gulari, E., Kim, J. (2008). Sensitive and selective label-free DNA detection by conjugated polymer-based microarrays and intercalating dye. *Chem. Mater.*, *20*, 2848–2850.
68. Lee, K., Rouillard, J. M., Kim, B. G., Gulari, E., Kim, J. (2009). Conjugated polymers combined with a molecular beacon for label-free and self-signal-amplifying DNA microarrays. *Adv. Funct. Mater.*, *19*, 3317–3325.

SUPRAMOLECULAR LIQUID CRYSTALS

CHAPTER 14

ADVANCED SYSTEMS OF SUPRAMOLECULAR LIQUID CRYSTALS

Takuma Yasuda and Takashi Kato

Department of Chemistry and Biotechnology, The University of Tokyo, Tokyo, Japan

14.1 INTRODUCTION

Liquid crystals are unique soft materials that combine molecular order and fluidity [1–3]. Liquid crystals are now widely applied to informational displays, in which the nematic phase, the simplest liquid-crystalline (LC) structure, has been utilized. Recently, functionalizations of nanostructured liquid crystals such as smectic, columnar, and cubic LC materials (Fig. 14-1a) have attracted attention [4–16]. New advanced soft materials for electronics [15–19], ionics [5–7,19,20], photonics [21,22], actuators [23], membranes [14], and biological functions [24] have been developed using the nanostructured liquid crystals. Self-assembled structures from nano- to micrometer scale are highly dependent on molecular shapes and molecular interactions of the nanostructured liquid crystals.

For further functionalization of liquid crystals, use of specific intermolecular interactions, such as hydrogen bonding, is one of promising and versatile approaches [5–7,25–30] because biological systems use hydrogen bonding to construct highly functional molecular complexes. The supramolecular design of liquid crystals is now widely accepted and studied [31–44]. In 1989, Kato and Fréchet [31,32], and Lehn and coworkers [33] reported that supramolecular mesogenic complexes with well-defined structures can be built by the formation of hydrogen bonds between different molecules (Fig. 14-1b). The first examples of such supramolecular liquid crystals were hydrogen-bonded complexes **1** and **2**. Complex **1** shows smectic and nematic LC phases [31], while complex **2** forms a columnar phase [33]. In these cases, the supramolecular mesogens with rigid structures are formed by the self-assembly of complementary molecular components. These early studies [31–33] have shown great possibilities as

Supramolecular Soft Matter: Applications in Materials and Organic Electronics, First Edition.
Edited by Takashi Nakanishi.
© 2011 John Wiley & Sons, Inc. Published 2011 by John Wiley & Sons, Inc.

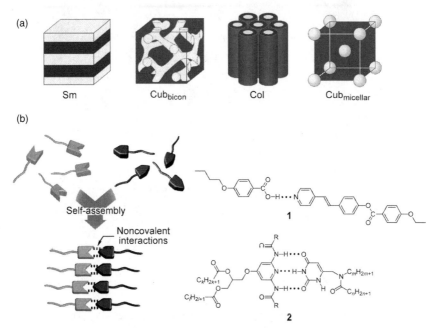

Figure 14-1 (a) Morphologies of phase-segregated LC nanostructures. Sm, smectic; Cub_bicon, bicontinuous cubic; Col, columnar; Cub_micellar, discontinuous (micellar) cubic. (b) Schematic illustration of supramolecular assembly of LC complexes through specific noncovalent interactions (left) and early examples of supramolecular hydrogen-bonded mesogenic complexes (right).

well as the importance of the supramolecular noncovalent approaches for the development of new dynamically functional LC materials.

In this chapter, we focus on functional nanostructured liquid crystals derived from supramolecular approaches. In Section 14.2, we show how we design materials structures of liquid crystals toward functionalization. In Section 14.3, we describe specific properties of the nanostructured LC materials exhibiting ionic and electronic functions.

14.2 DESIGN OF MATERIALS STRUCTURES

14.2.1 Specific Interactions for Supramolecular Self-Assembly

Since the discovery of the foregoing supramolecular liquid crystals consisting of complimentary molecular components [31,33], supramolecular approaches have been well-established as a new design strategy for the development of dynamically functional LC materials [5–7]. New molecular shapes, self-assembled structures, and intriguing properties have been found in a variety of nanostructured liquid crystals. Supramolecular side-chain [32] and main-chain [34] polymers that exhibit LC structures in bulk states have also been developed. One of the great

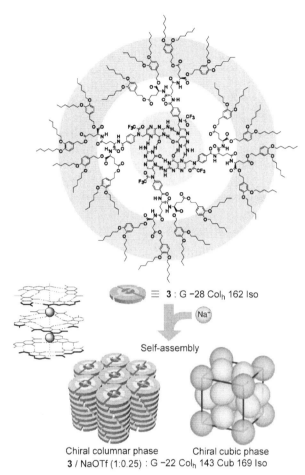

Figure 14-2 Formation of chiral cubic and chiral columnar phases induced by supramolecular self-assembly of a folic acid derivative with a sodium salt by hydrogen bonding and ion–dipolar interactions [37,38].

advantages of these supramolecular liquid crystals built using noncovalent interactions is their dynamic properties. For example, hydrogen-bonded LC polymer networks display reversible mesomorphic–isotropic phase transitions as a consequence of thermoinduced formation/dissociation (on/off switching) of the hydrogen bonds [35,36]. Thus, hydrogen bonds can be utilized as versatile structure-forming interactions to fabricate stimuli-responsive supramolecular LC materials.

One recent example of stimuli-responsive supramolecular liquid crystals is hydrogen-bonded folate rosettes [37–42]. The disklike tetramer is formed by folic acid derivative **3** containing oligo(glutamic acid) moieties and terminal alkoxy chains (Fig. 14-2). They are capable of forming hexagonal columnar (Col_h) and micellar cubic (Cub) LC phases over wide temperature ranges. The segmentation of the columns results in the transition to the micellar cubic phase. It is of interest that the addition of sodium ions induces the successive formation of chiral columnar and chiral micellar cubic phases, as schematically shown in Fig. 14-2 [37,38]. The circular dichroism (CD) spectra for the complexes of **3** with a sodium salt exhibit obvious Cotton effects in the LC states, whereas no induced CD signal is detected for the LC phases of **3** alone. The ion–dipolar interactions between sodium ions and the hydrogen-bonded folate rosettes lead to the stabilization of the helical ordering within the columnar stacks of **3**. It has been revealed that the columnar assemblies of the folic acid derivatives serve as an artificial ion channel material, showing selective ion-transporting properties [42].

In addition to hydrogen bonding, ionic interactions [20,45–51], charge-transfer (CT) interactions [52], and halogen bonding [53,54] have also been employed for the design of supramolecular LC assemblies. Bruce and coworkers have demonstrated that the complexation of nonmesomorphic 4-alkoxystilbazoles with iodopentafluorobenzene results in the formation of halogen-bonded liquid crystals exhibiting smectic and nematic phases [53,54].

14.2.2 New Molecular Shapes and Architectures

Significant research interest has been devoted to the design of new molecular shapes for the development of nanostructured functional liquid crystals. While conventional rod- and disk-shaped molecules are recognized as basic structures of thermotropic liquid crystals [1], a variety of unconventional shapes of mesogenic molecules such as compounds **4–8** (Fig. 14-3) have been designed as new motifs for functional liquid crystals. For example, bent-core [55–57], shuttlecocks [58–60], bowls [61,62], dendrimers/dendrons [10,11,43,44,63–66], cones [67–70], macrocyclic rings [71–74], and polycatenar [75–78] mesogens have been developed. Shape-driven self-assembly enables these exotic LC molecules to form hitherto unavailable supramolecular arrays possessing anisotropic properties.

Takezoe, Watanabe, and coworkers demonstrated that banana-shaped LC molecule **4** has a polar order and induces chiral superstructures in smectic phases [55]. Nakamura, Kato, and coworkers reported fullerene-based LC molecule **5** with the shape of a shuttlecock, which forms hexagonal columnar phases [58–60]. The attachment of five mesogenic substituents to one pentagon of fullerene produces conical molecules, which can stack into polar columnar arrays in the LC phases. Columnar liquid crystals featuring bowl-shaped distorted π-conjugated cores (buckybowl-like structures) have recently emerged [61,62]. The dicyanomethylene-appended truxene **6** having a distorted π-conjugated core exhibits unique multivalent redox and n-type semiconducting properties [61]. Liquid crystals based on shape-persistent macrocycles (e.g., compound **8**) are also valuable candidates for the formation of one-dimensional tubular superstructures [71–74]. The macrocyclic columnar liquid crystals with interior nanosized pores can act as host materials for guest compounds.

Goodby and coworkers reported a series of dendritic liquid crystals based on silsesquioxanes as the central part [10,64,65]. Dendrimer **9** (Fig. 14-4) tethering eight cyanobiphenyl mesogenic units can form smectic LC phases that

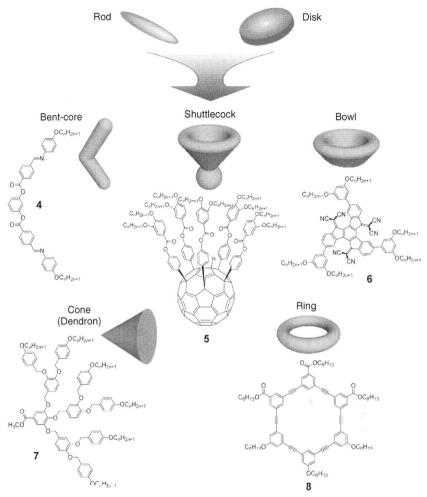

Figure 14-3 Molecular shapes of functional nanostructured liquid crystals: the progress from conventional rods and disks to unconventional structures such as bent-core [55], shuttlecock [58], bowl [61], cone [63], and ring [71].

comprise alternating organic and inorganic layers. For such dendritic liquid crystals, the incorporation of the multiple mesogenic units has an ability to organize nonmesogenic functional entities into nanostructured ordered LC states.

Kato, Stoddart, Sauvage, and coworkers developed [2]catenane- and [2]rotaxane-based dendritic molecules, **10** and **11** (Fig. 14-4), which include mobile, mechanically interlocked structures as the central core [79–82]. Catenane **10** and rotaxane **11** are capable of forming layered smectic A (SmA) phases in the temperature range of 83–117 and 10–150°C, respectively. The redox-driven mechanical movement (molecular shuttling) of the macrocycle within rotaxane **11** along its dumbbell component has been achieved both in

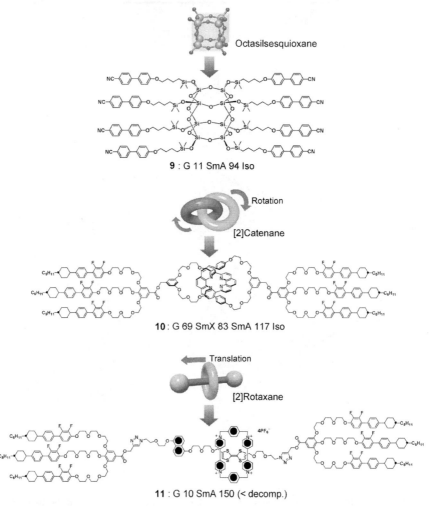

Figure 14-4 Design of dendritic liquid crystals based on octasilsesquioxane [65], catenane [79], and bistable rotaxane [80,81]. G, glassy; SmA, smectic A; SmX, unidentified smectic phase; Iso, isotropic; decomp., decomposed before isotropization.

solution and in the condensed LC phase [80,81]. The π-accepting macrocycle initially encircles the π-donating tetrathiafulvalene (TTF) recognition unit by CT interactions. When the TTF recognition unit is oxidized electrochemically, the cationic macrocycle moves to the 1,5-dioxynaphthalene (DNP) recognition unit as a result of Coulombic repulsion. Electrochromism has also been observed for the bistable rotaxane **11** as a result of the redox-driven mechanical movements. The appropriate combination of specific intermolecular interactions and the molecular shape of liquid crystals would yield a new generation of stimuli-responsive functional liquid crystals.

14.2.3 Nanosegregation in Liquid Crystals

Nanosegregation of incompatible parts in block molecules plays an essential role in the formation and stabilization of ordered LC nanostructures. It is well-known that in lyotropic liquid crystals, amphiphiles spontaneously form nanosegregated structures comprising hydrophilic and hydrophobic parts in the presence of water [1]. Thermotropic LC structures such as smectic, bicontinuous cubic, hexagonal columnar, and micellar cubic phases (Fig. 14-1a) are formed by nanosegregation. Recently, more complex nanosegregated LC structures have been reported [4–13]. Tschierske and coworkers have systematically studied the nanosegregation behavior of amphiphilic molecules combining polyhydroxy and alkyl moieties [83–85]. With the increase of the volume fraction of the polyhydroxy moiety, the interface curvature between two incompatible molecular parts should increase, giving rise to the structural variation from the smectic through columnar to micellar cubic nanostructures. The formation of intermolecular hydrogen-bonding networks as well as the nanosegregation between the hydrophilic and lipophilic parts cooperatively facilitate such self-organization of the polyhydroxy amphiphilic molecules.

Rod–coil block molecules (i.e., rodlike rigid molecules tethering flexible coiled chains) have been demonstrated to self-assemble into nanosegregated LC structures [13,86–88]. Lee and coworkers have reported a series of rod–coil diblock and coil–rod–coil triblock molecules based on aromatic mesogenic units and polyether coils [13,87]. In these rod–coil molecules, drastic changes in the self-assembled structures could be induced by controlling the volume fraction of molecular blocks and/or the rod–coil molecular length.

14.3 DESIGN OF MATERIALS FUNCTIONS

This section describes the design of LC assemblies toward specific functions and the control of their alignment. Several recent examples of ion- and electrofunctional nanostructured liquid crystals are mainly explained. A variety of mesomorphic superstructures has been constructed through noncovalent interactions, which affect ionic and electronic transport functions.

14.3.1 Ionic Functions

Self-organization of ion-conductive materials based on poly(ethylene oxide)s (PEOs) and incorporation of ionic liquids into ordered LC nanostructures is a useful strategy for the development of directionally dependent (anisotropic) ion-conductive materials [5–7,45–47,89–97]. It has been known that coordination complexes of PEOs with lithium salts serve as solid polymer electrolytes, in which the dissolved ions can move by the segmental motions of the disordered polyether chains. If these ion-active materials can be organized into well-aligned LC structures, efficient and anisotropic ion transportation would be accomplished through the formation of ion-conductive nanochannels. In this context, rod–coil–rod [89–92] and rod–coil [93] block molecules containing an oligo(ethylene oxide) segment have been prepared as two-dimensional (2D)

LC ion conductors. The lithium salt complexes of these block molecules have shown smectic LC phases in wide temperature ranges, and the anisotropic ion conduction along the layer structures has been observed for the LC complexes aligned vertically on substrates.

Macroscopically oriented LC nanostructures acting as low-dimensional ion-conductive pathways can be preserved in self-standing films by means of *in situ* photopolymerization of aligned LC complexes (Fig. 14-5a) [94,95]. The complexes of polymerizable LC monomer **12** with lithium triflate form layered smectic phases. UV irradiation of the homeotropically aligned complexes in the smectic A (SmA) phase results in the formation of free-standing film materials. It has been found that the polymer complex ($13/LiOSO_2CF_3$) forms a macroscopically aligned layered structure, which consists of ion-conductive layers and insulating mesogenic layers, as shown in Fig. 14-5a. The oriented photopolymerized film shows an ionic conductivity of 1.5×10^{-6} S/cm at 35°C along the direction parallel to the smectic layers, which is approximately 4500 times higher than that perpendicular to the layers [94]. The ionic conductivities of photopolymerized films can be enhanced to the order of 10^{-3} S/cm by introducing a tetra(ethylene oxide) segment at the terminal (instead of the inner part) of the mesogenic monomer [95]. It is conceivable that faster segmental motion of the terminal tetra(ethylene oxide) chain should result in the enhanced ionic conductivities.

One-dimensional (1D) ion conduction has been achieved in columnar LC materials based on ionic liquids [6,45–47,97]. Ionic liquids have great potential for application as electrolytes for energy devices, including batteries and capacitors. However, they have been commonly used as isotropic liquid states. The induction of aligned columnar nanostructures to ionic liquids can be expected to enhance anisotropic ion-transporting properties. Fan-shaped imidazolium salts **14a,b** (Fig. 14-5b) having a covalently linked 3,4,5-trialkoxybenzyl group have been developed as a new class of ionic columnar liquid crystals [45,46]. The self-assembly and phase segregation of these molecules in the nanometer scale give rise to the formation of Col_h LC structures possessing inner ion-conducting paths. In the macroscopically oriented monodomain materials of **14a,b** obtained by mechanical shearing, the ionic conductivities parallel to the columnar axis are one or two orders of magnitude higher than those perpendicular to the axis. In contrast, polydomain columnar materials prepared without shearing do not exhibit anisotropic ion conduction properties [45,46]. One-dimensional ion-conductive films have also been prepared from an imidazolium-based ionic columnar liquid crystal containing polymerizable acrylate groups at the terminal [47].

Supramolecular materials design can be applied to the fabrication of higher ion-conductive columnar LC assemblies. The complexation of dihydroxy-functionalized molecules **15a,b** and imidazolium salt **16** (Fig. 14-5b) via hydrogen bonding and nanosegregation provides supramolecular columnar liquid crystals incorporating mobile ions into the inner part of the columns [97]. It has been demonstrated that these noncovalent-type supramolecular LC materials exhibit ionic conductivities on the order of 10^{-3} S/cm at ambient temperature, which are about 700 times higher than those of the covalent-type columnar LC imidazolium salts **14a,b**.

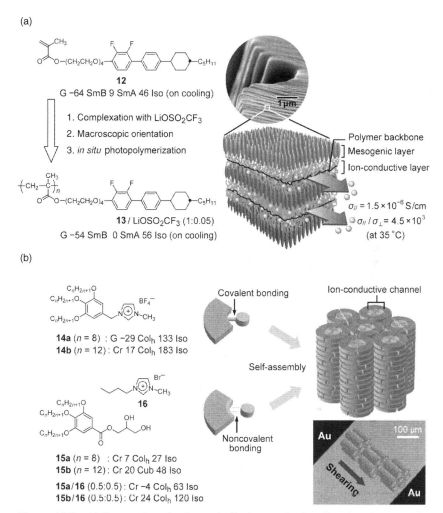

Figure 14-5 (a) Preparation of anisotropically ion-conductive films based on an LC polymer with a lithium salt [94]. (b) Design of 1D ion-conductive columnar LC materials: fan-shaped molecules bearing an imidazolium moiety [45,46] and molecular complexes of fan-shaped molecules with an ionic liquid [97]. *Source*: Reproduced with permission from Refs 45 and 94. Copyright 2003, 2004 American Chemical Society.

14.3.2 Electronic Functions

Use of nanostructured LC assemblies has attracted much attention for the development of new electronic and optoelectronic functional soft materials as well. The well-organized π-stacked structures of discotic (disklike) and calamitic (rodlike) aromatics in columnar and smectic LC phases can provide efficient pathways to promote electrical conduction [15–19,98–107]. Over the past two decades, π-conjugated columnar liquid crystals based on triphenylenes [98,99] and hexabenzocoronenes [15,100–102] and smectic liquid crystals based on oligothiophenes

[16,103–108] have been extensively studied as LC semiconductors exhibiting fast electronic conduction. High anisotropic field-effect transistor (FET) characteristics have been demonstrated for macroscopically oriented LC hexabenzocoronenes by Müllen and coworkers [102]. The hole mobility along the columnar axis is 75 times higher than that perpendicular to the column, indicating the predominant 1D electronic conduction within the columns.

A variety of LC nanostructures can be built from linearly extended π-conjugated oligomers by introducing multiple flexible chains into both the terminals (Fig. 14-6a) [77,78]. The polycatenar oligothiophenes **17** and **18** have a characteristic geometrical structure combining calamitic and discotic molecules and can self-assemble into smectic, columnar, and micellar cubic LC structures. It has been shown that LC properties of the polycatenar oligothiophenes are highly dependent on the number and length of the terminal alkoxy chains. While oligothiophene **17a** ($n = 12$) predominantly exhibits a tetragonal columnar (Col_{tet}) LC phase in the range of 110–151°C, oligothiophene **17b** ($n = 18$) with the elongated alkoxy chains induces the formation of a micellar cubic (Cub) phase in higher temperatures. As for oligothiophenes **18a,b**, the tetra-alkoxy substitution allows the molecules to adopt a calamitic shape, leading to the formation of highly ordered smectic phases over wide temperature ranges. This is one of the first attempts toward the development of functional π-conjugated nanostructured liquid crystals that are capable of forming layered, columnar, and globular electroactive superstructures by self-assembly processes.

It is noteworthy that the columnar and layered LC assemblies composed of π-conjugated oligothiophenes function as 1D and 2D hole transporters [77]. Figure 14-6b displays the temperature dependence of hole mobilities of the oligothiophene-based liquid crystals, evaluated by time-of-flight measurements. In the isotropic phase of **17a** ($n = 12$), low mobilities on the order of 10^{-6} cm^2/V $\cdot$ s are observed (circles in Fig. 14-6b). On cooling to the isotropic–Col_{tet} phase-transition temperature, the mobility discontinuously increases by two orders of magnitude, indicating that the 1D π-stacks of the oligothiophenes are responsible for the high hole mobility in the Col_{tet} phase. The hole mobilities of **17a** gradually increases up to ca 10^{-2} cm^2/V $\cdot$ s on further lowering the temperature. Meanwhile, oligothiophene **18a** shows almost temperature-independent hole transport properties over the entire temperature range of the smectic phase, giving mobilities of approximately 10^{-2} cm^2/V $\cdot$ s (squares in Fig. 14-6b). By contrast, no long-range hole transport behavior could be observed in the Cub phase of **17b** ($n = 18$). It is conceivable that in the Cub phase, the π-conjugated oligothiophene moieties are located inside the micelles. Therefore, they should be isolated from one another by the insulating alkoxy shells, preventing effective hole transport between the micelles.

14.4 SUMMARY AND OUTLOOK

We have described the materials design and the functions of some recently developed LC materials. To fabricate highly functional soft materials, the induction of

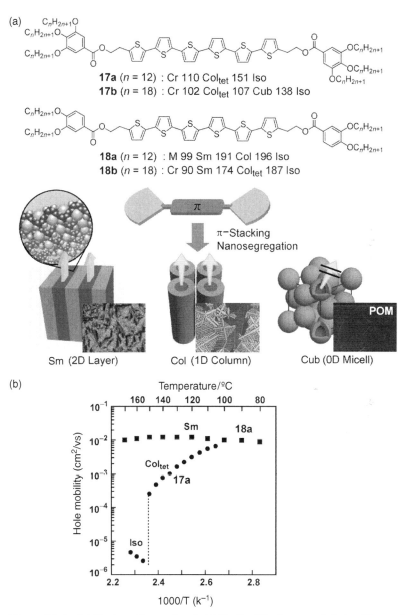

Figure 14-6 (a) Design of π-conjugated LC oligothiophenes and their self-organization into smectic, columnar, and micellar cubic nanostructures [77]. (b) Temperature dependence of hole mobilities of oligothiophenes **17a** (filled circles) and **18a** (filled squares) on cooling. The dashed line denotes the isotropic–Col_{tet} phase transition.

LC order within molecular condensed states is a promising and versatile approach. The use of LC ordering enables us to control molecular self-organization processes and to induce dynamic and anisotropic functions. We have shown that three important factors, that is, (i) control of specific intermolecular interactions, (ii) design of the shape and architecture of LC molecules, and (iii) use of nanosegregation, are essential for great advances in functional nanostructured LC materials with tailor-made properties. The development of composite materials based on liquid crystals and self-assembled fibers is also of great importance because the properties can be tuned and/or enhanced by the formation of heterogeneous structures [109,110]. We believe that these functional nanostructured LC materials featuring supramolecular design will be fascinating not only for fundamental research in chemistry, physics, and materials science but also for application as active entities in electronic and optoelectronic devices.

REFERENCES

1. Demus, D., Goodby, J. W., Gray, G. W., Spiess, H.-W., Vill, V., eds. (1998). *Handbook of Liquid Crystals*. Wiley-VCH, Weinheim.
2. Gray, G. W., ed. (1987). *Thermotropic Liquid Crystals*. Wiley, Chichester.
3. de Gennes, P. G. (1993). *The Physics of Liquid Crystals*. Oxford University Press, New York.
4. Kato, T., ed. (2008). Liquid crystalline functional assemblies and their supramolecular structures. *Struct. Bond.*, *128*, 1–222.
5. Kato, T. (2002). Self-assembly of phase-segregated liquid crystal structures. *Science*, *295*, 2414–2418.
6. Kato, T., Mizoshita, N., Kishimoto, K. (2006). Functional liquid-crystalline assemblies: self-organized soft materials. *Angew. Chem. Int. Ed.*, *45*, 38–68.
7. Kato, T., Yasuda, T., Kamikawa, Y., Yoshio, M. (2009). Self-assembly of functional columnar liquid crystals. *Chem. Commun.*, 729–739.
8. Tschierske, C. (2001). Micro-segregation, molecular shape and molecular topology—partners for the design of liquid crystalline materials with complex mesophase morphologies. *J. Mater. Chem.*, *11*, 2647–2671.
9. Laschat, S., Baro, A., Steinke, N., Giesselmann, F., Hägele, C., Scalia, G., Judele, R., Kapatsina, E., Sauer, S., Schreivogel, A., Tosoni, M. (2007). Discotic liquid crystals: tailor-made synthesis to plastic electronics. *Angew. Chem. Int. Ed.*, *46*, 4832–4887.
10. Goodby, J. W., Saez, I. M., Cowling, S. J., Görtz, V., Draper, M., Hall, A. W., Sia, S., Cosquer, G., Lee, S.-E., Raynes, E. P. (2008). Transmission and amplification of information and properties in nanostructured liquid crystals. *Angew. Chem. Int. Ed.*, *47*, 2754–2787.
11. Donnio, B., Buathong, S., Bury, I., Guillon, D. (2007). Liquid crystalline dendrimers. *Chem. Soc. Rev.*, *36*, 1495–1513.
12. Yoshizawa, A. (2008). Unconventional liquid crystal oligomers with a hierarchical structure. *J. Mater. Chem.*, *18*, 2877–2889.
13. Ryu, J.-H., Lee, M. (2008). Liquid crystalline assembly of rod–coil molecules. *Struct. Bond.*, *128*, 63–98.
14. Gin, D. L., Bara, J. E., Noble, R. D., Elliott, B. J. (2008). Polymerized lyotropic liquid crystal assemblies for membrane applications. *Macromol. Rapid Commun.*, *29*, 367–389.
15. Pisula, W., Zorn, M., Chang, J. Y., Müllen, K., Zentel, R. (2009). Liquid crystalline ordering and charge transport in semiconducting materials. *Macromol. Rapid Commun.*, *30*, 1179–1202.
16. Funahashi, M. (2009). Development of liquid-crystalline semiconductors with high carrier mobilities and their application to thin-film transistors. *Polym. J.*, *41*, 459–469.
17. Sergeyev, S., Pisula, W., Geerts, Y. H. (2007). Discotic liquid crystals: a new generation of organic semiconductors. *Chem. Soc. Rev.*, *36*, 1902–1929.

18. Hoeben, F. J. M., Jonkheijm, P., Meijer, E. W., Schenning, A. P. H. J. (2005). About supramolecular assemblies of π-conjugated systems. *Chem. Rev.*, *105*, 1491–1546.
19. Funahashi, M., Shimura, H., Yoshio, M., Kato, T. (2008). Functional liquid-crystalline polymers for ionic and electronic conduction. *Struct. Bond.*, *128*, 151–179.
20. Binnemans, K. (2005). Ionic liquid crystals. *Chem. Rev.*, *105*, 4148–4204.
21. O'Neill, M., Kelly, S. M. (2003). Liquid crystals for charge transport, luminescence, and photonics. *Adv. Mater.*, *15*, 1135–1146.
22. Sagara, Y., Kato, T. (2009). Mechanically induced luminescence changes in molecular assemblies. *Nat. Chem.*, *1*, 605–610.
23. Ikeda, T., Mamiya, J., Yu, Y. (2007). Photomechanics of liquid-crystalline elastomers and other polymers. *Angew. Chem. Int. Ed.*, *46*, 506–528.
24. Woltman, S. J., Jay, G. D., Crawford, G. P. (2007). Liquid-crystal materials find a new order in biomedical applications. *Nat. Mater.*, *6*, 929–938.
25. Kato, T., Mizoshita, N., Kanie, K. (2001). Hydrogen-bonded liquid crystalline materials: supramolecular polymeric assembly and the induction of dynamic function. *Macromol. Rapid Commun.*, *22*, 797–814.
26. Kato, T. (2000). Hydrogen-bonded liquid crystals: molecular self-assembly for dynamically functional materials. *Struct. Bond.*, *96*, 95–146.
27. Kato, T., Mizoshita, N. (2002). Self-assembly and phase segregation in functional liquid crystals. *Curr. Opin. Solid State Mater. Sci.*, *6*, 579–587.
28. Kato, T., Fréchet, J. M. J. (2006). Hydrogen-bonded liquid crystals built from hydrogen-bonding donors and acceptors. Infrared study on the stability of the hydrogen bond between carboxylic acid and pyridyl moieties. *Liq. Cryst.*, *33*, 1429–1437.
29. Ciferri, A., ed. (2005). *Supramolecular Polymers*, 2nd ed. Taylor & Francis, London.
30. Lehn, J.-M. (1995). *Supramolecular Chemistry: Concepts and Perspectives*. VCH, Weinheim.
31. Kato, T., Fréchet, J. M. J. (1989). A new approach to mesophase stabilization through hydrogen bonding molecular interactions in binary mixtures. *J. Am. Chem. Soc.*, *111*, 8533–8534.
32. Kato, T., Fréchet, J. M. J. (1989). Stabilization of a liquid-crystalline phase through noncovalent interaction with a polymer side chain. *Macromolecules*, *22*, 3818–3819.
33. Brienne, M.-J., Gabard, J., Lehn, J.-M., Stibor, I. (1989). Macroscopic expression of molecular recognition. Supramolecular liquid crystalline phases induced by association of complementary heterocyclic components. *J. Chem. Soc., Chem. Commun.*, 1868–1870.
34. Fouquey, C., Lehn, J.-M., Levelut, A.-M. (1990). Molecular recognition directed self-assembly of supramolecular liquid crystalline polymers from complementary chiral components. *Adv. Mater.*, *2*, 254–257.
35. Kato, T., Kihara, H., Kumar, U., Uryu, T., Fréchet, J. M. J. (1994). A liquid crystalline polymer network built by molecular self-assembly through intermolecular hydrogen bonding. *Angew. Chem. Int. Ed. Engl.*, *33*, 1644–1645.
36. Kihara, H., Kato, T., Uryu, T., Fréchet, J. M. J. (1996). Supramolecular liquid-crystalline networks built by self-assembly of multifunctional hydrogen-bonding molecules. *Chem. Mater.*, *8*, 961–968.
37. Kato, T., Matsuoka, T., Nishii, M., Kamikawa, Y., Kanie, K., Nishimura, T., Yashima, E., Ujiie, S. (1969). Supramolecular chirality of thermotropic liquid-crystalline folic acid derivatives. *Angew. Chem. Int. Ed. Engl.*, *43*, 1969–1972.
38. Kamikawa, Y., Nishii, M., Kato, T. (2004). Self-assembly of folic acid derivatives: induction of supramolecular chirality by hierarchical chiral structures. *Chem.—Eur. J.*, *10*, 5942–5951.
39. Kanie, K., Nishii, M., Yasuda, T., Taki, T., Ujiie, S., Kato, T. (2001). Self-assembly of thermotropic liquid-crystalline folic acid derivatives: hydrogen-bonded complexes forming layers and columns. *J. Mater. Chem.*, *11*, 2875–2886.
40. Kanie, K., Yasuda, T., Ujiie, S., Kato, T. (2000). Thermotropic liquid-crystalline folic acid derivatives: supramolecular discotic and smectic aggregation. *Chem. Commun.*, 1899–1900.
41. Kanie, K., Yasuda, T., Nishii, M., Ujiie, S., Kato, T. (2001). Hydrogen-bonded lyotropic liquid crystals of folic acids: responses to environment by exhibiting different complex patterns. *Chem. Lett.*, *30*, 480–481.

42. Sakai, N., Kamikawa, Y., Nishii, M., Matsuoka, T., Kato, T., Matile, S. (2006). Dendritic folate rosettes as ion channels in lipid bilayers. *J. Am. Chem. Soc.*, *128*, 2218–2219.

43. Sagara, Y., Kato, T. (2008). Stimuli-responsive luminescent liquid crystals: change of photoluminescent colors triggered by a shear-induced phase transition. *Angew. Chem. Int. Ed.*, *47*, 5175–5178.

44. Sagara, Y., Yamane, S., Mutai, T., Araki, K., Kato, T. (2009). A stimuli-responsive, photoluminescent, anthracene-based liquid crystal: emission color determined by thermal and mechanical processes. *Adv. Funct. Mater.*, *19*, 1869–1875.

45. Yoshio, M., Mukai, T., Ohno, H., Kato, T. One-dimensional ion transport in self-organized columnar ionic liquids. *J. Am. Chem. Soc.*, *126*, 994–995.

46. Yoshio, M., Ichikawa, T., Shimura, H., Kagata, T., Hamasaki, A., Mukai, T., Ohno, H., Kato, T. (2007). Columnar liquid-crystalline imidazolium salts: effects of anions and cations on mesomorphic properties and ionic conductivities. *Bull. Chem. Soc. Jpn.*, *80*, 1836–1841.

47. Yoshio, M., Kagata, T., Hoshino, K., Mukai, T., Ohno, H., Kato, T. (2006). One-dimensional ion-conductive polymer films: alignment and fixation of ionic channels formed by self-organization of polymerizable columnar liquid crystals. *J. Am. Chem. Soc.*, *128*, 5570–5577.

48. Tanabe, K., Yasuda, T., Yoshio, M., Kato, T. (2007). Viologen-based redox-active ionic liquid crystals forming columnar phases. *Org. Lett.*, *9*, 4271–4274.

49. Tanabe, K., Yasuda, T., Kato, T. (2008). Luminescent ionic liquid crystals based on tripodal pyridinium salts. *Chem. Lett.*, *37*, 1208–1209.

50. Ujiie, S., Tanaka, Y., Yano, Y., Mori, A., Iimura, K. (2006). Thermal and liquid crystalline properties of ionic liquid crystalline systems. *Kobunshi Ronbunshu*, *63*, 11–18.

51. Bazuin, C. G., Guillon, D., Skoulios, A., Nicoud, J.-F. (1986). The thermotropic mesophase structure of two long-chain alkyl pyridinium halides. *Liq. Cryst.*, *1*, 181–188.

52. Ringsdorf, H., Wüstefeld, R., Zerta, E., Ebert, M., Wendorff, J. H. (1989). Induction of liquid crystalline phases: formation of discotic systems by doping amorphous polymers with electron acceptors. *Angew. Chem. Int. Ed. Engl.*, *28*, 914–918.

53. Bruce, D. W. (2008). Halogen-bonded liquid crystals. *Struct. Bond.*, *126*, 161–180.

54. Nguyen, H. L., Horton, P. N., Hursthouse, M. B., Legon, A. C., Bruce, D. W. (2004). Halogen bonding: a new interaction for liquid crystal formation. *J. Am. Chem. Soc.*, *126*, 16–17.

55. Niori, T., Sekine, T., Watanabe, J. Furukawa, T., Takezoe, H. (1996). Distinct ferroelectric smectic liquid crystals consisting of banana shaped achiral molecules. *J. Mater. Chem.*, *6*, 1231–1233.

56. Pelzl, G., Diele, S., Weissflog, W. (1999). Banana-shaped compounds—a new field of liquid crystals. *Adv. Mater.*, *11*, 707–724.

57. Reddy, R. A., Tschierske, C. (2006). Bent-core liquid crystals: polar order, superstructural chirality and spontaneous desymmetrisation in soft matter systems. *J. Mater. Chem.*, *16*, 907–961.

58. Sawamura, M., Kawai, K., Matsuo, Y., Kanie, K., Kato, T., Nakamura, E. (2002). Stacking of conical molecules with a fullerene apex into polar columns in crystals and liquid crystals. *Nature*, *419*, 702–705.

59. Matsuo, Y., Muramatsu, A., Hamasaki, R., Mizoshita, N., Kato, T., Nakamura, E. (2004). Stacking of molecules possessing a fullerene apex and a cup-shaped cavity connected by a silicon connection. *J. Am. Chem. Soc.*, *126*, 432–433.

60. Matsuo, Y., Muramatsu, A., Kamikawa, Y., Kato, T., Nakamura, E. (2006). Synthesis and structural, electrochemical, and stacking properties of conical molecules possessing buckyferrocene on the apex. *J. Am. Chem. Soc.*, *128*, 9586–9587.

61. Isoda, K., Yasuda, T., Kato, T. (2009). Truxene-based columnar liquid crystals: self-assembled structures and electro-active properties. *Chem. Asian J.*, *4*, 1619–1625.

62. Miyajima, D., Tashiro, K., Araoka, F., Takezoe, H., Kim, J., Kato, K., Takata, M., Aida, T. (2009). Liquid crystalline corannulene responsive to electric field. *J. Am. Chem. Soc.*, *131*, 44–45.

63. Percec, V., Cho, W.-D., Ungar, G. (2000). Increasing the diameter of cylindrical and spherical supramolecular dendrimers by decreasing the solid angle of their monodendrons via periphery functionalization. *J. Am. Chem. Soc.*, *122*, 10273–10281.

64. Saez, I. M., Goodby, J. W. (2008). Supermolecular liquid crystals. *Struct. Bond.*, *128*, 1–62.

65. Mehl, G. H., Goodby, J. W. (1996). Liquid-crystalline, substituted octakis (dimethyl-siloxy)octasilsesquioxanes: oligomeric supermolecular materials with defined topology. *Angew. Chem. Int. Ed. Engl.*, *35*, 2641–2643.

66. Deschenaux, R., Donnio, B., Guillon, D. (2007). Liquid-crystalline fullerodendrimers. *New J. Chem.*, *31*, 1064–1073.

67. Xu, B., Swager, T. M. (1993). Rigid bowlic liquid crystals based on tungsten-oxo calix[4]arenes: host-guest effects and head-to-tail organization. *J. Am. Chem. Soc.*, *115*, 1159–1160.

68. Komori, T., Shinkai, S. (1993). Novel columnar liquid crystals designed from cone-shaped calix[4]arenes. The rigid bowl is essential for the formation of the liquid crystal phase. *Chem. Lett.*, *22*, 1455–1458.

69. Hatano, T., Kato, T. (2006). A columnar liquid crystal based on triphenylphosphine oxide—its structural changes upon interaction with alkaline metal cations. *Chem. Commun.*, 1277–1279.

70. Kimura, M., Hatano, T., Yasuda, T., Morita, J., Akama, Y., Minoura, K., Shimomura, T., Kato, T. (2009). Photoluminescent liquid crystals based on trithienylphosphine oxides. *Chem. Lett.*, *38*, 800–801.

71. Mindyuk, O. Y., Stetzer, M. R., Heiney, P. A., Nelson, J. C., Moore, J. S. (1998). High resolution X-ray diffraction study of a tubular liquid crystal. *Adv. Mater.*, *10*, 1363–1366.

72. Höger, S. (2004). Shape-persistent macrocycles: from molecules to materials. *Chem.—Eur. J.*, *10*, 1320–1329.

73. Shimura, H., Yoshio, M., Kato, T. (2009). A columnar liquid-crystalline shape-persistent macro-cycle having a nanosegregated structure. *Org. Biomol. Chem.*, *7*, 3205–3207.

74. Seo, S. H., Jones, T. V., Seyler, H., Peters, J. O., Kim, T. H., Chang, J. Y., Tew, G. N. (2006). Liquid crystalline order from *ortho*-phenylene ethynylene macrocycles. *J. Am. Chem. Soc.*, *128*, 9264–9265.

75. Nguyen, H.-T., Destrade, C., Malthête, J. (1997). Phasmids and polycatenar mesogens. *Adv. Mater.*, *9*, 375–388.

76. Fazio, D., Mongin, C., Donnio, B., Galerne, Y., Guillon, D., Bruce, D. W. (2001). Bending and shaping: cubics, calamitics and columnars. *J. Mater. Chem.*, *11*, 2852–2863.

77. Yasuda, T., Ooi, H., Morita, J., Akama, Y., Minoura, K., Funahashi, M., Shimomura, T., Kato, T. (2009). π-Conjugated oligothiophene-based polycatenar liquid crystals: self-organization and photoconductive, luminescent, and redox properties. *Adv. Funct. Mater.*, *19*, 411–419.

78. Yasuda, T., Kishimoto, K., Kato, T. (2006). Columnar liquid crystalline π-conjugated oligothio-phenes. *Chem. Commun.*, 3399–3401.

79. Baranoff, E. D., Voignier, J., Yasuda, T., Heitz, V., Sauvage, J.-P., Kato, T. (2007). A liquid-crystalline [2]catenane and its copper (I) complex. *Angew. Chem. Int. Ed.*, *46*, 4680–4683.

80. Aprahamian, I., Yasuda, T., Ikeda, T., Saha, S., Dichtel, W. R., Isoda, K., Kato, T., Stoddart, J. F. (2007). A liquid-crystalline bistable [2]rotaxane. *Angew. Chem. Int. Ed.*, *46*, 4675–4679.

81. Yasuda, T., Tanabe, K., Tsuji, T., Coti, K. K., Aprahamian, I., Stoddart, J. F., Kato, T. (2010). A redox-switchable [2]rotaxane in a liquid-crystalline state. *Chem. Commun.*, *46*, 1224–1226.

82. Aprahamian, I., Miljanić, O. Š., Dichtel, W. R., Isoda, K., Yasuda, T., Kato, T., Stoddart, J. F. (2007). Clicked interlocked molecules. *Bull. Chem. Soc. Jpn.*, *80*, 1856–1869.

83. Fuchs, P., Tschierske, C., Raith, K., Das, K., Diele, S. (2002). A thermotropic mesophase comprised of closed micellar aggregates of the normal type. *Angew. Chem. Int. Ed.*, *41*, 628–631.

84. Cheng, X., Das, M. K., Diele, S., Tschierske, C. (2002). Influence of semiperfluorinated chains on the liquid crystalline properties of amphiphilic polyols: novel materials with thermotropic lamellar, columnar, bicontinuous cubic, and micellar cubic mesophases. *Langmuir*, *18*, 6521–6529.

85. Tschierske, C. (2007). Liquid crystal engineering—new complex mesophase structures and their relations to polymer morphologies, nanoscale patterning and crystal engineering. *Chem. Soc. Rev.*, 36, 1930–1970.

86. Stupp, S. I., LeBonheur, V., Walker, K., Li, L. S., Huggins, K. E., Keser, M., Amstutz, A. (1997). Supramolecular materials: self-organized nanostructures. *Science*, *276*, 384–389.

87. Lee, M., Cho, B.-K., Zin, W.-C. (2001). Supramolecular structures from rod–coil block copoly-mers. *Chem. Rev.*, *101*, 3869–3892.

88. Yang, W.-Y., Ahn, J.-H., Yoo, Y.-S., Oh, N.-K., Lee, M. (2005). Supramolecular barrels from amphiphilic rigid-flexible macrocycles. *Nat. Mater.*, *4*, 399–402.

89. Ohtake, T., Ogasawara, M., Ito-Akita, K., Nishina, N., Ujiie, S., Ohno, H., Kato, T. (2000). Liquid-crystalline complexes of mesogenic dimers containing oxyethylene moieties with LiCF$_3$SO$_3$: self-organized ion conductive materials. *Chem. Mater.*, *12*, 782–789.
90. Ohtake, T., Takamitsu, Y. Ito-Akita, K., Kanie, K., Yoshizawa, M., Mukai, T., Ohno, H., Kato, T. (2000). Liquid-crystalline ion-conductive materials: self-organization behavior and ion-transporting properties of mesogenic dimers containing oxyethylene moieties complexed with metal salts. *Macromolecules*, *33*, 8109–8111.
91. Ohtake, T., Ito, K., Nishina, N., Kihara, H., Ohno, H., Kato, T. (1999). Liquid-crystalline complexes of a lithium salt with twin oligomers containing oxyethylene spacers. An approach to anisotropic ion conduction. *Polym. J.*, *31*, 1155–1158.
92. Hoshino, K., Kanie, K., Ohtake, T., Mukai, T., Yoshizawa, M., Ujiie, S., Ohno, H., Kato, T. (2002). *Macromol. Chem. Phys.*, *203*, 1547–1555.
93. Iinuma, Y., Kishimoto, K., Sagara, Y., Yoshio, M., Mukai, T., Kobayashi, I., Ohno, H., Kato, T. (2007). Uniaxially parallel alignment of a smectic A liquid-crystalline rod–coil molecule and its lithium salt complexes using rubbed polyimides. *Macromolecules*, *40*, 4874–4878.
94. Kishimoto, K., Yoshio, M., Mukai, T., Yoshizawa, M., Ohno, H., Kato, T. (2003). Nanostructured anisotropic ion-conductive films. *J. Am. Chem. Soc.*, *125*, 3196–3197.
95. Kishimoto, K., Suzawa, T., Yokota, T., Mukai, T., Ohno, H., Kato, T. (2005). Nano-segregated polymeric film exhibiting high ionic conductivities. *J. Am. Chem. Soc.*, *127*, 15618–15623.
96. Yoshio, M., Mukai, T., Kanie, K., Yoshizawa, M., Ohno, H., Kato, T. (2002). Layered ionic liquids: anisotropic ion conduction in new self-organized liquid-crystalline materials. *Adv. Mater.*, *14*, 351–354.
97. Shimura, H., Yoshio, M., Hoshino, K., Mukai, T., Ohno, H., Kato, T. (2008). Noncovalent approach to one-dimensional ion conductors: enhancement of ionic conductivities in nanostructured columnar liquid crystals. *J. Am. Chem. Soc.*, *130*, 1759–1765.
98. Boden, N., Bushby, R. J., Clements, J. (1993). Mechanism of quasi-one-dimensional electronic conductivity in discotic liquid crystals. *J. Chem. Phys.*, *98*, 5920–5931.
99. Adam, D., Schuhmacher, P., Simmerer, J., Häussling, L., Siemensmeyer, K., Etzbachi, K. H., Ringsdorf, H., Haarer, D. (1994). Fast photoconduction in the highly ordered columnar phase of a discotic liquid crystal. *Nature*, *371*, 141–143.
100. Herwig, P., Kayser, C. W., Müllen, K., Spiess, H. W. (1996). Columnar mesophases of alkylated hexa-*peri*-hexabenzocoronenes with remarkably large phase widths. *Adv. Mater.*, *8*, 510–513.
101. van de Craats, A. M., Warman, J. M., Fechtenkötter, A., Brand, J. D., Harbison, M. A., Müllen, K. (1999). Record charge carrier mobility in a room-temperature discotic liquid-crystalline derivative of hexabenzocoronene. *Adv. Mater.*, *11*, 1469–1472.
102. Shklyarevskiy, I. O., Jonkheijm, P., Stutzmann, N., Wasserberg, D., Wondergem, H. J., Christianen, P. C. M., Schenning, A. P. H. J., de Leeuw, D. M., Tomović, Ž, Wu, J., Müllen, K., Maan, J. C. (2005). High anisotropy of the field-effect transistor mobility in magnetically aligned discotic liquid-crystalline semiconductors. *J. Am. Chem. Soc.*, *127*, 16233–16237.
103. Funahashi, M., Hanna, J. (2000). High ambipolar carrier mobility in self-organizing terthiophene derivative. *Appl. Phys. Lett.*, *76*, 2574–2576.
104. Funahashi, M., Hanna, J. (2005). High carrier mobility up to 0.1cm^2 V^{-1} s^{-1} at ambient temperatures in thiophene-based smectic liquid crystals. *Adv. Mater.*, *17*, 594–598.
105. Funahashi, M., Zhang, F., Tamaoki, N. (2007). High ambipolar mobility in a highly ordered smectic phase of a dialkylphenylterthiophene derivative that can be applied to solution-processed organic field-effect transistors. *Adv. Mater.*, *19*, 353–358.
106. Oikawa, K., Monobe, H., Nakayama, K., Kimoto, T., Tsuchiya, K., Heinrich, B., Guillon, D., Shimizu, Y., Yokoyama, M. (2007). High carrier mobility of organic field-effect transistors with a thiophene-naphthalene mesomorphic semiconductor. *Adv. Mater.*, *19*, 1864–1868.
107. van Breemen, A. J. J. M., Herwig, P. T., Chlon, C. H. T., Sweelssen, J., Schoo, H. F. M., Setayesh, S., Hardeman, W. M., Martin, C. A., de Leeuw, D. M., Valeton, J. J. P., Bastiaansen, C. W. M., Broer, D. J., Popa-Merticaru, A. R., Meskers, S. C. J. (2006). Large area liquid crystal monodomain field-effect transistors. *J. Am. Chem. Soc.*, *128*, 2336–2345.

108. Kimura, M., Yasuda, T., Kishimoto, K., Götz, G., Bäuerle, P., Kato, T. (2006). Oligothiophene-based liquid crystals exhibiting smectic A phases in wider temperature ranges. *Chem. Lett.*, *35*, 1150–1151.

109. Kato, T., Hirai, Y., Nakaso, S., Moriyama, M. (2007). Liquid-crystalline physical gels. *Chem. Soc. Rev.*, *36*, 1857–1867.

110. Kato, T., Mizoshita, N., Moriyama, M., Kitamura, T. (2005). Gelation of liquid crystals with self-assembled fibers. *Top. Curr. Chem.*, *256*, 219–236.

CHAPTER 15

SUPRAMOLECULAR AND DENDRITIC LIQUID CRYSTALS

Isabel M. Saez and John W. Goodby

Department of Chemistry, University of York, York, UK

15.1 INTRODUCTION

The manipulation of the noncovalent forces that hold the constituents of a system together by self-assembly, leading to the design of "smart" functional supramolecular materials, has already become a major development in materials science and has a strong impact on the development of new systems that mimic the complexity found in the biological world. This area of research offers a powerful alternative for the development of materials and construction of devices at mesoscopic length scales, bridging the gap that exists between top-down miniaturization and bottom-up nanofabrication.

The performance of materials of complex biological function relies on the hierarchical organization of the self-assembled system. This self-assembly process on multiple length scales is the mechanism by which Nature transforms molecular systems into complex machines capable of performing a variety of tasks depending on their environment. Adaptability, and therefore reversible interactions, is one of the properties of such systems. The ability to organize the different parts of the molecular structure into precise regions in space and time to yield structures with different compartments, which in turn perform separate functions, is another important trademark of living systems. Through self-assembly and self-organization processes, liquid-crystalline phases have opened up new perspectives in materials science toward the design and engineering of hierarchical self-organizing supramolecular materials [1–3]. Self-organization in two- and three-dimensional space offered by the liquid-crystalline medium is an ideal vehicle to explore and control the organization of matter on the nanometer to the micrometer scale, which are fundamental in the emerging development of nanotechnology.

Shape and function are intimately related. In analogy with smart biological macromolecules, such as enzymes, proteins, and so on, which have precisely

Supramolecular Soft Matter: Applications in Materials and Organic Electronics, First Edition.
Edited by Takashi Nakanishi.

defined three-dimensional structures, a range of strategies and concepts has been sought in macromolecular chemistry, trying to emulate their complexity of form and function. Among them, the most successful so far is the development of self-assembling and self-organizing dendrimers.

At a molecular level, in very simple terms, the general molecular features of a mesogen include a dichotomous structure, composed of an anisotropic rigid core and flexible peripheral chains, a certain degree of amphiphilicity or segregation within the molecular structure, complementarity of molecular shape, and cooperativity. The formation of liquid-crystalline phases depends on the weak, reversible, and anisotropic intermolecular interactions (dipolar, electrostatic, van der Waals, hydrogen bonding) present between the molecules in such way that the mesophase has properties that cannot be attained by the individual molecules, thus expressing a hierarchy of function. The dynamic and reversible organization found in mesophases confers on liquid crystals (LCs) unique properties among soft materials, such as extraordinary responsiveness to external stimuli and self-repairing and self-healing while being able to support electronic, photonic, magnetic, and biological functional properties.

The basic molecular mesogenic design relies on an anisometric core, either rodlike or disklike in shape, with appended flexible chains that may possess some degree of phase separation from the core (aliphatic carbon chains or chains partially substituted by fluorine or silicon). In low-molar-mass thermotropic LCs, the type of mesophase formed depends first on basic molecular structural features. In a simplistic approximation, rodlike molecules form calamitic mesophases, whereas disklike molecules form columnar mesophases. Other molecular shapes have been explored with the aim of further controlling the macroscopic organization. For example, polycatenar (molecules bearing multiple peripheral chains) and sanidic (board shaped) systems and bent-core, bowlic (molecules with a rigid core that can stack up one inside another), and metallomesogens have been explored, and excellent reviews have covered the state of the art in their respective fields. Similarly, the design of amphotropic LCs, such as carbohydrate-based LCs, polyphilic LCs, and supramolecular H-bonded and halogen-bonded LCs, has been explored and expanded along the same lines.

The incorporation of a chiral center in the molecular structure confers the molecule with a subtle twist that translates as a helical arrangement at the macroscopic level. Undoubtedly, this modification at the molecular level furnishes one of the most profound impacts in mesophase type. Chirality in LCs continues to be a major area in the field of research and its applications, for example, in LCDs (liquid crystal displays) testify the effort dedicated to this subject.

The search for new applications for LCs has fuelled the development of nonconventional mesogenic materials. In many cases, these innovative new materials have molecular architectures that do not fall into the traditional categories of rodlike or disklike but are characterized by intermediate shapes, which in turn contain several elements capable of being independently manipulated, giving rise to perfectly controlled molecular geometries and topologies. Similarly, their mesogenic behavior is not restricted to either the thermotropic or the lyotropic type; they can display both types. This has led to the discovery of new types of mesophases with fascinating structures, which have been summarized [4].

As part of our ongoing work on nonconventional mesogens, we have sought to develop *functional mesogenic materials* in the form of *dendritic LCs and multipedes*. The structures of such materials possess large but discrete molecular systems created when many mesogenic units (i.e., molecular entities that promote mesophase formation) are tied covalently to a central core (scaffold), with precisely defined molecular architecture and topology [5–7].

15.1.1 Defining the Structures of Supermolecules

Several molecular architectures can be used to describe the gross structures of supramolecular LCs. For example, two mesogenic units, that is, molecular entities that favor mesophase formation, can be tied end to end to give linear supramolecular materials. If the mesogenic groups are the same, they are called *dimers*; if they are different they are referred to as *bimesogens*. The mesogenic units may also be tied together laterally rather than end to end, or they may have terminal units tied to lateral units to give T-shaped dimers or bimesogens. With trimers, the situation becomes more complicated because not only are there linearly and laterally linked possible structures but also structures in which the mesogenic units could be linked to a central point, creating a "molecular knot." Similarly, tetra-, penta-, and other substituted supermolecules can be created. Increased numbers of mesogenic units can be attached to a central point by introducing a central scaffold on which the supramolecular structures can be built. Thus, cyclic, caged, dendrimer, or hyperbranched scaffolds can be utilized (Fig. 15-1).

Thus, the simple strategy of linking mesogenic units covalently to a central scaffold produces liquid-crystalline materials with a potentially large structural diversity. This strategy leads to the concept of a supermolecule, a giant molecular entity that is made up of covalently bonded identifiable molecular units, each one of them with specific properties, which interact with each other in defined ways. Alternatively, different mesogens can be used to further the modulation of physical properties and thereby increase complexity.

The induction and stabilization of mesophases in dendritic LCs arise from the competition and overall balance between the tendency of the dendritic core to adopt a globular isotropic conformation leading to maximum entropy, the anisotropic intermolecular forces between the mesogenic units that afford the maximum enthalpic gain, and the microphase segregation resulting from chemical and structural incompatibilities of the dendritic core and the anisometric units.

The way in which supermolecules can self-organize is dependent to a large degree, apart from the nature of the constituent mesogenic units, on simple structural features, such as the density of the mesogenic units attached to the periphery of the central scaffold, their topology of attachment, and the degree to which they are decoupled from the scaffold. For example, the density of mesogenic groups attached to the periphery can effectively change the overall gross shape

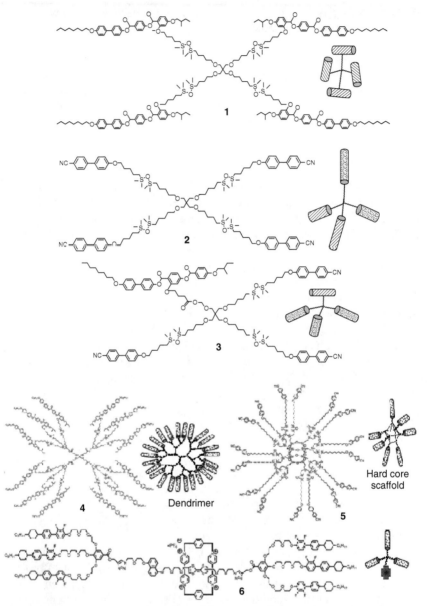

Figure 15-1 Templates for the design of liquid-crystalline supramolecular and dendritic materials and representative structures.

of the structure of the supermolecule from being rodlike to disklike to spherulitic. Therefore, supermolecular materials with an overall rodlike shape will support the formation of calamitic mesophases (including the various possibilities of smectic polymorphism), the disklike shape tends to support columnar mesophases, and spherulitic systems form cubic phases (Fig. 15-2).

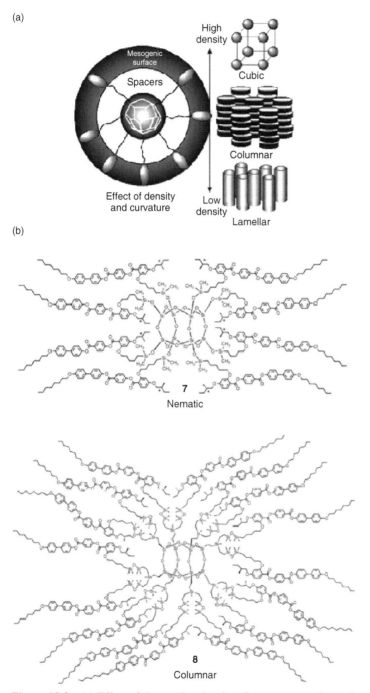

Figure 15-2   (a) Effect of the number density of mesogens on the surface of the supramolecular structure on the formation of various mesophases. (b) Representative examples of supermolecules.

Although there have not been exhaustive and systematic studies on the effects that the type (rodlike, disklike, or spherulitic) of mesogenic group attached to the central scaffold has on mesophase formation, it is clear for rodlike mesogenic groups that the topology of the attachment can markedly influence the type of mesophase formed and polymorphism exhibited. As shown in Fig. 15-3, in analogy with side-chain liquid crystal polymers (SCLCPs), the terminal attachment (end-on) of the mesogenic units tends to support a variety of smectic (lamellar) and/or nematic phases, whereas the lateral attachment (side-on) of the mesogenic units often leads to suppression of the smectic phases and domination of the nematic phase since the lateral chains disrupt the interactions between the mesogenic cores.

In addition to the number density and orientation of attachment, the degree to which the mesogenic units are decoupled from the central structure is important. The shorter the linking unit the more likely the material will act as a single supramolecular entity, whereas the longer the spacer the more likely it is that the properties of the individual mesogenic groups will dominate the overall properties of the material.

Apart from these coarse property–structure activity relationships, a secondary level of structure needs to be considered in defining the finer points of such relationships. Several topologies can be generated by the coupling of mesogenic units with dendritic scaffolds. For example, the central scaffolds may be considered to be soft (flexible) or hard (rigid), the different types of mesogens present may be randomly distributed on the surface or arranged in blocks to yield a segregated structure, microphase segregating units could be incorporated into the periphery or the scaffold of the material, mesogenic units may be incorporated into the scaffold as well as the periphery, metallomesogenic units as well as conventional organic mesogenic moieties may be incorporated, and chirality may be introduced into any part of the structure. Furthermore, by the incorporation of chemically or physically active moieties into the structures of supramolecular systems, "functional" or "smart" self-organizing materials can be created. Such materials systems are related to proteins in that they can be designed to have specific structures and properties. For example, in Fig. 15-1 (Compound **6**), functional moieties may interact with each other or with other chemical entities introduced into the supramolecular system, enhancing their response as a consequence of being incorporated into an ordered supramolecular assembly.

15.2 LIQUID-CRYSTALLINE DENDRIMERS

As described above, various topologies have been documented for dendritic and hyperbranched polymers, and this topic has received intense interest [8]. Several complementary reviews have covered the field of LC dendrimers and supramolecular LCs [9–15]. Although the exhaustive revision of this area is outside the scope of this article, several major types can be easily identified:

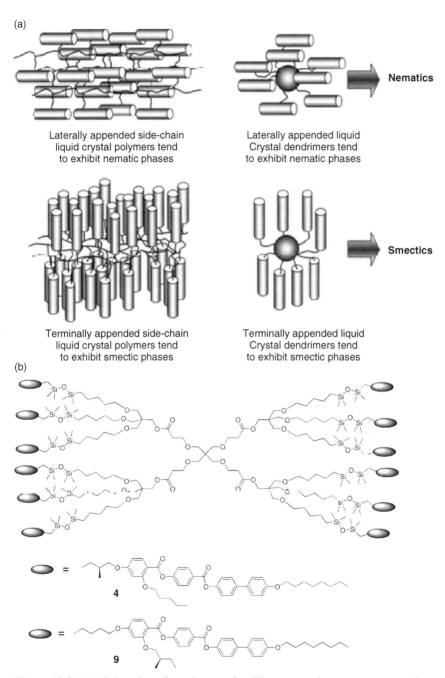

Figure 15-3 (a) Orientation of attachment of rodlike mesogenic groups to a central scaffold and mesophases formed for supramolecular systems, shown in comparison to the structures of side-chain liquid crystal polymers. (b) Representative examples of supermolecules.

1. Supramolecular LC dendrons and dendrimers, systems without mesogenic groups that form LC phases, in which the formation of the mesophases is due to self-assembly process of the constituent dendrons, aided by nanophase segregation.

2. LC dendrimers formed by the attachment of LC moieties to the periphery of a suitable core dendrimer, in which the mesogenic units are located on the surface of the dendritic core.

3. Hyperbranched polymers and LC dendrimers in which the mesogenic groups form part of each branching unit and the end groups of the dendrimer.

15.2.1 Supramolecular Self-Assembling Liquid Crystal Dendrons and Dendrimers

Within this vast group, the main representatives have been developed by Percec, based on the self-assembly process of the constituent polybenzyl ether dendrons, aided by microphase segregation, into supramolecular cylindrical, conical, and spherical dendrimers that self-organize into liquid-crystalline phases. The beauty and strength of this concept is that an almost infinite modulation of properties of the mesophase (type and lattice parameters) can be achieved from relatively simple molecular building blocks by analyzing extensive libraries of monodendrons with exquisite and very subtle variations in molecular design [16].

The structural motifs, AB_nX, are derived from 4-hydroxy, 3,4-dihydroxy, 3,5-dihydroxy, and 3,4,5-trihydroxy benzoates. The generation number, the substitution pattern, a diversity of focal point groups, and peripheral flexible chains are the structural parameters that allow subtle fine tuning of the rate of growth and deformation of the dendrons, microphase segregation and their self-assembling properties into mesophases. The overall shape of the dendrons varies from flat, tapered fan, semidiscoid, discoid, and conical to spherical, whereby lamellar, columnar, and cubic mesophases of various symmetries and huge diversity of structural parameters are formed through molecular shape control. Substitution of alkyl chains for partially fluorinated chains in these families of dendrons has the effect of changing the dendron conformation as well as the microphase segregation, modifying further the stability and the type of mesophase formed.

15.2.2 Side-Chain Liquid Crystal Dendrimers and Multipedes

Several families of dendrimers functionalized with mesogens have been extensively studied and provide the best correlation to date on structure and properties. Polypropyleneimine (PPI) and polyamidoamine (PAMAM) dendrimers have been functionalized with a large variety of mesogenic units. Several materials based on appending cyanobiphenyl (CB) moieties to PPI dendrimers have been prepared, containing 4, 16, and 64 end groups separated by pentyl or decyl spacers, with all exhibiting the smectic A phase [17]. The dendrimers orient into an antiparallel arrangement, yielding an interdigitated bilayer, with the spacing of the smectic layers almost independent of the dendrimer generation but dependent on the alkylene spacer length. The formation of the smectic phase indicates that the dendrimer scaffold is not globular but adopts a completely distorted conformation, even at higher generations, and resides at the interface between the layers composed of the mesogenic units. The change in conformation is driven by a phase separation between the anisometric rigid units and the flexible skeleton (of the spacer plus dendrimer) together with nanophase separation between the polar dendritic interior and the apolar mesogenic end groups.

This model of a polar interior and apolar surface units together with the anisotropy of the mesogenic units driving the supramolecular organization of the mesophase is repeated in other families of dendrimers. Serrano, among others, has described extensive series of dendrimers based on PPI and PAMAM cores functionalized with a variety of end-on and side-on mesogens [18]. When the cross-sectional area of the mesogens and dendrimer is similar, an elongated cylinder model is proposed, whose inner part is occupied by the dendrimer core and the outer part by the mesogens (since the core has a high degree of conformational freedom), which gives rise to the lamellar organization described above. However, the introduction of more terminal chains attached to the mesogens alters the cross-sectional area of the dendrimer and mesogenic units; the mesogenic units cannot be accommodated into a cylindrical structure, resulting in curvature at the interface between the dendrimer and mesogenic units, and prefer to be arranged into a disklike radial arrangement. The dendritic molecules fill up disks, or disk equivalents, ultimately leading to the formation of columnar mesophases.

From these examples, it is clear that the intermolecular interactions between the mesogenic units provide an enthalpic gain that is able to override the maximum entropic gain of the globular conformation of the dendritic core, and hence the distortion of the molecular shape is such that the flexible dendritic cores are able to segregate from the mesogenic units to yield nematic, smectic, and columnar mesophases. The specific type of mesomorphism found is dictated from then on by the density of mesogens attached to the dendritic scaffold, which affects the interfacial curvature. The flexibility of the scaffold is critical in allowing the distortions needed to support the formation of rodlike or disklike shapes of the supermolecule.

The effect of the shape and deformability of the dendritic scaffold is highlighted by the next group of materials. The scaffold in this example is a first-generation dendrimer based on pentaerythritol, with terminal C=C double bonds, suitable to carry out the attachment of the mesogens by hydrosilylation. The chiral S-2-methylbutyl group was incorporated onto the four-ring mesogens, known to strongly promote nematogenic behavior. The laterally appended dendrimer **4**, Fig. 15-3b (top), exhibits as expected a chiral nematic phase, with smectic mesophase formation being suppressed [6], whereas the terminally appended dendrimer **9**, Fig. 15-3b (bottom), exhibits smectic and chiral nematic phases. This demonstrates that the lateral appending of the mesogens causes disruption to the intermolecular packing, thereby destabilizing smectic mesomorphism and favoring the nematic phase. However, in both cases, the scaffold must deviate from the globular conformation to a collapsed one in order to fit within the mesogens.

However, if the dendritic core is not easily deformable, the competition between the tendency of the dendrimer to adopt a globular conformation and the ability of the mesogenic units to maximize their interactions determines the outcome of the type of mesophase formed (or indeed the lack of mesomorphism).

15.2.3   Effect of Hard Core Scaffolds — Silsesquioxane Scaffolds

A contrasting comparison with the dendrimers derived from flexible scaffolds described above is provided by dendrimers and supermolecules derived from rigid cage systems such as the octasilsesquioxane core. Cyanobiphenyl-substituted (CB) silsesquioxanes, such as **10**, exhibit lamellar mesophases [19]. The types of mesophases exhibited (SmC and SmA) demonstrate that the dendritic structure of the supermolecules is constrained to being rodlike and that the rodlike conformers tilt over at the smectic A to smectic C phase transition, as shown in Fig. 15-4. Thus, the structures of both the smectic A and smectic C phases have alternating organic and inorganic layers. As the inorganic and organic layers have differing refractive indices, the mesophase structures are essentially nanostructured birefringent slabs.

Other calamitic [20–22], bent-core [23], and discotic [24] mesogenic units have been attached to the silsesquioxane core to produce supramolecular materials that, in general, follow the trend shown here. That is, the systems are distorted from the globular geometry to yield rodlike conformers that better support the intermesogen interactions and hence in general preserve the type of mesophase originally shown by the mesogenic monomer. This rule is still followed when the density of mesogens in the periphery of the supermolecule is increased by the number of mesogens per functionalization site of the core (Fig. 15-1) [19].

As noted earlier for SCLCPs, in general, side-on attachment of the mesogenic subunits leads to the suppression of smectic phases and enhancement of the nematic phase (Fig. 15-3). By appending laterally attached chiral mesogenic units to the octasilsesquioxane scaffold, we targeted the induction of the chiral nematic phase into this class of materials [25] (Fig. 15-2b). **7** (Fig. 15-2b), exhibits an extraordinarily wide temperature range for the chiral nematic phase. A transition from a glassy state to the chiral nematic phase occurs near to room temperature, and then, the phase extends over $90°C$ before transforming to the isotropic liquid at $116.9°C$ [25]. For such a large molecular system, the mesophase is quite accessible and the melting point is very low.

The proposed local structure of the chiral nematic phase consists of rodlike/tubular structures in which the mesogenic units are expected to intermingle between the supermolecules. Thus, the surface of the polypede acts as a molecular recognition surface, in a way similar to recognition surfaces that are created in surfactant systems and Janus grains as described by de Gennes [26].

As in the CB systems, it was possible to increase the number of mesogens anchored to the core by dendritic growth of the central scaffold. Thus, the dendritic octasilsesquioxane hexadecamer **8** (Fig. 15-2b) exhibits chiral nematic,

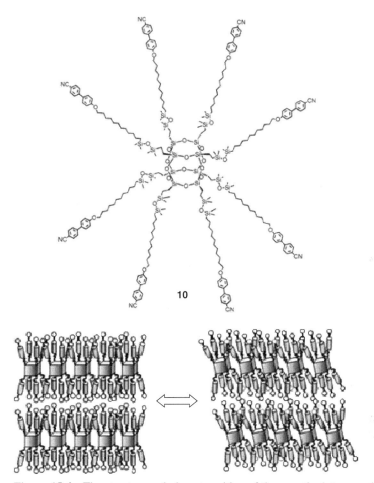

10

Figure 15-4   The structure and phase transition of the smectic A to smectic C phase of multipede **10**, where the spheres represent the cyano groups.

hexagonal disordered columnar, and rectangular disordered columnar phases, in the sequence g 5.4 Col*$_{rd}$ 30 Col*$_{hd}$ 102.3 N* 107.7°C Iso Liq [27].

Thus, the increase in the number density of the mesogens bound to the central scaffold transforms the situation from the octamers, which exhibit calamitic phases, to one where the hexadecamer exhibits columnar phases.

Similarly, the formation of the columnar phases in the dendrimer **8**, but not in the side-chain polysiloxanes, indicates that the silsesquioxane core assists in the interaction of neighboring units, resulting in the formation of the hexagonal and rectangular disordered structures, presumably through segregation of the siloxane cores from the mesogenic units in distinct columns. These remarkable new mesophase structures are favored by the spacer length used to attach the mesogen to the dendrimer core, since it has been kept relatively short (five methylene units) in order to prevent full decoupling of the mesogenic motions from the silsesquioxane core and, by forcing the mesogens to pack closely together around the dendritic core, to form "hexagonal tubular nematic-columnar" or "rectangular tubular nematic-columnar" phases.

The hybrid structures of such "tubular nematic-columnar LC phases" lend themselves as potential model systems for the development of photonic band-gap materials, whereby large differences in refractive indices between the inorganic and organic sections can be engineered into the system through design and synthesis. In addition, the chiral nematic phase of the hexadecamer shows the selective reflection of blue light, indicating that the pitch of the chiral nematic phase is submicron, $\sim$0.2–0.3 μm.

15.2.4   Effect of Hard Core Scaffolds — Fullerene Scaffolds

Increasing the size of the multifunctional core in supermolecules alters the balance between the free volume required by the core and the number of mesogens needed to occupy effectively the volume around the core in such a way that the interactions needed between the mesogens to support the formation of a LC phase can take place. The consequence of the need to maintain these interactions is that increasing the size of the core requires an increase in the number of mesogens around it.

Apart from octasilsesquioxane, other rigid cage structures can be used as the central building block, for example, [60]fullerene. The hexa-adducts of [60]fullerene can give a spherical distribution of mesogenic substituents about the central scaffold, and thus, these types of supramolecular material have topologies similar to those of the octasubstituted silsesquioxane. Not surprisingly, therefore, the fullerodendrimers exhibit mesophase behavior similar to the silsesquioxane dendrimers.

The incorporation of lateral mesogenic substituents onto [C60]Fullerene, particularly chiral mesogens, can lead to interesting structural behavior and properties. The Bingel reaction of malonates functionalized with laterally appended mesogens yields a supramolecular material with 12 mesogenic units symmetrically positioned about the C_{60} core, **11**, thereby creating a spherical architecture [28] (Fig. 15-5a). Again, because of the lateral attachment of the mesogenic units, it exhibits the chiral nematic phase. However, unlike the silsesquioxane dendrimer with the same mesogen, **11** does not exhibit the columnar phases.

Thus, the fullerene moiety is shielded effectively among the laterally attached mesogens, without disturbing the helical supramolecular organization of the mesophase. Furthermore, as the mesogenic units are symmetrically distributed all over the fullerene sphere, they effectively isolate it, thereby decreasing the possibility of aggregation of the C_{60} units, which is detrimental to mesophase formation.

It is also interesting to consider how the selection process for the helical organization of material **10** is generated on cooling from the isotropic liquid. Cooling into the chiral nematic phase, however, the helical organization was expected to be a result of the organized packing of the dendritic supermolecules, that is, they are considered to be no longer spherical in shape. However, it was

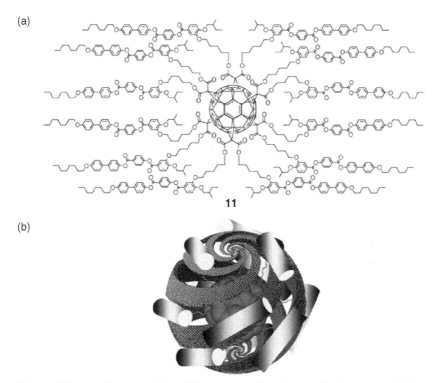

(a)

11

(b)

Figure 15-5   (a) Structure of the fullerene hexaadduct **11** and (b) the proposed helical structure of a nanomolecular "Boojum.".

found that, when the diameter of the C_{60} core is compared to the length of the mesogenic units, it is clear that flexible, random packing of the mesogenic units about the core in the LC state was not possible and that the mesogens are required to be organized in their packing arrangements relative to one another, both on the surface of the dendrimer and between individual dendrimer molecules. One possibility was postulated whereby the mesogenic units are oriented parallel to one another; thus, when the material cools into the liquid-crystalline phase, directional order of the mesogens is selected by the external environment, such as the surface. In doing so, this information is transmitted to the other mesogens associated with the spherical dendrimer and further to the neighboring dendritic supramolecular compounds. Alternatively, for an individual dendrimer, it was proposed that the direction of the mesogens would spiral around the C_{60} core to give poles at the top and bottom of the structure. Thus, the spherical dendrimer was projected to have a well-defined chiral surface, thereby resulting in the creation of a chiral nanoparticle, that is, a nanomolecular "Boojum" (Fig. 15-5b). When the chiral nanoparticles pack together, they were expected to do so through chiral surface recognition processes, resulting in the formation of a helical supramolecular structure. Consequently, the chiral supramolecular nanoparticles transmit their local organization through amplification to adjacent molecules, resulting in values of the helical twisting power that are higher and pitch that is shorter than might be expected for such large molecular entities.

15.3 MULTIPEDES

As noted in Section 15.1, the design of multipedes allows the incorporation of functional moieties into self-assembling and/or self-organizing states of matter. Of the two tetramer polypedes (Fig. 15-1), **1** has four laterally appended chiral mesogenic groups and therefore exhibits a chiral nematic phase as expected. Similarly, tetramer **2**, which has four terminally appended CB mesogenic groups, exhibits a smectic phase as expected [29].

When a laterally appended mesogen is introduced into a system where all the other mesogens are terminally appended, as in **3**, the laterally appended mesogen acts as a disrupter of the orderly packing of the molecules. This results in the suppression of the formation of smectic modifications and support for nematic phases. Thus, compound **2**, which is smectic A, is converted into a chiral nematic phase as in **3**. Remarkably, these high-molecular-mass materials have nematic phases at room temperature, but their crystallization is suppressed and they have low viscosity, indicating that the materials have rheological properties similar to those of low-molar-mass mesogens.

Fine tuning of material properties therefore can be achieved by using mixtures of mesogens attached to the periphery of the scaffold. In addition, the disordering induced can substantially lower melting points and widen temperature ranges of desirable LC phases. Furthermore, such materials are miscible with low-molar-mass materials and can be used to modify their physical properties.

15.3.1 "Janus" Liquid-Crystalline Multipedes

One of the more intriguing and challenging aspects in materials science is understanding the molecular recognition and self-assembling processes in materials with diversely functionalized faces or sides, which can yield supramolecular objects that may recognize and select left from right or top from bottom, as described by de Gennes [26].

A recent molecular design, "Janus" liquid-crystalline molecular materials in the form of segmented structures that contain two different types of mesogenic units has recently been reported [30]. Janus materials favour different types of mesophase structures, grafted onto the same scaffold, to create giant molecules that contain different hemispheres ("Janus" refers to materials with two faces, such as fluorocarbon/hydrocarbon or hydrophilic/hydrophobic, designed to enhance nanophase segregation) (Fig. 15-6).

The complementary materials, for example, **12** shown in Fig. 15-6, based on a central scaffold made up of pentaerythritol and tris(hydroxymethyl)aminomethane units linked together, where one unit carries three CB (smectic preferring) and the other carries three chiral phenyl benzoate (PB) (chiral nematic preferring) mesogenic moieties, were investigated [31].

(a)

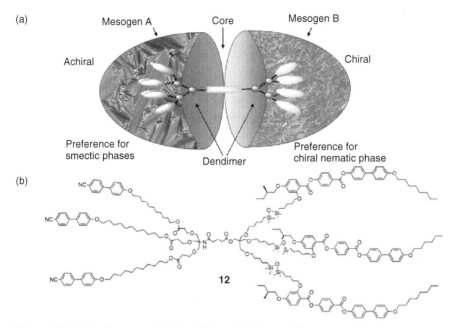

(b)

Figure 15-6 (a) Conceptual design of Janus liquid-crystalline supramolecular systems, composed of a smectic liquid crystal (left) and a chiral nematic liquid crystal (right) and (b) the structure of a Janus liquid crystal, **12**.

Comparison of the phase behavior of the two possible Janus compounds shows clearly that the overall topology of the molecule in respect to the inner core (i.e., which hemisphere carries what mesogen) plays a significant role in determining the type of mesophase formed, since simply placing them in different hemispheres changes the mesophase exhibited.

The manipulation of the structural fragments (mesogenic units, central scaffold, and linking units) in the molecular design of such supramolecular systems potentially allows one to vary the mesophase type and therefore the physical properties and potential applications of materials. Thus, the molecular design of these systems is flexible and potentially capable of incorporating functional units, thereby allowing us to take some steps toward the molecular and functional complexity found in living systems.

Other "Janus" amphiphilic dendrimers have been reported, the mesogenic behavior of which is induced not by the presence of promesogenic units attached to a central scaffold but simply by microphase segregation of the constituent segments of the dendrimer [32–34].

15.3.2 Functional Liquid-Crystalline Supermolecules

The concept of self-organizing functional materials has been realized by the incorporation, by covalent attachment, of the functional entity within a liquid-crystalline material. This approach to self-organization appears to be particularly interesting for functional groups that are not well-adapted to being organized in nanoscale architectures. For example, supramolecular fullerene assemblies [35] and supramolecular systems [12] have been described.

Let us first examine the proportion of mesomorphic groups that are required in the structure of a fullerene supermolecule for it to be mesogenic and then investigate the incorporation of a second functional group in the system, which might contribute or modify the activity of the fullerene unit. A mesogenic dendron of at least four CB-based groups is required to induce mesomorphism in the fulleropyrrolidine family. The parent multipede [36] exhibits a smectic A phase with a clearing point of 168°C. The introduction of the π-conducting phenylenevinylene arm at the N atom of the fulleropyrrolidine produces a material that clears at 171°C [37], and the introduction of ferrocene as the N side chain yields a material that clears at 169°C [38]. All three materials also exhibit glassy states with transitions to the smectic A phase at similar temperatures. The structural difference between the three arises from the substituent at the pyrrolidine N atom. The three materials differ greatly in structure; however, the LC properties are almost invariant with structure. These results indicate that the self-organizing properties of the three materials are dominated by the mesogenic dendron, which is attached to the C_{60} unit, and that the nonmesogenic parts of each material have little or no effect. These cumulative results demonstrate that the incorporation of mesogenic units into a supramolecular structure can be utilized in the self-organization and self-assembly properties, while still retaining the functionality of the supermolecule. These are properties not unlike those of a protein.

In the examples, the mesomorphic groups used for the self-organization process are CBs. As with many of the other examples used previously in this article, lateral substitution of the mesogens can be used to control the self-organization process to yield nematic phases. In the methanofullerene family, the introduction of only two side-on mesogens in the malonate precursor, as those used in **1**, does not provide enough "shielding" of the fullerene core, which disrupts the interactions between the mesogens and destroys the LC phase. Nematic phases can be induced to form, however, by bifurcation at the surface of the fullerene unit and the incorporation of mesogens that are laterally attached [39]. The values of the pitch of the helix of the chiral nematic phase obtained indicate that C_{60} fits within the helical structure formed by the mesogens themselves without causing any significant perturbation of the pitch. This, in turn, implies that, although the large C_{60} unit disturbs the mesogenic interactions, therefore lowering the clearing point, it can be effectively camouflaged within the self-organizing chiral nematic medium, provided that enough mesogenic subunits are available in the dendritic addend, without markedly suppressing the liquid-crystalline state. When the number density of the mesogenic subunits is increased, thereby reducing the weight fraction of the fullerene units in the supramolecular structure, the LC phases become increasingly more stable and the spherical fullerene unit becomes increasingly hidden within the LC matrix [12]. This result indicates that the LC units have the ability to incorporate and organize non-mesogenic functional units into the self-organizing state, without too much detriment to the LC itself.

15.3.3 Effect of the Chirality at the Core

As noted earlier, the incorporation of chiral groups in the LC moieties can have the effect of inducing a small twist on the molecular packing arrangement in the mesophase, resulting in a helical configuration. The effect on the mesophase properties is that, nonlinear properties, including thermochromism, ferroelectricity, antiferroelectricity, electrostriction, and flexoelectricity, appear.

However, another possibility is the introduction of chirality at the core of the supermolecule. A series of novel chiral star-shaped supermolecules incorporating carbohydrates, methyl α-D-glucopyranoside, or methyl α-D-mannopyranoside as chiral cores surrounded by four CB mesogenic groups linked by either C6 or C11 alkyl chain spacers have been examined [40]. These thermotropic carbohydrate-based chiral LCs exhibit the chiral nematic and smectic A mesophases over broad temperature ranges. The induction of the chiral nematic phase demonstrates that the transmission of structural information from the chiral core to the nonchiral mesogens takes place at the supramolecular level; the core imposes a twist on the packing of the molecules, inducing the formation of the helical superstructure. Comparison of the transition temperatures reveals a lower isotropization temperature for the mannosides than for the glucosides, indicating lower thermal stability of the mannoside mesophases. This effect can be attributed to the orientational and stereochemical properties associated with each inner chiral core.

When the CB mesogens are attached by a long C11 spacer, glycoside and mannoside tetramers display wide-temperature-range SmA mesophases. In contrast, incorporating a shorter spacer, C6, induces right-handed chiral nematic mesophases regardless of the chiral core structure. This effect, as observed with the octasilsesquioxanes presented before, is related to the lack of full decoupling of the mesogenic moieties with shorter spacers from the core, which allows the core to express its unique properties.

Similarly, chirality at the surface of the core unit can be introduced in fullerene supermolecules. The 1,3-dipolar cycloaddition reaction of azomethine ylides with C_{60} generates a stereogenic carbon atom at the pyrrolidine ring. This versatile reaction has been used successfully to prepare mesogenic chiral fulleropyrrolidines, derived from mesogenic nonchiral aldehyde dendrons to induce mesoscopic chirality in the materials. The difference between this fulleropyrrolidine and materials **7** and **8** is that the chirality is imparted in the latter ones by the chiral mesogens, whereas in the former one the chirality is imparted by the core. A second-generation dendron, which carries four laterally branched mesogens, was selected in order to observe the nematic phase. The ensuing four diastereoisomeric fullerodendrimers, (R,S), (R,R), (S,R), and (S,S), display the chiral nematic phase [41]. The absolute configurations of the stereogenic centers were determined by X-ray crystallography, 1D and 2D NMR experiments, and circular dichroism spectroscopy. The cholesteric pitch for all the stereoisomers, determined from the fingerprint texture, has values ranging from 2.8 to 4.0 μm, indicating that these materials have relatively long pitches. The handedness of the cholesteric helix was determined from optical observations, and so, a full correlation between the absolute configuration of the asymmetric carbons and the handedness of the macroscopic helix could be established. The fulleropyrrolidine derivatives exhibit supramolecular helical organizations with a right-handed helix for the (R,S) and (R,R) diastereoisomers and a left-handed helix for the (S,R) and (S,S) diastereoisomers. Furthermore, the configuration of the phenylamine carbon [PhCH(CH$_3$)] seems to dominate the chiral response in the mesophase; the isomers with the R configuration at the phenylamine carbon produce a right-handed helix, whereas the S configuration at the phenylamine carbon induces a left-handed helix.

15.3.4 Complex Functional Multipedes

It is highly desirable to obtain self-organized nanostructures based on functional materials because their organization may improve and/or induce new properties and be the basis for the creation of molecular machines. As it became apparent from the results demonstrated with the octasilsesquioxane and the fullerene cores, the design principle of attaching covalently mesogenic moieties to functional cores in order to induce mesogenic behavior is applicable to many other functional moieties, which, due to their isotropic molecular shape, bulkiness, and/or strong intermolecular interactions, are not ideally suited to support the formation of mesophases [42]. Thus, the periodic ordering of non-mesogenic functional units through the process of self-organization can be controlled by inducing liquid crystallinity and in turn can be tailored by the nature of the mesogens attached to it. This strategy has been applied successfully to the incorporation of the lanthanide metals as core in supermolecules, aimed at exploiting the optical properties brought in by the lanthanides, in particular, their luminescence.

The lanthanide nitrato complexes of the type [LnL$_1$(NO$_3$)$_3$], where L carries two CB-based mesogenic arms, indeed form smectic A mesophases in the $80-180^\circ$C temperature range with the heavier lanthanides (Ln = Eu–Lu) [43]. Destabilization of the smectic A phase in favor of the nematic phase was achieved by the introduction of a methyl group within the CB unit.

The magnetic properties of metal clusters and their behavior as molecular magnets have also attracted much attention, but these materials are difficult to integrate into macroscopic devices by self-assembly. The two-dimensional organization of molecular magnets based on the Mn$_{12}$ cluster [Mn$_{12}$O$_{12}$(AcO)$_{16}$(H$_2$O)$_4$] have been studied, inducing its self-assembling behavior by replacing the acetate ligands with gallate anions substituted with CB moieties. X-ray studies indicate that the fluid birefringent mesophase observed between 40 and 150°C corresponds to a smectic phase with modulation of the electronic density, due to the lamellar periodicity [44]. This result suggests an organization in which the peripheral mesogenic groups are equally distributed on either side of the metallic cluster in a compact manner (cylindrical molecular conformation), resembling the microsegregated smectic structures formed by mesogenic silsesquioxanes **5**, **6**, and **7**, and the clusters are further organized in a regular square lattice (similar to the so-called filled mesh phase).

Light emissive properties, in particular, emission in the red–NIR (near-infrared) region and electroluminescence, of metal clusters containing metal–metal multiple bonds have been demonstrated for the smectic phase of the molybdenum octahedral clusters $[Mo_6L_8Br_6]^{2-}$. The substitution reaction of the weakly bonded apical fluoride anions in the cluster $[Mo_6F_8Br_6]^{2-}$ by carboxylate anions based on gallic acid carrying CB moieties, L, as in the lanthanide complexes above, leads to the generation of a smectic phase, although it could not be assigned to a particular type [45]. The emissive properties of the native cluster are preserved in the mesomorphic state upon complexation with the CB-based ligands.

The same principle demonstrated for the self-organization of functional inorganic cores and nanoparticles by appending mesogenic moieties has also been applied to organic functional materials. Current research by Kato, Stoddart, and Sauvage has demonstrated how the self-organization properties of LCs can be used in the self-assembling of functional systems to give an oriented and organized bulk material [46]. A novel system, **6**, in which the properties of LCs have been combined with those of rotaxanes to create a self-organizing molecular switch has been described. This material self-organizes to give a smectic A phase, which has a temperature range from 7 to 146°C. Bistable switching was observed by examination of the electrochemical response, which was characterized by cyclic voltammetry.

15.4 SUMMARY

In this article, we have demonstrated the potential for creating a wide variety of novel supramolecular materials with designer structural and functional properties. In our own development of new materials, we have borne in mind the philosophy and approach used by living systems in the creation of monodisperse and discrete supramolecular materials with specific functionality. In particular, the design features employed in the creation of proteins have been used in the preparation of materials that do not bear any common atomistic relationship with proteins themselves. In this way, this new field lies open for the development of materials for which design is limited only by human imagination.

REFERENCES

1. Lehn, J.-M. (1995). *Supramolecular Chemistry: Concepts and Perspectives*. VCH, Weinheim.
2. Kato, T. (2002). Self-assembly of phase-segregated liquid crystal structures. *Science*, *295*, 2414–2418.
3. Elemans, J. E., Rowan, A. E., Nolte, R. J. M. (2003). Mastering molecular matter: supramolecular architectures by hierarchical self-assembly. *J. Mater. Chem.*, *13*, 2661–2671.
4. Demus, D., Goodby, J. W., Gray, G. W., Spiess, H.-W., Vill, V., eds. (1998). *Handbook of Liquid Crystals*, Vols. 1–3, Wiley-VCH, Weinheim.
5. Goodby, J. W., Mehl, G. H., Saez, I. M., Tuffin, R. P., Mackenzie, G., Auzely-Velty, R., Benvegnu, T., Plusquellec, D. (1998). Liquid crystals with restricted molecular topologies: supermolecules and supramolecular assemblies. *Chem. Commun.*, 2057–2070.
6. Saez, I. M., Goodby, J. W. (2005). Supermolecular liquid crystals. *J. Mater. Chem.*, *15*, 26–40.
7. Saez, I. M., Goodby, J. W. (2008). Supermolecular liquid crystals. *Struc. Bond.*, *15*, 26–40.

8. Newkome, G. R., Moorefield, C. N., Vögtle, F. (2000). *Dendrimers and Dendrons: Concepts, Syntheses, Applications*. Wiley-VCH, Weinheim.

9. Donnio, B., Guillon, D. (2006). Liquid crystalline dendrimers and polypedes. *Adv. Polym. Sci.*, *201*, 45–155.

10. Donnio, B., Buathong, S., Bury, I., Guillon, D. (2007). Liquid crystalline dendrimers. *Chem. Soc. Rev.*, *36*, 1495–1513.

11. Marcos, M., Martín-Rapún, R., Omenat, A., Serrano, J. L. (2007). Highly congested liquid crystal structures: dendrimers, dendrons, dendronized and hyperbranched polymers. *Chem. Soc. Rev.*, *36*, 1889–1901.

12. Deschenaux, R., Donnio, B., Guillon, D. (2007). Liquid-crystalline fullerodendrimers. *New J. Chem.*, *31*, 1064–1073.

13. Rosen, B. M., Wilson, C. J., Wilson, D. A., Peterca, M., Imam, M. R., Percec, V. (2009). Dendron-mediated self-assembly, disassembly, and self-organization of complex systems. *Chem. Rev.*, *109*, 6275–6540.

14. Tschierske, C. (2001). Micro-segregation, molecular shape and molecular topology, partners for the design of liquid crystalline materials with complex mesophase morphologies. *J. Mater. Chem.*, *11*, 2647–2671.

15. Kato, T., Mizoshita, N., Kishimoto, K. (2006). Functional liquid-crystalline assemblies: self-organized soft materials. *Angew. Chem. Int. Ed.*, *45*, 38–68.

16. Percec, V., Glodde, M., Bera, T. K., Miura, Y., Shiyanovskaya, I., Singer, K. D., Balagurusamy, V. S. K., Heiney, P. A., Schnell, I., Rapp, A., Spiess, H.-W, Hudson, S. D., Duank, H. (2002). Self-organization of supramolecular helical dendrimers into complex electronic materials. *Nature*, *419*, 384–387.

17. Baars, M. W. P. L., Söntjens, S. H. M., Fischer, H. M., Peerlings, H. W. I., Meijer, E. W. (1998). Liquid-crystalline properties of poly(propylene imine) dendrimers functionalized with cyanobiphenyl mesogens at the periphery. *Chem.—Eur. J.*, *4*, 2456–2466.

18. Marcos, M., Gimenez, R., Serrano, J. L., Donnio, B., Benoit, H., Guillon, D. (2001). Dendromesogens: liquid crystal organizations of poly (amidoamine) dendrimers versus starburst structures. *Chem.—Eur. J.*, *7*, 1006–1013.

19. Saez, I. M., Goodby, J. W. (1999). Supermolecular liquid crystal dendrimers based on the octasilsesquioxane core. *Liq. Cryst.*, *26*, 1101–1105.

20. Kreuzer, F.-H., Maurer, R., Spes, P. (1991). Liquid-crystalline silsesquioxanes. *Makromol. Chem., Macromol. Symp.*, *50*, 215–228.

21. Saez, I. M., Styring, P. (1996). Oligo(siloxane) rings and cages possessing nickel-containing liquid crystal side chains. *Adv. Mater.*, *8*, 1001–1005.

22. Mehl, G. H., Goodby, J. W. (1996). Liquid-crystalline substituted octakis (dimethylsiloxy) octasilsesquioxanes: Oligomeric supermolecular materials with defined topology. *Angew. Chem.*, *35*, 2641–2643.

23. Keith, C., Dantlgraber, G., Reddy, R. A., Baumeister, U., Prehm, M., Hahn, H., Lang, H., Tschierske, C. (2007). The influence of shape and size of silyl units on the properties of bent-core liquid crystals—from dimers via oligomers and dendrimers to polymers. *J. Mater. Chem.*, *17*, 3796–3805.

24. Miao, J., Lei Zhu, L. (2010). Topology controlled supramolecular self-assembly of octa triphenylene-substituted polyhedral oligomeric silsesquioxane hybrid supermolecules. *J. Phys. Chem. B*, *114*, 1879–1887.

25. Saez, I. M., Goodby, J. W. (2001). Chiral nematic octasilsesquioxanes. *J. Mater. Chem.*, *11*, 2845–2851.

26. de Gennes, P.-G. (1992). Soft matter (Nobel Lecture). *Angew. Chem. Int. Ed. Engl.*, *31*, 842–845.

27. Saez, I. M., Goodby, I. M., Richardson, R. M. (2001). A liquid-crystalline silsesquioxane dendrimer exhibiting chiral nematic and columnar mesophases. *Chem.—Eur. J.*, *7*, 2758–2764.

28. Campidelli, S., Brandmüller, T., Hirsch, A., Saez, I. M., Goodby, J. W., Deschenaux, R. (2006). An optically-active liquid-crystalline hexa-adduct of [60]fullerene which displays supramolecular helical organization. *Chem. Commun.*, 4282–4284.

29. Saez, I. M., Goodby, J. W. (2003). Segregated liquid crystalline dendritic supermolecules—multipedes based on pentaerythritol scaffolds. *J. Mater. Chem.*, *13*, 2727–2739.

30. Saez, I. M., Goodby, J. W. (2003). Design and properties of "Janus-like" supermolecular liquid crystals. *Chem. Commun.*, 1726–1727.

31. Saez, I. M., Goodby, J. W. (2003). "Janus" supermolecular liquid crystals—giant molecules with hemispherical architectures. *Chem.—Eur. J.*, *9*, 4869–4877.

32. Percec, V., Imam, M. R., Bera, T. K.,. Balagurusamy, V. S. K, Peterca, M., Heiney, P. A. (2005). Self-assembly of semifluorinated Janus-dendritic benzamides into bilayered pyramidal columns. *Angew. Chem. Int. Ed.*, *44*, 4739–4745.

33. Bury, I., Heinrich, B., Bourgogne, C., Guillon, D., Donnio, B. (2006). Supramolecular self-organization of "Janus-like" diblock codendrimers: synthesis, thermal behavior, and phase structure modeling. *Chem.—Eur. J.*, 12, 8396–8413.

34. Percec, V., Wilson, D. A., Leowanawat, P., Wilson, C. J., Hughes, A. D., Kaucher, M. S., Hammer, D. A., Levine, D. H., Kim, A. J., Bates, F. S., Davis, K. P., Lodge, T. P., Klein, M. L., DeVane, R. H., Aqad, E., Rosen, B. M., Argintaru, A. O., SienkowskaM. J., Rissanen, K., Nummelin, S., Ropponen, J. (2010). Self-assembly of Janus dendrimers into uniform dendrimersomes and other complex architectures. *Science*, *328*, 1009–1014.

35. Nakanishi, T., Shen, Y., Wang, J., Li, H., Fernandes, P., Yoshida, K., Yagai, S., Takeuchi, M., Ariga, K., Kurthf, D. J., Möhwald, H. (2010). Superstructures and superhydrophobic property in hierarchical organized architectures of fullerenes bearing long alkyl tails. *J. Mater. Chem.*, 20, 1253–1260.

36. Campidelli, S, Lenoble, J, Barberá, J, Paolucci, F, Marcaccio, M, Paolucci, D, Deschenaux, R. (2005). Supramolecular fullerene materials: dendritic liquid-crystalline fulleropyrrolidines. *Macromolecules*, *38*, 7915–7925.

37. Campidelli, S., Deschenaux, R., Eckert, J. F., Guillon, D., Nierengarten, J. F. (2002). Liquid-crystalline fullerene–oligophenylenevinylene conjugates. *Chem. Commun.*, 656.

38. Campidelli, S., Vázquez, E., Milic, D., Prato, M., Barberá, J., Guldi, D. M., Marcaccio, M., Paolucci, D., Paolucci, F., Deschenaux, R. (2008). Photophysical, electrochemical, and mesomorphic properties of a liquid-crystalline [60]fullerene–peralkylated ferrocene dyad. *J. Mater. Chem.*, *18*, 1504–1509.

39. Campidelli, S., Eng, C., Saez, I. M., Goodby, J. W., Deschenaux, R. (2003). Functional polypedes—chiral nematic fullerenes. *Chem. Commun.*, 1520–1521.

40. Belaissaoui, A., Cowling, S. J., Saez, I. M., Goodby, J. W. (2010). Core chirality based tailoring of the liquid crystalline properties of supermolecular tetrapedes. *Soft Matter*, *6*, 1958–1963.

41. Campidelli, S., Bourgun, P., Guintchin, B., Furrer, J., Stoeckli-Evans, H., Saez, I. M., Goodby, J. W., Deschenaux, R. (2010). Diastereoisomerically pure fulleropyrrolidines as chiral platforms for the design of optically active liquid crystals. *J. Am. Chem. Soc.*, *132*, 3574–3581.

42. Goodby, J. W., Saez, I. M., Cowling, S. J., Gasowska, J., MacDonald, R. A., Sia, S., Watson, P., Toune, K. J., Hird, M., Lewis, R. A., Lee, S.-E., Vaschenko, V. (2009). Molecular complexity and the control of self-organising processes. *Liq. Cryst.*, *36*, 567–605.

43. Terazzi, E., Bocquet, B., Campidelli, S., Donnio, B., Guillon, D., Deschenaux, R., Piguet, C. (2006). Encoding calamitic mesomorphism in thermotropic lanthanidomesogens. *Chem. Commun.*, 2922–2924.

44. Terazzi, E., Bourgogne, C., Welter, R., Gallani, J.-L., Guillon, D., Rogez, G., Donnio, B. (2008). Single-molecule magnets with mesomorphic lamellar ordering. *Angew. Chem. Int. Ed.*, *47*, 490–495.

45. Molard, Y., Dorson, F., Circu, V., Roisnel, T., Artzner, F., Cordier, S. (2010). Clustomesogens: liquid crystal materials containing transition-metal clusters. *Angew. Chem. Int. Ed.*, *49*, 3351–3355.

46. Aprahamian, I., Yasuda, T., Ikeda, T., Saha, S., Dichtel, W. R., Isoda, K., Kato, T., Stoddart, F. J. (2007). A liquid-crystalline bistable [2]Rotaxane. *Angew. Chem. Int. Ed.*, *46*, 4675–4679.

CHAPTER 16

PHOTORESPONSIVE CHIRAL LIQUID CRYSTALS

Ratheesh K. Vijayaraghavan and Suresh Das

Chemical Sciences and Technology Division, National Institute for Interdisciplinary Science and Technology (CSIR), Thiruvananthapuram, Kerala, India

16.1 INTRODUCTION

Liquid crystals (LCs) are a class of soft materials, which have been extensively exploited for the past several decades [1] for a variety of technological applications [3] ranging from display devices [2] to the development of high-strength polymers [4]. They also hold tremendous potential in the development of a wide range of advanced technologies such as telecommunication, information storage, and sensors. The term liquid crystal signifies a state of aggregation that is an intermediate between the solid state and isotropic liquid, by virtue of which they possess some unique properties such as the ability to be processed into uniform thin films with long-range molecular order, birefringence (i.e., anisotropy in refractive index and the ability to undergo switching of their long-range molecular order under the influence of external stimuli such as electric and magnetic fields), and so on. A combination of these properties has revolutionized the display industry. An alternative method for switching the long-range molecular order of liquid crystalline films and consequently their optical properties, which is being actively pursued, is the use of photons. Photochemical transformation in liquid-crystalline materials is known to bring about modifications in the alignment or phase transition properties of the liquid-crystalline host. This becomes possible because even a weak photochemical process occurring at the molecular level in liquid-crystalline materials are translated into macroscopic phenomena because of the cooperative interactions and long-range ordering of liquid-crystalline molecules, leading to amplification of signals. In photoactive liquid crystals, switching times can be several orders of magnitude higher compared to electrical- or magnetic-field-induced switching, making them potentially applicable in a number of advanced photonic applications, such as in high-speed

Supramolecular Soft Matter: Applications in Materials and Organic Electronics, First Edition.
Edited by Takashi Nakanishi.

rewritable recording devices [5,6], optical data storage [7,8], patterned nanostructures [9–11,13,14], fluorescence imaging [12,15], and band-gap materials [16,36]. Among the various classes of photoresponsive liquid crystals that have been investigated, those that exhibit macroscopic chirality are especially interesting. In chiral liquid crystals, the preferred orientation of molecules along a certain direction described by the unit vector (the director **n**) undergoes a helical twist from layer to layer. When the pitch of the helical arrangement of **n** in such liquid crystals is of the order of visible light, Bragg reflection results in the selective reflection of circularly polarized light, resulting in the films being brightly colored. This selective reflection of visible light makes chiral liquid crystals useful in a variety of applications such as noninvasive thermometers, polarizing mirrors, electro-optic displays, and optical storage. When such materials are designed to be photoactive, their optical properties can be tuned with light, leading to a wide range of photonic applications.

16.1.1 Liquid Crystals

Liquid-crystalline materials can be generated by a large variety of molecules, including organic and inorganic compounds, ions, zwitterions, elastomers, dendrimers [17,18], oligomeric systems, as well as metal-organic complexes [22–25]. Transitions to these intermediate mesomorphic states may be brought about by purely thermal processes (thermotropic mesomorphism) or by the influence of solvents (lyotropic mesomorphism). This chapter deals mainly with the photoresponsive aspects of chiral thermotropic liquid crystals. Typically, molecules that can exhibit liquid crystallinity are considered to belong to one of the two distinct classes of molecules: anisometric and amphiphilic (Fig. 16-1).

Anisometric (nonamphiphilic) molecules include rod- or disklike molecules or aggregates in most cases, giving exclusively thermotropic liquid crystals. The second class includes amphiphilic molecules such as detergents and lipids, leading to lyotropic and thermotropic mesophases.

Thermotropic liquid crystals are generally further distinguished with respect to the molecular shape of the constituent molecules [26], namely, calamitic (rodlike), discotic [27], and sanidic (board shaped). The vast majority of thermotropic liquid crystals are composed of rodlike molecules, and these can be further classified into two major classes, nematics (N) and smectics (Sm). The nematic phase is a one-dimensionally ordered fluid in which molecular long axes are oriented along a vector **n** (the *director*) but are otherwise completely disordered. In smectic phases, molecules are arranged in diffuse layers and show orientational and short-range positional order within the plane of the layers. Several smectic phases have been identified, but the two most commonly observed are the smectic A (SmA) and smectic C (SmC) phases. In the SmA phase, molecular long axes are oriented along a director **n**, which is parallel to the layer normal; in the SmC phase, the director **n** is tilted at a temperature-dependent angle with respect to the layer normal. Unlike the nematic phase, which completely lacks positional order, the diffused layer structure of the SmA and SmC phases results, on the time average, in a segregation of the rigid cores from the side chains.

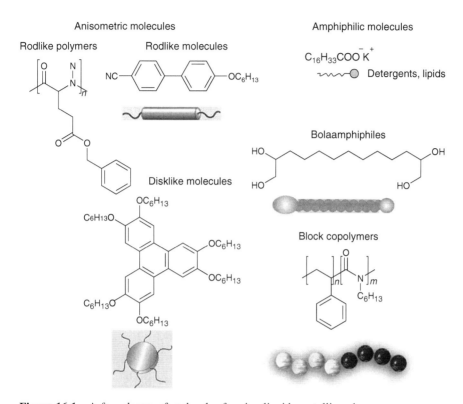

Figure 16-1   A few classes of molecules forming liquid-crystalline phases.

16.1.2   Chirality in Liquid Crystals

Chiral LCs can be obtained by doping of an achiral phase formed by achiral mesogens with a chiral dopant or by designing mesogens with suitably located chiral centers in the same molecule. Most studies on the mechanism by which the chirality of the molecule is mapped onto the liquid-crystalline phase deal with the mode of transfer of chirality from the chiral center of the molecule to the rest of the molecule (intramolecular transfer) and the subsequent transfer of chirality to the liquid-crystalline phase (intermolecular transfer) [28]. The question of how chirality can be transferred from one level to an adjacent level is of fundamental interest, and several recent studies have tried to address these issues in both host–guest systems and chiral molecules exhibiting liquid-crystalline phases. However, the transfer of chirality from the molecular level to the liquid-crystalline phase remains a complex phenomenon, which is still not well understood [29].

Two major classes of chiral liquid crystals are the chiral nematic [30–31] and the chiral smectic C phases. Chiral nematic (N*) LCs possess the inherent property of selective reflection of light because of a helical superstructure originating from their chirality [34,35]. Chiral smectic C (SmC*) phases [32] attract

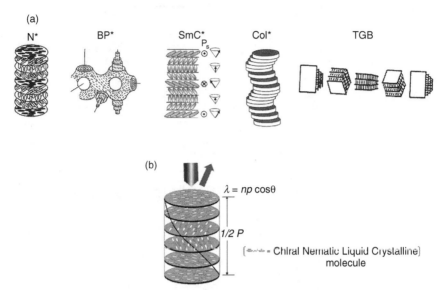

Figure 16-2 (a) Helical structures formed in chiral liquid crystals. Cholesteric phase (N*), blue phase (BP), chiral smectic C phase (SmC*), chiral columnar phase (Col*), and twist grain boundary phase (TGB). (b) Diagram of the helical structure of cholesteric liquid crystal with selective reflection. *Source*: Adapted from reference [28,34].

considerable attention, since their potential for application in display devices is next only to nematics, which are well known for their applications in flat panel liquid crystal displays (LCDs) [33]. The smectic C* liquid-crystalline state, due to its bistability, fast electro-optic switching (10^3 times faster than nematic LCs), and wide viewing angle, finds application in commercial high-resolution reflective microdisplays. In addition, molecules possessing the smectic C* phase are potential candidates for nonlinear optics [37,38,41,42] and photonic applications [39,40]. In view of these applications, the design and development of chiral LCs are of technological importance. Some of the other interesting examples of chiral liquid-crystalline phases are the chiral nematic (N*) phase, smectic C* (SmC*) phase, blue phase (BP), twist grain boundary (TGB) phases, and chiral columnar phase (Fig. 16-2a).

The chiral nematic liquid-crystalline phase (N*), also referred to as the cholesteric phase since this phase was initially observed in cholesterol derivatives, was the first thermotropic liquid-crystalline material discovered in 1888 by the Austrian botanist Reinitzer. Instead of the uniform alignment of the director field occurring in the nematic phase, the corresponding chiral nematic phase exhibits a helical structure that shows a uniform helix axis with a certain sign and certain pitch. The pitch, the distance needed for the director of the mesogens in each plane to rotate through $360°$, is a measure of chirality of the system. The cholesteric phase can reflect light of a specific wavelength depending on the helical pitch of the liquid crystal, $\lambda = np \cos\theta$, where n stands for the mean refractive index of the liquid crystal and θ is the incident angle (Fig. 16-2b). The wavelength of light reflected is divided into right and left circularly polarized components, and the component of light with the same sense of rotation of electric field vectors as the helical sense of the molecular ordering is transmitted and the other component is reflected.

The pitch of chiral nematic liquid crystals depends on the chemical structures of the liquid-crystalline molecules, the nature and concentration of the dopants, and other factors such as temperature and presence of an electric field. The addition of small amounts of optically active dopants to a nematic liquid-crystalline host can lead to the formation of a chiral nematic phase. The pitch of the induced chiral nematic phase is dependent on the concentration (c) of the dopant, the helical twisting power (β) of the dopant, and the enantiomeric excess (ee) of the dopant, as shown in the following equation:

$$p = [\beta \cdot c \cdot (ee)]^{-1}$$

The helical twisting power (β) is an intrinsic property of any chiral dopant, which indicates how efficient this molecule is in inducing a chiral orientation in the liquid-crystalline material. Molecules that bear a structural resemblance to the mesogenic host often possess a significant helical twisting power.

With regards to temperature, Gibson considered two possible effects on the helical pitch, namely, (i) the intermolecular distance along the helix axis increases, tending to increase the pitch and (ii) the displacement angle θ between adjacent molecules in the helical stack increases, tending to decrease the pitch. In fact, the pitch of pure single cholesteric compounds always decreases with increasing temperature, indicating that the second effect governing the displacement angle θ is more important than the longitudinal helical expansion [30].

The chiral smectic C (SmC*) phase is the chiral analog of the SmC phase. The SmC* ferroelectric liquid crystals (FLCs) exhibit spontaneous polarization (P_S) as a result of the presence of polar groups, which imparts the molecules with an electric dipole moment [28]. In the SmC* phase, the ferroelectric liquid-crystalline molecules are aligned parallel to each other to form a layer with a tilt between the direction of the long axis of the ferroelectric liquid crystalline molecule and the normal to the smectic layer. The average direction of the long axis (director, **n**) is defined in each layer, and owing to the chiral group in the ferroelectric liquid-crystalline molecules, the director adopts a helicoidal structure with a characteristic pitch. As a result, the polarization of each layer also spirals along the direction of the helical pitch and the overall polarization averages to zero.

In surface-stabilized alignments (Fig. 16-3), SmC* liquid crystals are contained in thin cells with a thickness less than the pitch of the helix. In such cells, the interaction of the ferroelectric liquid-crystalline molecules is strong enough to prevent the helical arrangement and the molecules align nearly perpendicular to the substrate. As a result, the polarization of each layer is also aligned in the same direction. Such a surface-stabilized ferroelectric liquid crystal (SSFLC) cell exhibits bistability wherein the polarization is aligned upward or downward and can be switched from one form to the other by application of an external electric field.

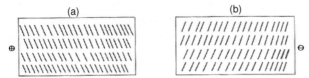

Figure 16-3   The SSFLC cell: The "bookshelf" geometry of a thin film of smectic C* sandwiched between two glass plates: (a) field "up" (normal to the plane of the diagram) and (b) field "down".

The remarkable feature of SSFLC is that the switch on and switch off times are just a few tens of microseconds, that is, about 10^3 times faster than the twisted nematic device. For a given cell thickness, the material parameters that determine the switching speed are polarization and viscosity [5].

BPs are an interesting class of chiral liquid crystals possessing a fluid lattice, which self-assemble into three-dimensional cubic defect structures. The periodicity of these defect structures is of the order of the wavelength of visible light, resulting in interesting optical properties such as selective reflection of circularly polarized light [45,46]. These properties combined with the ability to reorient the molecules in BPs using external electric fields make them very attractive for applications such as fast light modulators and tunable photonic crystals. One of the major factors that prevents the use of BP*s in technological applications is that they generally appear over a very narrow temperature range (ca 1 K), between the isotropic and chiral nematic (N*) phases of highly chiral liquid crystals. Enhancing the temperature range of the BPs is a major challenge, and several studies have been reported on efforts toward achieving this goal. Kikuchi *et al.* reported polymer-stabilized BPs for which the temperature range could be extended to more than 60 K [48]. Although fast electro-optic switching was observed in these systems, the switching was from the defect or disclination lines since the polymer matrix restricted deformation of the BP lattice, thereby preventing color switching. Coles *et al.* reported that eutectic mixtures of three homologs of a symmetric dimer doped with a small quantity of a highly twisting chiral additive showed a BP with a very large temperature gap, whose reflected light could be switched over a wide range in 10^{-2} s by applied electric fields [45]. More recently, Yang and coworkers have reported stable BPs in a H-bonded self assembly between a chiral donor and an achiral acceptor with the temperature range of the BP extending to 23 K [47].

TGB phases are another class of frustrated liquid-crystalline phase. At the normal chiral nematic to smectic A* (N* to SmA*) transition, the helical ordering of the chiral nematic phase collapses to give the layered structure of the smectic A* phase. However, for strongly chiral systems, there can be competition between the need of the molecules, because of their chiral packing requirements, to form a helical macrostructure and the requirement for the formation of a lower energy layered structure. This frustration is relieved by the formation of a helical structure, with the helical axis parallel to the planes of the layers. The helix and lamellar structures are, however, incompatible with one another and cannot coexist without the formation of defects. The coexistence is achieved by small blocks/sheets of molecules with a local smectic structure, which are rotated with respect to one another via sets of screw dislocations located at their interfaces, thereby giving a quasi-helical structure. In this way, the twist distortions are localized in defects that periodically punctuate the normal smectic state.

16.2   PHOTORESPONSIVE LIQUID CRYSTALS

Liquid-crystalline phases are highly susceptible to minor changes in their microenvironment, since the cooperative behavior of liquid-crystalline molecules and their long-range order help to amplify relatively weak processes occurring at the molecular level into macroscopic phenomena. Photochemically induced alignment or phase transitions can be effectively brought about in LCs doped with small amounts of photoresponsive molecules or inherently photoresponsive LCs. The use of light as an external stimulus in switching the alignment or phase of LCs has several advantages since such processes can be brought about in a rapid and precise manner with a high degree of spatial control. The choice of wavelengths, polarization, and intensity of light are some of the other variables that can be effectively used to bring about greater degrees of control in light-induced transformations [49,50]. A large variety of photoresponsive liquid crystals have been designed and investigated for optical data storage including holography, and these materials have been especially attractive from the point of view of developing ultrafast switching devices that can function in the millisecond to microsecond time domain [55,56]. The use of photoresponsive liquid-crystalline elastomers wherein light energy can be converted to mechanical energy is another area that is attracting considerable interest with a view of developing them for application in light-driven actuators [43,51].

Although both reversible and irreversible photochemical transformations can be utilized for such switching applications, those that are thermally or photochemically reversible are preferred for most of the applications mentioned above. In host–guest systems, the presence of extraneous molecules is known to reduce the isotropization temperatures of liquid crystals, and the extent to which this happens can vary with the shape of the dopant molecule. For example, when nematic or smectic liquid crystals are doped with the rod-shaped trans isomer of azobenzenes, the isotropization temperature of the liquid crystal is not significantly altered, whereas the bent cis isomer destabilizes the liquid-crystalline phase substantially [52]. Thus, photoisomerization of *trans*-azobenzene doped into liquid-crystalline samples held close to their isotropization temperature can result in photoinduced phase transition of the host liquid crystal to its isotropic phase. Since such phase transitions do not involve a change in temperature, they are termed as isothermal phase transitions. The trans–cis isomerization of azobenzene derivatives is thermally reversible, as a result of which the sample can be reverted back to the initial liquid-crystalline phase. The various parameters that control photoinduced isothermal phase transitions in host–guest systems, such as the nature of the host LCs, photochromic guest, concentration of the guest, as well as temperature, have been extensively investigated [53,54]. Photochemical phase transitions in host–guest systems in which the host is a liquid-crystalline polymer have also been widely investigated.

Photoisomerization of azobenzene derivatives is a very fast process and generally occurs in the picosecond time domain. The relatively long time required for photoinduced isothermal phase transitions in azobenzene-doped LCs can be attributed to the time required for the host liquid-crystalline molecules to react to the formation of the cis isomer of the azobenzene dopant. It would therefore be possible to reduce the response or switch times for photoinduced isothermal phase transitions by designing liquid-crystalline molecules that are intrinsically photoactive [51,52]. Ikeda and coworkers have developed monomeric and polymeric azobenzene derivatives that are nematic in the trans form and isotropic in the cis form. Time-resolved studies indicated that photoinduced isothermal phase transition occurred in about 200 μs, which is several orders of magnitude faster than photoswitching observed in host–guest systems. Apart from azobenzenes, photochromic molecules such as spiropyrans, fulgides, and alkenes have been investigated as dopants for bringing about photoinduced isothermal phase transitions. Recently, the use of diphenylbutadiene as a chromophore for the design of photoresponsive liquid crystals has been reported.

16.3 PHOTORESPONSIVE CHIRAL LIQUID CRYSTALS

16.3.1 Chiral Nematics

As described above, the pitch of chiral nematic LCs is sensitive to the nature and concentration of dopants. The helical twisting power of dopant molecules is closely related to the structure of the dopant molecule and its compatibility with the host liquid-crystalline molecules. Photochemically induced changes in the structure or stereochemistry of the dopant can therefore lead to significant changes in the pitch of the chiral nematic LCs.

Light-induced change in the pitch of cholesteric liquid crystals was first reported by Haas *et al.* in mixtures composed of cholesteryl bromide and other cholesteryl derivatives. The irreversible changes observed in these systems were attributed to photodecomposition of cholesteryl bromide. Reversible photoinduced changes in the reflection band of chiral nematics were first reported by Sackmann by making use of photoisomerization of azobenzene doped into cholesteric liquid-crystalline mixtures. Schuster *et al.* reported that the photochromism of fulgide could change the helical pitch of a cholesteric liquid crystal. Yokoyama *et al.* showed that binaphtol derivatives of indolylfulgides functioned as chiral dopants to generate chiral nematic phases on addition to nematic liquid crystals and that photoirradiation induced dramatic changes in the cholesteric pitch. The reverse effect, namely, conversion of a chiral nematic phase to a nematic phase has also been reported. Schuster and coworkers showed that photoracemization of optically active binapthyl derivatives in doped liquid crystals induced a chiral nematic to nematic phase transition. Subsequently, Feringa and coworkers reported that such transitions could be conducted in a reversible manner. Using photoisomerization of donor-acceptor-substituted inherently dissymmetric alkenes (**1**), between cis and trans forms using different wavelengths, reversible light-induced conversion of cholesteric to nematic liquid-crystalline phases as well as reversible alteration of the macroscopic helical pitch and screw sense in chiral nematics could be achieved [57].

(**P**)-*trans*-**1**   (**M**)-*cis*-**1**

The ability of such bistable asymmetrically substituted stilbenes labeled as molecular motors to change the pitch as well as the screw sense of the helix of chiral nematics has been cleverly used by Feringa and coworkers to perform mechanical work by rotating objects four orders of magnitude larger than the chiral dopants [30]. The fluorene-based molecular rotors (**2**) are highly efficient bistable dopants for calamitic liquid crystals, showing enormous helical twisting powers in some liquid-crystalline hosts.

2

Helical twisting powers of 144 μm^{-1} were obtained for the stable form of the motor, whereas the unstable form of the motor also displays a large helical twisting power. Moreover, the helical twisting powers of the stable and unstable form have opposite signs, allowing efficient switching between helicities. The rotational reorganizations brought about by photoirradiation of liquid-crystalline hosts doped with **2** was shown to generate a torque on embedded microscopic particles (Fig. 16-4).

With regard to use of photoresponsive chiral nematics for long-term storage of images, the fluidity of liquid crystals is disadvantageous, since slow diffusion can lead to a blurring of the images over a period of time. In order to overcome this problem, photoinduced changes in the pitch have been investigated in polymeric liquid crystals. Irradiation of cholesteryl polymer containing photoactive

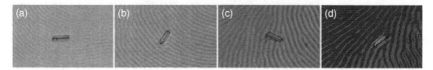

Figure 16-4   Features of a light-driven molecular motor. Glass rod rotating on the liquid crystal during irradiation with ultraviolet light. Frames (a)–(d) were taken at 15-s intervals and show clockwise rotations of 28° (b), 141° (c), and 226° (d) of the rod relative to the position in (a).

(−)-2-arylidene-*p*-menthane-3-one (**3**) as pendants resulted in elongation of the pitch, resulting in a red shift and broadening of the reflectance band.

$$hv$$

3

R = Ph, OMe, NMe$_2$,
Cl, CN, NO$_2$

A polymeric liquid-crystalline system was also reported wherein photo-induced switching of dopant azobenzene was accompanied by cross-linking of the polymeric LC induced by diacrylate and a polymerization initiator. Using this procedure, the pitch of the LC polymer could be shifted from 645 nm to a cross-linked state showing reflection at 750 nm. Feringa *et al.* used the photoisomerization of the donor-acceptor-substituted alkenes doped into a chiral polymerizable cholesteric acrylate mixture, in which the copolymerization could be used to lock the cholesteric helix to generate a stable polymer. Using pho-toisomerization of the dopant, the authors have shown that the reflected color of the film could be switched between red and green. The color change induced by light could be used for permanent storage of information by fixing it using pho-topolymerization. The use of polymeric LCs for imaging through photoinduced pitch changes suffers from several drawbacks, which include slow switching times, broadening of the reflectance bands, and difficulty in erasure of the stored images.

Tamaoki and coworkers have made elegant use of the glass-forming abil-ities of some cholesteric liquid crystals to overcome the problems associated with photoswitching in low-molecular-weight and polymeric LCs. A glassy liq-uid crystal could be obtained when the cholesteric phase is rapidly cooled without allowing it to recrystallize, that is, the macroscopic long-range helical ordering is preserved in the solid state [34]. The glassy materials so obtained retain all the optical characteristics such as selective reflection and selective transmittance of circularly polarized light. Glassy liquid crystals can also be obtained by polymerizing the cholesteric phase. Tamaoki and coworkers were able to design several dicholesteric esters with molecular weights of over 1000, which could form stable glasses on rapid cooling from the cholesteric liquid-crystalline phases [58,59]. Using these systems they could overcome some of the problems associated with photoswitching in low-molecular-weight and polymeric LCs. Glassy liquid crystals were obtained by rapidly cooling thin films of these materials held at high temperature at which they possess the cholesteric phase. In the process, the macroscopic long-range helical ordering is preserved in the solid state. As a result, the glassy materials retain all the optical characteristics such as selective reflection and selective transmittance of circularly polarized light observed in the cholesteric phase at high temperatures. The color of the glassy state could be tuned by rapid cooling of the cholesteric films held at different temperatures. Glassy LCs could also be obtained by spin coating the dicholesteryl esters dissolved in dichloromethane. To control the optical properties of glass-forming dimesogens by light, they were doped with azobenzene derivatives. The following figure shows the structures of azobenzene derivatives used for enabling photochemical control of the color of light reflected by the cholesteric glasses.

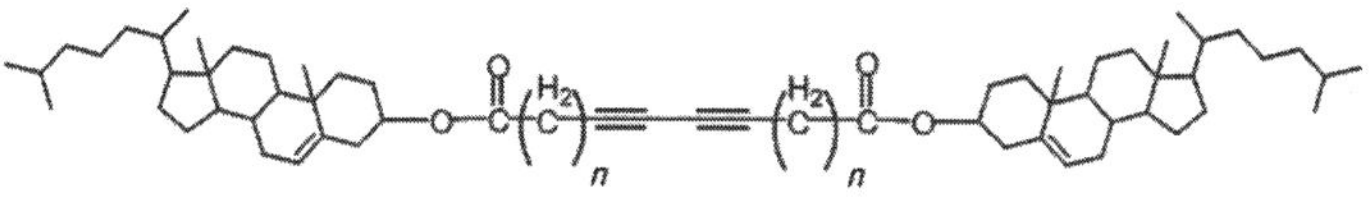

4-*n* (*n* = 2, 3, 4, 5, 6, 7, 8, 10)

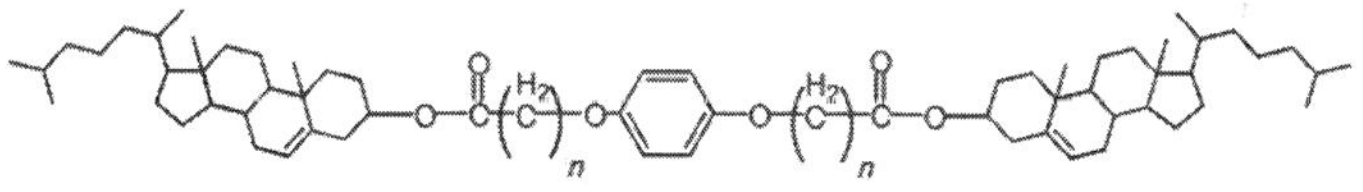

5-*n* (*n* = 1, 3, 4, 5, 6, 7, 9, 10, 11)

6

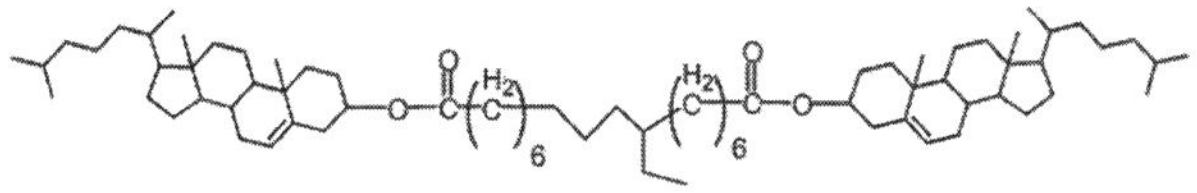

7

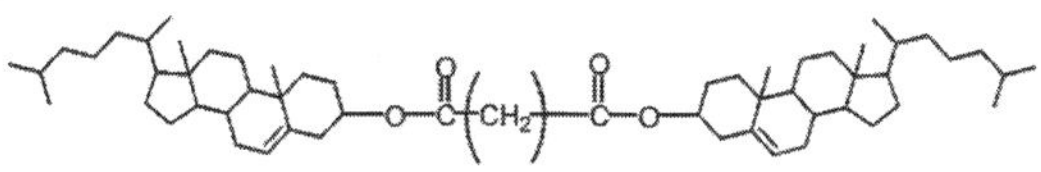

8

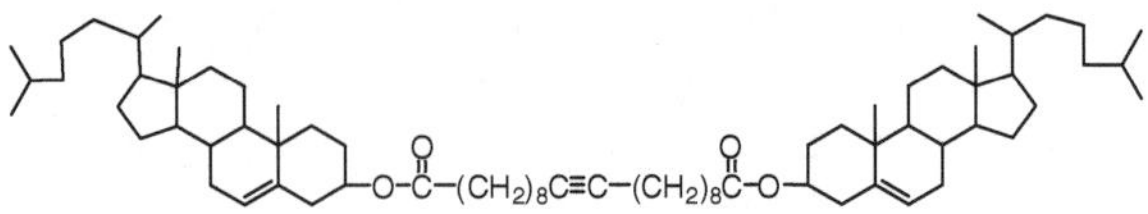

9

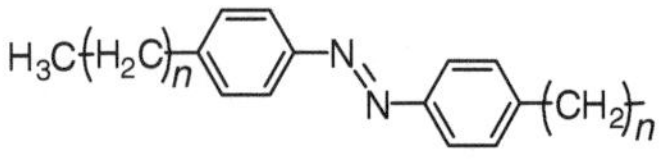

$H_3C(H_2C)_n$—⟨ ⟩—N=N—⟨ ⟩—$(CH_2)_n$

10 (n = 3,6,10–15)

A red reflection band was observed for a thin ($\sim$10 μm) film of a 98:2 weight ratio mixture of **9** and **10** ($n = 12$) kept between two quartz plates at 120°C. Irradiation of the film at this temperature with 366-nm UV light resulted in a change in its color from red to green to blue with increasing time of irradiation. This effect could be attributed to trans–cis photoisomerization of the azobenzene-based dopant. The thin films could be solidified by freezing them in an ice bath at any stage. The solidified glasses obtained in this manner retained the iridescent color of the cholesteric film from which it was quenched. Different colors could be fixed by changing the energy of the irradiated UV light since the pitch at the irradiated region was dependent on the cis/trans ratios of the azobenzene dopant at that site.

An interesting observation in this class of materials was that, although the *cis*-azobenzene begins to thermally revert back to the trans state at high temperatures, there was a slight time lag for the pitch to revert back. Because of this, the film could be rapidly quenched in ice without causing much blurring of the image. Once the glass was formed, the molecular alignment would not be disturbed by the reverse cis to trans thermal isomerization process. As a result, the pitch and hence the color reflected by the film could be fixed. Similar observations were made in dicholesteric esters doped with donor-acceptor-substituted diphenylbu-tadienes, namely, 1-(alkoxyphenyl)-4-(cyanophenyl)buta-1E,3E-dienes [60]. The photoisomers of these diphenylbutadienes were thermally stable even at elevated temperatures, making it easier to fix the photoinduced images into the glassy form. The images thus stored could be erased by irradiating these films under controlled conditions with 266-nm light, which led to complete conversion of the cis isomers to the all-trans form.

The extent of shift in the reflectance wavelength of the doped film compared to that of pure dicholesteryl mesogen depended on the length of the alkyl chains on the photoresponsive dopant. This dependence was attributed to the difference in the interaction between the azobenzene-derived dopant and the host cholesteric liquid crystal. On the basis of detailed X-ray analysis of irradiated films of the dicholesteryl esters doped with diphenylbutadiene derivatives it was proposed that the effect of the dopant on the pitch of the cholesteric host was due to induction of such smectic clusters.

The authors were able to show that the microscopic reorganization of smectic domains within the helical superstructure, induced by the dopant molecules and their photoisomers, resulted in changes in their macroscopic light-reflecting properties. Photoisomerization of these dopants led to a hypsochromic shift of the reflected light, and this effect could be used to tune the color of the light reflected by these mixtures over the entire visible region. The cholesteric pitch and hence the color of the light reflected by these materials could be fixed in a glassy state by rapidly cooling them from their cholesteric temperatures to 0°C. The efficiency of these materials for full-colour photoimaging has also been demonstrated.

Doping the glass-forming cholesteric liquid crystal dicholesteryl-10,12-docasadiynediote (CD8) with 1-(alkoxyphenyly)-4-(cyanophenyl)buta-$1E,3E$-dienes resulted in a redshift in the wavelength of light reflected by the host cholesteric liquid crystals. The extent of redshift was dependent on the concentration of the dopant, the temperature of the film, and also the molecular length of the dopant. Photoisomerization of these dopants lead to a blueshift of the reflected light, allowing color tuning of the films over a large part of the visible region. The cholesteric pitch and hence the light reflected by these materials could be fixed in a glassy state by cooling them from their cholesteric temperatures to 0°C.

X-ray diffraction (XRD) analyses of the BCn-CD8 mixtures exhibited two different diffraction peaks at 3.2° and 17.1°, corresponding to layer spacings of 28.0 and 5.2 Å. The broad large-angle peak was assigned to the distance between molecules in adjacent layers of the cholesteric phase, and the small-angle peak to the smectic domains. The full width at half maximum (FWHM) of the small-angle peak was significantly dependent on dopant parameters such as alkoxy chain length and concentration. FWHM was the least for mixtures containing BC8 and BC10, which exhibited a maximum redshift in the reflectance band. A similar reduction in FWHM was observed with increasing concentrations of the dopant, which also led to a redshift in the reflectance band. The decreased values of FWHM indicate improved ordering of the smectic domains, which is associated with an increase in the pitch of the cholesteric mixture. Comparison of the XRD patterns of the photolyzed BCn-CD8 mixtures with that of the unphotolysed mixtures showed that the blueshift in the reflectance band was associated with a broadening of the small-angle XRD peak, indicating that this effect could be attributed to the destruction of smectic cybotactic regions as a result of the bent structure of the photoproducts, namely, the E,Z and Z,E isomers.

Since inherently photoactive LCs possess several advantages over doped systems, dimesogenic liquid-crystalline molecules, consisting of a cholesterol moiety linked to a butadiene chromophore via flexible alkyl chains, were synthesized and investigated [61]. These dimesogenic liquid-crystalline molecules possessed the combined glass-forming properties of the cholesterol moiety and the photochromic properties of the butadiene moiety.

CBC, X = -O-

8CBC, X = -O-(CH$_2$)$_8$-O-

12CBC, X = -O-(CH$_2$)$_{12}$-O-

Photoinduced trans–cis isomerization of the butadiene chromophore in these materials could be utilized to bring about an isothermal smectic to cholesteric phase transition. By photochemically controlling the cis/trans isomer ratio, the pitch of the cholesteric phase could be continuously varied, making it possible to tune the color of the film over the entire visible region, and the color images thus generated could be stabilized by converting them to N* glasses by rapidly quenching the temperature of the irradiated film to ~0°C (Fig. 16-5a).

Figure 16-5b shows the dependence of the phase transition of CBC12 on the amount of cis isomers formed as a result of irradiation for different lengths of time. At 105°C, it was observed that just 1% conversion to the cis isomer resulted in SmA* to N* transition. The SmA* phase was not observed in samples containing more than 7.5% of the cis isomers. A schematic representation of the photoinduced effect is shown in Fig. 16-5c. On photoirradiation, formation of the cis isomers initially destabilizes the SmA* phase, bringing about a reduction in the SmA* to N* phase transition temperature. With increasing concentrations of the cis isomers, the pitch of the cholesteric phase decreases, resulting in a blueshift in its reflection band. At high cis concentrations, an isothermal phase transition to the isotropic phase occurs. The isotropization temperature was lowered by ~90°C, which is substantially larger than any of the previous reports on photoresponsive LCs. Although the photogenerated cis isomers were thermally stable, they could be reverted back photochemically. Thus, 266-nm photoirradiation of a film containing a mixture of the cis and trans isomers in the solid state resulted in near complete recovery of the all-trans isomer. Using this procedure, the images recorded using 366-nm light could be erased by irradiating them with 266-nm laser light.

The pitch of the cholesteric phase of the trimesogenic derivatives containing two cholesterol terminal moieties linked to a central diphenylbutadiene chromophore (DBC series) [62] could also be continuously varied thermally and photochemically to tune the color of the film over the entire visible region, and the color images thus generated could be stabilized by converting them to chiral nematic glasses.

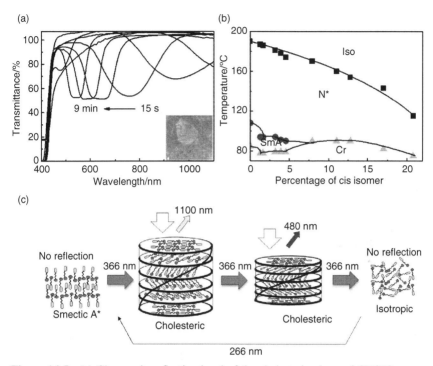

Figure 16-5   (a) Changes in reflection band of the cholesteric phase of CBC12 on photoirradiation of its smectic film held at 105°C. Inset shows an image stored in a cholesteric glass, obtained by energy-modulated irradiation of a smectic film at 105°C followed by quenching to ~0°C. (b) Effect of photoinduced cis isomers on phase transition of CBC12 in the cooling cycle. (c) Schematic representation of photoinduced isothermal phase transition from SmA* to isotropic phase via the N* phase.

DBC, X = -O-

8DBC, X = -O-(CH₂)₈-O-

12DBC, X = -O-(CH₂)₁₂-O-

These materials were also highly photoluminescent, exhibiting circularly polarized characteristics in the glassy liquid-crystalline state. Currently, there is substantial interest in the development of luminescent chiral nematic LCs because of their potential applications in organic light emitting diodes (OLEDs) and as lasing materials.

16.3.2 Chiral Smectic C (SmC*)

A drawback with photoinduced transformation in liquid crystals such as pitch change or phase transitions is that the switching speeds are usually in the millisecond time scale. In an effort to overcome this, photoresponsive ferroelectric liquid-crystalline systems were investigated. The first report on the photoactive SmC* was from the group of Ikeda in 1993. The chemical structures of the molecules investigated are given below (**11** and **12**) [63].

The ferroelectric liquid-crystalline films generated from compound **11** were doped with a chiral photochromic azo-compound, which undergoes trans–cis isomerization, inducing a change in the switching potential of the host liquid-crystalline film, thereby causing switching at the irradiated sites. Compound **12** exhibits an SmC* liquid-crystalline phase, in which the spontaneous polarization arises from the carbonyl group of the molecule, which is aligned perpendicular to the long molecular axis. In the ferroelectric liquid-crystalline matrix, the molecules are aligned parallel to each other to form a layer with a tilt between the direction of the long axis of the ferroelectric liquid-crystalline molecule and the normal to the smectic layer. FLCs show a much quicker response to the applied electric field owing to spontaneous polarization, P_s, and exhibit a hysteresis between the applied field across the cell and the polarization. The optical properties of the ferroelectric liquid-crystalline matrix doped with the photoresponsive azobenzene derivative (3 mol%) changed drastically on photoisomerization of the azobenzene unit, and the polarization direction flipped in the opposite direction on light illumination. The origin of this change was attributed to the disruption of the orientational order of the SmC* layer caused by the trans–cis photoisomerization of the azobenzene chromophore [64].

Subsequently, several reports suggested that spontaneous polarization of the SSFLCs can be photomodulated in the near-UV range using reversible trans–cis isomerization of chiral azobenzene dopants [65,66].

Lemieux and coworkers have extensively investigated the effect of photochromic molecules such as thioindigo on ferroelectric liquid-crystalline states [44]. The interesting aspect of thioindigo as a dopant is that the molecule maintains a rodlike shape in both isomeric forms. In the trans form, the thioindigo chromophore is nonpolar, whereas in the cis form, it possesses a transverse dipole moment, which can contribute to P_s provided that some degree of stereopolar coupling exists between the chromophore and a chiral structural unit.

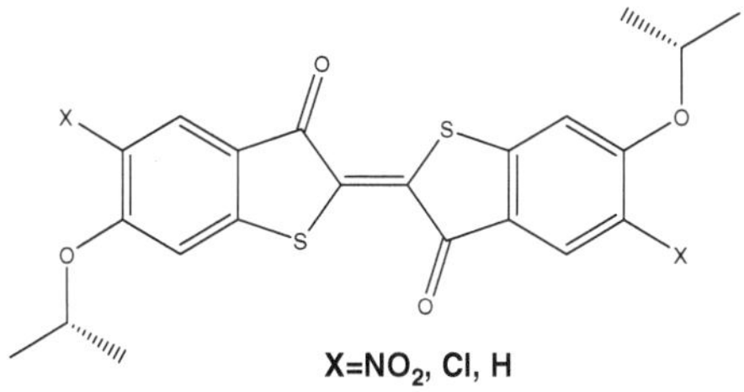

X=NO$_2$, Cl, H

13

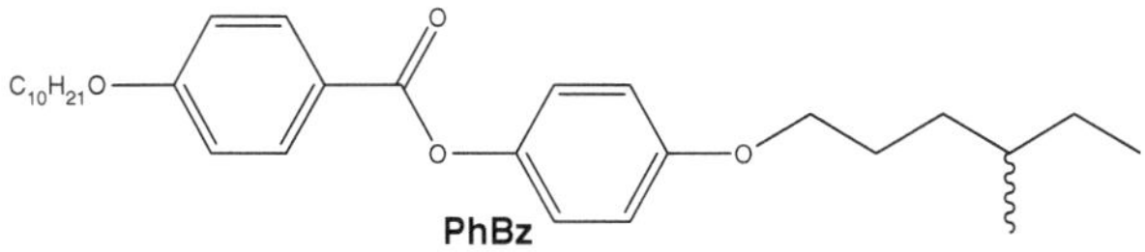

PhBz

14

Doping the SmC matrix with the photochromic material **13** followed by trans–cis photoisomerization of the dopant resulted in a 10% decrease of P_S, which is opposite in trend to that expected from the change in transverse dipole moment of the thioindigo core. This effect was attributed to the lack of stereopolar coupling between the core and the chiral side chains, and it was postulated that the modulation may be caused by a change in the conformational equilibrium of the chiral side chains imposed on the SmC lattice by photoisomerization to the cis-form. In order to reverse this trend and further enhance P_S photomodulation by harnessing the large change in transverse dipole moment of the thioindigo core on trans–cis photoisomerization, the authors have synthesized (R,R)-6,6'-bis(1-methylheptyloxy)-5,5'-dinitrothioindigo (**15**), in which the chiral side chains should be strongly coupled to the thioindigo core via both steric and dipole–dipole coupling with the adjacent nitro groups [62].

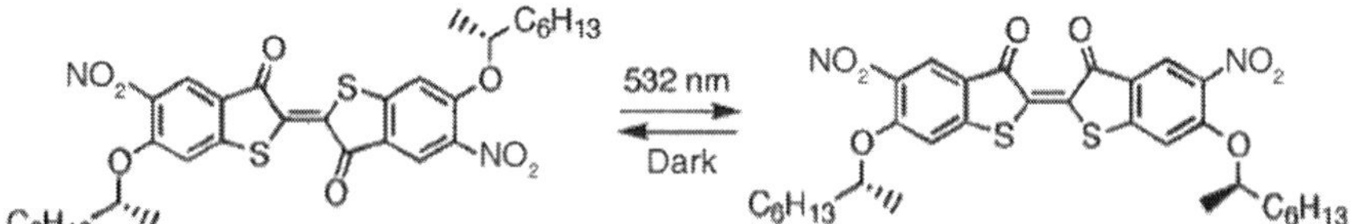

trans-**15**   *cis*-**15**

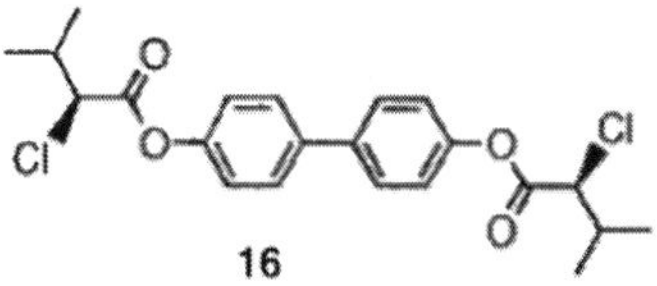

16

Compound **15** was doped into the racemic SmC host **16** in the mole fraction range of $0.005 < x\mathrm{d} < 0.03$ to give an induced SmC* liquid-crystalline phase. The P_S was measured with the sample under constant irradiation at two different wavelengths producing the cis-enriched photostationary state. An 82% *increase* in P_S was achieved on irradiation. Under these conditions, thermal relaxation to the trans isomer occurred within 10 s in the SmC* phase, a much faster rate than that observed in benzene solution at room temperature. The irradiation cycle was repeated several times without any decrease in the degree of photomodulation.

16.3.3 Blue Phases

Studies on photoresponsive frustrated liquid-crystalline phases such as BPs are relatively rare. In 2005, Chanishvili and coworkers reported the effect of photoisomerization of azobenzene derivative dopants on BPs [67]. The authors used two types of nematic hosts, the first consisting of ZhK-440 (NIOPIK), which is a mixture of two-thirds p-n-butyl-p-methoxyazoxybenzene and one-third p-n-butyl p-heptanoylazoxybenzene, and the second consisting of ZhK-537 (NIOPIK), which is the ester of 4-cyanophenyl and 4-heptylcinnamic acid. Both these host materials are capable of undergoing trans—cis photoisomerization. Two different chiral dopants were added to these nematic hosts, namely, the liquid-crystalline CB-15 (4-cyano-4-(2-methylbutyl)biphenyl, Merck) and non-liquid-crystalline MLC-6248 (2011R, Merck). Different compositions exhibiting chiral nematic liquid-crystalline phases were prepared, and the photoinduced pitch changes were investigated in these systems. One of the compositions consisting of 75% of ZHK440 and 25% MLC-6248 (by weight) showed an interesting formation of BP on trans to cis photoisomerization of the photochromic unit. This is the first time BPs have been induced by trans–cis isomerization. The BP was not present in the original thermal cycle of the mixture. Illumination of the mixture with 365-nm UV light resulted in the generation of the BPs as indicated by the selective reflection characteristic of BPs. The BP reflection could be tuned by further illumination.

Generation of stable BP in an inherently photoresponsive liquid-crystalline molecule was reported recently for the first time, in which a chiral diphenylbutadiene derivative (**17**) was found to exhibit photoinduced isothermal phase transition from SmA* to BP [68].

17

This molecule did not exhibit a BP in either heating or cooling cycles. The thermal phase sequence was observed as *Cr 120 SmA* 126 N* 132 Iso–Iso 131 N* 124 TGBA* 123 SmA* 110 Cr*. When the material held in the SmA* liquid-crystalline state at 124°C in the heating cycle was irradiated with 360-nm light for 100s, photoinduced formation of the cis isomer resulted in the formation of the BP with a characteristic classic BP texture and the reflection band was centred at 510nm. Subsequent irradiation led to a blueshift in the reflection band. The BP thus formed was thermodynamically stable and could be maintained at

Figure 16-6 Photoinduced isothermal phase transition of compound **17** from SmA* to blue phase at 124°C on irradiation with 360-nm light. (*See insert for color representation of the figure.*)

this state for several hours. On heating, this phase underwent isotropization at 128°C. Films drawn from inherently photoactive LCs can exhibit much faster switching times and enhanced stability compared to doped films. The BP that could be generated from the SmA* phase by irradiation with low-intensity 360-nm light (0.6 mW/cm^2) had a significantly wide temperature range, and its Bragg reflection could be tuned by varying the irradiation time. High-performance liquid chromatography (HPLC) analysis of the film irradiated for 100s indicated the total content of the cis isomers in this mixture to be 17%. Figure 16-6 shows the formation of the BP on photoirradiation of the film held at 124°C.

16.4 SUMMARY

Photoresponsive chiral liquid crystals can result in the amplification of a weak photochemical transformation, bringing out their bulk properties, including isothermal phase transformation, ferroelectric nature, and pitch of the cholesteric liquid-crystalline materials. Photochemical reactions of chromophores present as dopants or as an intrinsic part of the liquid-crystalline molecules can be utilized to bring about such transformations. A fine control over these transformations can be achieved by proper selection of the photoactive chromophore as well as by selective use of various parameters of light such as wavelength, intensity, and polarization. In view of this, design and study of novel photoresponsive liquid-crystalline systems continues to be an active area of research, where efforts are mainly concentrated on improving their response time, resolution, and ability.

REFERENCES

1. O'Neill, M., Kelly, S. M. (2003). Liquid crystals for charge transport, luminescence and photonics. *Adv. Mater.*, *15*, 1135–1146.
2. Adam, D., Schuhmacher, P., Simmerer, J., Haussling, L., Siemensmeyer, K., Etzbach, K. H., Ringsdorf, H., Haarer, D. (1994). Fast photoconduction in the highly ordered columnar phase of a discotic liquid crystal. *Nature*, *371*, 141–143.

3. Sirringhaus, H., Wilson, R. J., Friend, R. H., Inbasekaran, M., Wu, W., Woo, E. P., Grell, M., Bradley, D. D. C. (2000). Mobility enhancement in conjugated polymer field-effect transistors through chain alignment in a liquid-crystalline phase. *Appl. Phys. Lett.*, *77*, 406–408.

4. Jahromi, S., Kuipers, W. A. G., Norder, B., Mijs, W. J. (1995). Liquid crystalline epoxide thermosets. Dynamic mechanical and thermal properties. *Macromolecules*, *28*, 2201–2211.

5. Blumstein, A., Blumstein, R. B., Clough, S. B., Hsu, E. C. (1975). Oriented polymer growth in thermotropic mesophases. *Macromolecules*, *8*, 73–76.

6. Hoyle, C. E., Watanabe, T., Whitehead, J. B. (1994). Anisotropic network formation by photopolymerization of liquid crystal monomers in a low magnetic field. *Macromolecules*, 27, 6581–6588.

7. Tomasulo, M., Giordani, S., Raymo, F. M. (2005). Fluorescence modulation in polymer bilayers containing fluorescent and photochromic dopants. *Adv. Funct. Mater.*, *15*, 787–794.

8. Maly, K. E., Wand, M. D., Lemieux, R. P. (2002). Bistable ferroelectric liquid crystal photoswitch triggered by a dithienylethene dopant. *J. Am. Chem. Soc.*, *124*, 7898–7899.

9. Hikmet, R. A. M., Kemperman, H. (1998). Electrically switchable mirrors and optical components made from liquid-crystal gels. *Nature*, *392*, 476–479.

10. Kawata, S., Kawata, Y. (2000). Three-dimensional optical data storage using photochromic materials. *Chem. Rev.*, *100*, 1777–1788.

11. Tabe, Y., Yokoyama, H. (2003). Coherent collective precession of molecular rotors with chiral propellers. *Nat. Mater.*, *2*, 806–809.

12. Hubert, C., Rumyantseva, A., Lerondel, G., Grand, J., Kostcheev, S., Billot, L., Vial, A., Bachelot, R., Royer, P. (2005). Near-field photochemical imaging of noble metal nanostructures. *Nano. Lett.*, *5*, 615–619.

13. de Jong, J. J. D., Hania, P. R., Pugžlys, A., Lucas, L. N., de Loos, M., Kellog, R. M., Feringa, B. L., Duppen, K., van Esch, J. H. (2005). Light-driven dynamic pattern formation. *Angew. Chem. Int. Ed.*, *44*, 2373–2376.

14. Furumi, S., Janietz, D., Kidowaki, M., Nakagawa, M., Morino, S., Stump, J., Ichimura, K. (2001). Polarized photoluminescence from photo-patterned discotic liquid crystal films. *Chem. Mater.*, *13*, 1434–1437.

15. Willets, K. A., Ostroverkhova, O., He, M., Twieg, R. J., Moerner, W. E. (2003). Novel fluorophores for single-molecule imaging. *J. Am. Chem. Soc.*, *125*, 1174–1175.

16. Kubo, S., Gu, Z.-Z., Takahashi, K., Fujishima, A., Segawa, H., Sato, O. (2004). Tunable photonic band gap crystals based on a liquid crystal-infiltrated inverse opal structure. *J. Am. Chem. Soc.*, *126*, 8314–8319.

17. Binnemans, K. (2005). Ionic liquid crystals. *Chem. Rev.*, *105*, 4148–4204.

18. Martin-Rapun, R., Marcos, M., Omenat, A., Barbera, J., Romero, P., Serrano, J. L. (2005). Ionic thermotropic liquid crystal dendrimers. *J. Am. Chem. Soc.*, *127*, 7397–7403.

19. Kato, T. (2002). Self-assembly of phase-segregated liquid crystal structures. *Science*, *295*, 2414–2418.

20. Yoshio, M., Mukai, T., Kanie, K., Yoshizawa, M., Ohno, H., Kato, T. (2002). Layered ionic liquids: anisotropic ion conduction in new self-organized liquid-crystalline materials. *Adv. Mater.*, *14*, 351–354.

21. Nithyanandhan, J., Jayaraman, N., Davis, R., Das, S. (2004). Synthesis, fluorescence and photoisomerization studies of azobenzene functionalized poly(alkyl aryl ether) dendrimers. *Chem. Eur. J.*, *10*, 689–698.

22. Marcos, M., Martin-Rapún, R., Omenat, A., Barberá, J., Serrano, J. L. (2006). Ionic liquid crystal dendrimers with mono-, di- and trisubstituted benzoic acids. *Chem. Mater.*, *18*, 1206–1212.

23. Sessler, J. L., Callaway, W. B., Dudek, S. P., Date, R. W., Bruce D. W. (2004). Synthesis and characterization of a discotic uranium-containing liquid crystal. *Inorg. Chem.*, *43*, 6650–6653.

24. Elliott, J. M., Chipperfield, J. R., Clark, S., Teat, S. J., Sinn, E. (2002). Criteria for liquid crystal formation in 5-alkoxy-, 5-alkylamino-, and 5-alkanoyl-tropolone complexes of transition metals (Cu^{II}, Zn^{II}, Ni^{II}, Co^{II}, UO_2^{VI}, VO^{IV}). The first uranium metallomesogen. Crystal structure of bis(5-hexadecyloxytropolonato)copper(II). *Inorg. Chem.*, *41*, 293–299.

25. Clark, S., Elliott, J. M., Chipperfield, J. R., Styring, P., Sinn, E. (2002). The first uranium based liquid crystals. Uranyl metallomesogens from β-diketone and tropolone ligands. *Inorg. Chem. Comm.*, *5*, 249–251.

26. Dierking, I., Ed. *Textures of Liquid Crsystals*. Wiley VCH, Weinheim, 2003.

27. Chandrasekhar, S., Sadashiva, B. K., Suresh, K. A. (1977). Liquid crystals of disc-like molecules. *Pramana*, *9*, 471–480.

28. Kitzerow, H.-S., Bahr, C., Eds. *Chirality in Liquid Crystals*. Springer, New York, 2001.

29. di Matteo, A., Todd, S. M., Gottarelli, G., Solladié, G., Williams, V. E., Lemieux, R. P., Ferrarini, A., Spada, G. P. (2001). Correlation between molecular structure and helicity of induced chiral nematics in terms of short-range and electrostatic-induction interactions. The case of chiral biphenyls. *J. Am. Chem. Soc.*, *123*, 7842–7851.

30. Eelkema, R., Pollard, M. M., Katsonis, N., Vicario, J., Broer, D. J., Feringa, B. L. (2006). Rotational reorganization of doped cholesteric liquid crystalline films. *J. Am. Chem. Soc.*, *128*, 14397–14407.

31. Kato, T., Matsuoka, T., Nishii, M., Kamikawa, Y., Kanie, K., Nishimura, T., Yashima, E., Ujiie, S. (2004). Supramolecular chirality of thermotropic liquid-crystalline folic acid derivatives. *Angew. Chem. Int. Ed.*, *43*, 1969–1972.

32. Solladié, G., Hugelé, P., Bartsch, R. (1998). A new family of enantiomerically pure smectic C* liquid crystals with a bridged chiral biphenyl core. *J. Org. Chem.*, *63*, 3895–3898.

33. Lemieux, R. P. (2001). DNA-templated assembly of helical cyanine dye aggregates: A supramolecular chain polymerization. *Acc. Chem. Res.*, *34*, 845–853.

34. Tamaoki, N. (2001). Cholesteric liquid crystals for color information technology. *Adv. Mater.*, *13*, 1135–1147.

35. Yoshioka, T., Ogata, T., Nonaka, T., Moritsugu, M., Kim, S.-N., Kurihara, S. (2005). Reversible photon mode full color display by means of photochemical modulation of cholesteric structure. *Adv. Mater.*, *17*, 1226–1229.

36. Hwang, J., Song, M. H., Park, B., Nishimura, S., Toyooka, T., Wu, J. W., Takanishi, Y., Ishikawa, K., Takezoe, H. (2005). Electro-tunable optical diode based on photonic bandgap liquid-crystal heterojunctions. *Nat. Mater.*, *4*, 383–387.

37. Lagerwall, S. T. *Ferroelectric and Antiferroelectric Liquid Crystals*. Wiley-VCH, Weinheim, Germany, 1999.

38. Walba, D. M. (1995). Fast ferroelectric liquid-crystal electrooptics. *Science*, *270*, 250–251.

39. Clark, N. A., Lagerwall, S. T. In *Ferroelectric Liquid Crystals: Principles, Properties and Applications*. (Eds.: Goodby, J. W., Blinc, R., Clark, N. A., Lagerwall, S. T., Osipov, M. A., Pikin, S. A., Sakurai, T., Yoshino, K., Zeks, B.), Gordon and Breach, Philadelphia, 1991.

40. Walba, D. M., Xiao, L., Keller, P., Shao, R., Link, D., Clark, N. A. (1999). Ferroelectric liquid crystals for second order nonlinear optics. *Pure Appl. Chem.*, *71*, 2117–2123.

41. Walba, D. M., Dyer, D. J., Sierra, T., Cobben, P. L., Shao, R. F., Clark, N. A. (1996). Ferroelectric liquid crystals for nonlinear optics: Orientation of the disperse red 1 chromophore along the ferroelectric liquid crystal polar axis. *J. Am. Chem. Soc.*, *118*, 1211–1212.

42. Walba, D. M., Ros, M. B., Clark, N. A., Shao, R. F., Robinson, M. G., Liu, J. Y., Johnson, K. M., Doroski, D. (1991). Design and synthesis of new ferroelectric liquid crystals. An approach to the stereocontrolled synthesis of polar organic thin films for nonlinear optical applications. *J. Am. Chem. Soc.*, *113*, 5471–5474.

43. Ikeda, T., Kanazawa, A. *Molecular Switches*. (Ed.: Feringa, B. L.) Wiley-VCH, Weinheim, 2001; 363–1397.

44. Lemieux, R. P. (2004). Photoswitching of ferroelectric liquid crystals using chiral thioindigo dopants: the development of a photochemical Switch Hitter. *Chem. Rec.*, *3*, 288–295.

45. Coles, H. J., Pivnenko, M. N. (2005). Liquid crystal 'blue phases' with a wide temperature range. *Nature*, *436*, 997–1000.

46. Crooker, P. P. (1989). Plenary lecture. The blue phases. A review of experiments. *Liq. Cryst.*, *5*, 751–775.

47. He, W., Pan, G., Yang, Z., Zhao, D., Niu, G., Huang, W., Yuan, X., Guo, J., Cao, H., Yang, H. (2009). Wide blue phase range in a hydrogen-bonded self-assembled complex of chiral fluoro-substituted benzoic acid and pyridine derivative. *Adv. Mater.*, *21*, 2050–2053.

48. Kikuchi, H., Yokota, M., Hisakado, Y., Yang, H. (2002). Polymer-stabilized liquid crystal blue phases. *Nat. Mater.*, *1*, 64–68.

49. Ikeda, T., Tsutsumi, O. (1995). Optical switching and image storage by means of azobenzene liquid-crystal films. *Science*, *268*, 1873–1875.

50. Tsutsumi, O., Shiono, T., Ikeda, T., Galli, G. (1997). Photochemical phase transition behavior of nematic liquid crystals with azobenzene moieties as both mesogens and photosensitive chromophores. *J. Phys. Chem. B.*, *101*, 1332–1337.

51. Yu, Y., Ikeda, T. (2006). Soft actuators based on liquid-crystalline elastomers. *Angew. Chem. Int. Ed.*, *45*, 5416–5418.

52. Mallia, V. A., George, M., Das, S. (1999). Photochemical phase transition in hydrogen-bonded liquid crystals. *Chem. Mater.*, *11*, 207–208.

53. Kurihara, S., Ikeda, T., Tazuke, S. (1990). Photochemically induced isothermal phase transition in liquid crystals. Effect of interaction of photoresponsive molecules with matrix mesogens. *Mol. Cryst. Liq. Cryst.*, *178*, 117–132.

54. Ikeda, T., Miyamoto, T., Kurihara, S., Tazuke, S. (1990). Effect of structure of photoresponsive molecules on photochemical phase transition of liquid crystals IV. Photochemical phase tansition behaviors of photochromic azobenzene guest/polymer liquid crystal host mixtures. *Mol. Cryst. Liq. Cryst.*, *188*, 223–233.

55. Shishido, A., Tsutsumi, O., Kanazawa, A., Shiono, T., Ikeda, T. (1997). Distinct photochemical phase transition behavior of azobenzene liquid crystals evaluated by reflection-mode analysis. *J. Phys. Chem. B*, *101*, 2806–2810.

56. Shishido, A., Tsutsumi, O., Kanazawa, A., Shiono, T., Ikeda, T., Tamai, N. (1997). Rapid optical switching by means of photoinduced change in refractive index of azobenzene liquid crystals detected by reflection-mode analysis. *J. Am. Chem. Soc.*, *119*, 7791–7796.

57. Delden, R. A., Gelder, M. B., Huck, N. P. M., Feringa, B. L. (2003). Controlling the color of cholesteric liquid-crystalline films by photoirradiation of a chiroptical molecular switch used as dopant. *Adv. Funct. Mater.*, *13*, 319–324.

58. Moriyama, M., Song, S., Matsuda, H., Tamaoki, N. (2001). Effects of doped dialkylazobenzenes on helical pitch of cholesteric liquid crystal with medium molecular weight: utilization for full-colour image recording. *J. Mater. Chem.*, *11*, 1003–1006.

59. Mallia, V. A., Tamaoki, N. (2004). Design of chiral dimesogens containing cholesteryl groups; formation of new molecular organizations and their application to molecular photonics. *Chem. Soc. Rev.*, *33*, 76–84.

60. Davis, R., Mallia, V. A., Das, S., Tamaoki, N. (2004). Butadienes as a novel photo-chrome for tuning of cholesteric glasses: influence of microscopic molecular reorganization within the helical superstructure. *Adv. Funct. Mater.*, *14*, 743–748.

61. Abraham, S., Mallia, V. A., Ratheesh, K. V., Tamaoki, N., Das, S. (2006). Reversible thermal and photochemical switching of liquid crystalline phases and luminescence in diphenylbutadiene-based mesogenic dimers. *J. Am. Chem. Soc.*, *128*, 7692–7698.

62. Vijayaraghavan, R. K., Abraham, S., Akiyama, H., Furumi, S., Tamaoki, N., Das, S. (2008). Photoresponsive glass-forming butadiene-based chiral liquid crystals with circularly polarized photoluminescence. *Adv. Funct. Mater.*, *18*, 2510–2517.

63. Ikeda, T., Sasaki, T., Ichimura, K. (1993). Photochemical switching of polarization in ferroelectric liquid-crystal films. *Nature*, *361*, 428–430.

64. Sasaki, T., Ikeda, T. (1995). Photochemical control of properties of ferroelectric liquid crystals: Photochemically induced reversible change in spontaneous polarization and electro-optic property. *J. Phys. Chem. 99*, 13013–13018.

65. Sasaki, T., Ikeda, T., Ichimura, K. (1994). Photochemical control of properties of ferroelectric liquid crystals. Photochemical flip of polarization. *J. Am. Chem. Soc.*, *116*, 625–628.

66. Dinescu, L., Lemieux, R. P. (1999). Optical switching of a ferroelectric liquid crystal spatial light modulator by photoinduced polarization inversion. *Adv. Mater.*, *11*, 42–45.

67. Chanishvili, A., Chilaya, G., Petriashvili, G., Collings. P. J. (2005). *Trans-cis* isomerization and the blue phases. *Phys. Rev. E*, *71*, 051705–015710.

68. Ratheesh, K. V., Abraham, S., Shankara Rao, D. S.; Krishna Prasad S., Das, S. (2010). Light induced generation of stable blue phase in photoresponsive diphenylbutadiene based mesogen. *Chem. Commun.*, *46*, 2796–2798.

CHAPTER 17

LIQUID CRYSTALS TOWARD SOFT ORGANIC SEMICONDUCTORS

Yo Shimizu

National Institute of Advanced Industrial Science and Technology, Kansai Center, Ikeda, Osaka, Japan

17.1 INTRODUCTION

Organic electronics is now one of the most important research topics toward future technology of ubiquitous electronic devices. In particular, it should be really noticed that *Printed Electronics* has just launched to the future world of information technology involving a technology of energy resources such as solars cell and batteries, where one can fabricate a variety of electronic devices on flexible polymer substrates by printing technologies using *inks* of semiconductors and conductors for making electronic circuits [1,2]. These semiconducting and conducting materials, therefore, must have some "solubility" in common organic solvents, and in a special case, good solubility in water would be required for making aqueous inks.

Organic compounds have a long history of research, where chemical structures have been designed based on functions. Solubility has been one of the important factors for chemicals in industries. This fact essentially implies that liquid-crystalline materials are good candidates for semiconducting inks because of the flexible chains in the molecular structures, such as alkoxy and alkyl chains, rendering liquid crystallinity with a strong self-assembling nature of molecules [3]. For example, phthalocyanine, which is hardly soluble in organic solvents, could become more soluble when alkyl substituents are introduced into the phthalocyanine ring, and it is also required for facilitating liquid crystallinity. Furthermore, this leads to the fact that column chromatography could play an important role in the purification of the compound, which should be followed by recrystallization from solution.

Also, the self-assembling nature of liquid crystals is essentially interesting and important in terms of fabricating a thin film with a uniformity of molecular alignment. In *Printed Electronics*, thin films should be formed by the evaporation

Supramolecular Soft Matter: Applications in Materials and Organic Electronics, First Edition.
Edited by Takashi Nakanishi.

of solvent to give a uniform alignment of molecules, whereby one could bring out the potential ability of materials in a best way to provide higher performance as devices; in addition, this filming process should be executed in a spontaneous way. It has been well known that nematic liquid crystals have been applied to flat panel displays because of the easily controllable alignment of molecules that could provide a switching device as light valves. The result of field effect transistor (FET) studies on a calamitic liquid-crystalline system was reported to show that the easily controllable molecular alignment for the less viscous and lower ordered liquid crystals such as nematic phase leads to the easy fabrication of the large-area monodomain films with the more highly ordered structure to give good performance as a thin film transistor [4]. This indicates that it is possible to use mesogenic materials as a solid active layer in such devices if the domain state of liquid-crystalline phase is well maintained to give good quality solid thin film at room temperature by cooling for recrystallization [5].

Liquid crystal is a dynamic state of matter, and therefore, obtaining fast transport of charged carriers by electronic processes depends on the dynamics. In other words, mesogenic molecules fluctuate in position and orientation because of their dynamic state of matter. This allows us to treat liquid crystals as a disordered system of molecular aggregations. Therefore, the electronic hopping process profoundly concerns a coherent length of molecular array, where electrons easily and successively hop from molecule to molecule for a certain distance among the oriented molecules, and the molecular fluctuations disturb these arrays of molecules, leading to the slower charge transport as a macroscopic property, even though the electron transfer between molecules is a phenomenon of picosecond order. However, the present status of carrier mobility in mesophase has shown that the mobility in a level of amorphous silicon ($\sim 10^{-1}$ cm^2/V/s) could be easily obtained as its drift mobility (mobility observed for charge transport from electrode to electrode with a macroscopic distance), while it is strongly affected by its morphological properties [6–8]. In addition, self-healing of domain boundary may be good for the electronic hopping process in some liquid-crystalline phases, even though the compounds exhibiting fast mobility are always highly ordered systems, such as smectic E and B phases [9]. Therefore, these are highly viscous systems, implying a difficulty to control the molecular alignment to fabricate the homogeneous thin films in molecular orientation.

In this chapter, we shortly review the present status of liquid-crystal-based organic semiconductors with low molecular weights, and a perspective for soft organic semiconductors is mentioned, which are interesting and should be attractive for organic semiconductors applicable to flexible devices.

17.2 SMECTICS AND DISCOTICS AS ANISOTROPIC ORGANIC SEMICONDUCTORS

As shown in Fig. 17-1, liquid crystals are of two conventional categories based on the anisotropy of molecular shape, and rod-like and disk-like shapes are the

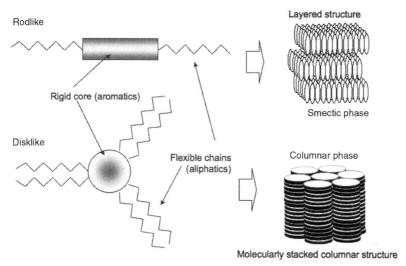

Figure 17-1   Two categories of liquid-crystalline molecules and important molecular orders of liquid crystals for electronic charge transport.

typical sets of anisotropic molecules for liquid crystals. The former is calamitic liquid crystal and the latter, discotic liquid crystal. In addition, as a characteristic chemical structure, the molecule consists of aromatics centered in the molecular structure forming the "core" of the molecule, with flexible chains attached to it at one or two edges of the molecule in calamitics and at the peripheral part of the molecule in discotics. These anisotropic characters of mesogenic molecules, therefore, form a variety of molecular packing in their mesophases as dynamic states of molecular aggregations. However, the important property of semiconductors, typically carrier mobility by electronic hopping process, is ascribed to how often the molecules meet in such an aggregation in an appropriate geometry of molecules for charge hopping, and thus, the static (time-averaged) orders of molecules in liquid-crystalline phases are primarily important for the efficient charge hopping among molecules. Therefore, smectic layered and discotic columnar structures are important for the charge transport pathway; the former potentially has a two-dimensional pathway, and the latter has a one dimensional pathway for charge transport.

The charge hopping rate, k, is described according to the semiclassical Marcus theory for charge transfer as follows [10]:

$$k = (4\pi/h)t^2(4\pi\lambda k_B T)^{-0.5}\exp(-\lambda/4k_B T) \qquad (17.1)$$

where t is transfer integral and λ is reorganization energy. t is a function of the splitting of HOMO/LUMO in energy level for hole/electron, and λ is the energy gap between the neutral and charged state of the molecule, which depends on the molecular structure. Therefore, for increasing the charge hopping rate, molecular design could be used to increase the value of t and decrease the value of λ.

The variables t and λ have been extensively studied for discotics in terms of not only translational, but also rotational displacements in columnar phase; these are more crucial factors than the periodical distance of stacking molecules within a column because it is, on an average, 3.5 Å [11–14]. Recent atomistic approaches involving dynamic aspects of molecular aggregation indicate that the dynamics could affect the efficiency of charge hopping by two orders of magnitude [15], indicating the importance of dynamics control in mesophase for more efficient charge transport in both nanoscopic and macroscopic scales.

17.2.1 Smectic Liquid-Crystalline Semiconductors

For calamitics, the translational displacement of molecules against the layers and rotational displacement within the layers are considered as crucial factors. For the smectic liquid-crystalline phases, highly ordered smectics such as the SmE phase which has a herringbone-type arrangement of molecules within a layer (as a time-averaged molecular order) exhibit fast mobility in the order of 10^{-1}–10^{-2} cm^2/Vs [16,17]. A herringbone-type packing is well known as an important packing motif for charge transport in solid organic semiconductors [18,19]. Therefore, it is quite reasonable to see that liquid crystals showing fast carrier mobility are highly ordered ones. According to the definition of a liquid crystal (orientational order of molecules remains after melting of the solid phase, while the positional order of molecular gravity center is destroyed), only five smectic phases, SmA, SmC, SmB$_{hex}$, SmI, and SmF, are real liquid crystals. However, so-called liquid crystals of smectics performing as fast mobility semiconductors are likely to have higher molecular orders as shown in Fig. 17-2, and thus, a "liquid-crystalline semiconductor" would be called to be a "mesophase semiconductor" as a recategorized area of research [20], if organic semiconductors should retain the self-assembling nature for organic electronics such as *Printed Electronics*. In this newly categorized area, one would have to find a relationship between crystalline solid and highly ordered mesophases in terms of dynamics control of molecules in mesophase. The empirical molecular designs of calamitic liquid crystals would be applied to the molecular design of mesophase materials to realize highly ordered smectics, and this might lead to the design of crystalline semiconductors with a rod-like shape of molecules probably in a way of its reduced consideration.

Simple molecular design to have a highly ordered smectic-like phase has been examined for 2,9-bis(2,5-dioctylthienyl)naphthalenes (8TNAT8) as shown in Fig. 17-3. In order to realize the densest packing for the molecular free volume, which allows the rod-like molecules to fluctuate in a rotational way around the longitudinal axis within a layer, the molecular shape should be simply rod-like without any lateral substituents because they would prevent the rotational fluctuations within the limited free volume. Also, the aromatics that take on a role of charge hopping should be placed in the center of the molecules, and the additional flexible chains having an appropriate length should be attached in a symmetrical way. In fact, a ditheinyl benzene analog with octyl chains exhibits three types of highly ordered mesophases with a smectic-like layered structure to give a fast

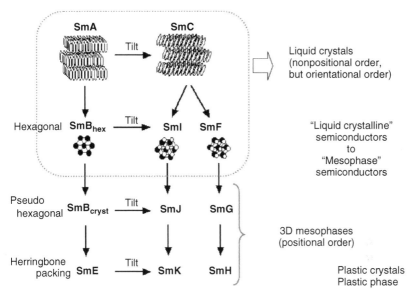

Figure 17-2 Liquid-crystalline semiconductors to mesophase semiconductors: highly ordered mesophase is necessary to realize highly efficient charge transport.

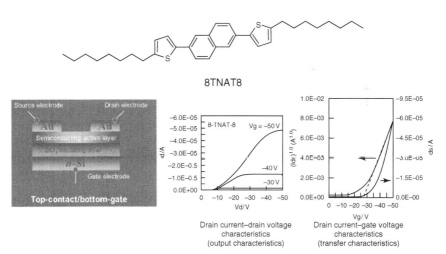

Figure 17-3 Thin film transistor performance of 8TNAT8 for the top-contact/bottom-gate geometry of device.

drift mobility in the order of $10^{-2}-10^{-1}$ cm²/Vs, which shows a typical behavior of smectic liquid-crystalline semiconductors, such as a temperature-independent mobility [21]. Further examination to get a mesophase semiconductor with a highly ordered smectic phase was continued to realize a simple phase transition; a simple phase sequence of "crystal−mesophase−isotropic" phases was attained by the replacement of the benzene ring with a naphthalene ring, leading to an increase in the lateral width of the rod-shaped molecule (8TNAT8). This increase in the lateral width could prevent the rotational fluctuations of molecules around the longitudinal axis in a layer. The results revealed that this compound exhibits only one highly ordered mesophase between 91 and 181°C, which has a three-dimensional lattice (triclinic with a smectic-like layered structure), while it is a viscous, not brittle solid (Unpublished results. The 3-dimenstional structure of the mesophase was determined by the combination of an XRD for nonaligned sample and a dilatometry technique studied by Dr Benoit Heinrich and Dr Daniel Guillon in IPCMS, France). The drift mobility measured by a time-of-flight (TOF) technique is in the order of 10^{-2} cm^2/Vs and 10^{-1}/cm^{2}Vs for the mesophase and crystalline phases, respectively. These were obtained for polydomain films sandwiched by two ITO electrodes. Single-crystal X-ray diffraction (XRD) studies revealed that the crystalline solid has a monoclinic lattice, in which the molecules are positioned in a way similar to that in the mesophase, and this result implies that such a small stepwise change of mobility could be observed at the transition between the mesophase and crystalline phases (Unpublished results. The single-crystalline XRD studies were carried out by Dr Motoo Shiro, Rigaku Corporation, Japan). In fact, this is justified by the result that only a small stepwise change of mobility is observed at the phase transition. Consequently, this similarity of molecular arrangements in the mesophase and crystalline solid would indicate that an empirical design concept that has been developed in chemical research on the structure-mesomorphism relationship of liquid crystals would be good toward a new concept of molecular design for organic solid semiconductors exhibiting fast mobility of charged carriers.

On the other hand, a simple thin film transistor was fabricated with 8TNAT8, which has a typical "top-contact/bottom-gate" geometry, to see how it behaves in a FET [22]. For the simple transistor shown in Fig. 17-3, high performance was attained at a mobility of 0.14 cm^2/Vs at room temperature (*in vacuo*), which is just comparable to the drift mobility of the material around room temperature. Considering that the thin film morphology deposited on the insulating layer of silicon substrate is not homogeneous (polydomain), the fact may indicate that such polydomain films for both drift and FET mobility measurements may involve a sort of self-healing effect among the domain boundaries. Also some attention is paid to a spontaneous alignment behavior of mesogenic molecules at the interface of the substrate. It is well known that, for open films of liquid-crystalline semiconductors on substrate, aromatics with aliphatic chains are likely to align in a way in which the aliphatic moieties, with the interface, are facing air because of the surface energy relationship. A similar situation also applies for mesogenic molecules. In the case of 8TNAT8, XRD studies on the active layers fabricated on a silicon substrate for FET indicates that the polydomain film in the transistor has high homogeneity of molecular alignment, of which the smectic-like layer plane lies on the substrate surface even for the polydomain films.

The analogous anthracene derivative with hexyl chains was reported for the FET device to show 0.50 cm^2/Vs mobility with a high on/off ratio (10^7) [23]. The authors simply describe that this compound exhibits a high-temperature liquid-crystalline mesophase, and it indicates that the self-assembling nature could contribute to the performance of devices in this case as well.

These results strongly suggest that the essential strategy to develop new organic semiconductors with liquid-crystalline nature is to control the molecular orders in both static and dynamic point of views, even though the thin film devices are with the crystalline films fabricated by use of liquid-crystalline properties [5].

Recent studies on calamitic mesophase semiconductors have found several new compounds exhibiting a fast mobility in the order of 10^{-1} cm²/Vs in their mesophases [8,16,17,21,22], although the mobility in the crystalline solids is not reported. In particular, perylene diimide derivatives as a n-type semiconductor are likely to show rather fast mobility both in the mesophase (classified as calamitic mesophase) and crystalline phase [24]. It is interesting to see that perylene itself could be a molecule with a lathlike shape, implying a shape that is intermediate between rod and disk. In fact, some chemical modifications lead to the variant mesomorphisms between calamitics and discotics (smectic and columnar mesophases, respectively) [25].

17.2.2 Discotic Liquid-Crystalline Semiconductors

For discotic liquid-crystalline semiconductors, columnar order is important for charge transport in an electronic process, where the molecules stack to each other in a face-to-face manner, and could provide an efficient pathway for charge hopping. The first discovery of fast mobility in mesophase as an epoch-making result pointed out that even in a dynamic aggregation of molecules such as the mesophase, charges so easily hop among the neighboring molecules in a macroscopic point of view [6], and this means one could set up a certain length of molecular order suitable for charge hopping in the columnar aggregations of disk-shaped molecules by molecular design. As mentioned earlier, the charge hopping rate is in proportion to the second order of magnitude for the transfer integral, and thus, the expansion of π-electronic conjugation system is mostly a right strategy to get the largest value of transfer integral when one considers that the rotational fluctuations of molecules are allowed around the columnar axis. Typical examples of discotic liquid-crystalline semiconductors are shown in Fig. 17-4.

Hexabenzocolonene is a typical example, and its derivatives exhibit fast mobility in the order of $10^{-2}-10^{-1}$ cm²/Vs for the columnar mesophases that were obtained by both TOF [26] and pulse-radiolysis time-resolved microwave conductivity (PR-TRMC) [27] techniques. The mobility obtained by the PR-TRMC method is an intrinsic mobility of charges within small domains, and thus, the charge transport is not significantly affected by its morphology. The large gap of observed mobilities between those techniques could be ascribed to the difference in morphology. The PR-TRMC technique has been used for easy evaluation of mobility of polydomain samples to reveal that a big molecule of metallomesogen is as interesting a compound as liquid-crystalline semiconductors [28].

This molecular design concept, in which the larger size of π-electronic conjugation systems should be fully extended, is also represented by graphene, which

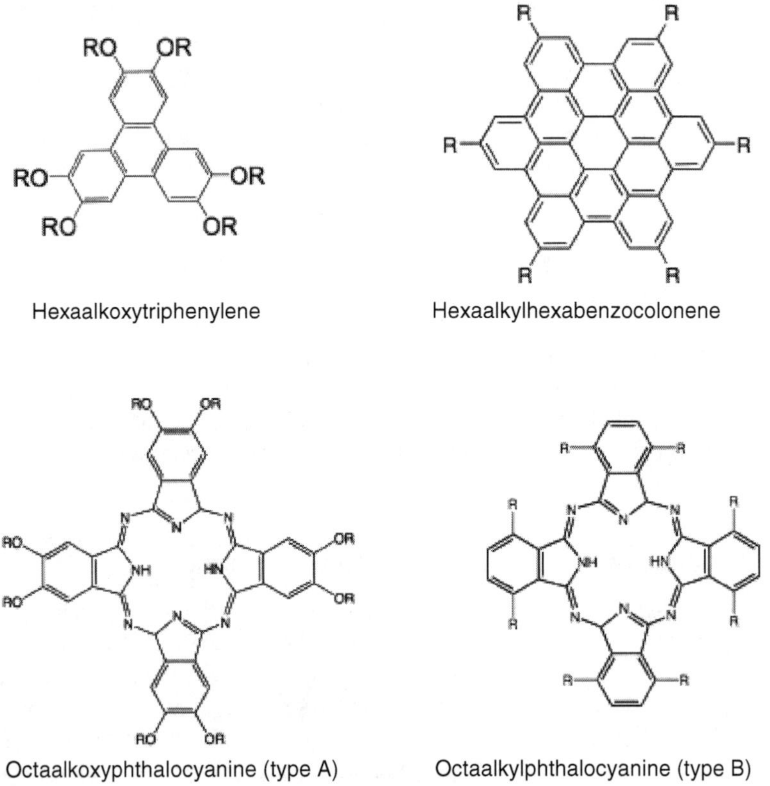

Hexaalkoxytriphenylene

Hexaalkylhexabenzocolonene

Octaalkoxyphthalocyanine (type A)

Octaalkylphthalocyanine (type B)

Figure 17-4 Chemical structures of discotic liquid-crystalline semiconductors.

is a topical material similar to semiconductors evolved from graphite [29,30]. In addition, a recent simulation study shows that the possible mobility is over 10 cm^2/Vs for defect-free samples of liquid-crystalline acenes [31]. Recent studies have also developed new techniques based on PR-TRMC to reveal fast mobility hole and electron for a variety of organic conductors [32].

As another strategy in discotics, one has to control the molecular fluctuations around and lateral to the columnar axis. In fact, even in the case of discotics, highly ordered columnar mesophases are likely to exhibit a fast mobility of carriers as shown in the case of triphenylenes, in which three-dimensional "helical" mesophase exhibits a fast mobility comparable to that of a-Si [6] (Fig. 17-5). In the "helical" mesophase, the triphenylenes stack to form columnar structures with a helical periodicity along the columnar axis because of the difficulty of molecular rotation. In other words, this situation can be obtained both by depressing the rotational fluctuations and by decreasing the rotation-allowed angle in which a higher transfer integral as a density of state and, consequently, a higher mobility could be expected. A qualitative discussion on the relationship between order–disorder states of columns and mobility was reported in the early days of this research area [33] and current studies are more quantitative as mentioned earlier.

The following is an example of other strategies to get more efficient charge transport for discotics. Phthalocyanine is one of the well-known compounds, which is a pigment and photoconductor and which has shown success in practical use. Therefore, phthalocyanine mesogens were already reported in the early 1980s and were synthesized in terms of liquid-crystalline conductors [34]. Currently, two types of phthalocyanine mesogens are known, as shown in Fig. 17-5. One has eight chains attached at the 2, 3, 9, 10, 16, 17, 23, and 24 positions and the other at 1, 4, 8, 11, 15, 18, 22, and 25 positions of a phthalocyanine ring. Several studies have been reported for the semiconducting properties of the former, although only two have reported those of the latter. The semiconducting properties of the former were reported to exhibit the mobility in the orders of 10^{-1} [35] and 10^{-2} cm^2/Vs [36] for the Col$_h$ mesophase, which were evaluated by PR-TRMC and TOF techniques, respectively. It is reasonable that the mobilities in the Col$_h$ mesophase are observed in such orders of magnitude because the maximum transfer integral arises every 45° with respect to the rotation of phthalocyanine planes around the columnar axis within the stacking column [37]. However, Iino et al. reported the ambipolar nature of charge transport for the latter mesogen of phthalocyanine with octyl chains, and the drift mobilities for the spontaneously aligned homeotropic film are almost temperature-independent to be in the order of 10^{-1} cm^2/Vs for the Col$_h$ and Col$_r$ mesophases [38,39]. This type of phthalocyanine mesogens has a steric problem for molecular stacking to form a stable and tight columnar structure because of the attaching position of alkyl chains. In fact, the homologs (C12–C6) were reported to exhibit disordered Col$_h$ and Col$_r$ phases where the molecules never tightly, thus periodically stack to each, and this is due to the protective repulsive force working among the stacking molecules derived from the chains [40]. More recently, while the hexyl homolog was found to show a drift mobility comparable to that of the octyl homolog in the Col$_h$ mesophase, the crystal phase exhibits a strong temperature-dependent hole mobility for polydomain films, which could be determined by their clear decay curves in TOF measurements for the Col$_h$ as well as crystal phases, and the hole mobility reaches 1.4 cm^2/Vs at room temperature [41]. This homolog does not show the tendency of spontaneous homeotropic alignment, unlike the octyl homolog. The crystal structure was reported to have a sort of interdigitated columnar arrangement and an extraordinarily large periodical stacking of molecules along the columnar axis in the crystal (8–9 Å) [42]. Therefore, charge transport might be realized not to be along the columnar axis. This indicates that when the central core has alkyl chains attached not at the peripheral part of molecular plane but at the inner part, subsequent contacts of molecules could be realistic, which could be a hopping site of electrons between molecules. In fact, the mobility in the Col$_h$ mesophase exhibits almost the same values for the homologs and is not significantly affected both by the alkyl chain length and the manner of molecular alignment in the cell (homeotropic and non-homeotropic films) [43]. These results imply other factors in the molecular geometry for efficient charge transport, considering that bulky tails peripherally attached to the central core is a negative factor for hexabenzocolonenes with branched chains [44].

On the other hand, an intermolecular interaction having a strong tendency to form molecularly segregated orders in mesophase was studied in terms of dynamic control of molecules. It is well known that perfluoroalkanes and hydrocarbon alkanes typically exhibit "fluorophilic and fluorophobic interactions" to form molecular segregations in the nanoscopic scale because of the weak van der Waals interaction of the former molecules. In fact, perfluoroalkylated triphenylenes have enhanced thermal stability of columnar mesophase [45], although the carrier mobility was evaluated to be in the same order of magnitude as those of the corresponding nonfluorinated homologs [46]. However, the introduction of perfluoroalkyl chains into the peripheral part of triphenylene surely leads to the induction of a strong tendency of spontaneous homeotropic alignment between a variety of substrate surfaces [47]. This indicates that one could facilitate mesogenic semiconductors with the spontaneous alignment property by chemical designs of mesogens to give an attractive property on the alignment control of molecules for thin film devices.

In the discotic columnar mesophases, intermolecular interactions would be strong and attractive tools for controlling the dynamics of molecules, although one has to consider that dipolar parts tend to work as a trap site for charged carriers.

Also, the control of molecular alignment is an important and interesting issue when considering the application to thin film devices such as transistors and solar cells. Columnar mesophase is normally a viscous state as a liquid-crystalline phase, although it sometimes exhibits a clear tendency to spontaneously align in a homeotropic manner between substrates. However, it is difficult to control the alignment when the mesophases show much higher viscosity derived from the highly ordered columns as well as the largely extended π-electronic conjugation system, which has a large polarizability leading to the stronger van der Waals forces.

However, this difficulty, which should be overcome for fabricating well-controlled domains with uniformly aligned molecules, has developed some interesting techniques for the alignment control of molecules. For surface modification, the usefulness of friction transfer technique was revealed in the use of polytetrafluoroethylene (PTFE) for a hexabenzocolonene mesogen. The well-aligned polymer backbones of PTFE induce the spontaneous alignment of the columnar axis, which is parallel to the direction of the polymers [48]. Furthermore, a novel technique to fabricate a film on a substrate with the uniform alignment of molecules was developed, in which single-crystal growth technique was evolved into two-dimensional thin films. Zone casting technique was found to exhibit useful performance as thin film fabrication method using the solution, where one can realize a single domain fabrication as a film by careful sliding of the substrate and adding the solution from a designed nozzle [49]. On the other hand, although a transient state of film is generated, an infrared technique was developed to fabricate a single domain with a line shape. In this technique, one has to select the wavelength of infrared to excite the corresponding mode of vibration for a chemical bond forming a molecular frame [50]. The combination of the directions between the incident polarization and transition moment of vibrational excitation leads to the formation of a new, uniformly aligned domain in the liquid-crystalline film, and its direction can be controlled, although a certain power of incidence is necessary depending on the viscosity of the mesophase [51].

However, most difficulties in controlling the molecular alignment are seen in the case of an open film on a substrate, and this is important for a sort of thin film devices such as transistors. Recently, a unique technique was reported, which is applicable to the open-film fabrication on a substrate, and the wider area could be obtained with a uniform alignment of molecules. One can fabricate such an open film on a substrate by using a "sacrificial layer," which is removed easily after the formation of a uniformly aligned film between the layer and substrate [52]. This technique would be useful for "*printed electronics*."

17.3 SUMMARY

Liquid crystal is a type of soft matter and possesses anisotropic orders of molecules. Also, the self-assembling nature should be utilized for the thin film device fabrication [53]. In particular, *Printed Electronics* have opened the door toward practical electronic devices combined with a variety of relevant materials. A soft substrate gives a soft electronic device, and all-printed devices are the features. The fabrication technologies have to be developed depending on the materials. The flexible devices should be tough enough to withstand a variety of shear forces, such as bending and twisting, coming from the external conditions. Funahashi *et al*. reported for the flexibility–performance relationship that a liquid-crystalline semiconductor in mesophase (not solid crystal) could exhibit a merit to use as the active layer [54]. The flexibility may need a material that would have, more or less, an elastomeric property, and the anisotropic fluid, such as liquid crystal, is surely a good candidate, where the molecules easily exchange their positional and orientational orders responding to the external force field.

On the other hand, it is easily understood that the molecular design for depressing motion leads to the formation of solid state, that is, a crystalline solid is obtained, where the molecules are arranged in an appropriate manner for charge hopping, and if any, it might be a bandlike structure. The present status unfortunately shows that it is not easy to predict the crystal structure only by the knowledge of chemical structure of molecules, although the Cambridge database of crystallography has elaborated to collect the huge number of published data of crystal structures. Therefore, molecular design of highly ordered liquid crystals is one of the strategic studies toward new organic semiconductors with self-assembling nature, and even now some empirical design concept for highly ordered mesophases of calamitic and discotic liquid crystals would be useful. In particular, calamitics is much more familiar to this issue because of the wide diversity of smectic phases, such as SmA, SmB, SmC, ... and SmK. Also, bicontinuous cubic mesophases formed by calamitic molecules are interesting in terms of the short range layered structure as well as its isotropic charge transport as macroscopic properties because the cubic phase consists of modified layered structures under a frustrated state of dynamic molecular aggregations [55]. This must hold good in the case of discotics as well. It was recently found that a triphenylene discotic liquid crystal exhibiting Col_h and bicontinuous cubic mesophases could show the high mobility of charged carriers determined by a TRMC technique [56].

For the chemical approach in the experimental studies on liquid-crystalline semiconductors, one has to consider sufficient purification of compounds as electronic materials. Repetitive purification of compounds sometimes drastically improves the carrier mobility characteristics. In other words, even a small amount of impurity could drastically depress efficient charge transport [57]. A typical example for the change of photocurrent decay curves in the TOF measurements for a type B phthalocyanine is shown in Fig. 17-5.

However, it is difficult to quantify the purity of organic compounds for the electronic materials, and it is surely a new issue to make a new standard of purity for organic materials, which would be different from the case of silicon. The elaboration of such repetitive purification is required for liquid-crystalline semiconductors as well. However, it is an advantage that the long tails improve the solubility of compounds into organic solvents, and this leads to easy access for column chromatography and recrystallization from solution.

From the practical point of view, it is also worthwhile to insist that photopolymerization after fabricating the liquid-crystalline film on a substrate to have a single domain and designed domain patterns is promising for its practical use if the mobility could be stable in a wide range of temperature without any depression of carrier transport efficiency [58].

Finally, recent studies on organic thin film solar cells by use of liquid-crystalline semiconductors exhibit interesting results. High performance could be attained for a simple bulk-heterojunction cell with a liquid-crystalline phthalocyanine and a non-mesogenic C_{60} derivative as seen that the power conversion efficiency could be over 3% with a high external quantum efficiency (>70%)

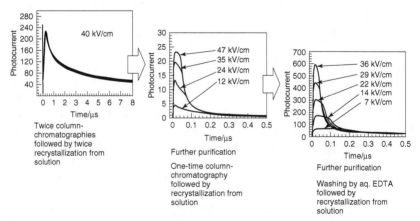

Figure 17-5 Change of the photocurrent decay curves observed for type B phthalocyanine with hexyl chains. It depends on the purification process.

[59]. This surely indicates the importance of mesomorphic fullerenes exhibiting efficient charge transport [60]. It may indicate the importance of liquid crystallinity for such a segregated structure in a mixture. Spontaneous formation of highly ordered periodical structures of molecular aggregation is an attractive property of mesophase materials, and it might be able to realize a hierarchical structure required for more effective charge separation and transport.

REFERENCES

1. Gamota, D. R., Brazis, P., Kalyanasundarum, K., Zhang, J., eds. (2004). *Printed Organic and Molecular Electronics*. Springer, Berlin.
2. Suganuma, K., ed. (2010). *Front Line of the Printed Electronics Engineering*. CMC, Tokyo.
3. Demus, D., Goodby, J., Gray, G. W., Spiess, H.-W., Vill, V., eds. (1998). *Handbook of Liquid Crystals*. Wiley-VCH, Weinheim.
4. van Breemen, A. J. J. M., Herwig, P. T., Chlon, C. H., Sweelssen, J., Schoo, H. F. M., Setayesh, S., Hardeman, W. M., Bastiaansen, C. W. M., Broer, D. J., Popa-Merticaru, A. R., Meskers, S. C. J. (2006). Large area liquid crystalline monodomain field-effect transistors. *J. Am. Chem. Soc.*, *128*, 2336–2345.
5. Iino, H., Hanna, J. (2006). Availability of liquid crystalline molecules for polycrystalline organic semiconductor thin films. *Jpn. J. Appl. Phys.*, *45*, L867–L870.
6. Adam, D., Schuhmacher, P., Simmerer, J., Häussling, L., Siemensmeyer, K., Etzbach, K. H., Ringsdorf, H., Haarer, D. (1994). Fast photoconduction in the highly ordered columnar phase of a discotic liquid crystal. *Nature*, *371*, 141–143.
7. Iino, H., Takayashiki, Y., Hanna, J., Bushby, R. J., Haarer, D. (2005). High electron mobility of $0.1 cm^2 V^{-1} s^{-1}$ in the highly ordered columnar phase of hexahexylthiotriphenylne. *Appl. Phys. Lett.*, *87*, 192105-1–192105-3.
8. Chen, L.-Y., Ke, T.-H., Wu, C.-C., Chao, T.-C, Wong, K.-T., Chang, C.-C. (2007). Anisotropic ambipolar carrier transport and high bipolar mobilities up to $0.1 cm^2 V^{-1} s^{-1}$ in aligned liquid-crystal glass films of oligofluorene. *Appl. Phys. Lett.*, *91*, 163509-1–163509-3.
9. Maeda, H., Funahashi, M., Hanna, J. (2000). Effect of domain boundary on carrier transport of calamitic liquid crystalline photoconductive materials. *Mol. Cryst. Liq. Cryst.*, *346*, 183–192.
10. Marcus, R. A. (1993). Electron transfer reactions in chemistry: theory and experiments. *Rev. Mod. Phys.*, *65*, 599–610.
11. Senthilkumar, K., Grozema, F. C., Bickelhaupt, F. M., Siebbeles, L. D. A. (2003). Charge transport in columnar stacked triphenylenes: effects of conformational fluctuations on charge transfer integrals and site energies. *J. Chem. Phys.*, *119*, 9809–9817.
12. Cornil, J., Lemaur, V., Calbert, J.-P., Brédas, J. –L. (2002). Charge transport in discotic liquid crystals: a molecular scale description. *Adv. Mater.*, *14*, 726–726.
13. Brédas, J. –L., Calbert, J. P., da Silva Filho, D. A., Cornil, J. (2002). Organic semiconductors: a theoretical characterization of the basic parameters governing charge transport. *Proc. Natl. Acad. Sci. U.S.A.* *99*, 5804–5809.
14. Lemaur, V., da Silva Filho, D. A., Coropceanu, V., Lehmann, M., Geerts, Y., Piris, J., Debije, M. G., van de Craats, A. M., Senthilkumar, K., Siebbeles, L. D. A., Warman, J. M., Brédas, J.-L., Cornil, J. (2004). Charge transport properties in discotic liquid crystals: a quantum-chemical insight into structure-property relationships. *J. Am. Chem. Soc.*, *126*, 3271–3279.
15. Olivier, Y., Muccioli, L., Lemaur, V., Geerts, Y. H., Zannoni, C., Cornil, J. (2009). Theoretical characterization of the structural and hole transport dynamics in liquid-crystalline phthalocyanine stacks. *J. Phys. Chem. B*, *113*, 14102–14111.
16. Funahashi, M., Zhang, F., Tamaoki, N. (2007). High ambipolar mobility in a highly ordered smectic phase of a dialkylphenylterthiophene derivative that can be applied to solution-processed organic field-effect transistors. *Adv. Mater.*, *19*, 353–358.

17. Funahashi, M., Hanna, J. (2005). High carrier mobility up to $0.1 cm^2 V^{-1} s^{-1}$ at ambient temperatures in thiophene-based smectic liquid crystals. *Adv. Mater.*, *17*, 594–598.

18. Fritz, S. E., Martin, S. M., Frisbie, C. D., Ward, M. D., Toney, M. F. (2004). Structural characterization of a pentacene monolayer on an amorphous SiO_2 substrate with grazing incidence X-ray diffraction. *J. Am. Chem. Soc. 126*, 4084–4085.

19. Cornil, J., Calbert, J. P., Bredas, J.-L. (2001). Electronic structure of the pentacene single crystal: relation to transport properties. *J. Am. Chem. Soc. 123*, 1250–1251.

20. Shimizu, Y., Oikawa, K., Nakayama, K., Guillon, D. (2007). Mesophase semiconductors in field effect transistors. *J. Mater. Chem.*, *17*, 4223–4229.

21. Oikawa, K., Monobe, H., Takahashi, J., Tsuchiya, K., Heinrich, B., Guillon, D., Shimizu, Y. (2005). A novel calamitic mesophase semiconductor with the fastest mobility of charged carriers: 1,4-di(5'-octyl-2'-thienyl)benzene. *Chem. Commun.*, 5337–5339.

22. Oikawa, K., Monobe, H., Nakayama, K., Kimoto, T., Tsuchiya, K., Heinrich, B., Guillon, D., Shimizu, Y., Yokoyama, M. (2007). High carrier mobility of organic field-effect transistors with a thiophene-naphthalene mesomorphic semiconductor. *Adv. Mater.*, *19*, 1864–1868.

23. Meng, H., Sun, F., Goldfinger, M. B., Jaycox, G. D., Li, Z., Marshall, W. J., Blackman, G. S. (2005). High-performance, stable organic thin-film field-effect transistors based on bis-5'-alkylthiphen-2'-yl-2,6-anthracene semiconductors. *J. Am. Chem. Soc.*, *127*, 2406–2407.

24. Struijk, C. W., Sieval, A. B., Dakhorst, J. E. J., van Dijk, M., Kimkes, P., Köhorst, R. B. M., Donker, H., Schaafsma, T. J., Picken, S. J., van de Craats, A. M., Warman, J. M., Zuilhof, H., Sudhölter, E. J. R. (2000). Liquid crystalline perylene diimides: architecture and charge carrier mobilities. *J. Am. Chem. Soc.*, *122*, 11057–11066.

25. An, Z., Yu, J., Jones, S. C., Barlow, S., Yoo, S., Domercq, B., Prins, P., Siebbeles, L. D. A., Kippelen, B., Marder, S. R. (2005). High electron mobility in room-temperature discotic liquid crystalline perylene diimides. *Adv. Mater.*, *17*, 2580–2583.

26. Kastler, M., Laquai, F., Müllen, K., Wegner, G. (2006). Room-temperature nondispersive hole transport in a discotic liquid crystal. *Appl. Phys. Lett.*, *89*, 252103.

27. van de Craats, A. M., Warman, J. M., Fechtenkötter, A., Brand, J. D., Harbison, M. A., Müllen, K. (1999). Record charge carrier mobility in a room-temperature discotic liquid-crystalline derivative of hexabenzocoronene. *Adv. Mater.*, *11*, 1469–1472.

28. Ban, K., Nishizawa, K., Ohta, K., van de Craats, A. M., Warman, J. M., Yamamoto, I., Shirai, H. (2001). Discotic liquid crystals of transition metal complexes 29: mesomorphism and charge transport properties of alkylthio-substituted phthalocyanine rare-earth metal sandwich complexes. *J. Mater. Chem.*, *11*, 321–331.

29. Novoselov, K. S., Geim, A. K., Morozov, S. V., Jiang, D., Zhang, Y., Dubonos, S. V., Grigorieva, I. V., Firsov, A. A. (2004). Electric field effect in atomically thin carbon films. *Science*, *306*, 666–669.

30. Hwang, E. H., Adam, S., Das Sarma, S. (2007). Carrier transport in two-dimensional graphene layers. *Phys. Rev. Lett.*, *98*, 186806–1–186806–4.

31. Feng, X., Macron, V., Pisula, W., Hansen, M. R., Lirkpatrick, J., Grozema, F., Andrienko, D., Kremer, K., Müllen, K. (2009). Towards high charge-carrier mobilities by rational design of the shape and peripherary of discotics. *Nat. Mater.*, *8*, 421–426.

32. Saeki, A., Seki, S., Takenobu, T., Iwasaa, Y., Tagawa, S. (2008). Mobility and dynamics of charge carriers in Rubrene single crystals studied by flash-photolysis microwave conductivity and optical spectroscopy. *Adv. Mater.*, *20*, 920–923.

33. Simmerer, J., Griesen, B., Paulus, W., Kettner, A., Shuhmacher, P., Etzbach, K. H., Simensmeiyer, K., Joachim, H. W., Ringsdorf, H., Haarer, D. (1996). Transient photoconductivityin a discotic hexagonal plastic crystal. *Adv. Mater.*, *8*, 815–819.

34. Piechocki, C., Simon, J., Skoulios, J., Guillon, D., Weber, P. (1982). Discotic mesophases obtained from substituted metallophthalocyanines. Toward liquid crystalline one-dimensional conductors. *J. Am. Chem. Soc. 104*, 5245–5247.

35. Warman, J. M., de Haas, M. P., Dicker, G., Grozema, F. C., Piris, J., Debije, M. G. (2004). Charge mobilities in organic semiconducting materials determined by pulse-radiolysis time-resolved microwave conductivity: π-bond-conjugated polymers versus π-π-stacked discotics. *Chem. Mater.*, *16*, 4600–4609.

36. Deibel, C., Janssen, D., Heremans, P., De Cupere, V., Geerts, Y., Benkhedir, M. L., Adriaenssens, G. J. (2006). Charge transport properties of a metal-free phthalocyanine discotic liquid crystal. *Org. Electron.*, *7*, 495–499.

37. Tant, J., Geerts, Y. H., Lehmann, M., De Cupere, V., Zucchi, G., Laursen, B. W., Bjørnholm, T., Lemaur, V., Marcq, V., Burquel, A., Hennebicq, E., Gardebien, F., Viville, P., Beljonne, D., Lazzaroni, R., Cornil, J. (2005). Liquid crystalline metal-free phthalocyanines designed for charge and exciton transport. *J. Phys. Chem. B*, *109*, 20315–20323.

38. Iino, H., Hanna, J., Bushby, R. J., Movaghar, B., Whitaker, B. J., Cook, M. J. (2005). Very high time-of-flight mobility in the columnar phases of a discotic liquid crystal. *Appl. Phys. Lett.*, *87*, 132102-1–132102-3.

39. Iino, H., Takayashiki, Y., Hanna, J., Bushby, R. J. (2005). Fast ambipolar carrier transport and easy homeotropic alignment in a metal-free phthalocyanine derivative. *Jpn. J. Appl. Phys.*, *44*, L1310–L1312.

40. Cook, M. J., Daniel, M. F., Harrison, K. J., McKeown, N. B., Thomson, A. J. (1987). 1,4,8,11,15,18,22,25-Octa-alkylphthalocyanines: new discotic liquid crystal materials. *J. Chem. Soc., Chem. Commun.*, 1086–1087.

41. Miyake, Y., Shiraiwa, Y., Okada, K., Monobe, H., Hori, T., Yamasaki, N., Yoshida, H., Cook, M. J., Fujii, A., Ozaki, M., Shimizu, Y. (2011). High carrier mobility up to 1.4 cm^2 V$-$1 s$-$1 in non-peripheral octahexyl phthalocyanine. *Appl. Phys. Express*, *4*, 021604-1-021604-3.

42. Chambrier, I., Cook, M. J., Helliwell, M., Powell, A. K. (1992). X-ray crystal structure of a mesogenic octa-substituted phthalocyanine. *J. Chem. Soc., Chem. Commun.*, 444–445.

43. Miyake, Y., Okada, K., Monobe, H., Hori, T., Yamasaki, N., Yoshida, H., Cook, M. J., Fujii, A., Ozaki, M., Shimizu, Y. Charge transport properties in a homologous series of 1,4,8,11,15,18,22,25-hexaalkylphthalocyanine, to be submitted.

44. Pisula, W., Kastler, M., Wasserfallen, D., Mondeshki, M., Piris, J., Schnell, I., Müllen, K. (2006). Relation between supramolecular order and charge carrier mobility of branched alkyl hexa-peri-hexabenzocoronenes. *Chem. Mater.*, *18*, 3634–3640.

45. Terasawa, N., Monobe, H., Kiyohara, K., Shimizu, Y. (2002). Fluorination effect of the peripheral chains on the mesomorphic properties in discotic liquid crystals of hexa-substituted triphenylene. *Chem. Lett.*, *32*, 214–215.

46. Terasawa, N., Monobe, H., Kiyohara, K., Shimizu, Y. (2003). Strong tendency towards homeotropic alignment in a hexagonal columnar mesophase of fluoroalkylated triphenylenes. *Chem. Commun.*, 1678–1679.

47. Miyake, Y., Fujii, A., Ozaki, M., Shimizu, Y. (2009). Carrier mobility of a columnar mesophase formed by a perfluoroalkylated triphenylene. *Synth. Met.*, *159*, 875–879.

48. van de Craats, A. M., Stutzmann, N., Bunk, O., Nielsen, M. M., Watson, M., Müllen, K., Chanzy, H. D., Sirringhaus, H., Friend, R. H. (2003). Meso-epitaxial solution-growth of self-organizing discotic liquid-crystalline semiconductors. *Adv. Mater.*, *15*, 495–499.

49. Tracz, A., Jeszka, J. K., Watson, M. D., Pisula, W., Müllen, K., Pakula, T. (2003). Uniaxial alignment of the columnar super-structure of a hexa (alkyl) hexa-peri-hexabenzocoronene on untreated glass by simple solution processing. *J. Am. Chem. Soc.*, *125*, 1682–1683.

50. Monobe, H., Awazu, K., Shimizu, Y. (2000). Change of liquid-crystal domains by vibrational excitation for a columnar mesophase. *Adv. Mater.*, *12* 1495–1499.

51. Monobe, H., Kiyohara, K., Heya, M., Awazu, K., Shimizu, Y. (2003). Anisotropic change of liquid-crystal domains by polarized infrared pulsed laser irradiation for a columnar mesophase. *Adv. Funct. Mater.*, *13*, 919–924.

52. Pouzet, E., De Cupere, V., Heintz, C., Anderesen, J. W., Breiby, D. W., Nielsen, M. M., Viville, P., Lazzaroni, R., Gbabode, G., Geerts, Y. H. (2009). Homeotropic alignment of a discotic liquid crystal induced by a sacrificial layer. *J. Phys. Chem. C*, *113*, 14398–14406.

53. Smits, E. C. P., Mathijssen, S. G. J., van Hal, P. A., Setayesh, S., Geuns, T. C. T., Mutsaers, K. A. H. A., Cantatore, E., Wondergem, H. J., Werzer, O., Resel, R., Kemerink, M., Kirchmeyer, S., Muzafarov, A. M., Ponomarenko, S. A., de Boer, B., Blom, P. W. M., de Leeuw, D. M. (2008). Bottom-up organic integrated circuits. *Nature*, *455*, 956–959.

54. Zhang, F., Funahashi, M., Tamaoki, N. (2010). Flexible field-effect transistors from a liquid crystalline semiconductor by solution processes. *Org. Electron.*, *11*, 363–368.

55. Suisse, J. -M., Mori, H., Monobe, H., Kutsumizu, S., Shimizu, Y., Charged carrier mobility in the cubic (Ia3d) mesophase of 1,2-bis(4'-n-nonyloxybenzoyl)-hydrazine (BABH-9), submitted for publication.

56. Alam, M. A., Motoyanagi, J., Yamamoto, Y., Fukushima, T., Kim, J., Kato, K., Tanaka, M., Saeki, A., Seki, S., Tagawa, S., Aida, T. (2009). 'Bicontinuous cubic' liquid crystalline materials from discotic molecules: a special effect of paraffinic side chains with ionic liquid pendants. *J. Am. Chem. Soc.*, *131*, 17722–17723.

57. Ahn, H., Ohno, A., Hanna, J. (2007). Impurity effects on charge carrier transport in various mesophases of smectic liquid crystals. *J. Appl. Phys.*, *102*, 093718–1–093718–6.

58. Inoue, M., Monobe, H., Ukon, M., Petrov, V. F., Watanabe, T., Kumano, A., Shimizu, Y. (2005). Fast charged carrier mobility of a triphenylene-based polymer film possessing nematic order. *Opto-Electron. Rev.*, *13*, 303–308.

59. Hori, T., Miyake, Y., Yamasaki, N., Yoshida, H., Fujii, A., Shimizu, Y., Ozaki, M. (2010). Solution processable organic solar cell based on bulk heterojunction utilizing phthalocyanine derivative. *Appl. Phys. Exp.*, *3*, 101602–1–101602–3.

60. Nakanishi, T., Shen, Y., Wang, J., Yagai, S., Funahashi, M., Kato, T., Fernandes, P., Möhwald, H., Kurth, D. G. (2008). Electron transport and electrochemistry of mesomorphic fullerenes with long-range ordered lamellae. *J. Am. Chem. Soc.*, *130*, 9236–9237.

SUPRAMOLECULAR COMPOSITES BASED ON CARBON NANOTUBES

CHAPTER 18

CNT/POLYMER COMPOSITE MATERIALS

Tsuyohiko Fujigaya,[1] *Yasuhiko Tanaka,*[1] *and*
Naotoshi Nakashima[1,2]

[1]Department of Applied Chemistry, Kyushu University, Fukuoka, Japan
[2]Japan Science and Technology Agency, CREST, Tokyo, Japan

18.1 INTRODUCTION

Carbon nanotubes (CNTs) are made of rolled up graphene sheets with one-dimensional extended π-conjugated structures, discovered in 1991 by Iijima [1]. They are classified into mainly three types in terms of the number of graphene layers within a CNT, that is, single-walled carbon nanotubes (SWNTs), double-walled carbon nanotubes (DWNTs), and multiwalled carbon nanotubes (MWNTs), which have one, two, and more than three walls, respectively. CNTs have been the central materials in the field of nanomaterials science and nanotechnology because of their remarkable electronic, mechanical, and thermal properties that far exceed those of existing materials. One of the key issues in the utilization of such seminal materials for basic research together with their potential applications is to develop a methodology to solubilize/disperse them in solvents [2–4] since as-synthesized CNTs form tight bundled structures [5] because of their strong van der Waals interactions (0.9 eV/nm) [6]. Solubilization/dispersion techniques can be categorized mainly into two methods, namely "chemical" and "physical" modification. Solubilization/dispersion of CNTs based on physical adsorption of dispersant molecules possesses several advantages, such as the ease of preparation process and maintaining intrinsic CNT properties, which are in sharp contrast to that based on chemical modification [7–9].

In this review, after a brief description of the strategy to solubilize CNTs in solution, we summarize our recent study to design and fabricate novel CNT/polymer nanohybrids and their applications.

Supramolecular Soft Matter: Applications in Materials and Organic Electronics, First Edition.
Edited by Takashi Nakanishi.
© 2011 John Wiley & Sons, Inc. Published 2011 by John Wiley & Sons, Inc.

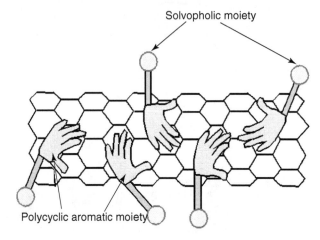

Figure 18-1 Schematic illustration of the solubilization of CNTs through physical adsorption of dispersant molecules on the surfaces of CNTs.

18.2 STRATEGY FOR CARBON NANOTUBE SOLUBILIZATION AND FUNCTIONALIZATION

Figure 18-1 shows the general idea for the solubilization of CNTs in solution [10–12], in which the solubilizers are functional compounds that are composed of an aromatic moiety and a solvophilic moiety. According to this concept, solubilizers carrying a hydrophilic or hydrophobic moiety are expected to dissolve CNTs in dipolar solvents such as water and alcohols or in nonpolar solvents, respectively. This model is simple and is applicable to a variety of aromatic amphiphiles, including porphyrins, polyphenols, compounds carrying a pyrene, anthracene [13,14], terphenyl [14,15], perylene [16], triphenylene [17], phenanthrene [18], and pentacene [19]. Furthermore, a large number of aromatic polymers, polysaccharides, and surfactants are reported to solubilize CNTs in solution [4,20].

18.3 REDOX REACTION AND DETERMINATION OF ELECTRONIC STATES OF CARBON NANOTUBES

In this section, we describe how to determine the electronic states (redox potentials and Fermi levels) of (n,m) SWNTs having their own chirality. The electronic states of CNTs, one of the most fundamental features of nanotubes, strongly depend on the chirality of the CNTs [21]. The electric states of SWNTs are closely related to their structures, which are identified with a specified diameter and chiral angle uniquely relating to a pair of integers (n,m), the chirality indices. The first-principles calculation well explains the electronic properties of SWNTs that are characterized by van Hove singularities [22–24]. Several papers describing solution redox chemistry of SWNTs have been published [25–28]. Kazaoui and coauthors [25] reported the absorption spectral changes of SWNTs induced by chemical oxidation. Zheng and Diner [26] reported the oxidation potentials of the (6,5)-enriched SWNT using chemical oxidation with an oxidant (K_2IrCl_6). They estimated the oxidation potential of (6,5)SWNTs to be approximately 0.8 V versus SHE (standard hydrogen electrode potential).

In situ spectroelectrochemistry of SWNTs is a powerful technique for investigating the potential-controlled electronic structures of the nanotubes, and many spectroelectrochemical studies have been carried out using bundled CNTs [28–32]. For example, Kavan and coauthors [29,31] reported electrochemical - potential-dependent Raman spectra of SWNT films on electrodes, and Murakoshi and coauthors reported chirality-dependent redox potentials and Fermi levels of semiconducting SWNTs on an electrode using electrochemical Raman spectroscopy [32].

Although many groups have endeavored to understand the fundamental properties of the CNTs and many attempts have been made to determine the electronic properties of SWNTs, it was not easy to determine the redox potentials of isolated (n,m)SWNTs. In order to determine the electronic states of isolated (n,m)SWNTs, we fabricated a transparent indium tin oxide (ITO) electrode modified with a film of carboxymethylcellulose sodium salt (CMC)/poly(diallyldimethylammonium chloride) [33,34]. From the near-infrared (NIR) photoluminescence (PL) spectrum of the film (Fig. 18-2a), it was found that the film contains isolated SWNTs whose chiralities are (6,5), (8,3), (7,5), (8,4), (10,2), (7,6), (9,4), (10,3), (8,6), (9,5), (12,1), (11,3), (8,7), (10,5), and (9,7).

We carried out *in situ* NIR absorption spectroelectrochemistry using the modified electrode. The NIR absorption spectra of the individually solubilized SWNTs were found to bleach when the external potential applied to the electrode was arbitrarily stepped up from 0.0 to −1.0 V and from 0.0 to +1.1 V. Unfortunately, the determination of the redox potentials of individual SWNTs having their own chirality indices was difficult because NIR absorption peaks from nanotubes with several different chirality indices have bandgaps that overlap one another. The PL of SWNTs is consistent with that of isolated SWNTs, and therefore characterization of isolated SWNTs is possible. We carried out *in situ* NIR PL spectroelectrochemistry in a way similar to *in situ* NIR absorbance spectroelectrochemistry [33,34]. As a typical example, in Fig. 18-2b, we show the PL spectra of (7,5), (7,6), and (10,3)SWNTs at given applied potentials from 0.0 to −1.0 V and then from 0.0 to +1.1 V. It is evident that the PL spectra show a strong applied potential dependence. Figure 18-2c shows normalized PL intensity of (7,5), (7,6), and (10,3)SWNTs as a function of external applied potential and the Nernst analysis curves (solid line) of the experimental results. From the observed inflection point, we can determine the oxidation and reduction potentials of (7,5), (7,6), and (10,3)SWNTs. In a similar way, the redox potentials of the above-mentioned 15 isolated (n,m)SWNTs together with Fermi levels and bandgaps were determined. The finding advances the basic science and understanding of CNTs and opens the way to a new breed of nanoscience.

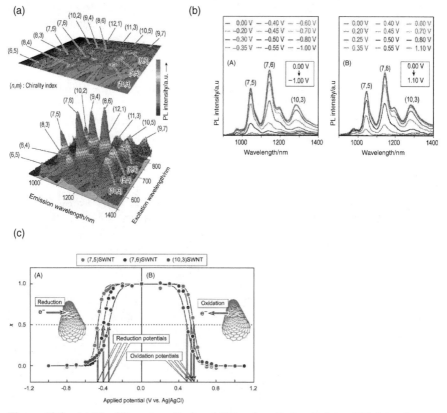

Figure 18-2 (a) The 2D contour (top) and 3D surface (bottom) plots of PL intensity as a function of emission and excitation wavelengths. The data were obtained for a film containing isolated SWNTs coated on an ITO electrode. (b) Dependence of external applied potential on the PL spectra of the film containing (7,5), (7,6), and (10,3)SWNTs on an ITO electrode. (c) Normalized PL intensity of the film containing isolated SWNTs on an ITO electrode as a function of external applied potential and the Nernst analysis curves (each solid line) of the experimental results. (*See insert for color representation of the figure.*)

18.4 DNA/CARBON NANOTUBE HYBRIDS

The combination of CNTs and DNA (or RNA) is of fundamental and applied interest in many chemical and biochemical areas. We have reported the finding that double-stranded DNA (dsDNA) molecules dissolve SWNTs in aqueous solutions [33]. Barisci *et al.* [35] fabricated SWNTs/dsDNA fibers that were mechanically strong and conductive and exhibited useful capacitance values up to 7.2 F/g. Iijima *et al.* [36] showed high-resolution transmission electron microscopic (TEM) and scanning tunneling microscopic (STM) images of dsDNA/multiwalled CNTs. Gladchenko *et al.* [37] characterized fragmented dsDNA-wrapped SWNTs in aqueous solutions. He and Bayachou [38] described the layer-by-layer fabrication and characterization of dsDNA-wrapped SWNTs particles. Prato *et al.* [39]

described binding of plasmid DNA onto functionalized CNTs in the construction of CNT-based gene transfer vector systems. Dordick *et al.* [40] reported *in vitro* transcription and protein translation from CNTs/DNA assemblies. Coleman *et al.* [41] reported spontaneous dispersion of SWNTs by dsDNA. At the same time as our report that dsDNA solubilizes SWNTs, Zheng *et al.* [42,43] showed that single-stranded DNA (ssDNA) solubilizes SWNTs. After this report, many papers describing the properties of SWNTs/ssDNA have been published.

Optical absorbance spectra of SWNTs dispersed by ssDNA homopolymers show anisotropic absorbance of SWNTs [44]. We have constructed multilayer assemblies with alternating monolayers of poly(G)-wrapped SWNTs and poly(C)-wrapped SWNTs on quartz based on the complementary base pairing between nucleic acid bases G and C, which would be applicable in the fields of nanoscience and technology [45]. SWNTs/ssDNA (RNA) solutions can be used as materials for gene and protein delivery and nanotherapy [46]. DNA (RNA)–CNT hybrids play important roles in the rapid development of nanotechnologies [47]. SWNTs/ssDNA induce the alignment of membrane proteins [48] and can serve as rigid templates for the self-assembly of gold nanoparticles [49]. We described the modulation of NIR optical properties of individually dissolved SWNTs in aqueous solutions of dsDNA [50]. NIR PL mapping of the individually solubilized SWNTs in pure water gave only one spot from the (6,5) SWNTs, and the PL behaviors dramatically changed with pH. A possible mechanism for the tunable NIR optical behaviors is also reported.

Thermodynamic stability of the DNA/SWNT is another feature of the complex. By using a gel permeation chromatography (GPC) technique, we have proved that the binding of dsDNA and SWNTs is highly stable, namely, the detachment of the dsDNA from the surface of SWNTs is ignorable at least at one month [51]. The formation of a stable complex between DNA and CNTs has initiated a wide range of research in view of the biological applications, such as the conformation transition monitoring of DNA [52], redox sensing of glucose and hydrogen peroxide [53], hybridization detection between ssDNA and the complimentary DNA [54], and uptake estimation of DNA/SWNTs into the cell [55]. With the increased possibility of the use of DNA/CNTs as gene delivery carriers, a strong demand to avoid DNA damage during sonication arises. Modified DNA-wrapping protocol [56] using surfactant-dissolved SWNTs followed by the exchange of DNA in a dialysis membrane realized the sonication-free process [37,53].

We also examined the stability of oligo DNA/SWNT hybrids and reached the clear-cut answer for the stability of size-sorted ds and ss oligo DNA-solubilized SWNTs without containing the free unbound oligo DNA: we have discovered that the fractionated oligo dsDNA and ssDNA/SWNT hybrids are stable without desorption from the nanobiohybrids for at least one month, indicating that both the oligo dsDNA and ssDNA possess strong affinities to the SWNTs [57]. These results indicate that we need not consider desorption of the bound oligo dsDNA or oligo ssDNA from their nanohybrids with the SWNTs in water, which is of significant advantage to the utilization of the nanobiohybrids in wide areas of science.

18.5 CURABLE MONOMERS AND NANOIMPRINTING

Heat- and photocurable resins have been interesting as a matrix for the CNT composite owing to several advantages.

1. Most of these monomers are viscous liquids, and, principally, there is no need to add any solvent to obtain polymer/CNT composites.

2. Mixing with small monomers is expected to have lower entropic barriers to disperse compared to polymer melt mixing.

3. Quick solidification, especially in photocurable systems, can avoid the reaggregation often occurring during solvent evaporation process.

Epoxy/CNTs are one of the most extensively researched thermoset composites so far [58–65]. The combination of rheological study [66] and small angle nuclear scattering measurements [67] is a strong tool to understand the degree of dispersion in the composite [66]. Quick solidification without solvent removal was utilized to keep the CNT alignment formed prior to polymerization via magnetic- or electronic-induced orientation [68,69]. The combination has been demonstrated to yield good processability [70]. We have reported mold-assisted photolithography of bis-acrylate/SWNT composites (**UV-1** in Fig. 18-3a) by using a poly(dimethylsiloxane) (PDMS) stamp [71], and clear 2D patterns with a submicron scale were easily fabricated on a silicon wafer in few seconds (Fig. 18-3b). As expected, the degree of dispersion showed no change on polymerization, which was proved by the monitoring of vis-NIR absorption spectroscopy. The composite presented an extremely low electric percolation threshold (0.05–0.1 wt%) as well as low surface resistance accompanied by good dispersion compared to the other systems (in the order of 10^{-2} ohm/square), suggesting effective dispersion of the SWNTs in the matrix. These patterned polymer/SWNT composites with high conductivity may offer novel potential applications, including an optical waveguide utilizing the nonlinear response of SWNTs [72], a scaffold for cell culture media [73,74], a thin film transistor composed of an SWNT network in the insulating resin, a separator for fuel cell, and a chemical/biological sensor [75,76].

18.6 CONDUCTIVE NANOTUBE HONEYCOMB FILM

Honeycomb structures from organic (polymer) and organic/inorganic hybrid materials are of interest because of their unique structures and functions. Since the first report by François et al. [77] that self-organized honeycomb structures are formed from star-shaped polystyrene or poly(styrene)-poly(paraphenylene) block copolymers in carbon disulfide under flowing moist gas, many papers have been published describing the formation of similar honeycomb structures using different kinds of organic (polymer) materials, including symmetric diblock copolymers [78], rod–coil diblock copolymers [79], a coil-like

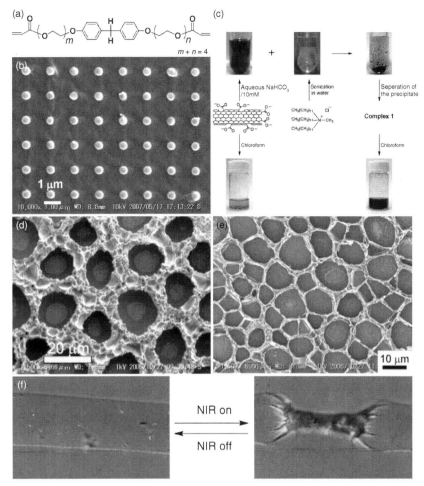

Figure 18-3 (a) Chemical structure of UV-1. (b) SEM image of a nanoimprinted patterns formed by UV-1/SWNT composite. (c) Preparation of complex 1. (d) SEM images of complex 1 on a glass substrate before the ion exchange. (e) SEM images of complex 1 on a glass substrate after the ion exchange. (f) Microscopic images of a PNIPAM/SWNT gel before (left) and after (right) near-IR laser irradiation.

polymer [80], ion-complexed polymers [81], lipid-packaged Pt complexes [82], poly(D,L-lactic-*co*-glycolic acid) [83], polysulfone [84], amphiphilic poly(*p*-phenylenes) [85], and a poly(ε-caprolactone)/amphiphilic copolymer [86].

We have reported the discovery that self-assembly of SWNTs into a honeycomb structure is spontaneous on glass substrates [87] and transparent plastic films, such as those of poly(ethylene terephthalate) (PET), which is a widely used engineering plastic in the industrial field. This structure is formed by a simple solution casting method using an SWNT/lipid conjugate (Complex **1**, Fig. 18-3c), which is an ion complex of shortened SWNTs and tridodecylmethylammonium chloride, a molecular-bilayer-forming amphiphile, and available from our previous study. Complex **1** is soluble in several organic solvents, including dichloromethane, chloroform, benzene, and toluene. We recently reported the formation of (semi)conducting SWNT honeycomb structures on flexible transparent polymer films [88]. We chose the PET film, which is a widely used engineering plastic in the industrial field. This study should be important from the viewpoint of potential applications of conducting SWNTs with honeycomb structures for the fabrication of conducting plastic films with transparent flexible properties. Such films might be useful in many areas of applications that require flexible conducting films. The typical scanning electron microscopic (SEM) images of a honeycomb structure are shown in Fig. 18-3d. The sizes of the unit cells are controllable by changing the experimental conditions.

The surface resistivity (R_s) of the cast films of complex **1** with honeycomb structures is insulating ($R_s > 10^8$ ohm/square) because of the coating of the tube surfaces with the ammonium lipid. We developed a method to remove the lipid from the films by employing the "ion-exchange method." The experimental procedure is simple, namely, each cast film is immersed overnight in a p-toluenesulfonic acid methanol solution and then rinsed with methanol followed by air drying. By this procedure, the methylene stretching vibrations in the Fourier transform infrared (FT-IR) spectroscopy of the film almost disappear. The Raman spectra of complex **1** before and after the ion exchange are virtually identical. The SWNTs remain intact during all the processes. The SEM images of the cast films after ion exchange are shown in Fig. 18-3e. After ion exchange, the skeletons with the honeycomb structures become thin because of the removal of the lipid. Higher magnification SEM measurements show oriented nanotubes along the honeycomb skeletons.

After ion exchange, a dramatic change in the R_s values is observed. The R_s values decrease with increasing concentration of complex **1** because of the formation of network structures in larger areas on the films. When the film is prepared from complex **1**, 3.0 mg/mL in chloroform, the R_s reached a high conducting value, 3.2×10^2 ohm/square. A similar behavior is observed when dichloromethane and benzene are used in place of chloroform. Interestingly, after ion exchange, the R_s values of the films decrease more than 10^4- to 10^6-fold compared to the original values.

The conductive SWNT honeycomb films on glass substrates and plastic films fabricated by the self-organization from nanotube solutions are useful in many areas of nanoscience and technology.

18.7 NANOTUBE/POLYMER GEL NEAR IR-RESPONSIVE MATERIALS

CNTs are characterized by their intense absorption in the NIR region, and this absorption gives a potential use for the NIR functional materials. Mainly two NIR-responsive materials have been explored. One is the NIR-saturable absorber necessary for solid state lasers based on the saturable absorption property of CNTs [89–91]. Sakakibara et al. demonstrated that the SWNT composites dispersed in a polyimide matrix are well suited for the reproducible construction of mode-locked fiber lasers and the generation of extremely short pulse durations [72,92,93]. Homogeneous dispersion of SWNTs in the polyimide matrix serves to minimize the loss of light caused by scattering and to realize such an excellent property. This application is quite unique and gives requisite optical devices such as laser and optical switches for NIR high-speed optical communication systems.

Another unique application of CNTs working in the NIR region is the photon-to-heat convertor utilizing efficient photoabsorption and photothermal conversions of CNTs in the NIR region. Boldor *et al.* reported that MWNTs showed a higher photothermal conversion efficiency than that of graphite [94]. Among the various light sources, NIR laser light is a fascinating stimulus, especially from a biomedical point of view, because tissues have only a slight absorption in the NIR region, which enables remote stimulation of NIR absorbance in the body from outside. Dai *et al.* reported an NIR-induced release of ssDNA from an ssDNA/SWNT composite dispersed in an aqueous medium [46]. Photothermal conversion occurred because of the effective nonradiative process of excited SWNTs generating intense heat in a very short period. As a result, the wrapped polymer is dissociated from the composites and the SWNTs start to aggregate through strong van der Waals interactions. They demonstrated that the photothermal conversion of CNTs irradiated by NIR light is effective to kill cancer cells stained with CNTs [46]. Clear unwrapping of the dispersant polymer induced by NIR photothermal conversion was reported by our group [95]. We described that NIR light irradiation of the SWNTs solubilized with an anthracene-carrying vinyl polymer caused flocculation of the SWNTs [95].

Furthermore, we have proposed the utilization of photothermal conversion of CNTs to thermoresponsive polymer materials. Poly(N-isopropylacrylamide) (PNIPAM) [96] and its derivatives are well-known thermoresponsive materials, which show a phase transition triggered by external stimuli, such as the solvent composition [97], pH [98], ionic strength [98], electric field [99], and light [100]. On irradiation with NIR light centered at 1064 nm, the PNIPAM/SWNT composite gel (200 μm in diameter) containing the SWNTs in the PNIPAM matrix immediately shrunk to a narrower gel (Fig. 18-3f) after 15 s. After turning off the irradiation, the shrunken gel gradually swells and becomes around 200 μm in diameter after about 67 s. The response time of the volume change is controllable by changing the concentration of the SWNTs as well as the power of the NIR laser light. Amazingly, no notable deterioration of the gel actuation is observed even after the 1200-cycle operation, namely, the SWNT composite gels are highly durable because of the toughness of the CNTs. In fact, the Raman spectra of the gels before and after the endurance test supports exhibit virtually identical G/D (Graphite/Defect) ratios, which guarantee that the SWNTs remain structurally intact. Very recently, Miyako *et al.* [101] reported two different kinds of smart polymer gels (agarose and PNIPAM gels) containing SWNTs and single-walled nanohorns (SWNHs) that show marked phase transitions on NIR irradiation; they found that, on NIR laser irradiation (1064 nm), the nanocarbon–agarose gel hybrids exhibit a gel-to-sol transition, whereas control agarose gel (without the nanocarbons) does not show any phase transition. Such NIR actuation of the polymer/CNT composites covers both a soft gel-type and solid film materials [102–108]. A wide range of absorption of CNTs provides an opportunity for a "molecular heater" to work at the various wavelengths of the light source.

18.8 ELECTROCATALYST FOR FUEL CELL USING SOLUBLE CNTs

Polybenzimidazole (PBI, Fig. 18-4a) is able to individually dissolve SWNTs based on the $\pi - \pi$ interaction between SWNTs and PBI [109]. PBI is a proton-conducting material for polymer electrolyte fuel cells (PEFCs) that can operate even under dry conditions above $100°C$ [110,111] and is a promising candidate as a substitution for a Nafion, widely used in the low-temperature PEFC systems [112–114]. The PEFC operations at higher temperatures afford many benefits such as decreased carbon monoxide poisoning on the catalyst metal particles, increase in the catalytic reaction rate, easy removal of generated water, and so on [115].

We newly designed and developed a novel PBI nanocomposite, in which CNTs and platinum (Pt) nanoparticles are employed as carbon support and redox catalyst, respectively. Especially, in our system, PBI is expected to act as (i) a Pt adsorbing material via the coordination of the Pt ion with the aromatic nitrogen on PBI, (ii) an MWNT-solubilizing material, and (iii) a proton conductor. Here, we focus on the design and fabrication of an interfacial nanostructure formed from a MWNT/**PBI**/Pt nanocomposite. Such a study is highly important since the catalyst efficiency largely depends on the catalyst morphology at their interfaces [116].

TEM image (Fig. 18-4b) of the composite shows that the Pt nanoparticles are rather uniformly deposited on the MWNT/**PBI**. From the TEM image, the mean diameter (d_{TEM}) of the Pt nanoparticles is 4.0 ± 1.8 nm. The capability of the composites as an electrocatalyst can be evaluated by their electrochemical activities. The amounts of Pt loaded on the electrode were determined by the amounts of composite solutions casted on the electrodes. Electrochemically active surface area (ECSA) of the MWNT/**PBI**/Pt composite was calculated to be 44.0 m^2/g of Pt, which is much greater than that of MWNT/Pt (22.5 m^2/g of Pt). This higher ECSA value for the MWNT/**PBI**/Pt composite might be due to the more homogeneous loading of the Pt particles on MWNTs than on MWNT/Pt. The Cyclic voltammomtry (CV) result obtained strongly suggests that the Pt nanoparticles immobilized on the PBI are close enough to the MWNTs to undergo facile electronic communication with the MWNTs and that the Pt particles are exposed to the electrolyte phase (Fig. 18-4c). This interface structure is most likely the structure that has been widely recognized as the "ideal" triple-phase boundary structure (Fig. 18-4d) that enables an excellent catalyst efficiency [117]. The formation mechanism of the above-mentioned structure is the growth of the Pt nanoparticles on the PBI layer, followed by penetration into the PBI layer to form a structure in close contact with the MWNT surfaces.

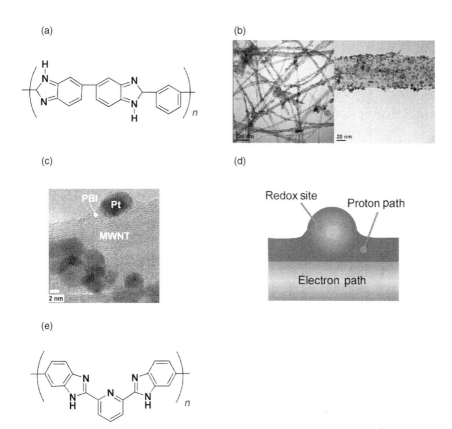

Figure 18-4 (a) Chemical structure of **PBI**. (b) Typical TEM image of a MWNT/**PBI**/Pt composite. (c) High-resolution TEM image of the MWNT/**PBI**/Pt. (d) Illustration of an "ideal" triple-phase boundary structure, in which all phases (electron path, proton path, and gas phases) are attached to the redox site. (e) Chemical structure of **PyPBI**.

Pt utilization efficiency (η_{Pt}) is the essential parameter to describe the catalyst performance and is calculated by dividing ECSA by the chemical surface area (CSA). CSA of the MWNT/**PBI**/Pt and MWNT/Pt were determined to be 71.1 and 58.1 m^2/g, respectively, when we used Eq. 18.1

$$CSA = 6/\rho d \tag{18.1}$$

where ρ is the density of Pt (21.09 g/cm^3) and d is the mean diameter of the Pt nanoparticles obtained by X-ray diffraction (XRD) (d_{XRD} and d'_{XRD}). Thus, the Pt utilization efficiency, η_{Pt}, (ECSA/CSA) of MWNT/**PBI**/Pt was determined to be as high as 74%, while that of MWNT/Pt was 39%. The Pt utilization of the commercial catalyst (Carbon black/Pt) is reported to be 54.8% [118]. The observed higher utilization efficiency of MWNT/**PBI**/Pt is explained by the formation of the "ideal" triple-phase boundary structure of the MWNT/**PBI**/Pt realized by the PBI adsorption onto MWNT and adsorption of Pt.

We have also used pyridine-containing polybenzimidazole (**PyPBI**, Fig. 18-4e) in place of PBI [119]. **PyPBI** acts as an efficient dispersant for CNT wrapping and produces a stable complex after removal of the unbound **PyPBI**. The wrapped **PyPBI** serves as glue for immobilizing Pt nanoparticles onto the surface of MWNTs without any strong oxidation process for the MWNTs. A highly homogeneous and remarkably efficient Pt loading onto the surface of MWNTs through a coordination reaction between Pt and **PyPBI** is achieved. CV measurements have revealed that the Pt nanoparticles deposited on the **PyPBI**-wrapped MWNTs have a high ECSA. The characteristic peaks in the negative region (from -0.13 to 0.2 V vs Ag/AgCl) attributable to atomic hydrogen adsorption and desorption on the Pt nanoparticle surfaces together with the Pt oxidation and Pt oxide reduction in the positive region (from 0.6 to 1.0 V vs Ag/AgCl) are apparently detected, which are similar to the reported profiles [120,121]. ECSA of the MWNT/**PyPBI**/Pt hybrid is calculated to be 51.6 m^2/g of Pt. The obtained ECSA value is higher than our previously reported value (44.0 m^2/g of Pt) for MWNT/**PBI**/Pt [122] and as high as the conventional CB/Pt system (55.8 m^2/g of Pt). Such a high ECSA value is explained by the excellent accessibility of the reactant on the surfaces of the homogeneously dispersed Pt and efficient electron conduction of the MWNTs lying beneath the Pt nanoparticles.

18.9   CONCLUDING REMARKS

In this review, we first described a strategy to solubilize CNTs based on noncovalent modification. Individual solubilization of CNTs is necessary for a wide range of scientific and technological applications of CNTs, and the preparation of individually dissolved SWNTs is the first step to afford them to practical use as well as fundamental studies. Individual solubilization based on noncovalent modification maintains the intrinsic properties of SWNTs. A large number of papers describing the applications of soluble CNTs has been reported, and some of them are unique for the CNT properties. We like to emphasize that a combination of soluble nanotubes and spectroelectrochemistry makes possible direct determination of the precise electronic properties of SWNTs. We highlighted the fabrication of novel CNT-based nanohybrids, including DNA/SWNTs, nanotube imprinting, self-assembled conducting nanotube honeycomb films, and photoresponsible polymer/SWNT gels. Furthermore, the design and fabrication of an electrocatalyst for fuel cells using PBI-solubilized CNTs that can operate in non-humid atmosphere is presented. CNT composites using soluble CNTs have high potential for application in the wide areas of nanomaterials science and technology and bioscience.

REFERENCES

1. Iijima, S. (1991). Helical microtubules of graphitic carbon. *Nature*, *354*, 56–58.
2. Nakashima, N. (2005). Soluble carbon nanotubes: fundamentals and applications. *Int. J. Nanosci*., *4*, 119–137.

3. Murakami, H., Nakashima, N. (2006). Soluble carbon nanotubes and their applications. *J. Nanosci. Nanotechnol.*, *6*, 16–27.

4. Nakashima, N., Fujigaya, T. (2007). Fundamentals and applications of soluble carbon nanotubes. *Chem. Lett.*, *36*, 692–697.

5. Dyke, C. A., Tour, J. M. (2004). Covalent functionalization of single-walled carbon nanotubes for materials applications. *J. Phys. Chem. A*, *108*, 11151–11159.

6. Girifalco, L. A., Hodak, M., Lee, R. S. (2000). Carbon nanotubes, buckyballs, ropes, and a universal graphitic potential. *Phys. Rev. B*, *62*, 13104.

7. Tasis, D., Tagmatarchis, N., Bianco, A., Prato, M. (2006). Chemistry of carbon nanotubes. *Chem. Rev.*, *106*, 1105–1136.

8. Balasubramanian, K., Burghard, M. (2005). Chemically functionalized carbon nanotubes. *Small*, *1*, 180–192.

9. Tasis, D., Tagmatarchis, N., Georgakilas, V., Prato, M. (2003). Soluble carbon nanotubes. *Chem.—Eur. J.*, *9*, 4000–4008.

10. Nakashima, N., Tomonari, Y., Murakami, H. (2002). Water-soluble single-walled carbon nanotubes via noncovalent sidewall-functionalization with a pyrene-carrying ammonium ion. *Chem. Lett.*, *6*, 638–639.

11. Murakami, H., Nomura, T., Nakashima, N. (2003). Noncovalent porphyrin-functionalized single-walled carbon nanotubes in solution and the formation of porphyrin-nanotube nanocomposites. *Chem. Phys. Lett.*, *378*, 481–485.

12. Tomonari, Y., Murakami, H., Nakashima, N. (2006). Solubilization of single-walled carbon nanotubes by using polycyclic aromatic ammonium amphiphiles in water—strategy for the design of high-performance solubilizers. *Chem.—Eur. J.*, *12*, 4027–4034.

13. Zhang, J., Lee, J. K., Wu, Y., Murray, R. W. (2003). Photoluminescence and electronic interaction of anthracene derivatives adsorbed on sidewalls of single-walled carbon nanotubes. *Nano Lett.*, *3*, 403–407.

14. Gregan, E., Keogh, S. M., Maguire, A., Hedderman, T. G., Neill, L. O., Chambers, G., Byrne, H. J. (2004). Purification and isolation of SWNTs. *Carbon*, *42*, 1031–1035.

15. Hedderman, T. G., Keogh, S. M., Chambers, G., Byrne, H. J. (2004). Solubilization of SWNTs with organic dye molecules. *J. Phys. Chem. B*, *108*, 18860–18865.

16. Feng, W., Fujii, A., Ozaki, M., Yoshino, K. (2005). Perylene derivative sensitized multi-walled carbon nanotube thin film. *Carbon*, *43*, 2501–2507.

17. Yamamoto, T., Miyauchi, Y., Motoyanagi, J., Fukushima, T., Aida, T., Kato, M., Maruyama, S. (2008). Improved bath sonication method for dispersion of individual single-walled carbon nanotubes using new triphenylene-based surfactant. *Jpn. J. Appl. Phys.*, *47*, 2000–2004.

18. Gotovac, S., Hattori, Y., Noguchi, D., Miyamoto, J.-I., Kanamaru, M., Utsumi, S., Kanoh, H., Kaneko, K. (2006). Phenanthrene adsorption from solution on single wall carbon nanotubes. *J. Phys. Chem. B*, *110*, 16219–16224.

19. Gotovac, S., Honda, H., Hattori, Y., Takahashi, K., Kanoh, H., Kaneko, K. (2007). Effect of nanoscale curvature of single-walled carbon nanotubes on adsorption of polycyclic aromatic hydrocarbons. *Nano Lett.*, *7*, 583–587.

20. Fujigaya, T., Nakashima, N. (2008). Methodology for homogeneous dispersion of single-walled carbon nanotubes by physical modification. *Polym. J.*, *40*, 577–589.

21. Saito, R., Dresselhaus, G., Dresselhaus, M. S. (1998). *Physical Properties of Carbon Nanotubes*. Imperial College Press, London.

22. Kataura, H., Kumazawa, Y., Maniwa, Y., Umezu, I., Suzuki, S., Ohtsuka, Y., Achiba, Y. (1999). Optical properties of single-wall carbon nanotubes. *Synth. Met.*, *103*, 2555–2558.

23. Saito, R., Dresselhaus, G., Dresselhaus, M. S. (2000). Trigonal warping effect of carbon nanotubes. *Phys. Rev. B: Condens. Matter Mater. Phys.*, *61*, 2981–2990.

24. Hamada, N., Sawada, S. I., Oshiyama, A. (1992). New one-dimensional conductors: graphitic microtubules. *Phys. Rev. Lett.*, *68*, 1579–1581.

25. Kazaoui, S., Minami, N., Kataura, H., Achiba, Y. (2001). Absorption spectroscopy of single-wall carbon nanotubes: effects of chemical and electrochemical doping. *Synth. Met.*, *121*, 1201–1202.

26. Zheng, M., Diner, B. A. (2004). Solution redox chemistry of carbon nanotubes. *J. Am. Chem. Soc.*, *126*, 15490–15494.

27. O'Connell, M. J., Eibergen, E. E., Doorn, S. K. (2005). Chiral selectivity in the charge-transfer bleaching of single-walled carbon-nanotube spectra. *Nat. Mater.*, *4*, 412–418.

28. Paolucci, D., Franco, M. M., Iurlo, M., Marcaccio, M., Prato, M., Zerbetto, F., Pénicaud, A., Paolucci, F. (2008). Singling out the electrochemistry of individual single-walled carbon nanotubes in solution. *J. Am. Chem. Soc.*, *130*, 7393–7399.

29. Kazaoui, S., Minami, N., Matsuda, N., Kataura, H., Achiba, Y. (2001). Electrochemical tuning of electronic states in single-wall carbon nanotubes studied by *in situ* absorption spectroscopy and AC resistance. *Appl. Phys. Lett.*, *78*, 3433–3435.

30. Kavan, L., Rapta, P., Dunsch, L., Bronikowski, M. J., Willis, P., Smalley, R. E. (2001). Electrochemical tuning of electronic structure of single-walled carbon nanotubes: in-situ Raman and vis-NIR study. *J. Phys. Chem. B*, *105*, 10764–10771.

31. Okazaki, K.-I., Nakato, Y., Murakoshi, K. (2003). Absolute potential of the Fermi level of isolated single-walled carbon nanotubes. *Phys. Rev. B: Condens. Matter Mater. Phys.*, *68*, 354341–354345.

32. Tanaka, Y., Hirana, Y., Niidome, Y., Nakashima, N. (2009). Experimentally Determined Redox Potentials of Individual (n,m) Single-Walled Carbon Nanotubes. *Angew. Chem. Int. Ed.*, *48*, 7655–7659.

33. Hirana, Y., Tanaka, Y., Niidome, Y., Nakashima, N. (2010). Strong Micro-Dielectric Environment Effect on the Band Gaps of (n,m) Single-Walled Carbon Nanotubes. *J. Am. Chem. Soc.*, *132*, 13072–13077.

34. Nakashima, N., Okuzono, S., Murakami, H., Nakai, T., Yoshikawa, K. (2003). DNA dissolves single-walled carbon nanotubes in water. *Chem. Lett.*, *32*, 456–457.

35. Barisci, J. N., Tahhan, M., Wallace, G. G., Badaire, S., Vaugien, T., Maugey, M., Poulin, P. (2004). Properties of carbon nanotube fibers spun from DNA-stabilized dispersions. *Adv. Funct. Mater.*, *14*, 133–138.

36. Iijima, M., Watabe, T., Ishii, S., Koshio, A., Yamaguchi, T., Bandow, S., Iijima, S., Suzuki, K., Maruyama, Y. (2005). Fabrication and STM-characterization of novel hybrid materials of DNA/carbon nanotube. *Chem. Phys. Lett.*, *414*, 520–524.

37. Gladchenko, G. O., Karachevtsev, M. V., Leontiev, V. S., Valeev, V. A., Glamazda, A. Y., Plokhotnichenko, A. M., Stepanian, S. G. (2006). Interaction of fragmented double-stranded DNA with carbon nanotubes in aqueous solution. *Mol. Phys.*, *104*, 3193–3201.

38. He, P., Bayachou, M. (2005). Layer-by-layer fabrication and characterization of DNA-wrapped single-walled carbon nanotube particles. *Langmuir*, *21*, 6086–6092.

39. Pantarotto, D., Singh, R., McCarthy, D., Erhardt, M., Briand, J.-P., Prato, M., Kostarelos, K., Bianco, A. (2004). Functionalized carbon nanotubes for plasmid DNA gene delivery. *Angew. Chem. Int. Ed.*, *43*, 5242–5246.

40. Rege, K., Viswanathan, G., Zhu, G., Vijayaraghavan, A., Ajayan, P. M., Dordick, J. S. (2006). *In vitro* transcription and protein translation from carbon nanotube-DNA assemblies. *Small*, *2*, 718–722.

41. Cathcart, H., Quinn, S., Nicolosi, V., Kelly, J. M., Blau, W. J., Coleman, J. N. (2007). Spontaneous debundling of single-walled carbon nanotubes in DNA-based dispersions. *J. Phys. Chem. C*, *111*, 66–74.

42. Zheng, M., Jagota, A., Semke Ellen, D., Diner Bruce, A., McLean Robert, S., Lustig Steve, R., Richardson Raymond, E., Tassi Nancy, G. (2003). DNA-assisted dispersion and separation of carbon nanotubes. *Nat. Mater.*, *2*, 338–342.

43. Zheng, M., Jagota, A., Strano, M. S., Santos, A. P., Barone, P., Chou, S. G., Diner, B. A., Dresselhaus, M. S., McLean, R. S., Onoa, G. B., *et al.* (2003). Structure-based carbon nanotube sorting by sequence-dependent DNA assembly. *Science*, *302*, 1545–1548.

44. Hughes, M. E., Brandin, E., Golovchenko, J. A. (2007). Optical absorption of DNA-carbon nanotube structures. *Nano Lett.*, *7*, 1191–1194.

45. Ishibashi, A., Yamaguchi, Y., Murakami, H., Nakashima, N. (2006). Layer-by-layer assembly of RNA/single-walled carbon nanotube nanocomposites. *Chem. Phys. Lett.*, *419*, 574–577.

46. Kam, N., Wong, S., O'Connell, M., Wisdom, J. A., Dai, H. (2005). Carbon nanotubes as multifunctional biological transporters and near-infrared agents for selective cancer cell destruction. *Proc. Natl. Acad. Sci. U.S.A.*, *102*, 11600–11605.

47. Onoa, B., Zheng, M., Dresselhaus, M. S., Diner, B. A. (2006). Carbon nanotubes and nucleic acids: tools and targets. *Phys. Status Solidi A: Appl. Mater. Sci.*, *203*, 1124–1131.

48. Douglas, S. M., Chou, J. J., Shih, W. M. (2007). DNA-nanotube-induced alignment of membrane proteins for NMR structure determination. *Proc. Natl. Acad. Sci. U.S.A.*, *104*, 6644–6648.

49. Han, X., Li, Y., Deng, Z. (2007). DNA-wrapped single-walled carbon nanotubes as rigid templates for assembling linear gold nanoparticle arrays. *Adv. Mater.*, *19*, 1518–1522.

50. Noguchi, Y., Fujigaya, T., Niidome, Y., Nakashima, N. (2008). Regulation of the near-IR spectral properties of individually dissolved single-walled carbon nanotubes in aqueous solutions of dsDNA. *Chem.—Eur. J.*, *14*, 5966–5973.

51. Noguchi, Y., Fujigaya, T., Niidome, Y., Nakashima, N. (2008). Single-walled carbon nanotubes/DNA hybrids in water are highly stable. *Chem. Phys. Lett.*, *455*, 249–251.

52. Heller, D. A., Jeng, E. S., Yeung, T.-K., Martinez, B. M., Moll, A. E., Gastala, J. B., Strano, M. S. (2006). Optical detection of DNA conformational polymorphism on single-walled carbon nanotubes. *Science*, *311*, 508–511.

53. Xu, Y., Pehrsson, P. E., Chen, L., Zhang, R., Zhao, W. (2007). Double-stranded DNA single-walled carbon nanotube hybrids for optical hydrogen peroxide and glucose sensing. *J. Phys. Chem. C*, *111*, 8638–8643.

54. Jeng, E. S., Moll, A. E., Roy, A. C., Gastala, J. B., Strano, M. S. (2006). Detection of DNA hybridization using the near-infrared band-gap fluorescence of single-walled carbon nanotubes. *Nano Lett.*, *6*, 371–375.

55. Kam, N., Wong, S., Liu, Z., Dai, H. (2006). Carbon nanotubes as intracellular transporters for proteins and DNA: an investigation of the uptake mechanism and pathway. *Angew. Chem. Int. Ed.*, *45*, 577–581.

56. Jeng, E. S., Barone, P. W., Nelson, J. D., Strano, M. S. (2007). Hybridization kinetics and thermodynamics of DNA adsorbed to individually dispersed single-walled carbon nanotubes. *Small*, *3*, 1602–1609.

57. Yamamoto, Y., Fujigaya, T., Niidome, Y., Nakashima, N. (2010). Fundamental properties of oligo double-stranded DNA/single-walled carbon nanotube nanobiohybrids. *Nanoscale*, *2*, 1767–1772.

58. dos Santos, A. S., Leite, T. D. O. N., Furtado, C. A., Welter, C., Pardini, L. C., Silva, G. G. (2008). Morphology, thermal expansion, and electrical conductivity of multiwalled carbon nanotube/epoxy composites. *J. Appl. Polym. Sci.*, *108*, 979–986.

59. Liu, L.-Q., Wagner, H. D. (2007). A comparison of the mechanical strength and stiffness of MWNT-PMMA and MWNT-epoxy nanocomposites. *Compos. Interfaces*, *14*, 285–297.

60. Bekyarova, E., Thostenson, E. T., Yu, A., Kim, H., Gao, J., Tang, J., Hahn, H. T., Chou, T. W., Itkis, M. E., Haddon, R. C. (2007). Multiscale carbon nanotube-carbon fiber reinforcement for advanced epoxy composites. *Langmuir*, *23*, 3970–3974.

61. Moisala, A., Li, Q., Kinloch, I. A., Windle, A. H. (2006). Thermal and electrical conductivity of single- and multi-walled carbon nanotube-epoxy composites. *Compos. Sci. Technol.*, *66*, 1285–1288.

62. Guzman de Villoria, R., Miravete, A., Cuartero, J., Chiminelli, A., Tolosana, N. (2006). Mechanical properties of SWNT/epoxy composites using two different curing cycles. *Compos. Part B: Eng.*, *37B*, 273–277.

63. Li, N., Huang, Y., Du, F., He, X., Lin, X., Gao, H., Ma, Y., Li, F., Chen, Y., Eklund, P. C. (2006). Electromagnetic interference (EMI) shielding of single-walled carbon nanotube epoxy composites. *Nano Lett.*, *6*, 1141–1145.

64. Zhu, J., Peng, H., Rodriguez-Macias, F., Margrave, J. L., Khabashesku, V. N., Imam, A. M., Lozano, K., Barrera, E. V. (2004). Reinforcing epoxy polymer composites through covalent integration of functionalized nanotubes. *Adv. Funct. Mater.*, *14*, 643–648.

65. Zhu, J., Kim, J., Peng, H., Margrave, J. L., Khabashesku, V. N., Barrera, E. V. (2003). Improving the dispersion and integration of single-walled carbon nanotubes in epoxy composites through functionalization. *Nano Lett.*, *3*, 1107–1113.

66. Song, Y. S., Youn, J. R. (2005). Influence of dispersion states of carbon nanotubes on physical properties of epoxy nanocomposites. *Carbon*, *43*, 1378–1385.

67. Chatterjee, T., Jackson, A., Krishnamoorti, R. (2008). Hierarchical structure of carbon nanotube networks. *J. Am. Chem. Soc.*, *130*, 6934–6935.

68. Bliznyuk, V. N., Singamaneni, S., Sanford, R. L., Chiappetta, D., Crooker, B., Shibaev, P. V. (2006). Matrix mediated alignment of single wall carbon nanotubes in polymer composite films. *Polymer*, *47*, 3915–3921.

69. Park, C., Wilkinson, J., Banda, S., Ounaies, Z., Wise, K. E., Sauti, G., Lillehei, P. T., Harrison, J. S. (2006). Aligned single-wall carbon nanotube polymer composites using an electric field. *J. Polym. Sci., Part B: Polym. Phys.*, *44*, 1751–1762.

70. Sandoval, J., Soto, K., Murr, L., Wicker, R. (2007). Nanotailoring photocrosslinkable epoxy resins with multi-walled carbon nanotubes for stereolithography layered manufacturing. *J. Mater. Sci.*, *42*, 156.

71. Fujigaya, T., Haraguchi, S., Fukumaru, T., Nakashima, N. (2008). Development of novel carbon nanotube/photopolymer nanocomposites with high conductivity and their application to nanoimprint photolithography. *Adv. Mater.*, *20*, 2151–2155.

72. Schibli, T. R., Minoshima, K., Kataura, H., Itoga, E., Minami, N., Kazaoui, S., Miyashita, K., Tokumoto, M., Sakakibara, Y. (2005). Ultrashort pulse-generation by saturable absorber mirrors based on polymer-embedded carbon nanotubes. *Opt. Express*, *13*, 8025–8031.

73. Aoki, N., Yokoyama, A., Nodasaka, Y., Akasaka, T., Uo, M., Sato, Y., Tohji, K., Watari, F. (2005). Cell culture on a carbon nanotube scaffold. *J. Biomed. Nanotechnol.*, *1*, 402–405.

74. Aoki, N., Yokoyama, A., Nodasaka, Y., Akasaka, T., Uo, M., Sato, Y., Tohji, K., Watari, F. (2006). Strikingly extended morphology of cells grown on carbon nanotubes. *Chem. Lett.*, *35*, 508–509.

75. Allen, B. L., Kichambare, P. D., Star, A. (2007). Carbon nanotube field-effect-transistor-based biosensors. *Adv. Mater.*, *19*, 1439–1451.

76. Balasubramanian, K., Burghard, M. (2006). Biosensors based on carbon nanotubes. *Anal. Bioanal. Chem.*, *385*, 452–468.

77. Widawski, G., Rawiso, M., Francois, B. (1994). Self-organized honeycomb morphology of star-polymer polystyrene films. *Nature*, *369*, 387–389.

78. Li, Z., Zhao, W., Liu, Y., Rafailovich, M. H., Sokolov, J., Khougaz, K., Eisenberg, A., Lennox, R. B., Krausch, G. (1996). Self-ordering of diblock copolymers from solution. *J. Am. Chem. Soc.*, *118*, 10892–10893.

79. Jenekhe, S. A., Chen, X. L. (1999). Self-assembly of ordered microporous materials from rod-coil block copolymers. *Science*, *283*, 372–375.

80. Srinivasarao, M., Collings, D., Philips, A., Patel, S. (2001). Three-dimensionally ordered array of air bubbles in a polymer film. *Science*, *292*, 79–82.

81. Yabu, H., Tanaka, M., Ijiro, K., Shimomura, M. (2003). Preparation of honeycomb-patterned polyimide films by self-organization. *Langmuir*, *19*, 6297–6300.

82. Lee, C.-S., Kimizuka, N. (2002). Pillared honeycomb nanoarchitectures formed on solid surfaces by the self-assembly of lipid-packaged one-dimensional Pt complexes. *Proc. Natl. Acad. Sci. U.S.A.*, *99*, 4922–4926.

83. Zhao, X., Cai, Q., Shi, G., Shi, Y., Chen, G. (2003). Formation of ordered microporous films with water as templates from poly(D,L-lactic-co-glycolic acid) solution. *J. Appl. Polym. Sci.*, *90*, 1846–1850.

84. Xu, Y., Zhu, B., Xu, Y. (2005). A study on formation of regular honeycomb pattern in polysulfone film. *Polymer*, *46*, 713–717.

85. Nurmawati, M. H., Renu, R., Ajikumar, P. K., Sindhu, S., Cheong, F. C., Sow, C. H., Valiyaveettil, S. (2006). Amphiphilic poly(*p*-phenylene)s for self-organized porous blue-light-emitting thin films. *Adv. Funct. Mater.*, *16*, 2340–2345.

86. Arai, K., Tanaka, M., Yamamoto, S., Shimomura, M. (2008). Effect of pore size of honeycomb films on the morphology, adhesion and cytoskeletal organization of cardiac myocytes. *Colloids Surf., A*, *313–314*, 530.

87. Takamori, H., Fujigaya, T., Yamaguchi, Y., Nakashima, N. (2007). Simple preparation of self-organized single-walled carbon nanotubes with honeycomb structures. *Adv. Mater.*, *19*, 2535–2539.

88. Wakamatsu, N., Takamori, H., Fujigaya, T., Nakashima, N. (2009). Self-organized single-walled carbon nanotube conducting thin films with honeycomb structures on flexible plastic films. *Adv. Funct. Mater.*, *19*, 311–316.

89. Chen, Y. C., Raravikar, N. R., Schadler, L. S., Ajayan, P. M., Zhao, Y. P., Lu, T. M., Wang, G. C., Zhang, X. C. (2002). Ultrafast optical switching properties of single-wall carbon nanotube polymer composites at 1.55mm. *Appl. Phys. Lett.*, *81*, 975–977.

90. Tatsuura, S., Furuki, M., Sato, Y., Iwasa, I., Tian, M., Mitsu, H. (2003). Semiconductor carbon nanotubes as ultrafast switching materials for optical telecommunications. *Adv. Mater.*, *15*, 534–537.

91. Sakakibara, Y., Tatsuura, S., Kataura, H., Tokumoto, M., Achiba, Y. (2003). Near-infrared saturable absorption of single-wall carbon nanotubes prepared by laser ablation method. *Jpn. J. Appl. Phys., Part 2*, *42*, L494–L496.

92. Sakakibara, Y., Rozhin, A. G., Kataura, H., Achiba, Y., Tokumoto, M. (2005). Carbon nanotube-poly(vinylalcohol) nanocomposite film devices: applications for femtosecond fiber laser mode lockers and optical amplifier noise suppressors. *Jpn. J. Appl. Phys. Part 1*, *44*, 1621–1625.

93. Nishizawa, N., Seno, Y., Sumimura, K., Sakakibara, Y., Itoga, E., Kataura, H., Itoh, K. (2008). All-polarization-maintaining Er-doped ultrashort-pulse fiber laser using carbon nanotube saturable absorber. *Opt. Express*, *16*, 9429–9435.

94. Boldor, D., Gerbo, N. M., Monroe, W. T., Palmer, J. H., Li, Z., Biris, A. S. (2008). Temperature measurement of carbon nanotubes using infrared thermography. *Chem. Mater.*, *20*, 4011–4016.

95. Narimatsu, K., Niidome, Y., Nakashima, N. (2006). Pulsed-laser induced flocculation of carbon nanotubes solubilized by an anthracene-carrying polymer. *Chem. Phys. Lett.*, *429*, 488–491.

96. Hirotsu, S., Hirokawa, Y., Tanaka, T. (1987). Volume-phase transitions of ionized N-isopropyl-acrylamide gels. *J. Chem. Phys.*, *87*, 1392–1395.

97. Katayama, S., Hirokawa, Y., Tanaka, T. (1984). Reentrant phase transition in acrylamide-derivative copolymer gels. *Macromolecules*, *17*, 2641–2643.

98. Tanaka, T., Fillmore, D., Sun, S.-T., Nishio, I., Swislow, G., Shah, A. (1980). Phase transitions in ionic gels. *Phys. Rev. Lett.*, *45*, 1636–1639.

99. Tanaka, T., Nishio, I., Sun, S. T., Ueno-Nishio, S. (1982). Collapse of gels in an electric field. *Science*, *218*, 467–469.

100. Suzuki, A., Tanaka, T. (1990). Phase transition in polymer gels induced by visible light. *Nature*, *346*, 345–347.

101. Miyako, E., Nagata, H., Hirano, K., Hirotsu, T. (2008). Photodynamic thermoresponsive nanocarbon-polymer gel hybrids. *Small*, *4*, 1711–1715.

102. Ahir, S. V., Terentjev, E. M. (2005). Photomechanical actuation in polymer-nanotube composites. *Nat. Mater.*, *4*, 491–495.

103. Ahir, S. V., Terentjev, E. M. (2006). Fast relaxation of carbon nanotubes in polymer composite actuators. *Phys. Rev. Lett.*, *96*, 133902/1–133902/4.

104. Ahir, S. V., Squires, A. M., Tajbakhsh, A. R., Terentjev, E. M. (2006). Infrared actuation in aligned polymer-nanotube composites. *Phys. Rev. B: Condens. Matter Mater. Phys.*, *73*, 085420/1–085420/12.

105. Lu, S., Panchapakesan, B. (2005). Optically driven nanotube actuators. *Nanotechnology*, *16*, 2548–2554.

106. Lu, S., Panchapakesan, B. (2007). Photomechanical responses of carbon nanotube/polymer actuators. *Nanotechnology*, *18*, 305502.

107. Yang, L., Setyowati, K., Li, A., Gong, S., Chen, J. (2008). Reversible infrared actuation of carbon nanotube-liquid crystalline elastomer nanocomposites. *Adv. Mater.*, *20*, 2271–2275.

108. Koerner, H., Price, G., Pearce, N. A., Alexander, M., Vaia, R. A. (2004). Remotely actuated polymer nanocomposites' stress-recovery of carbon-nanotube-filled thermoplastic elastomers. *Nat. Mater.*, *3*, 115.

109. Okamoto, M., Fujigaya, T., Nakashima, N. (2008). Individual dissolution of single-walled carbon nanotubes (SWNTs) using polybenzimidazole (PBI) and high effective reinforcement of SWNTs/PBI composite films. *Adv. Funct. Mater.*, *18*, 1776–1782.

110. Wang, J. T., Savinell, R. F., Wainright, J., Litt, M., Yu, H. (1996). A H2/O2 fuel cell using acid doped polybenzimidazole as polymer electrolyte. *Electrochim. Acta*, *41*, 193–197.

111. Li, Q., He, R., Jensen, J. O., Bjerrum, N. J. (2004). PBI-based polymer membranes for high temperature fuel cells—preparation, characterization and fuel cell demonstration. *Fuel Cells*, *4*, 147–159.

112. Heitner-Wirguin, C. (1996). Recent advances in perfluorinated ionomer membranes: Structure, properties and applications. *J. Memb. Sci.*, *120*, 1–33.
113. Kerres, J. A. (2001). Development of ionomer membranes for fuel cells. *J. Memb. Sci.*, *185*, 3–27.
114. Deluca, N. W., Elabd, Y. A. (2006). Polymer electrolyte membranes for the direct methanol fuel cell: A review. *J. Polym. Sci., Part B: Polym. Phys.*, *44*, 2201.
115. Li, Q., He, R., Gao, J.-A., Jensen, J. O., Bjerrum, N. J. (2003). The CO poisoning effect in PEMFCs operational at temperatures up to 200°C. *J. Electrochem. Soc.*, *150*, A1599–A1605.
116. Lee, S. J., Mukerjee, S., McBreen, J., Rho, Y. W., Kho, Y. T., Lee, T. H. (1998). Effects of Nafion impregnation on performances of PEMFC electrodes. *Electrochim. Acta*, *43*, 3693–3701.
117. Munakata, H., Ishida, T., Kanamura, K. (2007). Electrophoretic deposition for nanostructural design of catalyst layers on Nafion membrane. *J. Electrochem. Soc.*, *154*, B1368–B1372.
118. Wang, J. J., Yin, G. P., Zhang, J., Wang, Z. B., Gao, Y. Z. (2007). High utilization platinum deposition on single-walled carbon nanotubes as catalysts for direct methanol fuel cell. *Electrochim. Acta*, *52*, 7042–7050.
119. Fujigaya, T., Okamoto, M., Nakashima, N. (2009). Design of an assembly of pyridine-containing polybenzimidazole, carbon nanotubes and Pt nanoparticles for a fuel cell electrocatalyst with a high electrochemically active surface area. *Carbon*, *47*, 3227–3232.
120. Solla-Gullon, J., Lafuente, E., Aldaz, A., Martinez, M. T., Feliu, J. M. (2007). Electrochemical characterization and reactivity of Pt nanoparticles supported on single-walled carbon nanotubes. *Electrochim. Acta*, *52*, 5582.
121. Colon-Mercado, H. R., Popov, B. N. (2006). Stability of platinum based alloy cathode catalysts in PEM fuel cells. *J. Power Sources*, *155*, 253–263.
122. Okamoto, M., Fujigaya, T., Nakashima, N. (2009). Design of an assembly of poly(benzimidazole), carbon nanotubes, and Pt nanoparticles for a fuel-cell electrocatalyst with an ideal interfacial nanostructure. *Small*, *5*, 735–740.

CHAPTER 19

INTERACTION OF CARBON NANOTUBES AND SMALL MOLECULES

Sampath Srinivasan and Ayyappanpillai Ajayaghosh

Photosciences and Photonics Group, National Institute for Interdisciplinary Science and Technology (NIIST), CSIR, Trivandrum, India

19.1 INTRODUCTION

Carbon nanotubes (CNTs), particularly single-walled carbon nanotubes (SWNTs), by virtue of their high mechanical and electronic properties have attracted considerable interest for a wide range of applications in the field of advanced materials and biology [1–10]. SWNTs were first experimentally identified by Iijima [11,12]. Details of the structural and physical properties of CNTs and their application are discussed in various chapters of this book and hence need not be discussed here. Rather, the focus is on the physical interaction of CNTs with small molecules. This topic is extremely important in the context of the processability of CNTs for different applications without compromising the electronic properties. In order to put the topic of this chapter into perspective, we must emphasise that CNTs are highly insoluble materials because of the strong intertubular interaction leading to bundling. However, solubilization of CNTs in solvents is necessary to improve the processability and thereby application [13–15]. In particular, the solubilization in aqueous media is necessary to improve the biocompatibility of the nanotubes and enable their environment friendly characterization, separation, and self-assembly. In recent years, researchers have introduced several methods for improving the solubility of CNTs [16–18]. One of the efficient methods is chemical functionalization. However, this strategy strongly perturbs the electronic properties of CNTs because of the disturbance of the extended π–network of the nanotube [19–25]. Noncovalent modifications, on the other hand, not only improve the solubility of SWNTs in both organic solvents and water but also constitute nondestructive processes that preserve the primary structures of CNTs,

Supramolecular Soft Matter: Applications in Materials and Organic Electronics, First Edition.
Edited by Takashi Nakanishi.

along with their unique mechanical and electronic properties [26–31]. Noncovalent functionalization involves van der Waals, $\pi-\pi$, CH$-\pi$, or electrostatic interactions between molecules and the surface of CNTs.

A strategy to exfoliate the bundles and stabilize isolated SWNTs in aqueous solution is to overcome the hydrophobic interactions between the tubes by repulsive forces. Surfactants and water-soluble polymers are known to facilitate the dissolution of SWNTs in water. Aromatic molecules such as porphyrins, pyrene, and anthracene derivatives have been reported to interact with nanotube side walls by $\pi-\pi$ interaction. Interaction of CNTs with aromatic molecules having ionic groups facilitates solubilization of the composites in water. The adsorption of aromatic molecules is assumed to take place in a coplanar geometry with the aid of $\pi-\pi$ interactions, in analogy to the graphene surface. The strength of the $\pi-\pi$ stacking is expected to scale up with the size of the aromatic moiety, but owing to the curvature of the nanotube surface, the shape of the ring system may also control the affinity. In a polar environment, the major contributions to the $\pi-\pi$ interaction result from the electrostatic interactions and hydrophobic effects. Charge-transfer (CT) interactions with aromatic molecules have also been suggested to be efficient in promoting solubilization of CNTs. The following sections review recent reports pertaining to the interaction of CNTs with different types of organic molecules.

19.2 INTERACTION OF POLYAROMATICS WITH CNTs

The interaction of small molecules having different aromaticity with CNTs allows the dispersion or dissolution of the latter in a variety of solvents, which is governed by the property of the interacting molecules. An interesting study of gas phase adsorption of organic molecules such as cyclohexane (**1**), cyclohexene (**2**), cyclohexadiene (**3**), and benzene (**4**), in which the number of π–electrons monotonically increases from zero to six, has revealed that $\pi-\pi$ stacking is important for adsorption on CNTs [32]. This study clearly demonstrates that the molecule–CNT interactions in this series are controlled by coupling of π–electrons of the molecules with the electronic system of the nanotube. This phenomenon becomes important for interactions of organic molecules with nanotubes in the solution phase, leading to the solubilization of individual nanotubes.

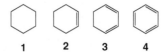

1　　**2**　　**3**　　**4**

Nakashima and coworkers have carried out extensive studies on the solubilization of CNTs in solvents with a variety of functional molecules that are composed of an aromatic unit and a solvophilic unit [33–36]. Solubilizers carrying a hydrophilic or hydrophobic unit can dissolve CNTs in polar solvents, such as water and alcohols, and in nonpolar solvents, respectively. Ammonium amphiphiles carrying a phenyl (**5**) or naphthyl (**6**) group did not act as CNT solubilizers; instead, the amphiphilic systems having phenanthrayl (**7**) and pyrenyl (**8**) groups play an important role in solubilization. The molecule **8**, an amphiphile that does not form micelles, is an excellent solubilizer for SWNTs. Fluorescence spectra provided direct evidence for the interaction as the pyrene excimer emission of the composite is significantly quenched when compared to that of the original solution because of the energy transfer from the pyrene moiety in **8** to the SWNTs. It has been reported that an aqueous solution of **8** has a tendency to dissolve semiconducting SWNTs with diameters in the range of 0.89–1.0 nm. This fine discrimination in the diameters of the SWNTs is of interest in the design of nanotube solubilizers that recognize a single nanotube chiral index, as the synthesis of SWNTs with a single chiral index is presently difficult, and for their potential application in many fields of science and technology.

Recently, Paloniemi *et al.* have reported that naphthalene derivatives with an amino group could disperse SWNTs in solvents owing to the contribution of a CT interaction between the amino group and the SWNTs [37]. They have prepared water-soluble anionic and cationic polyelectrolyte composites of CNTs via noncovalent interaction using the ionic pyrenes (**11** and **12**) and naphthalenes (**9** and **10**). CT from the adsorbates to the nanotubes is revealed by the positive shifts in the C 1s binding energy. Adsorbates containing an amino group show enhanced interaction with the nanotubes. Specific interactions between the substituents and the nanotube surface (CT, cation–π interactions) are important, especially with small aromatic molecules such as naphthalenes. The stability and solubility of the SWNT/polyanions and SWNT/polycations are high enough to enable their processing in an aqueous environment and introduction into self-assembled multilayer structures.

The electronic interaction of anthracene derivatives with SWNT has been studied by Murray *et al.* [38]. They have used different functional groups having various electrophilic capabilities and volumes. The effects of substituents were investigated by spectroscopic techniques. The absorption spectra of anthracene derivatives of SWNT/**13–17** showed weak absorbance shifts relative to the corresponding unattached compounds. The fluorescence maxima of chemisorbed anthracenes showed a redshift relative to the free compounds. The shift of emission to lower energy varied with the anthracene substituents, in the order **14** > **13, 16** > **15, 17**. The small energy shifts (redshift) may be attributed to the solvent-related excited state relaxations that lower excited/ground-state dipole moment. The emission energy shift suggests structure-dependent interaction, which might be correlated with electron donor or acceptor properties. The face-to-face $\pi - \pi$ interaction is accompanied by an electron donor–acceptor CT interaction between the aromatic adsorbate and the SWNT sidewall. Hence the SWNT adsorption process favors the anthracene derivatives based on their electron affinity, in the order $- CN > -Br > -CH_2OH > -H > -OH$, except for **17**. The affinity of **17** for SWNT may include the interaction of hydroxyl groups with the SWNT.

Electrochemical measurements of the relative electron affinity of the anthracene derivatives offer a clear correlation and support the CT interaction during chemisorption. Electrochemical potentials are well known to track electron affinity energetics. Voltammetric studies have revealed electron transfer reductions. A less negative reduction potential corresponds to a lower lowest unoccupied molecular orbital (LUMO) state, which is expected to electronically interact more strongly with the SWNT. Anthracene derivatives with more positive reduction potentials should exhibit the largest chemisorption coverages. Formal potentials for the first one-electron reduction of the anthracene derivatives fall in an order of **16** > **15** > **17** – **13** > **14**. This order is indeed approximately the same as that of the coverage of the anthracene derivatives on SWNTs.

The reversibility of an adsorbed molecule on CNT could be studied using a stronger adsorber, which displaces the previously adsorbed molecule. For example, a mixture of SWNT and **13** with a 100-fold molar excess of pyrene in tetrahydrofuran (THF) when stirred at room temperature for 72 h replaces the attached anthracene molecules. The emission and absorption spectra of the product correspond to those of pyrene, indicating that anthracene had been quantitatively displaced by pyrene. The fluorescence of anthracene adsorbed on the SWNTs is not much different from that observed in the absence of SWNTs [39].

Tromp *et al.* have functionalized CNTs by utilizing a size matching approach [40]. For example, molecular tweezers or "anchor molecules" were used to selectively interact with CNTs having a specific diameter. This approach is based on the $\pi-\pi$ interaction of CNTs with condensed aromatic compounds to form a host–guest pair. The anchor molecule **18** is shaped like a folded ribbon such that it fits well with CNTs at an optimum diameter but not efficient either with larger or smaller diameter tubes. The tail of the anchor molecule provides solubility, as well as the potential for attachment of functional groups that allow selective interaction with other moieties such as proteins and DNA sequences.

There have been few attempts to separate semiconducting and metallic nanotubes. Physical methods such as dielectrophoresis and density gradient centrifugation as well as chemical methods including covalent and noncovalent functionalization such as adsorption of bromine and diazotization have been employed to separate metallic and semiconducting SWNTs [41]. Recently, Rao *et al.* have separated semiconducting and metallic SWNTs from a mixture by the CT interaction with potassium salt of coronene tetracarboxylic acid (**19**) [42]. Optical absorption and Raman spectroscopic studies indicate the separation of metallic and semiconducting SWNTs. This method depends on the concentration of **19** and the time of interaction.

Interaction of *N,N'*-diphenyl glyoxaline-3,4,9,10-perylene tetracarboxylic acid diacidamide (**20**) with multiwalled carbon nanotube (MWNT) results in the dispersion of the latter in organic solvents, allowing the preparation of nanocomposite thin films at the molecular level [43]. Since **20** is an aromatic molecule with extensive π–conjugation and considerable planarity, irreversible adsorption occurs onto the surfaces of MWNTs in chloroform and dimethylformamide. The nanocomposites of **20**/MWNT have been characterized by using UV–vis absorption, photoluminescence (PL), and photocurrent spectral data. The redshift in the absorption spectra and the quenching in PL indicate π-stacking between **20** and MWNT. The enhanced photoconductivity and strongly quenched PL are explained in terms of efficient photoinduced charge separation between MWNT and **20**. This property of the **20**/MWNT composite has been used for the design of photovoltaic devices with enhanced efficiency.

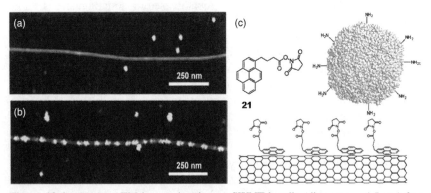

19   **20**

A simple and general approach to noncovalent functionalization of SWNTs and subsequent immobilization of various biological molecules onto nanotubes with a high degree of control and specificity has been reported (Fig. 19-1). This could be achieved with a bifunctional molecule, 1-pyrenebutanoic acid succinimidyl ester (**21**), which irreversibly adsorbs onto the inherently hydrophobic surface of SWNTs in dimethylformamide or methanol [44]. The anchored molecules on SWNTs are highly stable against desorption in aqueous solutions. This leads to further functionalization of SWNTs with the help of the succinimidyl ester groups that are reactive to nucleophiles such as primary and secondary amines that exist in abundance on the surface of most proteins. In this way, a required protein, ferritin, could be attached to SWNT. Atomic force microscopic (AFM) images have revealed the presence of immobilized ferritin along the SWNT.

Figure 19-1 (a) An AFM image showing an SWNT bundle (diameter = 4.5 nm) free of ferritin after incubation in a ferritin solution. (b) An image showing ferritin molecules (apparent height 10 nm) adsorbed on an SWNT bundle (diameter 2.5 nm) functionalized by **21** and incubated in a ferritin solution. (c) 1-Pyrenebutanoic acid succinimidyl ester **21** irreversibly adsorbed onto the sidewall of an SWNT via π-stacking. The amine groups on the protein react with the anchored succinimidyl ester to form amide bonds for protein immobilization.

As a control, SWNTs were exposed to the solution of ferritin alone, without treatment with **21**. In this case, the nanotubes were found to be free of adsorbed ferritin (Fig. 19-1). This technique enables the immobilization of a wide range of biomolecules on the sidewalls of SWNTs with high specificity and efficiency.

Ring opening metathesis polymerization (ROMP) reaction has been utilized for the preparation of polynorbornene-coated SWNTs using pyrene-anchored ruthenium alkylidenes molecule [45]. SWNTs are deposited onto SiO_2 substrates utilizing ferritin-derived catalysts and then functionalized by using norbornene solution. AFM imaging (tapping mode) of SWNT before and after polynorbornene coating showed polymer coating of ca 7.7 nm with an SWNT of 1.4-nm diameter. AFM studies revealed that the SiO_2 surface was essentially free of polynorborene, indicating the high selectivity of the functionalization step using polycyclic aromatics.

Harada *et al.* have reported a chemically responsive supramolecular SWNT hydrogel by the interaction of pyrene-attached cyclodextrin (CD) moieties on the SWNT surface (Scheme 19-1) [46]. Since CD is soluble in water [47–49], interaction between pyrene-modified CDs (**22**) and SWNTs resulted in **22**/SWNT hybrids. The vacant cavities of **22** around SWNT are able to capture guest molecules on the SWNT surface. By utilizing host–guest interactions between β-CDs of **22**/SWNT hybrids and polymers carrying guest moieties, supramolecular SWNT hydrogels were prepared. The host–guest complexes between β-CD moieties immobilized on the SWNT surface and dodecyl groups in poly(acrylic acid) (**23**, carrying 2 mol% of dodecyl groups) act as cross-links to form hybrid gel network structures. When sodium adamantane carboxylate (**24**, 100 equivalents to dodecyl moieties of **23**) was added to the hydrogel as a competitive guest, there was a gel to sol transition, indicating the dissociation of the host–guest complexes. Similarly, the gel changes to a solution when α-CD (**25**) (100 equivalents to dodecyl groups of **23**) was added as a competitive host. Gel–sol transition occurs since the dodecyl moieties form complexes with α-CD more favorably than with β-CD.

Stupp *et al.* have developed a strategy to create electronically active molecular coatings on CNTs that allowed the preparation of hybrid materials useful for photovoltaic applications [50]. The authors have used a rational molecular design approach to assemble oligothiophenes on the surface of CNTs. For this purpose, a quinquethiophene unit was functionalized with a pyrene moiety, which covalently interacts with CNTs. Photovoltaic devices were fabricated by depositing the CNT–thiophene composite on indium tin oxide (ITO)-coated glass, which showed a power conversion efficiency of 0.02%.

19.3   INTERACTION OF EXTENDED π–SYSTEMS WITH CNTs

Linear π–conjugated molecules are known to strongly interact with each other through π–stacking. Similarly, CNTs are also known to bundle together as a

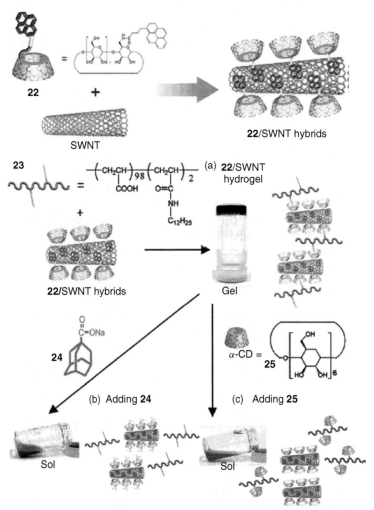

Scheme 19-1   (a) Scheme showing the reversible gelation of **22**/SWNT with **23**. Gel to sol transition on addition of a competitive guest **24** (b) and a competitive host **25** (c).

result of intertube interaction. It is reasonable to expect that these two systems may interact strongly when mixed together thereby allowing the dispersion of the CNTs. Thus, extended π-systems such as poly(phenylenevinylene) (PPV), poly(phenylacetylene) (PPA), and poly(*m*-phenylenevinylene-*co*-2,5-dioctyloxy-*p*-phenylenevinylene) (PmPV) have been shown to strongly interact with CNTs [51–54]. Stoddart and coworkers have prepared composites of nanotubes with alkoxy-modified phenylenevinylene-type polymers and studied the performance in a photovoltaic device [55]. In a subsequent work, the same researchers have studied the interaction of CNT with PmPV and poly(2,6-pyridinylenevinylene-*co*-2,5-dioctoxy-*p*-phenylenevinylene) (PPyPV). In the case of PPyPV-wrapped SWNTs, interaction is dominated by the protonated polymer [56–58]. PmPV, on the other hand, interacts with SWNTs in its charge-neutral form. Helical wrapping of the polymer backbone with the SWNT is responsible for the dispersion of the SWNT. An alternative strategy for solubilizing CNT using short rigid oligo(phyneleneethynylene)s (OPEs) was reported by Chen and coworkers [59]. Since the oligomers are short, no helical wrapping over the CNT is proposed. In this case, the interaction allowed a 20-fold solubility enhancement for small-diameter nanotubes. It is also possible to homogeneously disperse small-diameter CNTs within the matrices of polystyrene and polycarbonate [60]. These composites show dramatic improvements in the electrical conductivity at low filler loading (percolation threshold at 0.045 wt%).

Interaction of SWNTs with short linear π-conjugated oligo(p-phenylenevinylene) (OPV)-based gelators allows the preparation of hybrid organogels with improved stability (Fig. 19-2) [61]. Molecules **26–28** are known to form organogels above a critical concentration in nonpolar solvents. However, in relatively polar solvents, higher concentrations of OPVs are required for gelation, and the resulting gels are mechanically unstable (thixotropic) [62–64]. It is therefore necessary to improve the stability and mechanical properties of OPV gels for any potential application. It has been found that the addition of CNTs accelerates self-assembly and gelation of OPVs below their normal critical gelation concentration (CGC), resulting in CNT-dispersed stable hybrid gels. Spectroscopic studies indicated strong physical interaction between the self-assembled OPVs and CNTs. Rheological parameters of the composite gel showed an enhanced solid-like character relative to the OPV gel. The ratio of the storage modulus (G') to loss modulus (G'') is greater than unity for the hybrid gels. The physical reinforcement of the self-assembled OPV tapes by isolated CNTs is considered to be responsible for the improved rheological properties. Microscopic investigation has revealed the presence of isolated CNTs covered by self-assembled OPVs, thereby reinforcing the OPV supramolecular tapes. In this case, CNTs act as physical cross-links between the tapes, enhancing gel stability. This strategy to make CNT–OPV hybrid π-conjugated gels allowed the CNTs to retain their long aspect ratio as well as their electronic properties in the gel state.

Interaction of CNTs with the nongelling OPV molecules (**29–31**) resulted in the dispersion of the former in a variety of nonpolar solvents [65]. The well-dispersed nanocomposite can be coated on glass, metal, and mica surfaces, resulting in water-repellent self-cleaning surfaces with high water contact angles (CA) of about $165–170°$ and a sliding angle (SA) of less than $2°$ (Fig. 19-2). The fraction of air on the composite surface estimated by using the CA values indicated that the fraction of air trapped beneath the water droplets is very high. Water droplets could roll on the composite surface much more easily than on the CNT and OPV surfaces. It has been found that the force required for the water droplets to move on the composite is about 44 and 24 times less when compared to that of pure CNT and OPV coatings, respectively. Microscopic analysis of these composite coatings has revealed that they have a binary surface topography consisting of micrometer-sized hills and valleys with a nanoscale paraffin coating of hairy hydrocarbon chains akin to the two-tier surface morphology

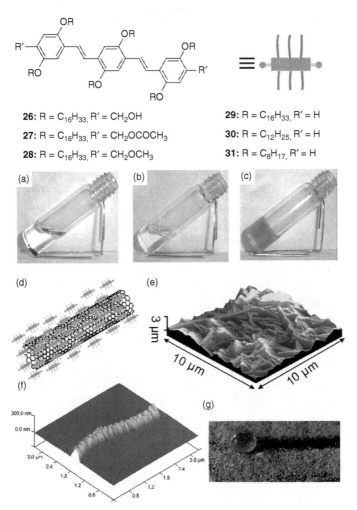

26: R = C$_{16}$H$_{33}$, R' = CH$_2$OH

27: R = C$_{16}$H$_{33}$, R' = CH$_2$OCOCH$_3$

28: R = C$_{16}$H$_{33}$, R' = CH$_2$OCH$_3$

29: R = C$_{16}$H$_{33}$, R' = H

30: R = C$_{12}$H$_{25}$, R' = H

31: R = C$_8$H$_{17}$, R' = H

Figure 19-2 Chemical structures of OPV derivatives. Photographs of immiscible SWNTs in toluene (a), solution of **26** in toluene (b), and gel of **26** in toluene with SWNTs (c). Schematic representation of the OPV-CNT π–π interaction (d). AFM images of **29**-MWNT composite coating (e) and of a single CNT coated with **29** (f). Photograph showing the self-cleaning ability of a dusted composite surface (g). (*See insert for color representation of the figure.*)

of lotus leaves. The CNT–OPV nanocomposite surface gives a high CA and very small SA with liquids having varying pH values and ionic strengths. Most importantly, it has been demonstrated that, in place of the regular micro- and nanostructured topography of natural systems, an irregular microstructure created by a nanostructured material is sufficient to mimic the superhydrophobic character of natural self-cleaning surfaces. The fact that superhydrophobicity is exhibited by the relatively inexpensive MWNTs, when compared to SWNTs, broadens the scope for potential applications of the composite.

19.4 INTERACTION OF CONJUGATED MACROCYCLES WITH CNTs

Macrocyclic compounds having extended π–conjugation, such as porphyrins and related systems, are known to solubilize CNTs in solvents [35,66–68]. For example, an appreciable solubility of SWNT has been achieved on interaction with zinc protoporphyrin (ZnPP) in dimethylformamide (DMF) [35]. Microscopic investigations revealed the presence of unbundled SWNTs with diameters ranging from 0.9 to 1.5 nm. UV–vis–NIR (near-infrared) spectra of the dispersed SWNT solutions revealed the characteristic absorption bands for ZnPP and SWNTs in the visible and NIR regions. The amount of SWNTs dissolved/dispersed in DMF is estimated to be ~10–20 mg/mL. Direct confirmation for ZnPP/SWNTs inter-actions was acquired from fluorescence studies. The porphyrin emission intensity in ZnPP/SWNTs solution is considerably decreased when compared with that of ZnPP. This quenching has been ascribed to the energy transfer from the photoex-cited porphyrin to SWNT. Sun *et al.* have reported the dispersion of SWNT in the presence of H_2-5,10,15,20-tetrakis(hexadecyloxyphenyl)-21H,23H-porphyrin (**32**) in a variety of organic solvents [66]. The selective interactions between **32** and semiconducting SWNT have been utilized to separate metallic (bulk conductivity = 1.1 S/cm) and semiconducting nanotubes (bulk conductivity = 0.007 S/cm). The interactions of **32** with SWNT are specific to the porphyrin free base, and the interaction decreased on complexation with the metal cation. For example, when **33** was used instead of **32** under the same experimental conditions, no interaction leading to the solubilization of SWNTs was observed.

$$M = 2H\ (\mathbf{32})$$
$$Zn^{2+}\ (\mathbf{33})$$

Interaction of CNTs with porphyrins having positively or negatively charged ionic head groups has been used to attach oppositely charged molecules on the CNT surface. Water-soluble porphyrins such as octapyridinium ZnP/H$_2$P salts or octacarboxylate ZnP/H$_2$P salts have been utilized to form electron donor–acceptor nanohybrids [69–72]. They were characterized by spectroscopic and microscopic techniques. Complex formation between CNT and **34** was confirmed by the redshifted Soret- and Q-bands in the absorption spectrum through the formation of isosbestic points. Photoexcitation of the nanohybrids with visible light resulted in the reduction of the CNT and oxidation of ZnP or H$_2$P.

34

Noncovalent functionalization of SWNT with pyrene-bearing nitrogenous bases can be used to self-assemble the electron/energy donor molecules, such as zinc tetraphenylporphyrin via axial coordination, which might result in a stable hybrid system. These hybrid systems may show maximum retainment of electronic and mechanical properties of CNT, which in turn, might provide better charge-separated states with efficient lifetimes. Self-assembled zinc napthalocyanine and SWNTs-based donor–acceptor nanohybrids have been prepared by treating the latter with pyrene-functionalized imidazole moiety (**35**). UV–vis–NIR spectra of the dispersed solutions revealed the characteristic peaks of the SWNT from 700 to 1100 nm. Using the imidazole ligand of the soluble **35**-SWNT, a zinc naphthalocyanine (**36**) was axially coordinated to yield **36**-**35**-SWNT donor–acceptor nanohybrids [73]. The structure and photophysical behavior of these nanohybrids were studied using TEM, UV–visible–NIR, and electrochemical methods. Steady-state and time-resolved emission studies have revealed efficient fluorescence quenching of the donor, **36**, in the nanohybrids. Nanosecond transient absorption spectra have revealed that the photoexcitation of **36** resulted in one-electron oxidation with simultaneous one-electron reduction of SWNT, thus generating a charge-separated state with a lifetime of 100–120 ns. Utilization of **36** as the electron donor in this system enabled observation of the donor cation radical (**36$^+$**), which acts as a direct evidence for the photoinduced electron transfer within these systems.

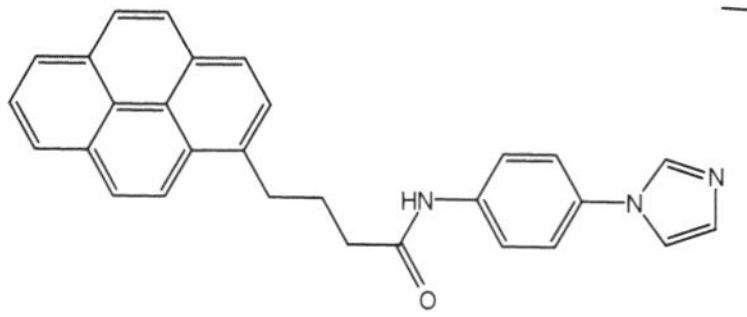

35

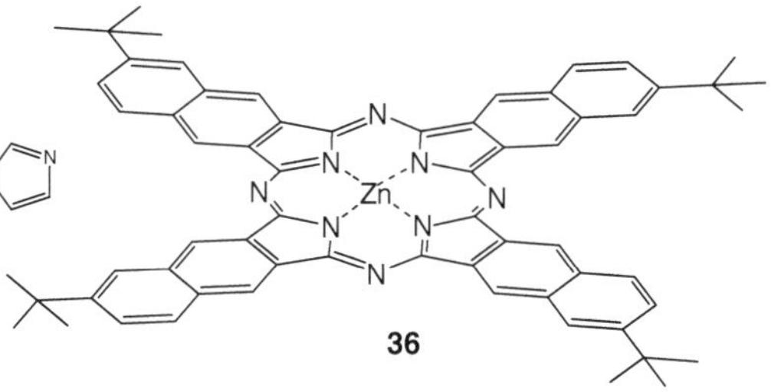

36

Among the various noncovalent functionalizations reported so far, the interaction of ammonium ion–crown ether with CNTs is considered as one of the potential methods because of the high degree of directionality with high binding energies. For example, pyrene functionalized with an alkyl ammonium cation resulted in the formation of **37**-SWNT. The alkyl ammonium cations of the soluble **37**-SWNT can be self-assembled with donor porphyrins that contain one or four 18-crown-6 moieties via alkyl ammonium ion–crown ether interactions. The resulting porphyrin-SWNT hybrids, **37**-SWNT-**38**, were stable and soluble in DMF [74]. Charge-stabilization was also observed for the nanohybrids in which the lifetime of the radical ion pairs was around 100 ns. This approach of nanohybrid formation utilizing the concept of crown ether–ammonium cation binding interaction has also been extended to the photo- and redox-active entity, fullerene [75]. The nanohybrids were isolated and characterized by TEM, UV–visible–NIR, and electrochemical methods. Further studies involving nanosecond transient absorption studies have confirmed electron transfer, and it was possible to spectrally observe the formation of fullerene anion radical. This study has demonstrated that hydrogen-bonding motif is useful to prepare SWNTs bearing donor–acceptor nanohybrids, which are useful for light-energy harvesting and photovoltaic applications.

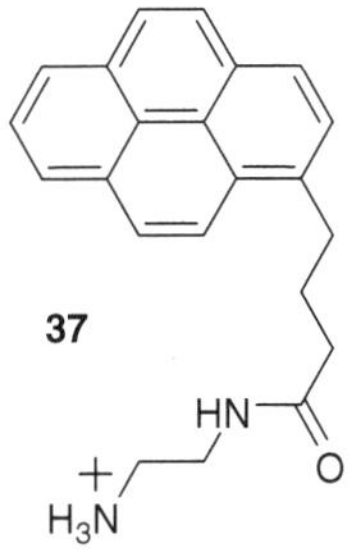

37

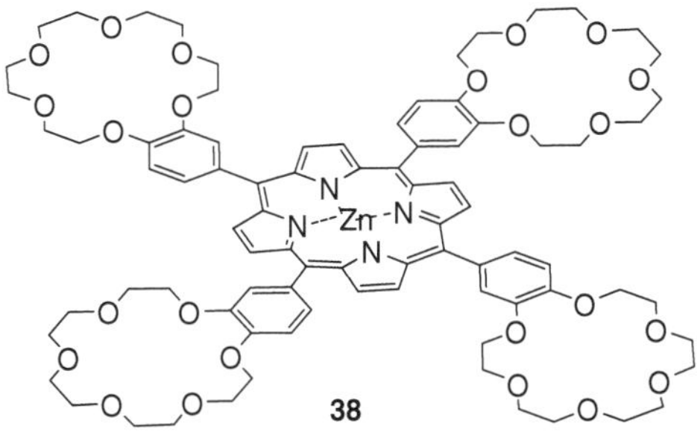

38

Recently, Sessler and coworkers have reported a sapphyrin-CNT-based ensemble in which SWNTs are strongly bound to pentapyrrolic expanded porphyrin macrocyles, such as sapphyrins (**39**) [76]. The noncovalent interaction in water and ionic liquid was characterized by UV–vis, fluorescence, and cyclic voltammetric studies. UV–vis studies of the aqueous solutions of sapphyrin and

SWNTs showed a redshift of the Soret band, indicating the presence of CT interactions within the ensemble. Photoirradiation of the sample resulted in electron transfer from the SWNTs to sapphyrin.

A dynamic coordination-directed solubilization of SWNTs in aqueous solutions was reported by Stoddart *et al.* through a combination of a free-base porphyrin (**40**) or Zn^{II} metalloporphyrin complex (**41**) carrying two pyridine ligands and a cis-protected Pd^{II} complex (**42**), which forms charged acyclic and/or cyclic adducts [77]. The complexes were characterized using ultraviolet and visible absorption spectroscopy and by high-resolution transmission electron microscopy (TEM). Solubilization occurs through a stepwise interaction of the flat π–surfaces of the metalloporphyrins with the curved π–surfaces of SWNT bundles. The pyridyl ligands of **41** are in an orientation suitable to coordinate with the charged **42** complexes under dynamic bond-forming conditions, resulting in the production of a cyclic or acyclic product on the SWNT bundles. From these interactions, the hydrophobic surfaces of the bundled SWNTs are eventually transformed into surfaces that are more hydrophilic in nature, rendering the bundles and ropes more soluble in aqueous solutions.

Chiral SWNTs have left- and right-handed helical structures, M and P, respectively. Peng *et al.* have reported the preferential extraction of optically active (*M*)- or (*P*)-SWNTs with 2,6-pyridylene-bridged chiral diporphyrins **43** [78–80]. In the circular dichroism spectra, the SWNTs extracted with **43** exhibited much larger intensity than those extracted with 1,3-phenylene-bridged chiral diporphyrins **44**, indicating an improved chiral discrimination ability of **43**. In particular, (6,5)-SWNTs display the most intensified circular dichroism signals among the SWNTs extracted with **43**. In addition, the SWNT extraction ability of **43** is shown to be considerably enhanced in comparison to **44**. These improved discrimination and extraction abilities of **43** are attributed to the formation of a more stable SWNT complex. Energy minimized structures for **43**:(6,5)-SWNT complexes have shown that (*R*)- and (*S*)-**43** form complexes preferentially with (*M*)- and (*P*)-(6,5)-SWNTs, respectively. These calculations have predicted that the **43**:(6,5)-SWNT complex is ~1.6 kcal/mol more stable than the corresponding complex of **44**, accounting for the improved abilities of **43** in the chiral discrimination and extraction of CNTs.

43: X = N
44: X = CH

19.5   INTERACTION OF IONIC LIQUIDS WITH CNTs

Aida and coworkers have shown that ionic liquids can covalently interact with CNTs to form hybrid gels. Ionic liquids have unique properties such as negligible vapor pressure, high chemical and thermal stabilities, high ionic conductivity, and broad electrochemical potential window. Moreover, ionic liquids have the ability to dissolve a wide variety of organic and inorganic substances. They have drawn increasing attention not only as a new class of electrolytes but also as recyclable alternatives to conventional organic solvents for wet processes, including chemical syntheses, catalyses, liquid/liquid extractions, and so forth. Imidazolium-ion-based ionic liquids are a new class of dispersants for CNTs [81]. Since ionic liquids are fluid at room temperature, they behave both as a solvent and dispersant, affording readily processable gel-like substances. This feature gives great advantage over the known examples, since one can "design" soft composite materials directly from the resulting gels, readily modify their physical properties, incorporate certain functionalities, and even transfer them into other fluid media or solid matrices. This method allows for large-scale processing of CNTs.

Furthermore, ion-conductive properties of the ionic liquid media can be used for electrochemical applications. Ionic liquids interact with the π-electronic surface of the SWNTs by means of cation–π and/or π–π interactions and form bucky gels. The gelation occurs in a variety of imidazolium-ion-based ionic liquids upon grinding with 0.5–1 wt% (critical gel concentration) of SWNTs. When SWNTs are ground into ionic liquid (**45**) in excess with respect to the critical gel concentration, the gel and ionic liquid phases are clearly separated from each other by centrifugation, indicating that the gel can trap a limited amount of the ionic liquid. Bucky gels thus obtained are easy to process into any shape. For example, through extrusion from a needle, one can fabricate a cable-like material that is not easily torn apart even when suspended. Owing to the negligible volatility of ionic liquids, bucky gels, in sharp contrast with ordinary organogels and hydrogels, are highly stable and can retain their physical properties even under reduced pressure. SWNTs in the bucky gels were not chemically disrupted, indicating that the gelation is induced only physically. Rheological properties have suggested that entanglement of SWNT bundles does not principally govern the elastic properties. Presumably, the system is ruled by a great number of weak physical cross-links among the SWNT bundles, for which a long-range ordering of ionic liquid molecules may be responsible.

Dong and coworkers have fabricated gold and glassy carbon electrodes coated with a bucky gel prepared from MWNTs and **46** [82]. Importantly, the redox potentials of $K_3Fe(CN)_6$, as evaluated by the modified and unmodified glassy carbon electrodes, are comparable to one another, demonstrating that the interfacial potential, possibly arising from the modification, is negligibly small. As supported by ac impedance and cyclic voltammogram measurements, electrical conduction governs the conductive properties of the bucky-gel layer deposited on the electrodes. Analogous to bucky gels with SWNTs, those with MWNTs most likely contain a well-developed nanotube network, which possibly serves as a conduction pathway from the electrodes.

Glassy carbon electrode modified with an MWNTs/**47** gel is capable of selectively detecting dopamine even in the presence of competing substrates such as ascorbic acid [83]. Baughman and coworkers have reported the first CNT nanotube-based actuators, in which two SWNT sheets (bucky paper) attached together with a double-sided Scotch tape show a bending motion in aqueous electrolytes [84]. In contrast, a bucky-gel-based actuator works in air without any support of external electrolytes. Furthermore, unlike conventional polymer actuators [85,86], this actuator operates without deposition of a metallic electrode layer. The bucky gels formed from polymerizable ionic liquids (**48**) and SWNTs, on *in situ* free-radical polymerization, give rise to highly reinforced, electroconductive polymer composites of SWNTs [81–83,87,88]. Examples of polymerizable ionic liquids include acrylate, methacrylate, and vinyl groups appended with imidazolium ion functionalities. Similar to ordinary imidazolium-ion-based ionic liquids, the polymerizable ionic liquids form gels with SWNTs. On heating in the presence of radical initiators such as azobisisobutyronitrile, these gels are transformed into black, homogeneous polymer materials (bucky plastics). A bucky plastic film, prepared from the methacrylate-appended ionic liquid monomer at

7 wt% content of SWNTs, displays an electrical conductivity as large as 1 S/cm and a 120-fold enhancement of the Young's modulus [87]. Such a large enhancement of the tensile modulus by SWNTs has rarely been reported. Furthermore, the observed conductivity is one of the highest values so far reported for SWNT-doped polymers ($\sigma < 10^{-2}$ S/cm). Because ionic liquids are ion conductive, their composites with electronically conducting CNTs are expected to be promising components for electrochemical devices, including solar cells [88]. Bucky gels may further enhance the potentials of CNTs, considering that they can solubilize a variety of synthetic and natural substances including redox-active proteins.

45: R = C_4H_9, BF_4
46: R = C_4H_9, PF_6
47 : R = C_8H_{17}, PF_6
48 : R = C_4H_8 $OCOCHCH_2$, PF_6

19.6 INTERACTION OF METAL ION COMPLEXES WITH CNTs

Bonifazi *et al.* have reported a simple method to prepare luminescent SWNTs by adsorbing an europium(III) complex (**49**) onto the surface of oxidized SWNTs (ox-SWNT) [89]. Z-contrast scanning transmission electron microscopic (STEM) images have revealed that the Eu^{III}-containing complex adheres to the SWNTs surface both as aggregates and monomolecular species, exhibiting a bright red emission. PL investigations conducted in solution and in polystyrene (PS) revealed that the presence of the SWNT units does not affect the Eu^{III}-centered luminescence, therefore leading to light emission in **49**·ox-SWNT materials, thus maintaining the fundamental structural and electronic properties of the ox SWNT.

49

Ikeda *et al.* have reported the solubilization of SWNTs in chloroform by using a Cu^{II} 2,2′-bipyridine derivative bearing two cholesteryl groups (**50**) [90]. However, **50** alone has only a weak ability to solubilize SWNTs. These results confirm the importance of the expansion of the π–conjugated planar structure. Furthermore, the addition of a solution of ascorbic acid in methanol to the $[Cu^{II}(50)_2]$-SWNT mixture immediately resulted in precipitation of the SWNTs from the chloroform solution, although the precipitate could be redissolved by bubbling air through it. $[Cu^{II}(50)_2]$ complex has a planar geometry, which makes it capable of interacting with the π surface of SWNT, while the complex $[Cu^{I}(50)_2]$ exists in a tetrahedral geometry, which decreases the interaction with SWNT. Controlling the solubilization of SWNTs using the redox properties of the solubilizing agents would be useful for the purification of metallic and semiconducting SWNTs and for the electrodeposition of SWNTs on electrode surfaces. Scherman *et al.* have reported that the CNTs were noncovalently functionalized by using amphiphilic GdL (III) chelates [91]. The functionalized nanotubes proved to be simultaneously powerful positive and negative paramagnetic contrast agents and were used as magnetic resonance imaging (MRI) contrast agents for *in vivo* imaging.

19.7   INTERACTION OF SURFACTANTS WITH CNTs

An alternative nondestructive method for the solubilization of CNTs in water is possible by the noncovalent interactions of amphiphilic molecules (surfactants). The hydrophilic part of surfactants interacts with water, and the hydrophobic part is adsorbed onto the nanotube surface, thus solubilizing CNTs by preventing them from aggregating into bundles. In the case of charged surfactants such as sodium dodecylsulfate (SDS) or tetraalkylammonium bromide, the dispersion of nanotubes is stabilized by electrostatic repulsion between the micelles [92]. In the case of charge-neutral surfactants, such as polyvinylpyrrolidone (PVP), the dispersion is mainly due to the large solvation shell created by the hydrophilic moieties of surfactants around the nanotube [93]. Surfactant molecules containing aromatic groups are capable of forming more specific and more directional $\pi-\pi$

stacking interactions with the graphitic surface of nanotubes. Interactions of SDS and structurally related sodium dodecylbenzene sulfonate (SDBS) with SWNTs have been compared to demonstrate the role of the aromatic groups. The alkyl chains of SDS and SDBS are of the same length, but the latter has a phenyl ring attached between the alkyl chain and the hydrophilic group. The presence of the phenyl ring makes SDBS surfactant more effective for solubilization of nanotubes than SDS because of the aromatic stacking between the SWNT and the phenyl rings of SDBS within the micelle [94].

Recently, Hirsch *et al.* have designed a variety of SWNT surfactants based on perylene bisimide derivatives [95]. They are capable of dispersing and exfoliating SWNTs, yielding higher populations of individualized nanotubes than SWNTs dispersed in a solution of SDBS with similar experimental parameters, although at much lower surfactant concentrations. Bolaamphiphilic perylene derivatives, as well as amphiphilic derivatives, interact with the nanotubes via $\pi-\pi$-stacking interaction (π-surfactants), as revealed by optical spectroscopy. A $\pi-\pi$-stacking interaction is reflected by a redshift of the optical transitions as well as a strong alteration of the nanotube emission pattern and a significant quenching of the nanotube- and perylene-fluorescence intensity. This strong interaction between the electron-poor aromatic system of the dispersing agent and the nanotube renders water-soluble perylene derivatives for nanotube exfoliation. Interaction of the amphiphilic perylene dye (**51**) not only facilitates the dispersion of SWNTs in water but also allows p-doping, resulting in radical ion-pairs of the CNTs [96]. Interaction of SWNTs with electron donors or acceptors generates active materials capable of producing electrical energy when irradiated. This approach is expected to be useful for the creation of electronic and optoelectronic devices in which regulated electrical transport properties become important.

51

19.8 INTERACTION OF FUNCTIONAL DYES WITH CNTs

Mao *et al.* have demonstrated that methylene blue (MB), a polynuclear aromatic electroactive dye, can be adsorbed onto SWNTs to form an electrochemically functional MB-SWNT nanostructure [97]. UV–visible, IR spectroscopic, and voltammetric studies have been used to show the CT interaction between them, in which MB acts as the electron acceptor and SWNT as the electron donor. Furthermore, the adsorption of MB onto SWNTs is found to solubilize the formed MB-SWNT nanostructure in water and facilitate layer-by-layer assembling of the prepared electrochemically functional nanostructure into a nanocomposite. Hu *et al.* have reported a noncovalent approach for the dissolution and surface modification of SWNTs by a commercially available diazo dye, Congo red [98]. On the basis of the π–stacking interaction between the sidewall and Congo red, SWNTs can be dissolved in water with a solubility as high as 3.5 mg/mL. Microscopic studies showed that most of the nanotube bundles were exfoliated into individual SWNTs or small ropes. Similarly, fluorescein and Prussian blue ($Fe_4^{3+}[Fe^{2+} (CN)_6]_3$) were used to modify the surface of CNTs through π–π interaction or electrostatic interaction [99,100]. Alternatively, small dye molecules can be filled inside the CNT. For example, Squarylium (**52**) dye was encapsulated in SWNT, and the filling rate of the molecules strongly depends on the nanotube diameter. A chloroform solution of **52** and SWNT was refluxed, filtered, and washed with chloroform several times to remove any nonencapsulated molecules. PL studies revealed efficient energy transfer from the encapsulated **52** to the SWNTs [101].

52

19.9 COMPOSITES OF LIQUID CRYSTALS AND CNTs

Liquid crystals (LCs) exhibit long-range orientational order along a preferred direction, which can be conveniently reoriented by the application of external fields. Thus, the self-organizing properties of anisotropic LCs could be easily exploited to obtain large-scale alignment of exfoliated SWNTs, as accomplished by a few research groups. The self-organizing properties of anisotropic, calamitic, and nematic LCs have been exploited to align CNTs, and these composites have been studied for various properties [102–106]. For example, Dierking *et al.* have shown that the parallel alignment of nanotubes can be obtained by dispersion in a self-organizing anisotropic fluid such as a nematic LC [103]. By utilizing the cooperative reorientation of LCs, the overall direction of the nanotube alignment was controlled both statically and dynamically by the application of external fields.

On the basis of continuum-based density functional theory, it has been proposed that CNTs should form a columnar phase or a lyotropic liquid-crystalline phase in the presence and absence of van der Waals interactions [107]. Later, the formation of lyotropic liquid-crystalline phases by acid-functionalized CNTs in water [108], DNA-stabilized CNTs in water [109], and functionalized CNTs in acids [110] were reported. The integration of SWNTs in a hexagonal lyotropic liquid-crystalline phase of the well-known surfactant Triton X 100 has been described by Weiss *et al.* [111], and the preparation of nematic nanotube gels was reported by Islam *et al* [112]. Recently, Kumar and Bisoyi have reported the functionalization of SWNTs with a discotic triphenylene moiety and their alignment in a columnar mesophase [113]. SWNT–LC hybrid systems may be important for many device applications, such as photoconductors, light-emitting diodes, photovoltaic solar cells, sensors, optical data storage, and thin-film transistors, as the dispersed SWNTs can be aligned in the desired direction using well-established LC alignment technologies.

19.10 CONCLUSIONS

Functionalized CNTs are progressively playing important roles in the research, development, and application of nanotube-based advanced materials. In this chapter, we have described briefly the noncovalent interaction of a variety of small molecules with CNTs. The solubility of SWNTs modified by noncovalent functionalization offers excellent opportunities not only in the understanding of the structural characteristics of SWNTs in solution but also in their use in various nanomaterials and devices. The methodologies for noncovalent functionalization of SWNTs, discussed in this chapter, promise a new class of nanocomposites for various applications. Noncovalent interaction extends the scope of the application of CNTs in biology as well as in materials science. Even though a large amount of information is available, there is still room for further development, which may lead to new materials with intriguing properties.

REFERENCES

1. Ajayan, P. M. (1999). Nanotubes from carbon. *Chem. Rev.*, *99*, 1787–1799.
2. Dai, L., Mau, A. W. H. (2001). Controlled synthesis and modification of carbon nanotubes and C_{60}: carbon nanostructures for advanced polymeric composite materials. *Adv. Mater.*, *13*, 899–913.
3. Hirsch, A. (2002). Functionalization of single-walled carbon nanotubes. *Angew. Chem. Int. Ed.*, *41*, 1853–1859.
4. Niyogi, S., Hamon, M. A., Hu, H., Zhao, B., Bhowmik, P., Sen, R., Itkis, M. E., Haddon, R. C. (2002). Chemistry of single-walled carbon nanotubes. *Acc. Chem. Res.*, *35*, 1105–1113.
5. Sun, Y.-P., Fu, K., Lin, Y., Huang, W. (2002). Functionalized carbon nanotubes: properties and applications. *Acc. Chem. Res.*, *35*, 1096–1104.
6. Dyke, C. A., Tour, J. M., (2004). Overcoming the insolubility of carbon nanotubes through high degrees of sidewall functionalization. *Chem.—Eur. J.*, *10*, 812–817.
7. Guldi, D. M., Rahman, G. M. A., Zerbetto, F., Prato, M. (2005). Carbon nanotubes in electron donor-acceptor nanocomposites. *Acc. Chem. Res.*, *38*, 871–878.
8. Xie, X.-L., Mai, Y.-W., Zhou, X.-P. (2005). Dispersion and alignment of carbon nanotubes in polymer matrix: a review. *Mater. Sci. Eng. R*, *49*, 89–112.
9. Tasis, D., Tagmatarchis, N., Bianco, A., Prato, M. (2006). Chemistry of carbon nanotubes. *Chem. Rev.*, *106*, 1105–1136.

10. Park, T.-J., Banerjee, S., Hemraj-Benny, T., Wong, S. S. (2006). Purification strategies and purity visualization techniques for single-walled carbon nanotubes. *J. Mater. Chem.*, *16*, 141–145.

11. Iijima, S. (1991). Helical microtubules of graphitic carbon. *Nature*, *354*, 56–58.

12. Iijima, S., Ichihashi, T. (1993). Single-shell carbon nanotubes of 1-nm diameter. *Nature*, *363*, 603–605.

13. Zheng, M., Jagota, A., Strano, M. S., Santos, A. P., Barone, P., Chou, S. G., Diner, B. A., Dresselhaus, M. S., Mclean, R. S., Onoa, G. B., Samsonidze, G. G., Semke, E. D., Usrey, M., Walls, D. J. (2003). Structure-based carbon nanotube sorting by sequence-dependent DNA assembly. *Science*, *302*, 1545–1548.

14. Li, S., He, P., Dong, J., Guo, Z., Dai, L. (2005). DNA-directed self-assembling of carbon nanotubes. *J. Am. Chem. Soc.*, *127*, 14–15.

15. Zorbas, V., Smith, A. L., Xie, H., Ortiz-Acevedo, A., Dalton, A. B., Dieckmann, G. R., Draper, R. K., Baughman, R. H., Musselman, I. H. (2005). Importance of aromatic content for peptide/single-walled carbon nanotube interactions. *J. Am. Chem. Soc.*, *127*, 12323–12328.

16. Hasobe, T., Fukuzumi, S., Kamat, P. V. (2005). Ordered assembly of protonated porphyrin driven by single-wall carbon nanotubes. J- and H-aggregates to nanorods. *J. Am. Chem. Soc.*, *127*, 11884–11885.

17. Karajanagi, S. S., Yang, H., Asuri, P., Sellitto, E., Dordick, J. S., Kane, R. S. (2006). Protein-assisted solubilization of single-walled carbon nanotubes. *Langmuir*, *22*, 1392–1395.

18. Nakanishi, T., Michinobu, T., Yoshida, K., Shirahata, N., Ariga, K., Moehwald, H., Kurth, D. G. (2008). Nanocarbon superhydrophobic surfaces created from fullerene-based hierarchical supramolecular assemblies. *Adv. Mater.*, *20*, 443–446.

19. Fischer, J. E. (2002). Chemical doping of single-wall carbon nanotubes. *Acc. Chem. Res.*, *35*, 1079–1086.

20. Charlier, J.-C. (2002). Defects in carbon nanotubes. *Acc. Chem. Res.*, *35*, 1063–1069.

21. in het Panhuis, M., Sainz, R., Innis, P. C., Kane-Maguire, L. A. P., Benito, A. M., Martinez, M. T., Moulton, S. E., Wallace, G. G., Maser, W. K. (2005). Optically active polymer carbon nanotube composite. *J. Phys. Chem. B*, *109*, 22725–22729.

22. White, C. T., Mintmire, J. W. (2005). Fundamental properties of single-wall carbon nanotubes. *J. Phys. Chem. B*, *109*, 52–65.

23. Yang, D.-Q., Rochette, J.-F., Sacher, E. (2005). Spectroscopic evidence for π-π interaction between poly(diallyl dimethylammonium) chloride and multiwalled carbon nanotubes. *J. Phys. Chem. B*, *109*, 4481–4484.

24. Tan, Y., Resasco, D. E. (2005). Dispersion of single-walled carbon nanotubes of narrow diameter distribution. *J. Phys. Chem. B*, *109*, 14454–14460.

25. Saini, R. K., Chiang, I. W., Peng, H., Smalley, R. E., Billups, W. E., Hauge, R. H., Margrave, J. L. (2003). Covalent sidewall functionalization of single wall carbon nanotubes. *J. Am. Chem. Soc.*, *125*, 3617–3621.

26. Kovtyukhova, N. I., Mallouk, T. E., Pan, L., Dickey, E. C., (2003). Individual single-walled nanotubes and hydrogels made by oxidative exfoliation of carbon nanotube ropes. *J. Am. Chem. Soc.*, *125*, 9761–9769.

27. Asai, M., Sugiyasu, K., Fujita, N., Shinkai, S. (2004). Facile and stable dispersion of carbon nanotubes into a hydrogel composed of a low molecular-weight gelator bearing a tautomeric dye group. *Chem. Lett.*, *33*, 120–121.

28. Nabeta, M., Sano, M. (2005). Nanotube foam prepared by gelatin gel as a template. *Langmuir*, *21*, 1706–1708.

29. Sabba, Y., Thomas, E. L. (2004). High-concentration dispersion of single-wall carbon nanotubes. *Macromolecules*, *37*, 4815–4820.

30. Yoshida, M., Koumura, N., Misawa, Y., Tamaoki, N., Matsumoto, H., Kawanami, H., Kazaoui, S., Minami, N. (2007). Oligomeric electrolyte as a multifunctional gelator. *J. Am. Chem. Soc.*, *129*, 11039–11041.

31. Bhattacharyya, S., Guillot, S., Dabboue, H., Tranchant, J.-F., Salvetat, J.-P. (2008). Carbon nanotubes as structural nanofibers for hyaluronic acid hydrogel scaffolds. *Biomacromolecules*, *9*, 505–509.

32. Sumanasekera, G. U., Pradhan, B. K., Romero, H. E., Adu, K. W., Eklund, P. C. (2002). Giant thermopower effects from molecular physisorption on carbon nanotubes. *Phys. Rev. Lett.*, *89*, 166801–166804.

33. Nakashima, N., Tomonari, Y., Murakami, H. (2002). Water-soluble single-walled carbon nanotubes via noncovalent sidewall-functionalization with a pyrene-carrying ammonium ion. *Chem. Lett.*, *31*, 638–639.

34. Nakashima, N., Kobae, H., Sagara, T., Murakami, H. (2002). Formation of single-walled carbon nanotube thin films on electrodes monitored by an electrochemical quartz crystal microbalance. *ChemPhysChem*, *3*, 456–458.

35. Murakami, H., Nomura, T., Nakashima, N. (2003). Noncovalent porphyrin-functionalized single-walled carbon nanotubes in solution and the formation of porphyrin–nanotube nanocomposites. *Chem. Phys. Lett.*, *378*, 481–485.

36. Tomonari, Y., Murakami, H., Nakashima, N., (2006). Solubilization of single-walled carbon nanotubes by using polycyclic aromatic ammonium amphiphiles in water—strategy for the design of high-performance solubilizers. *Chem.—Eur. J.*, *12*, 4027–4034.

37. Paloniemi, H., Ääritalo, T., Laiho, T., Like, H., Kocharova, N., Haapakka, K., Terzi, F., Seeber, R., Lukkari, J. (2005). Water-soluble full-length single-wall carbon nanotube polyelectrolytes: preparation and characterization. *J. Phys. Chem. B*, *109*, 8634–8642.

38. Zhang, J., Lee, J.-K., Wu, Y., Murray, R. W. (2003). Photoluminescence and electronic interaction of anthracene derivatives adsorbed on sidewalls of single-walled carbon nanotubes. *Nano Lett.*, *3*, 403–407.

39. Hedderman, T. G., Keogh, S. M., Chambers, G., Byrne, H. J. (2004). Solubilization of SWNTs with organic dye molecules. *J. Phys. Chem. B*, *108*, 18860–18865.

40. Tromp, R. M., Afzali, A., Freitag, M., Mitzi, D. B., Chen, Zh. (2008). Novel strategy for diameter-selective separation and functionalization of single-wall carbon nanotubes. *Nano Lett.*, *8*, 469–472.

41. Rao, C. N. R., Govindaraj, A. (2005). *Nanotubes and Nanowires*. Royal Society of Chemistry, Cambridge.

42. Voggu, R., Rao, K. V., George, S. J., Rao, C. N. R. (2010). A simple method of separating metallic and semiconducting single-walled carbon nanotubes based on molecular charge transfer. *J. Am. Chem. Soc.*, *132*, 5560.

43. Feng, W., Fujii, A., Ozaki, M., Yoshino, K. (2005). Perylene derivative sensitized multi-walled carbon nanotube thin film. *Carbon*, *43*, 2501–2507.

44. Chen, R. J., Zhang, Y., Wang, D., Dai, H., (2001). Noncovalent sidewall functionalization of single-walled carbon nanotubes for protein immobilization. *J. Am. Chem. Soc.*, *123*, 3838–3839.

45. Gómez, F. J., Chen, R. J., Wang, D., Waymouth, R. M., Dai, H. (2003). Ring opening metathesis polymerization on non-covalently functionalized single-walled carbon nanotubes. *Chem. Commun.*, 190–191.

46. Ogoshi, T., Takashima, Y., Yamaguchi, H., Harada, A. (2007). Chemically-responsive sol–gel transition of supramolecular single-walled carbon nanotubes (SWNTs) hydrogel made by hybrids of SWNTs and cyclodextrins. *J. Am. Chem. Soc.*, *129*, 4878–4879.

47. Chen, J., Dyer, M. J., Yu, M. F. (2001). Cyclodextrin-mediated soft cutting of single-walled carbon nanotubes. *J. Am. Chem. Soc.*, *123*, 6201–6202.

48. Chambers, G., Carroll, C., Farrell, G. F., Dalton, A. B., McNamara, M., Panhuis, M. I. H., Byrne, H. J. (2003). Characterization of the interaction of gamma cyclodextrin with single-walled carbon nanotubes. *Nano Lett.*, *3*, 843–846.

49. Liu, K., Fu, H., Xie, Y., Zhang, L., Pan, K., Zhou, W. (2008). Assembly of β-cyclodextrins acting as molecular bricks onto multiwall carbon nanotubes. *J. Phys. Chem. C*, *112*, 951–957.

50. Klare, J. E., Murray, I. P., Goldberger, J., Stupp, S. I. (2009). Assembling p-type molecules on single wall carbon nanotubes for photovoltaic devices. *Chem. Commun.*, 3705–3707.

51. Tang, B. Z., Xu, H. (1999). Preparation, alignment, and optical properties of soluble poly(phenylacetylene)-wrapped carbon nanotubes. *Macromolecules*, *32*, 2569–2576.

52. Ago, H., Petritsch, K., Shaffer, M. S. P., Windle, A. H., Friend, R. H. (1999). Composites of carbon nanotubes and conjugated polymers for photovoltaic devices. *Adv. Mater.*, *11*, 1281–1285.

53. Curran, S. A., Ajayan, P. M., Blau, W. J., Carroll, D. L., Coleman, J. N., Dalton, A. B., Davey, A. P., Drury, A., McCarthy, B., Maier, S., Strevens, A., (1998). A composite from poly(*m*-phenylenevinylene-*co*-2,5-dioctoxy-*p*-phenylenevinylene) and carbon nanotubes: a novel material for molecular optoelectronics. *Adv. Mater.*, *10*, 1091–1093.

54. Star, A., Lu, Y., Bradley, K., Gruner, G. (2004). Nanotube optoelectronic memory devices. *Nano Lett.*, *4*, 1587–1591.

55. Star, A., Stoddart, J. F., Steuerman, D., Diehl, M., Boukai, A., Wong, E. W., Yang, X., Chung, S. W., Choi, H., Heath, J. R. (2001). Preparation and properties of polymer-wrapped single-walled carbon nanotubes. *Angew. Chem. Int. Ed.*, *40*, 1721–1725.

56. Steuerman, D. W., Star, A., Narizzano, R., Choi, H., Ries, R. S., Nicolini, C., Stoddart, J. F. J., Heath, R. (2002). Interactions between conjugated polymers and single-walled carbon nanotubes. *J. Phys. Chem. B*, *106*, 3124–3130.

57. Star, A., Stoddart, J. F. (2002). Dispersion and solubilization of single-walled carbon nanotubes with a hyperbranched polymer. *Macromolecules*, *35*, 7516–7520.

58. Star, A., Liu, Y., Grant, K., Ridvan, L., Stoddart, J. F., Steuerman, D. W., Diehl, M. R., Boukai, A., Heath, J. R. (2003). Noncovalent side-wall functionalization of single-walled carbon nanotubes. *Macromolecules*, *36*, 553–560.

59. Chen, J., Liu, H., Weimer, W. A., Halls, M. D., Waldeck, D. H., Walker, G. C. (2002). Noncovalent engineering of carbon nanotube surfaces by rigid, functional conjugated polymers. *J. Am. Chem. Soc.*, *124*, 9034–9035.

60. Ramasubramaniam, R., Chen, J., Liu, H. (2003). Homogeneous carbon nanotube/polymer composites for electrical applications. *Appl. Phys. Lett.*, *83*, 2928–2930.

61. Srinivasan, S., Babu, S. S., Praveen, V. K., Ajayaghosh, A. (2008). Carbon nanotube triggered self-assembly of oligo(*p*-phenylenevinylene)s to stable hybrid π–gels. *Angew. Chem. Int. Ed.*, *47*, 5746–5749.

62. Ajayaghosh, A., Praveen, V. K. (2007). π–Organogels of self-assembled *p*-phenylenevinylenes: soft materials with distinct size, shape, and functions. *Acc. Chem. Res.*, *40*, 644–656.

63. George, S. J., Ajayaghosh, A. (2005). Self-assembled nanotapes of oligo(*p*-phenylene vinylene)s: sol–gel-controlled optical properties in fluorescent π–electronic gels. *Chem.—Eur. J.*, *11*, 3217–3227.

64. Ajayaghosh, A., Praveen, V. K., Srinivasan, S., Varghese, R. (2007). A composite from poly(*m*-phenylenevinylene-*co*-2,5-dioctoxy-*p*-phenylenevinylene) and carbon nanotubes: a novel material for molecular optoelectronics. *Adv. Mater.*, *19*, 411–415.

65. Srinivasan, S., Praveen, V. K., Philip, R., Ajayaghosh, A. (2008). Bioinspired superhydrophobic coatings of carbon nanotubes and linear π systems based on the "Bottom-up" self-assembly approach. *Angew. Chem. Int. Ed.*, *47*, 5750–5754.

66. Li, H., Zhou, B., Lin, Y., Gu, L., Wang, W., Fernando, K. A. S., Kumar, S., Allard, L. F., Sun, Y.-P. (2004). Selective interactions of porphyrins with semiconducting single-walled carbon nanotubes. *J. Am. Chem. Soc.*, *126*, 1014–1015.

67. Chen, J., Collier, C. P. (2005). Noncovalent functionalization of single-walled carbon nanotubes with water-soluble porphyrins. *J. Phys. Chem. B*, *109*, 7605–7609.

68. Rahman, G. M. A., Guldi, D. M., Campidelli, S., Prato, M. (2006). Electronically interacting single wall carbon nanotube–porphyrin nanohybrids. *J. Mater. Chem.*, *10*, 62–65.

69. Guldi, D. M., Rahman, G. M. A., Jux, N., Balbinot, D., Tagmatarchis, N., Prato, M. (2005). Multiwalled carbon nanotubes in donor–acceptor nanohybrids—towards long-lived electron transfer products. *Chem. Commun.*, 2038–2040.

70. Guldi, D. M., Prato, M., (2004). Electrostatic interactions by design. Versatile methodology towards multifunctional assemblies/nanostructured electrodes. *Chem. Commun.*, 2517–2525.

71. Guldi, D. M., Rahman, G. M. A., Jux, N., Balbinot, D., Hartnagel, U., Tagmatarchis, N., Prato, M. (2005). Functional single-wall carbon nanotube nanohybrids—associating SWNTs with water-soluble enzyme model systems. *J. Am. Chem. Soc.*, *127*, 9830–9838.

72. Guldi, D. M., Rahman, G. M. A., Sgobba, V., Kotov, N. A., Bonifazi, D., Prato, M. (2006). CNT-CdTe versatile donor-acceptor nanohybrids. *J. Am. Chem. Soc.*, *128*, 2315–2323.

73. Chitta, R., Sandanayaka, A. S. D., Schumacher, A. L., D'Souza, L., Araki, Y., Ito, O., D'Souza, F. (2007). Donor-acceptor nanohybrids of zinc naphthalocyanine or zinc porphyrin noncovalently

linked to single-wall carbon nanotubes for photoinduced electron transfer. *J. Phys. Chem. C*, *111*, 6947–6955.

74. D'Souza, F., Chitta, R., Sandanayaka, A. S. D., Subbaiyan, N. K., D'Souza, L., Araki, Y., Ito, O. (2007). Self-assembled single-walled carbon nanotube: zinc–porphyrin hybrids through ammonium ion–crown ether interaction: construction and electron transfer. *Chem.—Eur. J.*, *13*, 8277–8284.

75. D'Souza, F., Chitta, R., Sandanayaka, A. S. D., Subbaiyan, N. K., D'Souza, L., Araki, Y., Ito, O. (2007). Supramolecular carbon nanotube-fullerene donor-acceptor hybrids for photoinduced electron transfer. *J. Am. Chem. Soc.*, *129*, 15865–15871.

76. Boul, P. J., Cho, D.-G., Rahman, G. M. A., Marquez, M., Ou, Z., Kadish, K. M., Guldi, D. M., Sessler, J. L. (2007). Sapphyrin-nanotube assemblies. *J. Am. Chem. Soc.*, *129*, 5683–5687.

77. Chichak, K. S., Star, A., Alto, M. V. P., Stoddart, J. F. (2005). Single-walled carbon nanotubes under the influence of dynamic coordination and supramolecular chemistry. *Small*, *1*, 452–461.

78. Peng, X., Komatsu, N., Kimura, T., Osuka, A. (2007). Improved optical enrichment of SWNTs through extraction with chiral nanotweezers of 2,6-pyridylene-bridged diporphyrins. *J. Am. Chem. Soc.*, *129*, 15947–15953.

79. Dukovic, G., Balaz, M., Doak, P., Berova, N. D., Zheng, M., Mclean, R. S., Brus, L. E. (2006). Racemic single-walled carbon nanotubes exhibit circular dichroism when wrapped with DNA. *J. Am. Chem. Soc.*, *128*, 9004–9005.

80. Peng, X., Komatsu, N., Bhattacharya, S., Shimawaki, T., Aonuma, S., Kimura, T., Osuka, A. (2007). Optically active single-walled carbon nanotubes. *Nat. Nanotechnol.*, *2*, 361–365.

81. Fukushima, T., Kosaka, A., Ishimura, Y., Yamamoto, T., Takigawa, T., Ishii, N., Aida, T. (2003). Molecular ordering of organic molten salts triggered by single-walled carbon nanotubes. *Science*, *300*, 2072–2074.

82. Zhao, F., Wu, X., Wang, M., Liu, Y., Gao, L., Dong, S. (2004). Electrochemical and bioelectrochemistry properties of room-temperature ionic liquids and carbon composite materials. *Anal. Chem.*, *76*, 4960–4967.

83. Zhao, Y., Gao, Y., Zhan, D., Liu, H., Zhao, Q., Kou, Y., Shao, Y., Li, M., Zhuang, Q., Zhu, Z. (2005). Selective detection of dopamine in the presence of ascorbic acid and uric acid by a carbon nanotubes-ionic liquid gel modified electrode. *Talanta*, *66*, 51–57.

84. Baughman, R. H., Cui, C., Zakhidov, A. A., Iqbal, Z., Barisci, J. N., Spinks, G. M., Wallace, G. G., Mazzoldi, A., De Rossi, D., Rinzler, A. G., Jaschinski, O., Roth, S., Kertesz, M. (1999). Carbon nanotube actuators. *Science*, *284*, 1340–1344.

85. Electroactive Polymer (EAP) (2004). Actuators as artificial muscles, reality, potential, and challenges, 2nd ed. (Ed.: Bar-Cohen, Y.). SPIE Press, Washington, DC.

86. Smela, E. (2003). Conjugated polymer actuators for biomedical applications. *Adv. Mater.*, *15*, 481–494.

87. Fukushima, T., Kosaka, A., Yamamoto, Y., Aimiya, T., Notazawa, S., Takigawa, T., Inabe, T., Aida, T. (2006). Dramatic effect of dispersed carbon nanotubes on the mechanical and electroconductive properties of polymers derived from ionic liquids. *Small*, *2*, 554–560.

88. Fukushima, T., Aida, T. (2007). Ionic liquids for soft functional materials with carbon nanotubes. *Chem. Eur. J.*, *13*, 5048–5058.

89. Accorsi, G., Armaroli, N., Parisini, A., Meneghetti, M., Marega, R., Prato, M., Bonifazi, D. (2007). Wet adsorption of a luminescent Eu[III] complex on carbon nanotubes sidewalls. *Adv. Funct. Mater.*, *17*, 2975–2982.

90. Nobusawa, K., Ikeda, A., Kikuchi, J-I., Kawano, S-I., Fujita, N., Shinkai, S. (2008). Reversible solubilization and precipitation of carbon nanotubes through oxidation–reduction reactions of a solubilizing agent. *Angew. Chem. Int. Ed.*, *47*, 4577–4580.

91. Richard, C., Doan, B.-T., Beloeil, J.-C., Bessodes, M., Toth, E., Scherman, D. (2008). Noncovalent functionalization of carbon nanotubes with amphiphilic Gd[3+] chelates: toward powerful T_1 and T_2 MRI contrast agents. *Nano Lett.*, *8*, 232–236.

92. O'Connell, M. J., Bachilo, S. M., Huffman, C. B., Moore, V. C., Strano, M. S., Haroz, E. H., Rialon, K. L., Boul, P. J., Noon, W. H., Kittrell, C., Ma, J. P., Hauge, R. H., Weisman, R. B., Smalley, R. E. (2002). Band gap fluorescence from individual single-walled carbon nanotubes. *Science*, *297*, 593–596.

93. Moore, V. C., Strano, M. S., Haroz, E. H., Hauge, R. H., Smalley, R. E., Schmidt, J., Talmon, Y. (2003). Noncovalent functionalization of carbon nanotubes with amphiphilic Gd^{3+} chelates: toward powerful T_1 and T_2 MRI contrast agents. *Nano Lett.*, *3*, 1379–1382.

94. Islam, M. F., Rojas, E., Bergey, D. M., Johnson, A. T., Yodh, A. G. (2003). High weight fraction surfactant solubilization of single-wall carbon nanotubes in water. *Nano Lett.*, *3*, 269–273.

95. Backes, C., Schmidt, C. D., Rosenlehner, K., Hauke, F., Coleman, J. N., Hirsch, A. (2010). Nanotube surfactant design: the versatility of water-soluble perylene bisimides. *Adv. Mater.*, *22*, 788–802.

96. Ehli, C., Oelsner, C., Guldi, D. M., Alonso, A. M., Prato, M., Schmidt, C., Backes, C., Hauke, F., Hirsch, A. (2009). Adsorption of methylene blue dye onto carbon nanotubes: a route to an electrochemically functional nanostructure and its layer-by-layer assembled nanocomposite. *Nat. Chem.*, *1*, 243–249.

97. Yan, Y., Zhang, M., Gong, K., Su, L., Guo, Z., Mao, L. (2005). Adsorption of methylene blue dye onto carbon nanotubes: a route to an electrochemically functional nanostructure and its layer-by-layer assembled nanocomposite. *Chem. Mater.*, *17*, 3457–3463.

98. Hu, C., Chen, Z., Shen, A., Shen, X., Li, J., Hu, S. (2006). Water-soluble single-walled carbon nanotubes via noncovalent functionalization by a rigid, planar and conjugated diazo dye. *Carbon*, *44*, 428–434.

99. Ratchford, N. N., Bangsaruntip, S., Sun, X., Welsher, K., Dai, H. (2007). Noncovalent functionalization of carbon nanotubes by fluorescein-polyethylene glycol: supramolecular conjugates with pH-dependent absorbance and fluorescence. *J. Am. Chem. Soc.*, *129*, 2448–2449.

100. Han, S., Chen, Y., Pang, R., Wan, P., Fan, M. (2007). Fabrication of prussian blue/multiwalled carbon nanotubes/glass carbon electrode through sequential deposition. *Ind. Eng. Chem. Res.*, *46*, 6847–6851.

101. Yanagi, K., Iakoubovskii, K., Matsui, H., Matsuzaki, H., Okamoto, H., Miyata, Y., Maniwa, Y., Kazaoui, S., Minami, N., Kataura, H. (2007). Photosensitive function of encapsulated dye in carbon nanotubes. *J. Am. Chem. Soc.*, *129*, 4992–4997.

102. Lynch, M. D., Patrick, D. L. (2002). Organizing carbon nanotubes with liquid crystals. *Nano Lett.*, *2*, 1197–1201.

103. Dierking, I., Scalia, G., Morales, P. (2005). Liquid crystal–carbon nanotube dispersions. *J. Appl. Phys.*, *97*, 044309–5.

104. Dierking, I., Scalia, G., Morales, P., LeClere, D. (2004). Aligning and reorienting carbon nanotubes with nematic liquid crystals. *Adv. Mater.*, *16*, 865–869.

105. Baik, I.-S., Jeon, S. Y., Lee, S. H., Park, K. A., Jeong, S. H., An, K. H., Lee, Y. H. (2005). Electrical-field effect on carbon nanotubes in a twisted nematic liquid crystal cell. *Appl. Phys. Lett.*, *87*, 263110–3.

106. Mizoshita, N., Suzuki, Y., Hanabusa, K., Kato, T. (2005). Bistable nematic liquid crystals with self-assembled fibers. *Adv. Mater.*, *17*, 692–696.

107. Somoza, A. M., Sagui, C., Roland, C. (2001). Liquid-crystal phases of capped carbon nanotubes. *Phys. Rev. B*, *63*, 081403–4.

108. Song, W., Kinloch, I. A., Windle, A. H. (2003). Nematic liquid crystallinity of multiwall carbon nanotubes. *Science*, *302*, 1363–1363.

109. Badaire, S., Zakri, C., Maugey, M., Derre, A., Barisci, J. N., Wallace, G., Poulin, P. (2005). Liquid crystals of DNA-stabilized carbon nanotubes. *Adv. Mater.*, *17*, 1673–1676.

110. Rai, P. K., Pinnick, R. A., Parra-Vasquez, A. N. G., Davis, V. A., Schmidt, H. K., Hauge, R. H., Smalley, R. E., Pasquali, M. (2006). Isotropic-nematic phase transition of single-walled carbon nanotubes in strong acids. *J. Am. Chem. Soc.*, *128*, 591–595.

111. Weiss, V., Thiruvengadathan, R., Regev, O. (2006). Preparation and characterization of a carbon nanotube-lyotropic liquid crystal composite. *Langmuir*, *22*, 854–856.

112. Islam, M. F., Alsayed, A. M., Dogic, Z., Zhang, J., Lubensky, T. C., Yodh, A. G. (2004). Nematic nanotube gels. *Phys. Rev. Lett.*, *92*, 088303–4.

113. Kumar, S., Bisoyi, H. K. (2007). Aligned carbon nanotubes in the supramolecular order of discotic liquid crystals. *Angew. Chem. Int. Ed.*, *46*, 1501–1503.

CHAPTER 20

THE TUNING OF CNT DEVICES USING SELF-ASSEMBLING ORGANIC AND BIOLOGICAL MOLECULES

Jeong-O. Lee[1] and Ju-Jin Kim[2]

[1]NanoBio Fusion Research Center, Korea Research Institute of Chemical Technology, Daejeon, Korea
[2]Department of Physics, Chonbuk National University, Jeonju, Korea

20.1 INTRODUCTION

Self-assembled monolayers (SAMs) could potentially have a decisive influence on the development of organic and inorganic electronic devices [1–5]. SAMs are widely used as surface modifiers to accommodate the growth or adhesion of substances and can enhance the performance of organic electronic devices. The importance of SAMs to the production of nanoscale devices may prove to be even more dramatic; SAMs not only can potentially enhance device performance but also can alter the device carrier type [6 9]. In single walled carbon nanotube (SWNT) devices that are known to operate as Schottky transistors [10], SAMs may enhance the hole mobility or produce diodes and n-type CNT (carbon nanotube) transistors [6–9,11]. Furthermore, SAMs composed of organics or biomolecules can afford special functionalities to CNT devices. For example, CNT devices that include self-assembled organic moieties or are wrapped with DNA can operate as sensitive gas sensors [12,13]. In this chapter, we describe a variety of techniques for tuning CNT devices using SAMs composed of organic or biological molecules. In the first section, we focus on tuning the electrical properties of CNT devices using SAMs, and the second section describes applications of SWNT devices functionalized by SAMs.

Supramolecular Soft Matter: Applications in Materials and Organic Electronics, First Edition.
Edited by Takashi Nakanishi.
© 2011 John Wiley & Sons, Inc. Published 2011 by John Wiley & Sons, Inc.

20.2   TUNING SWCNT DEVICES USING SAMs COMPOSED OF ORGANIC AND BIOLOGICAL MOLECULES

SAMs have been used in organic electronic devices to tune device performance [2,5]. For example, de Boer *et al.* [14] and Khodabakhsh *et al.* [15] used SAMs to control the Schottky barrier heights in organic field-effect transistors (FETs). Recently, Hong *et al.* [5] demonstrated tuning of the hole injection barriers in pentacene organic film transistors by assembling a variety of SAMs on pristine Ag electrodes. By changing the type of molecule used for SAM production, they changed the work function of the SAM-treated Ag electrode from 4.14 to 5.35 eV, a range of more than 1 eV. In the case of small band gap semiconductors, such as CNTs with diameters >2 nm, the semiconductor carrier type may be converted from electron to hole simply by changing the Schottky barrier height without impurity doping [6,8].

The Schottky barrier in metal/semiconductor junctions is determined by the energy difference between the metal work function and the electron affinity of the semiconducting channel [16]. Hence, metal electrodes with low work functions are typically used for electron injection into the conduction band, whereas metal electrodes with high work functions are adequate for hole injection into the valence band [17–19]. In practice, however, it is not easy to find appropriate electrode metals with optimal work functions for use in particular conducting channels [19,20]. If the work function of the metal electrode could be varied systematically, it would be possible to precisely tailor the electrical properties of a device. Because SAMs attach to metal surfaces, such as Au and Ag, by a self-ordering mechanism that orients the inherent molecular dipoles of the SAM molecules relative to the metal surface, this molecular ordering alters the work function of metal electrodes, thereby modifying energy level alignment near the electrode/semiconductor junction [2,9]. SAM-modified electrodes have been widely used to tune organic electronic devices, maximizing performance.

A diverse array of nanoscale electronic devices that employ CNTs has been suggested. The FET, in which semiconducting CNTs are employed as the electron channel, is one of the most widely investigated nanodevice structures [21]. Substitutional impurity doping techniques used in the conventional bulk semiconductor industry encounter many challenges when applied to CNT devices because of the small volume of the channel, the nonuniformity of doping, and the perfect covalent bonding structure of CNTs. Precise control over the electrical properties of CNT devices is essential for future electronic device development [22]. As an alternative to impurity doping, contact doping may be used to modify the Schottky barrier height between CNTs and metal electrodes [22]. Charge transport in most carbon nanotube field-effect transistors (CNT-FETs) is strongly influenced by the Schottky barrier between the nanotubes and the contact electrode [10]. Under normal conditions, the fabricated CNT-FETs have shown only p-type operation. The characteristic p-type behavior of CNT-FETs has long been a source of debate [23–26]. Previously, hole-doping by environmental oxygen molecules was believed to be the principal cause of the p-type behavior of CNT-FETs [27,28]. However, it is generally agreed that p-type behavior arises from the low-lying Schottky barrier for hole transport at the high work function metal–CNT interface [29,30]. The position of the Fermi level at the interface between the metal electrode and the CNT channel should be the key parameter governing the transport characteristics. Electron injection into the conduction band from the metal electrode, that is, fabrication of an n-type CNT-FET, has been challenging, and much research has been conducted toward this goal.

Researchers have tried to fabricate CNT-FET electrodes using chemically active low work function metals, such as K and Ca [17,31]. Other attempts have used nitrogen doping to increase the electron carrier efficiency. However, chemical vapor deposition (CVD) experiments showed that nitrogen dopants tended to have a pyridine-like structure that could never be an electron donor to CNTs. Fabrication of electrodes using low work function metals is also not practical because these metals are very reactive and are not stable under ambient conditions. In this regard, precise control over the electrical properties of CNT-FETs remains a challenging and technically important research goal.

In recent years, our group has investigated the transport properties of CNT-FETs. Our main concern in this field has been the control of the Schottky barrier at the metal–nanotube contact interface, with the goal of achieving low-resistance contact or adjusting the conduction pattern. Our studies have been directed toward developing practical methods for tailoring conduction properties without using impurity doping techniques. We found that adsorption of SAMs at the metal–nanotube contact region significantly affected the transport properties of CNT-FETs. We have also attempted to tune the work functions of metal electrodes via postprocessing adsorption of various biomolecules during the fabrication of CNT-FETs. In this chapter, we review various studies regarding the engineering of electronic properties of CNT-FETs using SAMs, and applications of SAM-modified CNT-FETs.

20.2.1   Tuning of the Contact Barrier in Single-Walled CNT-FETs with SAMs

This section focuses on tuning of the Schottky barrier in SWNT devices using chemically tailored electrodes. Several research groups have attempted to modify the electrical properties of SWNT-based devices, either through conventional approaches, such as selective doping in portions of the nanotubes [32,33], or by employing local split gates to modify the carrier concentration of the nanotubes [26,34]. However, exploitation of some well-known characteristics of the Schottky barrier in SWNT-FETs may enable tuning of the conduction properties simply by adjusting the alignment of the Fermi level at the nanotube–metal contact interface.

CNT p-type operation mainly arises from the low Schottky barrier for hole transport at the metal–nanotube contact interface. Metals suitable for p-type operation, such as Au and Pd, usually have a higher work function than do CNTs. In this respect, fabrication of n-type SWNT-FETs has proved challenging and interesting. Nosho *et al.* [17], Oh *et al.* [35] and Kim *et al.* [18] constructed n-type FETs using a low work function metal electrode. Researchers have attempted construction of an asymmetric contact electrode to achieve Schottky diode operation in SWNT-FETs [36–38]. Yang *et al.* developed such an SWNT-based Schottky diode by placing Pd at the anode and Al at the cathode [38].

In our work, we fabricated symmetric p-type SWNT-FETs using a high work function metal and then systematically converted these devices by modifying the contact barriers using a SAM composed of thiolated molecules. The use of a SAM to control the Fermi level alignment between the SWNTs and the electrode metal can be advantageous, because SAM modifications are introduced under ambient conditions. In principle, the conduction type can be tuned by appropriate choice of the SAM molecule.

The SWNT-FETs used in this study were fabricated by the patterned growth method [39], conventional photolithography, and liftoff. Figure 20-1b shows an AFM (atomic force microscopy) image of a bare device used in our experiments (i.e., prior to SAM treatment). Onto this bare device, patterns were deposited using an SU-8 negative photoresist such that one or both of the contact electrodes were exposed to chemical treatment, as shown in the schematic in Fig. 20-1a. Two self-assembling molecules, 2-amino-ethanethiol (HSCH$_2$CH$_2$NH, M_w 77.15) and 3-mercaptopropionic acid (HSCH$_2$CH$_2$CO$_2$H, M_w 106.14), were purchased from Sigma–Aldrich and used without further treatment. These molecules were selected because they contained a thiol group, which favors the formation of highly ordered monolayers on Au surfaces, and opposite dipole moments [40]. Figure 20-1c shows the molecular structures of 2-aminoethanethiol and 3-mercaptopropionic acid and the predicted direction of the molecular dipole moments. First, to determine the SAM's effect on the contact area, we exposed 2-aminoethanethiol molecules to one of the contact regions of the bare SWNT-FET device, as shown. The electrical transport characteristics of the device were recorded in the absence of SAM treatment and the device was incubated in a 10 mM ethanol solution of SAM precursors. Figure 20-2a and 20-2b shows the current–voltage ($I-V$) characteristics and

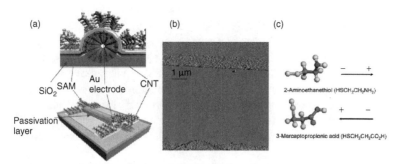

Figure 20-1   (a) Schematic diagram showing a SAM-tuned SWNT-FET device. (b) Atomic force microscopy image of a device used in contact engineering experiment. (c) Molecular structure of the two self-assembly molecules used in this study. (*See insert for color representation of the figure.*)

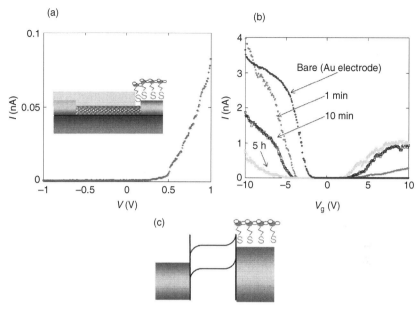

Figure 20-2 (a) $I-V$ curves measured after treatment of only one of the contacts with 2-aminoethanethiol. (b) Evolution of the electrical transfer characteristics during the formation of a 2-aminoethanethiol SAM on the exposed contacts: bare device, after 10 min of reaction, and after 5 h of reaction. (c) A suggested band diagram for this device.

transfer characteristics $(I-V_g)$ of the SWNT-FET after exposure of one of the contacts to 2-aminoethanethiol (reaction time of 10 min). The original device had symmetric and ohmic $I-V$ curves (data not shown) before treatment of the SAM. After treatment with 2-aminoethanethiol, the device exhibited highly asymmetric diode-like $I-V$ characteristics as shown in Fig. 20-2a.

For comparison, we fabricated a device in which both the contact electrodes were covered with SU-8, exposing only the nanotube channel region. After exposing the SU-8-covered electrodes to 2-aminoethanethiol, the device showed symmetric $I-V$ characteristics with reduced conductance. This was consistent with the observations of Kong *et al.*, who attributed the decrease in conductance to the electron doping characteristics of the amine groups [41]. When a 3-mercaptopropionic acid SAM was used in similar experiments, no significant changes in the transport characteristics were observed.

The diode-like operation of the SWNT-FET featuring a 2-aminoethanethiol SAM resulted from the formation of asymmetric contacts after the treatment of the cathode surface, as depicted in Fig. 20-2c. As discussed above, researchers have attempted to create diode-like SWNT-FETs by modifying the Schottky barrier at the metal–nanotube contact regions through construction of asymmetric contacts [38]. We used self-assembled surface functionalization instead of metals with different work functions to adjust the Fermi level alignment in a prefabricated nanotube transistor. For comparison, we used two molecules, 2-aminoethanethiol and 3-mercaptopropionic acid, which were predicted to have different dipole orientations (Fig. 20-1c). We found that, of these two molecules, only treatment with 2-aminoethanethiol produced a diode-like operation.

Ab initio electronic structure calculations were used to investigate the effects of the 2-aminoethanethiol and 3-mercaptopropionic acid molecules on bare Au surfaces. A slab with six gold layers was modeled, in which the two layers in the center were held fixed during geometric optimization calculation. Plotting the local potential with respect to the Fermi level, we obtained the work function for the bare Au surface as well as for the 2-aminoethanethiol- and 3-mercaptopropionic-acid-adsorbed Au surfaces, as shown in Fig. 20-3a–c, respectively. The calculated work function for the bare Au surface (5.2 eV) was very close to the previously reported experimental value [42]. The work functions for the 2-aminoethanethiol- and 3-mercaptopropionic-acid-adsorbed Au surfaces (Fig. 20-3b and c) exhibited obvious deviations from this value. Specifically, adsorption of 2-aminoethanethiol molecules substantially decreased the work function (3.9 eV), whereas adsorption of 3-mercaptopropionic acid had little effect (4.7 eV). This difference between the electronic structures confirmed that the asymmetric $I-V$ pattern observed after treating one electrode of the

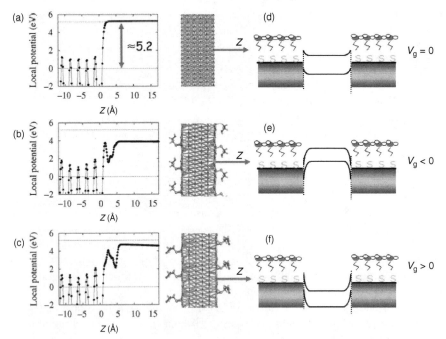

Figure 20-3 Plots of the local potentials with respect to the Fermi level of the (a) six layers of bare Au, (b) six layers of Au with adsorbed 2-aminoethanethiol molecules, and (c) six Au layers with adsorbed 3-mercaptopropionic acid molecules. *Source*: Reused with permission from Applied Physics Letters, 89, 243115 (2006). Copyright 2006, American Institute of Physics. (d) Band diagram for the SWNT-FET after SAM treatment. The suggested band diagrams for the SAM-treated SWNT-FET device under the gate bias conditions $V_g < 0$ (e) and $V_g > 0$ (f).

SWNT-FET with 2-aminoethanethiol was due to an adjustment of the Schottky barrier at the treated contact.

To investigate the effects of 2-aminoethanethiol treatment on the transfer characteristics, we measured the evolution of the $I-V_g$ characteristics for the SWNT-FETs as a function of treatment time as shown in the Fig. 20-2b. Prior to reaction with 2-aminoethanethiol, the device showed a typical p-type transistor behavior. After treatment with 2-aminoethanethiol, with incubation times of 10 min and 5 h, p-channel conduction decreased and n-type conduction became dominant. The 2-aminoethanethiol SAM showed a saturation behavior at a reaction time of 5 h, at which point the $I-V_g$ characteristics remained almost unchanged with further treatment. As a result, the SWNT-FET with a 2-aminoethanethiol SAM showed an ambipolar gate transfer character. The gate transfer character could be explained by considering the band diagram generated by SAM treatment of one contact. Before SAM treatment, the Fermi levels of the electrodes were aligned with the valence band of the CNTs, leading to p-type behavior. After the formation of a 2-aminoethanethiol SAM, however, the Fermi level shifted toward the conduction band (Fig. 20-3d). The hole current in the SWNT-FET with a treated electrode dominated under conditions of $V_g < 0$, as shown in Fig. 20-3e. However, significant electron conduction occurred in the treated devices when $V_g > 0$, as shown in Fig. 20-3f.

It should be noted that the threshold voltage of the device did not change significantly after SAM treatment of the contact electrodes, as shown in Fig. 20-2b. Shifts in the threshold voltage are, in general, largely due to variations in the doping level in the semiconductor channel region, whereas changes in the slope of the $I-V_g$ curves were derived from variations in the Schottky barrier at the contact region [22]. As shown in Fig. 20-2b, 2-aminoethanethiol treatment induced slope changes in the $I-V_g$ curve, eventually resulting in an ambipolar behavior with no prominent shifts in the threshold voltage. These findings confirmed that introduction of the SAM composed of 2-aminoethanethiol modified the Schottky barrier at the engineered contact. These results suggested that tuning the Schottky barrier by introducing a SAM composed of a selected molecular species may be a practical method for controlling conduction patterns in nanotube-based electronic devices.

20.2.2 Modification of the Electrical Properties in Double-Walled Nanotube Devices using SAMs

In this section, we focus on the tuning of the conduction properties in double-walled carbon nanotube FETs (DWNT-FETs) by SAM surface modification and the effect of ambient air on the electrical properties of these devices. The diameters of DWNTs are normally larger than those of SWNTs. Because the energy gap of a semiconducting nanotube is inversely proportional to the diameter of the tube, $E_g \sim 1/d$, where d is the tube diameter [43], DWNTs should have a smaller energy gap than SWNTs. Thus, the change in Fermi level alignment induced by a SAM could be stronger in a DWNT. For device fabrication, we dispersed CVD-grown DWNTs in methanol and deposited them onto degenerately doped Si wafers (with 300 nm SiO$_2$) at predefined coordinates. After locating suitable DWNTs, patterns for the electrical leads were generated using e-beam lithography. Thin adhesion layers of Ti (<5 nm) and Au (30 nm) were subsequently deposited by thermal evaporation to form electrodes. The thin titanium layer (<5 nm) acted as a glue to provide strong binding between the metal electrode and substrate. Since the thickness of adhesion layer is comparable to the diameter of DWNT, CNTs had many opportunities for regional contact with the gold layer. In addition, the edge of the metal electrode was more likely to be covered by the gold layer.

To ensure that the molecules reacted only with the Au electrodes and not with the sidewalls of the DWNT, the channel region of the DWNT-FET was covered using a negative e-beam resist (micro resist technology, ma-N2410). The diameter of the DWNT device used in this experiment was about 4.5 nm. To change the carrier type in the DWNT-FET, 2-aminoethanethiol molecules were introduced, which self-assembled at the interface between the Au electrodes and the CNTs, changing the transport properties of the Au metal/CNT metal contacts.

In general, oxygen significantly influenced the electrical properties of the CNT devices; electrical properties in vacuum are different from those at ambient atmosphere. At the metal–CNT interface, the oxygen adatoms increased the metal work function. In the CNT channel region, oxygen adsorbators increased the hole concentration. Under high vacuum, device character was most likely to be governed by the Fermi level alignment at the metal–CNT contact, since hole-donating oxygen adsorbates are removed in vacuum. Alignment favored the p-type character because of the high work function contact of the Au in our device. The removal of oxygen in the metal/CNT contacts resulted in a slight increase in the n-channel current at a positive gate bias voltage because of the reduction of the work function. However, the device did not transform from p- to n-type entirely via oxygen removal during the pumping process.

Contact-opened DWNT-FETs were immersed in 1 mM 2-aminoethanethiol in ethanol and allowed to react for more than 12 h. Samples were subsequently washed with ample quantities of clean ethanol to remove any physisorbed molecules and were dried with N$_2$ gas. Figure 20-4a and b presents the gate transfer characteristics of a bare DWNT-FET and a SAM-modified DWNT-FET, respectively, under vacuum. Prior to assembly of the SAM, the DWNT-FET showed a clear p-type transport behavior with a gate threshold voltage of −6.5 V. After SAM functionalization, the p-channel current decreased significantly, and a substantial increase in the n-channel current was observed in the positive gate bias region, in which low conduction was observed prior to SAM self-assembly. Although a small p-channel current existed, clear n-type transport was observed. The SAM-treated DWNT-FET underwent a clear transition from p-type to n-type transistor via contact modifications. For comparison, we performed a similar experiment in which both contact electrodes were covered with the resist, exposing only the nanotube body region. The device showed decreased conductance due to the electron doping nature of the amine groups as in the case of SWNT-FET.

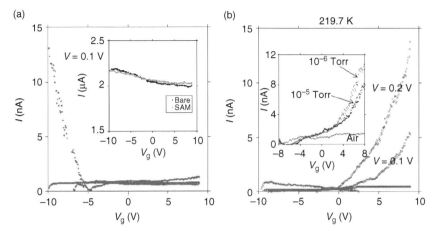

Figure 20-4 (a) $I-V_g$ curves of a bare DWNT-FET. Inset shows the gate transfer characteristics from metallic DWNT-FET before and after contact engineering. The source–drain bias voltage was 100 mV. (b) SAM-modified DWNT-FET under vacuum. The inset shows the $I-V_g$ curves of a SAM-modified DWNT-FET under different pressure conditions. The source–drain bias voltage was set to 100 mV.

If the metal film layer was sufficiently thin, the work function of the system was influenced by the overlaying metal or molecules [44]. According to *ab initio* electronic structure calculations, for a model system containing six gold layers on which 2-aminoethanethiol molecules were adsorbed, the work function of the SAM/Au system was substantially lowered, to 3.9 eV, relative to that of the pure Au electrode (5.2 eV) [9]. In our case, the thin Au metal layer at the edge of the electrode near the CNT/Au contact may have been affected by the SAM process such that the work function of the Au electrode near the interface was lowered. AFM measurements showed the presence of gradual slopes on the edges of the electrodes near the CNT/Au contacts. SAMs probably attached to the very thin layer of the Au electrode, effectively changing the electrical properties.

The reduction in the work function and the assembly of adsorbed polar molecules near the CNT–metal contact region rearranged the charge distribution and, consequently, modified the surface electrical properties. We investigated the effect of vacuum on contact-engineered devices in the vacuum chamber with an electrical feedthrough. The inset in Fig. 20-4b shows the $I-V_g$ curve of the SAM-modified DWNT-FET in air and under vacuum. As shown in the figure, enhanced n-channel current was observed for the SAM-modified DWNT-FET in vacuum. It should be noted that only minor changes were observed in the gate threshold voltages on evacuation. Such behavior was consistent with previous findings that oxygen adsorption on the contact electrodes changed the pinning condition at the metal–nanotube interface [22]. The n-type transport in SAM-modified DWNT-FETs could be further enhanced under higher vacuum ($\sim 10^{-6}$ Torr). Because the edge of the metal electrode (contact) was exposed, molecules could assemble in the interface region. In addition to lowering the work function of the Au electrode by SAM molecules, desorption of oxygen molecules in high vacuum assisted with enhancing the n-type transport.

Finally, we investigated the effects of energy level alignment using metallic DWNTs. The inset in Fig. 20-4a shows the effects of 2-aminoethanethiol SAM on a semimetallic DWNT device. The device showed very weak gate transfer characteristics with a channel resistance of 46 kΩ, suggesting semimetallic behavior. In semimetallic DWNT devices, electrical transfer characteristics showed negligible changes on evacuation or self-assembly. This property implied that work function control using SAMs was not possible in semimetallic DWNTs, in which the Schottky barrier did not exist.

In conclusion, self-assembly of polar molecules at the metal–nanotube interface affects the Schottky barrier height greatly, which appears as a dramatic change in semiconducting nanotube devices. Such phenomena could be related to lowering of the metal work function due to self-assembly, and charge redistribution at the interface results from the polarity of self-assembled molecules.

20.2.3   Modification of the Electrical Properties of Single-Walled Nanotube Devices using Nanoparticles

In this section, we demonstrate that the electrical properties of SWCNT devices can be changed dramatically by decorating CNT devices with nanosized particles, such as Au, polystyrene, or protein nanoparticles. In experiments performed by Na et al. [45], SWCNT devices were fabricated using laser ablation-grown CNTs and e-beam lithography. After fabrication, devices were allowed to react with dilute solutions of streptavidin-coated 10-nm Au nanoparticles or streptavidin-coated polystyrene nanoparticles. After reaction, devices were rinsed with deionized (DI) water and dried with N_2. The electrical properties and topographic images were recorded before and after nanoparticle decoration. Before decoration, each device showed a poor on/off ratio ($\sim$6) and a large subthreshold swing (>5 V/decade). However, after assembly of protein nanoparticles near the contact electrodes, devices exhibited enhanced performance: on/off ratios increased up to 10^4, and the subthreshold swing decreased to 0.15 V/decade. Interestingly, devices with nanoparticles assembled on the sidewalls of SWNTs did not show enhanced performances postdecoration. SWNTs with metallic transport capabilities showed much more dramatic changes after decoration. Before decoration, devices showed typical metallic behavior, and the conductance could not be modulated by the gate bias voltage. After samples were reacted with streptavidin-coated 10-nm Au nanoparticles, however, an FET-like gate response was observed in the metallic nanotubes. The conductance saturated immediately after the p-channel turned on, which was unusual for conventional metal-oxide–semiconductor field-effect transistorSemiconductor Field-Effect Transistors (MOSFETs) or SWNT-FETs. The authors posited that the contact barrier was modulated by the self-assembled protein nanoparticles. As shown in Fig. 20-5, proteins immobilized at the contact interface could be charged either negatively or positively, according to the gate bias, which effectively changed the

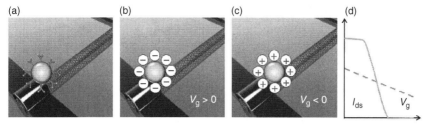

Figure 20-5 (a) Schematics of an SWNT-FET in which one contact was modified by protein-immobilized nanoparticles. (b) Application of a positive gate voltage induced a negative charge on the proteins, which blocked the tunneling of electrons from the contact electrode to the SWNT. (c) Electrons may tunnel through the barrier more easily when positively charged nanoparticles are present at the contact electrode. (d) Schematic diagram of the transfer characteristics of a nanoparticle-decorated SWNT-FET. The solid line denotes the transfer curve after decoration, and the dotted line indicates the transfer curve before decoration.

contact tunnel barrier height. *Ab initio* calculations supported the assertion that the molecular adsorbates could indeed modulate the potential barrier height.

Apart from this work, Wei *et al.* [6] showed that it was possible to transform unipolar SWNT-FETs to ambipolar SWNT-FETs using a polystyrene nanosphere assembly. SWNTs were grown by chemical vapor deposition techniques, and the authors used molybdenum as the contact electrode. Polystyrene nanospheres were assembled near the contact electrode as well as near the nanotube body. Devices that contained polystyrene near the contact electrodes showed enhanced sub-threshold swings and on/off ratios, in agreement with the observations of Na *et al.*, and enhanced ambipolar characteristics. The authors screened the contact region with polymethylmethacrylate (PMMA) to ensure that the contact electrodes were free from polystyrene nanospheres. In these devices, no improved performance or ambipolar behavior was observed, and the effects of the polystyrene nanoparticles arose mainly from the contact region. Because the relative permittivity of polystyrene is significantly higher than that of air, decoration of the contact region with nanospheres resulted in increased gate coupling. The authors attributed the observed improvements in device performance to a reduced Schottky barrier as a result of increased gate capacitance. The transition from unipolar to ambipolar transport may be explained by the increased gate coupling.

20.2.4 Sensors Based on Contact Barrier Modulation

Electronic sensors based on SWNT transistors have been extensively studied because of their potential uses in diagnostics, environmental monitoring, and the detection of chemical or biological warfare agents. Electronic sensors provide fast reaction times and point-of-care capabilities as well as cost-effectiveness resulting from sensor miniaturization. Unlike sensors based on semiconductor nanowires, the operational sensing mechanism in SWNT transistor sensors is still under debate [46]. A doping effect that stems from charge transfer from the analyte, producing electrostatic gating effects, is one potential mechanism. However, because SWCNT transistors function as Schottky transistors, modulation of the Schottky barrier could also form the basis of the sensing mechanism [46]. The change in effective capacitance in SWNT-FET sensors with completely packed molecular recognition elements may explain the observed conductance changes after analyte binding [47]. Here, we focus on SWNT-FET sensors that utilize Schottky barriers as the effective sensing area. In 2003, Chen *et al.* showed that major changes in the conductance of SWNT-FET sensors arose from the contact area, not from the nanotube body itself [46]. In their experiment, Pd/Au electrodes were patterned, SWNTs were grown via chemical vapor deposition, and, to discriminate between the effects of the contacts and nanotube channels, metal electrodes were obscured using SAMs composed of methoxy(poly(ethylene glycol))thiol (mPEG-SH). Although bare SWNT-FETs showed a change in conductance in the presence of bovine serum albumin (BSA), human serum albumin (HSA), human chorionic gonadotropin (HCG), polyclonal human IgG (hIgG), and avidin, no conductance changes were observed in mPEG-SH-passivated devices, with the exception of the avidin detection device. The authors confirmed that the majority of the conductance changes arose from the metal-SWNT contact area. Avidin detection was unique, because avidin molecules have a large net charge under the experimental pH conditions. (The isoelectric point of avidin is $10-11$; therefore, it is strongly positively charged at neutral pH.)

Recently, Tang *et al.* [48] demonstrated the detection of hybridization between a target single-stranded DNA (ssDNA) and a probe ssDNA that had been self-assembled on the metal electrode of an SWNT-FET. The probe ssDNA, functionalized with thiol groups, was self-assembled into an ordered monolayer on the Au electrode of an SWNT-FET. On hybridization with the target DNA, the conductance of the SWNT-FET decreased, whereas mismatched target oligo sequences yielded much smaller conductance changes (Fig. 20-6). Thiolated molecules have been shown to align energy levels [48] and shift the work function. Hybridization between a target ssDNA and a thiolated probe ssDNA on the Au electrodes reduced the Au work function, thereby decreasing the conductance of the p-type SWNT-FET.

20.3 FUNCTIONAL SWNT DEVICES COATED WITH SELF-ASSEMBLED ORGANIC OR BIOLOGICAL MOLECULES

By combining self-assembly with SWNT devices, it has been possible to create functional SWNT devices. For example, SWNT-FETs functionalized with pyrenecyclodextrin [13] and DNA-wrapped SWNT-FETs [12] were used as chemical sensors. Combinations of photoactive molecules and SWNT-FETs can yield SWNT optoelectronic devices [49]. In these structures, SWNT-FETs function as high-performance transducers that translate chemical or photo signals into an electrical readout. Several methods for functionalizing SWNT-FETs have been developed for each type of application. In this chapter, only self-assembly

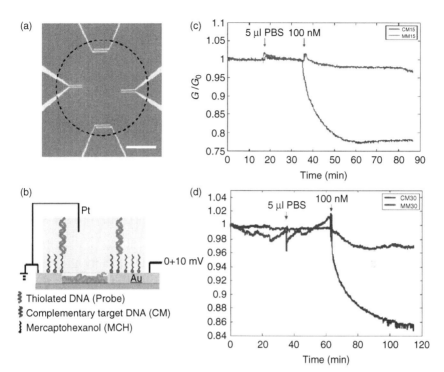

Figure 20-6   DNA detection using SWNT-FETs. (a) Optical microscopic image of a typical SWNT-FET. (b) Experimental layout. Au electrodes on the SWNT-FET were functionalized with mercaptohexanol and immobilized using thiolated probe DNA. The conductance change induced by reaction was measured in real time using a Pt reference electrode. (c) Real-time conductance from the device. A solution containing 100 nM of a 15-mer matching target decreased the conductance, whereas a mutant target produced only minor conductance changes. (d) Real-time conductance measurements on hybridization of 30-mer matching and mutant targets. *Source*: Adapted from Ref. [48], reprinted with permission, Copyright 2006, American Chemical Society.

approaches are discussed because the process of self-assembly is relatively simple and scalable and the perturbations to the electrical properties of SWNTs are minimal.

20.3.1   SWNT Chemical Sensors using Self-Assembled Molecules

In 2000, Kong *et al.* and Collins *et al.* showed that bare SWNT devices could be used as highly sensitive chemical sensors for NO_2, NH_3, and O_2 [28,50]. In the presence of NO_2 gas, electrons were transferred from the SWNTs to the NO_2 molecules, which increased the conductance. The NH_3 detection mechanism is under debate; the sensing mechanism may originate from either channel doping or Schottky barrier effects. Peng *et al.* showed that Schottky barrier modulation was the dominant mechanism at temperatures less than $150°C$ [51]. Sensor selectivity, which permits differentiation between chemical species, is accomplished through the design of molecular recognition layers that are specific to the analyte. Decoration of Pd nanoparticles on the surface of an SWNT-FET produced a hydrogen detector [52], and the combination of polyethyleneimine (PEI) and Nafion permitted selective detection of NO_2 and NH_3 [53]. Starch films were used as the recognition layer for detection of CO_2 using SWNT-FETs [54]. Chemoselective polymers were combined to produce sensors for nerve agent detection [55]. Zhang *et al.* showed that SWNT devices with electrochemically decorated polyaniline (PANI) films yielded high sensitivity toward NH_3 [56]. In most reported chemical sensors, chemoselective polymer films were used to confer selectivity on sensors. In film-based sensors, the thickness of a film played a crucial role in sensor performance. In most cases, sensitivity increased and the response time decreased with decreasing film thickness, because diffusion of target molecules was much faster in thinner films [57]. The recovery time for sensors was much longer in thick films because of trapping of the target molecules in the polymer matrix. Therefore, very thin molecular recognition layers are optimal for chemical sensing, and SAMs may provide a solution for fabricating such layers. Several methods have been reported for the formation of SAMs composed of receptor molecules adsorbed in SWNT channels. COOH groups may be added to SWNT sidewalls by acid treatment and subsequently linked to receptor molecules [58]. However, covalent modifications of CNT surfaces usually proceed at high temperatures and require toxic chemical treatments. Therefore, it is preferable to use molecules that can self-assemble or adsorb onto the sidewalls of SWNTs. Interactions such as van der Waals, $\pi-\pi$ stacking, and $CH-\pi$ interactions contribute to self-assembly and adsorption. Surfactants such as sodium dodecyl sulfate (SDS), sodium dodecylbenzene sulfonate (SDBS), cetyltrimethylammonium bromide (CTAB), Tween, Triton X, and siloxane polyether copolymer (PSPEO) have been demonstrated to solubilize and disperse CNTs [59]. Polymer wrapping with biopolymers is widely used to solubilize CNTs [60]. In polymer wrapping, the polymers helically wrap around CNTs to reduce strain. In principle, surfactants or polymers could be used to tether functional moieties to CNTs, thereby conferring specific functionality. In this chapter, we have restricted ourselves to discussion of those molecules that readily assemble on SWNTs and provide functionality, in and of themselves.

20.3.1.1 *SWCNT Assembly Using Pyrene-Based Molecules*
Binding of polyaromatic molecules to SWNTs is rather strong and stable. In fact, even among surfactants, molecules with aromatic groups interact better with SWNTs through $\pi-\pi$ stacking than molecules with alkyl chains [59]. As the number of aromatic groups increases, the binding affinity between SWNTs strengthens. Anthracene, a molecule containing three benzene rings, is known to have a strong binding affinity for SWNTs, but it can be replaced with pyrene [59]. Porphyrins and phthalocyanines also bind strongly to SWNTs [59]. Pyrene-based molecules can be used to lend functionality to SWNTs. Pyrene linked to an N-hydroxysuccinimide ester group can be used to immobilize biomolecules for the fabrication of SWNT-based

biosensors [61]. Detection of analytes, such as proteins and small molecules, and of microorganisms, such as *Escherichia coli*, was performed using SWNT sensors functionalized with pyrene-N-hydroxysuccinimide ester-immobilized antibodies or aptamers [62–64]. It is possible to confer functionality using the pyrene moiety itself. Simmons *et al.* used 1-pyrenecarboxylic acid to disperse SWNTs without harsh oxidation processes [65]. Pyrene derivatives with polar end groups can disperse SWNTs in solution or can be used to form functional layers on top of SWNTs. Zhao *et al.* synthesized pyrenecyclodextrin for fabricating sensors of organic molecules in aqueous solutions [13]. Pyrenecyclodextrin was synthesized by the condensation of mono(6-aminoethylamino-6-deoxy)-β-cyclodextrin with 1-pyreneacetic acid in N,N-dimethyl formamide. A solution containing 1.13×10^{-3} mol/L pyrenecyclodextrin in DMF (dimethylformamide) was reacted with SWCNTs, and pyrenecyclodextrin was specifically self-assembled onto the surfaces of SWNTs via π–π interactions. Organic compounds, such as 1-adamantanol, sodium cholate, and sodium deoxycholate, which bind within the cavities formed by cyclodextrin, could be detected via changes in the SWNT electrical signal.

20.3.2 Functional SWNT Devices Functionalized with Self-Assembled Biopolymers

It is an intriguing possibility that a variety of biopolymers, including DNA and polypeptides, can lend functionality to SWNTs. Nucleic acids, such as DNA or peptide nucleic acids (PNAs), may be used as biomolecular glue to control the site-specific positioning of SWNTs. Keren and coworkers showed that it was possible to fabricate SWNT devices using DNA and DNA-binding enzymes [66]. In Keren's approach, SWNTs were assembled onto DNA strands using the streptavidin–biotin interaction, and DNA wires were metallized by silver reduction and subsequent metallization of Au. Recently, Maune *et al.* self-assembled SWNT devices using DNA origami templates [67]. In this case, an SWNT binding sequence (poly T) was used to disperse the SWNTs. In 2003, Zheng *et al.* discovered that noncovalent binding of ssDNA to CNT surfaces was an effective means for dispersing CNTs [68,69]. Aromatic bases in ssDNA may interact with the sidewalls of SWNTs through π–π stacking, and a variety of ssDNA conformations on CNTs is possible. ssDNA may helically wrap around CNTs or simply adsorb onto the surface. Extensive theoretical and experimental studies have investigated the interaction of DNA with CNTs [70,71]. The binding strength between ssDNA and CNTs is sequence-specific. Poly(A) and poly(C) sequences have low binding free energies on nanotube surfaces and, therefore, are less efficient at dispersing CNTs than are poly(T) sequences [69]. Sequence-dependent binding efficiencies of ssDNA permit separation of metallic SWNTs from semiconducting SWNTs and SWNT separation by diameter. Assemblies of poly(GT) on CNTs produce a DNA–CNT hybrid that carries a net negative charge because of the negatively charged phosphate groups on the DNA backbone. The net charge of the DNA–CNT hybrid can vary according to the polarizability of the CNT. Metallic CNTs, with higher polarizability, screen the negative charges of DNA better than semiconducting CNTs. Therefore, in the presence of an anion exchange resin, the elution times for metallic and semiconducting DNA–CNT hybrids vary because of the differences in the strength of interaction between the hybrids and the solid phase [69]. On the other hand, in the absence of ssDNA, metallic and semiconducting SWNTs can be separated using agarose gel elec-trophoresis [72]. SWNTs dispersed using SDS showed clear separation between the metallic and semiconducting SWNTs by agarose gel electrophoresis. Tanaka and colleagues attributed the selective separation to affinity binding of semicon-ducting SWNTs to the agarose gel. Recently, Tu *et al.* described DNA sequence motifs designed for the structure-specific recognition and separation of CNTs [73]. They found several chiral-specific sequences from a library and attributed the sequence-specificity to the 2D form of the DNA. Most of the sequences identified appeared to form stable 2D sheets with adjacent strands, and these 2D sheets could be wrapped around the surfaces of specific types of SWNTs. The purity of the (10,5) SWNT species was electronically measured to be 99%, and the yield of separation could be optimized by controlling the ionic conditions and pH.

20.3.3 Applications of DNA-Wrapped SWNTs

Wrapping SWNTs with ssDNA is a method for dispersing SWNTs in aqueous solutions and for separating metallic SWNTs from semiconducting SWNTs simul-taneously. A number of other special applications for DNA-wrapped SWNTs have been described. In this section, we discuss a variety of applications of DNA-wrapped SWNTs, from sensors to electronic devices.

20.3.3.1 Electronic Materials Based on DNA-Wrapped CNTs DNA-wrapped CNTs are ideal candidates for electrode materials. DNA wraps around CNTs via $\pi-\pi$ stacking and produces debundled hybrid SWNTs. Electrolytes were shown to penetrate the interstitial spaces between SWNTs, leading to higher capacitance and actuation strain [74]. In electronic devices, DNA wrapping has provided improved adhesion to dielectric materials. Lu *et al.* showed that DNA-wrapping accommodated atomic layer deposition (ALD) of SiO_2 and HfO_2, which enabled the fabrication of high-performance SWNT-FETs [75]. While SWNT-FETs without DNA wrapping showed gate leakage for HfO_2 layers 5 nm thick, DNA-wrapped SWNT-FETs showed high performances with 3 nm thick HfO_2. The authors achieved an ideal limit of 60 mV/decade swing using DNA-wrapped SWNT-FETs.

20.3.3.2 Sensors Based on DNA-Wrapped CNTs Biosensing and DNA sequencing are obvious applications for DNA-wrapped CNTs in sensors. Both electronic and optical sensors have been shown to detect hybridization of DNA [75,76]. Star *et al.* immobilized capture probe ssDNA on the sidewalls of CNTs in an SWNT-FET to measure the hybridization of target DNA [76]. Probe ssDNAs were nonspecifically adsorbed on the SWNTs, as was shown by Zheng *et al.* Immobilization of capture ssDNA decreased the measured conductance, which was explained by an electron doping mechanism. This phenomenon is similar to the interaction between aromatic molecules and SWNTs, which also has qualities consistent with electron doping. The aromatic bases of capture ssDNA apparently were bound to the sidewalls of SWNTs. Hybridization of target DNA further decreased the conductance, whereas mutant target DNAs containing single nucleotide polymorphisms showed smaller conductance changes.

Hybridization between the probe and target DNAs was detected by optical methods. Jeng *et al.* demonstrated that hybridization to target DNA could be measured through its effect on the near-IR band gap fluorescence of SWNTs [77]. As was observed in the transistor experiments, the authors functionalized SWNT sidewalls using probe ssDNA and measured the hypsochromic shift in the near-IR SWNT fluorescence. After 48 h incubation with the target DNA, the energy of the (6,5) SWNT peak increased to ~ 2 meV, indicating a decrease in the effective dielectric constant, which the authors interpreted as a higher coverage of ssDNA. Noncomplementary target ssDNA did not yield such large changes in the fluorescence peak positions. The authors also measured the hybridization kinetics on the surfaces of SWNTs. A steady state was reached only after incubation for 13 h at $25°C$, whereas free DNA hybridization occurs on the timescale of 10 min.

DNA-wrapped SWNTs have been used not only for the detection of complementary DNA hybridization but also to monitor gaseous species and heavy metal ions. Staii *et al.* decorated the sidewalls of SWNT-FETs with ssDNA to develop chemical sensors [12]. The ability of single-stranded nucleotide oligomers to selectively bind various chemical and biological species has been known since 1980s, when the presence of functional nucleic acids that could specifically bind and regulate viral or cellular proteins were discovered [78]. Functional nucleic acids were called *aptamers* by Ellington and Szostak [79], and they are currently used in a variety of applications, such as therapeutic drugs, sensors, and affinity columns. Using this concept, ssDNA oligomers that could be immobilized onto SWNT surfaces to yield selective sensors were selected from a library. Staii *et al.* immobilized ssDNA on the sidewalls of SWNT-FETs to fabricate sensors for methanol, propionic acid, trimethylamine, 2,6-dinitrotoluene, and dimethylphosphonate [12].

Heavy metal ion detection using DNA-wrapped SWNTs has also been described. Heller *et al.* measured the near-IR emission energies of double-stranded DNA (dsDNA) immobilized on SWNT surfaces for the detection of Hg^{2+} ions in whole blood, tissues, and inside a living cell [80]. A 30-base ssDNA immobilized on an SWNT surface was shown to undergo a B to Z structural transition on exposure to heavy metal ions, because divalent metal cations bind to DNA and stabilize the Z form. The DNA conformal change altered the dielectric environment around the SWNT, which, as in the case of DNA hybridization, decreased the near-IR emission energy.

20.4 CONCLUSIONS

In this chapter, we reviewed the properties and applications of SWNT devices modified with SAMs composed of organic or biological molecules.

Contact engineering using SAMs of organic or biological molecules is very effective in tailoring SWNT devices. SAMs were used to create SWNT Schottky diodes by modifying one contact electrode with the SAM, and n-type SWNT-FETs were fabricated by lowering the work functions of both electrodes using SAMs. SAMs enhanced the performance of SWNT-FETs and enabled formation of nanoparticle-decorated SWNT-FETs, which showed improved on/off ratios as well as higher mobilities. SAMs can lend functionality to SWNT devices. Highly sensitive and selective chemical or biological sensors were developed by combining SAMs with SWNT devices, and optoelectronic devices were fabricated using self-assembly of photoactive materials.

On the basis of these findings, SAMs offer a versatile and simple route to fabricating functional SWNT devices, and we expect that the tremendous potential of this technique will be realized in the near future.

REFERENCES

1. Evans, S. D., Ulman, A. (1990). Surface potential studies of alkyl-thiol monolayers adsorbed on gold. *Chem. Phys. Lett.*, *170*, 462–466.
2. Campbell, I. H., Rubin, S., Zawodzinski, T. A., Kress, J. D., Martin, R. L., Smith, D. L., Barashkov, N. N., Ferraris, J. P. (1996). Controlling Schottky energy barriers in organic electronic devices using self-assembled monolayers. *Phys. Rev. B*, *54*, R14321–R14324.
3. Gundlach, D. J., Jia, L. L., Jackson, T. N. (2001). Pentacene TFT with improved linear region characteristics using chemically modified source and drain electrodes. *IEEE Electron Device Lett.*, *22*, 571–573.
4. Hamadani, B. H., Corley, D. A., Ciszek, J. W., Tour, J. M., Natelson, D. (2006). Controlling charge injection in organic field-effect transistors using self-assembled monolayers. *Nano Lett.*, *6*, 1303–1306.
5. Hong, J. P., Park, A. Y., Lee, S., Kang, J., Shin, N., Yoon, D. Y. (2008). Tuning of Ag work functions by self-assembled monolayers of aromatic thiols for an efficient hole injection for solution processed triisopropylsilylethynyl pentacene organic thin film transistors. *Appl. Phys. Lett.*, *92*, 143311–3.
6. Wei, D., Zhang, Y., Yang, Y., Hasko, D. G., Chu, D., Teo, K. B. K., Amaratunga, G. A. J., Milne, W. I. (2009). Transformation of unipolar single-walled carbon nanotube field effect transistors to ambipolar induced by polystyrene nanosphere assembly. *ACS Nano*, *2*, 2526–2530.
7. Lee, C. W., Zhang, K., Tantang, H., Lohani, A., Mhaisalkar, S. G., Lia, L. J., Nagahiro, T., Tamada, T., Chen, Y. (2007). Tuning of electrical characteristics in networked carbon nanotube field-effect transistors using thiolated molecules. *Appl. Phys. Lett.*, *91*, 103515–3.
8. Jeon, E. K., Kim, H. S., Kim, B. K., Kim, J. J., Lee, J. O., Lee, C. J. (2008). The modification of the electrical property in double-walled nanotube devices with a self-assembled monolayer of molecules. *J. Nanosci. Nanotechnol.*, *8*, 4349–4351.
9. Cui, X., Freitag, M., Martel, R., Brus, L., Avouris, Ph. (2003). Controlling energy-level alignments at carbon nanotube/Au contacts. *Nano Lett.*, *3*, 783–787.
10. Heinze, S., Tersoff, J., Martel, R., Derycke, V., Appenzeller, J., Avouris, Ph. (2002). Carbon nanotubes as Schottky barrier transistors. *Phys. Rev. Lett.*, *89*, 106801–4.
11. Kim, B. K., Kim, J. J., So, H. M., Kong, K., Chang, H., Lee, J. O., Park, N. (2006). Carbon nanotube diode fabricated by contact engineering with self-assembled molecules. *Appl. Phys. Lett.*, *89*, 243115–3.

12. Staii, C., Johnson, A. T. Jr. (2005). DNA-decorated carbon nanotubes for chemical sensing. *Nano Lett.*, *5*, 1774–1778.

13. Zhao, Y.-L., Hu, L., Stoddart, J. F., Gruner, G. (2008). Pyrenecyclodextrin-decorated single-walled carbon nanotube field-effect transistors as chemical sensors. *Adv. Mater.*, *20*, 1910–1915.

14. de Boer, B., Hadipour, A., Mandoc, M. M., Woudenbergh, T., Blom, P. W. M. (2005). Tuning of metal work functions with self-assembled monolayers. *Adv. Mater.*, *17*, 621–625.

15. Khodabakhsh, S., Poplavskyy, D., Heutz, S., Nelson, J., Bradley, D. D. C., Murata, H., Jones, T. S. (2004). Using self-assembling dipole molecules to improve hole injection in conjugated polymers. *Adv. Funct. Mater.*, *14*, 1205–1210.

16. Campbell, I. H., Hagler, T. W., Smith, D. L., Ferraris, J. P. (1996). Direct measurement of conjugated polymer electronic excitation energies using metal/polymer/metal structures. *Phys. Rev. Lett.*, *76*, 1900–1903.

17. Nosho, Y., Ohno, Y., Kishimoto, S., Mizutani, T. (2005). n-Type carbon nanotube field-effect transistors fabricated by using Ca contact electrodes. *Appl. Phys. Lett.*, *86*, 073105–073107.

18. Kim, H. S., Kim, B. K., Kim, J. J., Lee, J. O., Park, N. (2007). Controllable modification of transport properties of single-walled carbon nanotube field effect transistors with *in situ* Al decoration. *Appl. Phys. Lett.*, *91*, 153113–3.

19. Kim, H. S., Jeon, E. K., Kim, J. J., So, H. M., Chang, H., Lee, J. O., Park, N. (2008). Air-stable n-type operation of Gd-contacted carbon nanotube field effect transistors, *Appl. Phys. Lett.*, *93*, 123106–3.

20. Zhang, Z., Liang, X., Wang, S., Yao, K., Hu, Y., Zhu, Y., Chen, Q., Zhou, W., Li, Y., Yao, Y., Zhang, J., Peng, L. M. (2007). Doping-free fabrication of carbon nanotube based ballistic CMOS devices and circuits. *Nano Lett.*, *7*, 3603–3607.

21. Tans, S. J., Verschueren, A. R. M., Dekker, C. (1998). Room-temperature transistor based on a single carbon nanotube. *Nature*, *393*, 49–52.

22. Derycke, V., Martel, R., Appenzeller, J., Avouris, Ph. (2002). Controlling doping and carrier injection in carbon nanotube transistors. *Appl. Phys. Lett.*, *80*, 2773–2775.

23. Heinze, S., Tersoff, J., Avouris, Ph. (2003). Electrostatic engineering of nanotube transistors for improved performance. *Appl. Phys. Lett.*, *83*, 5038–5040.

24. Martel, R., Schmidt, T., Shea, H. R., Hertel, T., Avouris, Ph. (1998). Single- and multi-wall carbon nanotube field-effect transistors. *Appl. Phys. Lett.*, *73*, 2447–2449.

25. Appenzeller, J., Knoch, J., Derycke, V., Martel, R., Wind, S., Avouris, Ph. (2002). Field-modulated carrier transport in carbon nanotube transistors. *Phys. Rev. Lett.*, *89*, 126801–126804.

26. Freitag, M., Radosavljevic, M., Zhou, Y., Johnson, A. T., Smith, W. F. (2001). Controlled creation of a carbon nanotube diode by a scanned gate. *Appl. Phys. Lett.*, *79*, 3326–3328.

27. Sumanasekera, G. U., Adu, C. K. W., Fang, S., Eklund, P. C. (2000). Effects of gas adsorption and collisions on electrical transport in single-walled carbon nanotubes. *Phys. Rev. Lett.*, *85*, 1096–1099.

28. Collins, P. G., Bradley, K., Ishigami, M., Zettl, A. (2000). Extreme oxygen sensitivity of electronic properties of carbon nanotubes. *Science*, *287*, 1801–1804.

29. Bachtold, A., Hadley, P., Nakanishi, T., Dekker, C. (2001). Logic circuits with carbon nanotube transistors. *Science*, *294*, 1317–1320.

30. Martel, R., Derycke, V., Lavoie, C., Appenzeller, J., Chan, K. K., Tersoff, J., Avouris, Ph. (2001). Ambipolar electrical transport in semiconducting single-wall carbon nanotubes. *Phys. Rev. Lett.*, *87*, 256805–256808.

31. Javey, A., Tu, R., Farmer, D. B., Guo, J., Gordon, R. G., Dai, H. (2005). High performance n-type carbon nanotube field-effect transistors with chemically doped contacts. *Nano Lett.*, *5*, 345–348.

32. Zhou, C., Kong, J., Yenilmez, E., Dai, H. (2002). Modulated chemical doping of individual carbon nanotubes. *Science*, *290*, 1552–1555.

33. Antonov, R. D., Johnson, A. T. (1999). Subband population in a single-wall carbon nanotube diode. *Phys. Rev. Lett.*, *83*, 3274–3276.

34. Lee, J. U. (2005). Photovoltaic effect in ideal carbon nanotube diodes. *Appl. Phys. Lett.*, *87*, 073101–3.

35. Oh, H., Kim, J. J., Song, W., Moon, S., Kim, N., Kim, J., Park, N. (2006). Fabrication of n-type carbon nanotube field-effect transistors by Al doping. *Appl. Phys. Lett.*, *88*, 103503–103505.

36. Lu, C., An, L., Fu, Q., Zhang, H., Murduck, J., Liu, J. (2006). Schottky diodes from asymmetric metal-nanotube contacts. *Appl. Phys. Lett.*, *88*, 133501–3.

37. Manohara, H. M., Wong, E. W., Schlecht, E., Hunt, B. D., Siegel, P. H. (2005). Carbon nanotube Schottky diodes using Ti-Schottky and Pt-Ohmic contacts for high frequency applications. *Nano Lett.*, *5*, 1469–1474.

38. Yang, M. H., Teo, K. B. K., Milne, W. I., Hasko, D. G. (2005). Carbon nanotube Schottky diode and directionally dependent field-effect transistor using asymmetrical contacts. *Appl. Phys. Lett.*, *87*, 253116–3.

39. Kong, J., Soh, H. T., Cassell, A. M., Quate, C. F., Dai, H. (1998). Synthesis of individual single-walled carbon nanotubes on patterned silicon wafers. *Nature*, *395*, 878–881.

40. Ulman, A. (1991). *An Introduction to Ultrathin Organic Films; from Langmuir–Blodgett to Self-Assembly*. Academic Press, San Diego, CA.

41. Kong, J., Dai, H. (2001). Full and modulated chemical gating of individual carbon nanotubes by organic amine compounds. *J. Phys. Chem. B*, *105*, 2890–2893.

42. Michaelson, H. B. (1977). The work function of the elements and its periodicity. *J. Appl. Phys.*, *48*, 4729–4733.

43. Dresselhaus, M. S., Dresselhaus, G., Avouris, Ph. (2001). *Carbon Nanotubes: Synthesis, Structure, Properties, and Applications*, Springer-Verlag, Berlin.

44. Park, S., Colombo, L., Nishi, Y., Cho, K. (2005). *Ab initio* study of metal gate electrode work function. *Appl. Phys. Lett.*, *86*, 073118–3.

45. Na, P. S., Park, N., Kim, J., Kim, H., Kong, K., Chang, H., Lee, J.-O. (2006). A field effect transistor fabricated with metallic single walled carbon nanotubes. *Fuller. Nanotub. Car. N.*, *14*, 141–149.

46. Chen, R. J., Choi, H. C., Bangsaruntip, S., Yenilmez, E., Tang, X., Wang, Q., Chang, Y. L., Dai, H. (2004). An investigation of the mechanisms of electronic sensing of protein adsorption on carbon. *J. Am. Chem. Soc.*, *126*, 1563–1568.

47. Heller, I., Janssens, A. M., Männik, J., Minot, E. D., Lemay, S. G., Dekker, C. (2008). *Nano Lett.*, *8*, 591.

48. Tang, X., Bangsaruntip, S., Nakayama, N., Yenilmez, E., Chang, Wang, Q. (2006). Carbon nanotube DNA sensor and sensing mechanism. *Nano Lett.*, *6*, 1632.

49. Lu, C., Akey, A., Wang, W., Herman, I. P. (2009). Versatile formation of CdSe nanoparticle-single-walled carbon nanotube hybrid structure. *J. Am. Chem. Soc.*, *131*, 3446–3447.

50. Kong, J., Franklin, N. R., Zhou, C., Chapline, M. G., Peng, S., Cho, K., Dai, H. (2000). Nanotube molecular wires as chemical sensors. *Science*, *287*, 622–625.

51. Peng, N., Zhang, Q., Chow, C. L., Tan, O. K., Marzari, N. (2009). Sensing mechanism for carbon nanotube based NH3 gas detection. *Nano Lett.*, *9*, 1626–1630.

52. Sun, Q., Wang, H. H., Xia, M. (2008). Single-walled carbon nanotubes modified with Pd nanoparticles: unique building blocks for high-performance, flexible hydrogen sensors. *J. Phys. Chem. C*, *112*, 1250–1259.

53. Qi, P., Vermesh, O., Grecu, M., Javey, A., Wang, Q., Dai, H. (2003). Toward large arrays of multiplex functionalized carbon nanotube sensors for highly sensitive and selective molecular detection. *Nano Lett.*, *3*, 347–351.

54. Star, A., Han, T.-R., Joshi, V., Gabriel, J.-C. P., Gruner, G. (2004). Nanoelectronic carbon dioxide sensors. *Adv. Mater.*, *16*, 2049–2052.

55. Novak, J. P., Snow, E. S., Houser, E. J., Park, D., Stepnowski, J. L., McGill, R. A. (2003). Nerve agent detection using networks of single-walled carbon nanotubes. *Appl. Phys. Lett.*, *83*, 4026–4028.

56. Zhang, T., Nix, M. B., Yoo, B.-Y., Deshusses, M. A., Myung, N. V. (2006). Electrochemically functionalized single-walled carbon nanotube gas sensor. *Electroanalysis*, *18*, 1153–1158.

57. Ellis, D. L., Zakin, M. R., Bernstein, L. S., Rubner, M. F. (1996). Conductive polymer films as ultrasensitive chemical sensors for hydrazine and monomethylhydrazine vapour. *Anal. Chem.*, *68*, 817–822.

58. Hirsch, A. (2002). Functionalization of single-walled carbon nanotubes. *Angew. Chem. Int. Ed.*, *41*, 1853–1859.

59. Hu, C.-Y., Xu, Y.-J., Duo, S.-W., Zhang, R.-F., Li, M.-S. (2009). Non-covalent functionalization of carbon nanotubes with surfactants and polymers. *J. Chin. Chem. Soc.*, *56*, 234–239.

60. Witus, L. S., Rocha, J.-D. R., Yuwono, V. M., Paramonov, S. E., Weisman, R. B., Hartgerink, J. D. (2007). Peptides that non-covalently functionalize single-walled carbon nanotubes to give controlled solubility characteristics. *J. Mater. Chem.*, *17*, 1909–1915.

61. Chen, R. J., Zhang, Y., Wang, D., Dai, H. (2001). Noncovalent sidewall functionalization of single-walled carbon nanotubes for protein immobilization. *J. Am. Chem. Soc.*, *123*, 3838–3839.

62. Besteman, K., Lee, J.-O., Wiertz, F. G. M., Heering, H. A., Dekker, C. (2003). Enzyme-coated carbon nanotubes as single-molecule biosensors. *Nano Lett.*, *3*, 727–730.

63. Li, C., Currelli, M., Lin, H., Lei, B., Ishikawa, F. N., Datar, R., Cote, R., Thomson, M., Zhou, C. (2005). Complementary detection of prostate-specific antigen using In_2O_3 nanowires and carbon nanotubes. *J. Am. Chem. Soc.*, *127*, 12484–12485.

64. So, H.-M., Park, D.-W., Kim, Y.-H., Choi, S. Y., Kim, S. C., Chang, H., Lee, J.-O. (2008). Detection and titer estimation of *Escherichia coli* using aptamer-functionalized single-walled carbon nanotube field effect transistors. *Small*, *4*, 197–201.

65. Simmons, T. J., Bult, J., Hashim, D. P., Linhardt, R. J., Ajayan, P. M. (2009). Non-covalent functionalization as an alternative to oxidative acid treatment of single wall carbon nanotubes with applications for polymer composites. *ACS Nano*, *3*, 865–870.

66. Keren, K., Berman, R. S., Buchstab, E., Sivan, U., Braun, E. (2003). DNA-templated carbon nanotube field effect transistor. *Science*, *302*, 1380–1382.

67. Maune, H. T., Han, S., Barish, R. D., Bockrath, M., Godard, W. A. III, Rothemund, P. W. K., Winfree, E. (2010). Self-assembly of carbon nanotubes into two-dimensional geometries using DNA origami templates. *Nat. Nanotechnol.*, *5*, 61–66.

68. Zheng, M., Jagota, A., Semke, E. D., Diner, B. A., Mclean, R. S., Lustig, S. R., Richardson, R. E., Tassi, N. G. (2003). DNA-assisted dispersion and separation of carbon nanotubes. *Nat. Mater.*, *2*, 338–342.

69. Zheng, M., Jagota, A., Strano, M. S., Santos, A. P., Barone, P., Chou, S. G., Diner, B. A., Dresselhaus, M. S., Mclean, R. S., Onoa, G. B., Samsonidze, G. G., Semke, E. D., Usrey, M., Walls, D. J. (2003). Structure-based carbon nanotube sorting by sequence-dependent DNA assembly. *Science*, *302*, 1545–1548.

70. Meng, S., Kaxiras, E. (2009). Interaction of DNA with CNTs: properties and prospects for electronic sequencing. In *Biosensing using Nanomaterials* (Ed.: Merkoçi, A.), John Wiley & Sons, Inc., Hoboken, New Jersey.

71. Daniel, S., Rao, T. P., Rao, K. S., Rani, S. U., Naidu, G. R. K., Lee, H.-Y., Kawai, T. (2007). A review of DNA functionalized/grafted carbon nanotubes and their characterization. *Sens. Actuators, B*, *122*, 672–682.

72. Tanaka, T., Jin, H., Miyata, Y., Kataura, H. (2008). High-yield separation of metallic and semiconducting single-wall carbon nanotubes by agarose gel electrophoresis. *Appl. Phys. Express*, *1*, 114001–114001-3.

73. Tu, X., Manohar, S., Jagota, A., Zheng, M. (2009). DNA sequence motifs for structure-specific recognition and separation of carbon nanotubes. *Nature*, *460*, 250–253.

74. Shin, S. R., Lee, C. K., So, I., Jeon, J.-H., Kang, T. M., Kee, C., Kim, S. I., Spinks, G. M., Wallace, G. G., Kim, S. J. (2007). DNA-wrapped single-walled carbon nanotube hybrid fibers for supercapacitors and artificial muscles. *Adv. Mater.*, *20*, 466–470.

75. Lu, Y., Bangsaruntip, S., Wang, X., Zhang, L., Nishi, Y., Dai, H. (2006). DNA functionalization of carbon nanotubes for ultrathin atomic layer deposition of high k dielectrics for nanotube transistors with 60 mV/decade switching. *J. Am. Chem. Soc.*, *128*, 3518–3519.

76. Star, A., Tu, E., Niemann, J., Gabriel, J. P., Joiner, C. S., Valcke, C. (2006). Label-free detection of DNA hybridization using carbon nanotube network field effect transistors. *Proc. Natl. Acad. Sci. U.S.A.*, *103*, 921–926.

77. Jeng, E. S., Moll, A. E., Roy, A. C., Gastala, J. B., Strano, M. S. (2006). Detection of DNA hybridization using the near-infrared band-gap fluorescence of single-walled carbon nanotubes. *Nano Lett.*, *6*, 371–375.

78. Marchniak, R. A., Garcia-Blanco, M. A., Sharp, P. A. (1990). Identification and characterization of a HeLa nuclear protein that specifically binds to trans-activation-reponse (TAR) element of human immunodeficiency virus. *Proc. Natl. Acad. Sci. U.S.A.*, 87, 3624–3628.
79. Ellington, A. D., Szostak, J. W. (1990). *In vitro* selection of RNA molecules that bind specific ligands. *Nature*, *346*, 818–822.
80. Heller, D. A., Jeng, E. S., Yeung, T.-K., Martinez, B. M., Moll, A. E., Gastala, J. B., Strano, M. S. (2006). Optical detection of DNA conformational polymorphism on single-walled carbon nanotubes. *Science*, *311*, 508–511.

OPTOELECTRONICS BASED ON SUPRAMOLECULAR ASSEMBLIES

CHAPTER 21

MIMICKING PHOTOSYNTHESIS WITH FULLERENE-BASED SYSTEMS

Juan Luis Delgado,[1,2] *Dirk M. Guldi,*[3] *and Nazario Martín*[1,2]

[1]Departamento de Química Orgánica, Universidad Complutense de Madrid, Madrid, Spain
[2]IMDEA-Nanociencia, Ciudad Universitaria de Cantoblanco, Madrid, Spain
[3]Department of Chemistry and Pharmacy Interdisciplinary Center for Molecular Materials (ICMM), Friedrich-Alexander-Universitaet Erlangen-Nuenberg Egerlandstr. 3, Erlangen, Germany

21.1 INTRODUCTION

An increasing demand for energy is currently the most important problem facing mankind. The "fire age," on which our civilization has been based from the very beginning, is approaching its end. The rapid consumption of fossil fuel is expected to cause unacceptable environmental problems such as the greenhouse effect, which could lead to disastrous climatic consequences [1,2]. Thus, renewable and clean energy resources are definitely required in order to stop global warming [1,2]. Among renewable energy resources, solar energy is by far the largest exploitable resource. Nature harnesses solar energy for its production by photosynthesis, and fossil fuel is the product of photosynthesis [1,2]. Thus, extensive efforts have been devoted to develop artificial systems for the efficient and economical conversion of solar energy into stored chemical fuels [3–7]. Obviously, it is not easy to achieve a solution to all global environment and energy resource problems [8,9]. However, the importance and complexity of energy transfer and electron transfer processes in photosynthesis have inspired design and synthesis of a large number of donor–acceptor ensembles, including nanocomposites, that can mimic the energy transfer or electron transfer process in photosynthesis. The specific objective of this chapter is to describe recent developments of electron donor–acceptor nanocomposites and their applications aiming at efficient and economical conversion of solar energy into stored chemical fuels.

Supramolecular Soft Matter: Applications in Materials and Organic Electronics, First Edition.
Edited by Takashi Nakanishi.
© 2011 John Wiley & Sons, Inc. Published 2011 by John Wiley & Sons, Inc.

21.2 FULLERENES IN ELECTRON TRANSFER

Fullerenes are now readily available and exhibit exciting characteristics. For example, the delocalization of charges within the giant, spherical carbon framework together with the rigid, confined structure of the aromatic π-sphere offers unique opportunities for stabilizing charged entities [10–28]. Above all, the small reorganization energies of fullerenes in charge transfer reactions have led to a notable breakthrough in synthetic electron donor–acceptor systems by providing accelerated charge separation and decelerated charge recombination [29–34]. In light of the above, fullerenes have emerged as a particular promising electron-accepting building block in electron donor–acceptor conjugates/hybrids that give rise to long-lived radical ion pair states (Fig. 21-1) [10–28]. Studying multifunctional fullerene nanostructures—their design, synthesis, characterization, and performance—mandates a careful control over composition, interchromophore separation/angular relationship, overall dynamical and stimulus-induced reorganization, and electronic coupling. The overall goal is to achieve control over the organization of the assemblies and their physical and chemical properties (i.e., electronic structures, etc.), thereby enhancing desired functionalities through simple external parameters or variables, with the intent of creating new tailored materials.

21.2.1 Fullerene-Based Electron Donor–Acceptor Conjugates

The simplest possible combination of an electron-accepting fullerene is with an electron donor linked by a charge-mediating bridge. As electron donors, a wide

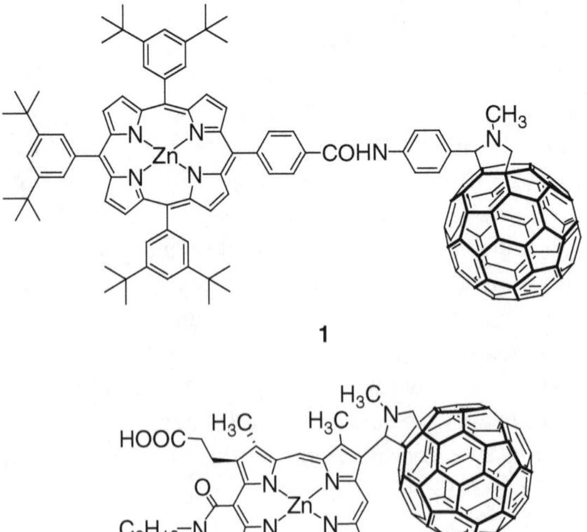

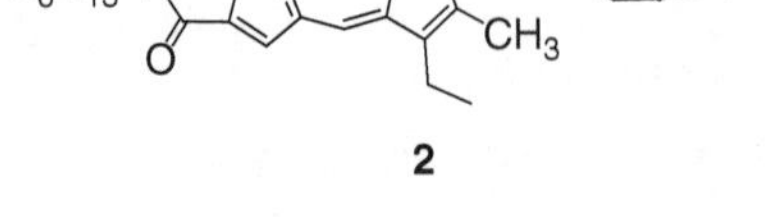

Figure 21-1 Leading examples of donor-fullerene-linked dyads exhibiting the formation of a long-lived charge-separated state.

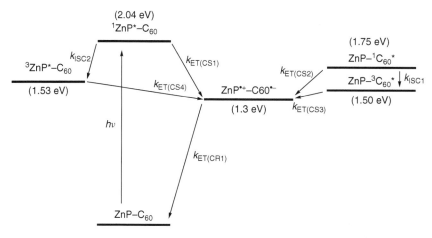

Scheme 21-1   Reaction scheme and energy diagram for **1** in PhCN.

range of building blocks—porphyrins [35–45], phthalocyanines [46,47], amines [48], polycondensed aromatics [49], transition complexes [50–52], carotenoids [53], ferrocenes [54,55], tetrathiafulvalenes (TTFs) [56], and others [57]—have been employed.

Scheme 21-1 summarizes the photophysical processes taking place in a prototype electron donor–acceptor conjugate, that is, $ZnP-C_{60}$ (**1**), and the corresponding energetics in benzonitrile [37]. In particular, photoexcitation of **1** in polar solvents is followed by charge transfer events that commence from the ZnP singlet excited state (9.5×10^9 s^{-1}), the ZnP triplet excited state ($> 1.5 \times 10^7$ s^{-1}), the C_{60} singlet excited state (5.5×10^9 s^{-1}), or, the C_{60} triplet excited state (1.5×10^7 s^{-1}). Common to these events is the formation of a radical ion pair state, namely, the one-electron-oxidized ZnP and the one-electron-reduced C_{60}, $ZnP^{\bullet+}-C_{60}^{\bullet-}$. The charge-separation efficiencies evolving from the ZnP singlet excited state and the C_{60} singlet excited state were determined as 95% and 23%, respectively. In a mechanistic view, the unquenched fractions of the singlet excited state precursor are subject to intersystem crossing processes to form the corresponding triplet manifolds. The latter are then the inception to the quantitative radical ion pair state formation [37].

Overall, the efficiency of $ZnP^{\bullet+}-C_{60}^{\bullet-}$ formation reaches a remarkable 99% in benzonitrile. Regardless of the formation pathway, $ZnP^{\bullet+}-C_{60}^{\bullet-}$ recombines to regenerate the singlet ground state with a remarkable lifetime of 0.77 μs (1.3×10^6 s^{-1}). The rate constant is also remarkable, especially when considering that it is nearly four orders of magnitude smaller than the dominating charge transfer event that starts from the ZnP singlet excited state [37]. Such a synergy of charge transfer events, that is, fast charge separation and slow charge recombination, for **1** in polar solvents is in marked contrast to non-fullerene-containing electron donor–acceptor conjugates, in which the charge recombination is faster than the charge separation [37].

Energy transfer between the energetically high-lying ZnP singlet excited state and the low-lying C_{60} singlet excited state competes with the direct charge transfer [10–28]. As a matter of fact, the outcome of these processes—energy versus charge transfer—is influenced by the specific combination of fullerene and donors as well as environmental aspects such as the solvent and the bridge that connects the fullerene with the donor. Moreover, the resulting radical ion pair state may decay, besides reinstating the singlet ground state, to afford singlet and triplet excited states of either the fullerene or the donor. The latter is a function of the relative energy levels of the states involved—radical ion pair state, singlet excited state, and triplet excited state—and is driven by the small reorganization energies of fullerene-containing electron donor–acceptor conjugates [10–28].

When considering chlorin-C_{60} (**2**), a radical ion pair state lifetime of 120 s (8.3×10^{-3} s^{-1}) was observed in frozen benzonitrile (123 K) [45], which constitutes the slowest charge recombination ever reported in a simple electron donor–acceptor conjugate [10–28]. The quantum yield of radical ion pair state formation was, however, as low as 12%, which is significantly inferior to the charge transfer performance seen on photoinduced charge separation in analogous ZnP systems—close to 100% when considering the fluorescence quenching of ZnP [45].

Using short bridges as connectors between ZnP and C_{60} is often accompanied by the formation of a short-lived exciplex [42,43]. Importantly, in such electron donor–acceptor conjugates, some fraction of the exciplex is converted into a long-lived radical ion pair state, whereas the remaining fraction of the exciplex decays rapidly to the ground state [42,43]. The low quantum yield may arise from the predominant decay of the exciplex state to the ground state rather than affording the radical ion pair state.

In summary, fullerene-containing electron donor–acceptor conjugates give rise to the successful mimicking of the primary events in photosynthesis, that is, light harvesting, unidirectional energy transfer, charge transfer, and so on. This led to the exploitation of fullerenes in multistep charge transfer systems in the form of triads [37,58,59], tetrads [60,61], pentads [62,63], and even hexads [64,65]. Leading examples of such multichromophore conjugates (**3–6**) are illustrated in Fig. 21-2.

This paragraph focuses on a carotenoid system (**3**), which is linked covalently to the free base analog of the aforementioned ZnP–C_{60} system [58]. A built-in energy gradient facilitates in C–H_2P–C_{60} the stepwise sequence of initial H_2P singlet excited state, charge separation to yield the one-electron-oxidized H_2P and the one-electron-reduced C_{60} (C–H_2P$^{\bullet+}$–$C_{60}$$^{\bullet-}$), and charge shift to afford the one-electron-oxidized C and the one-electron-reduced C_{60} (C$^{\bullet+}$–H_2P–$C_{60}$$^{\bullet-}$). Interestingly, in 2-methyltetrahydrofuran, two processes lead to C–H_2P$^{\bullet+}$–$C_{60}$$^{\bullet-}$. One is the process that evolves from the H_2P singlet excited state, and the other is the counterpart that starts with the C_{60} singlet excited state. A subsequent charge shift produces C$^{\bullet+}$–H_2P–$C_{60}$$^{\bullet-}$ with a total quantum yield of 88%. The final charge-separated state decays by charge recombination (0.34 μs) to yield the C triplet excited state rather than the ground state because of the small reorganization energy [58]. Similar multichromophore conjugates were

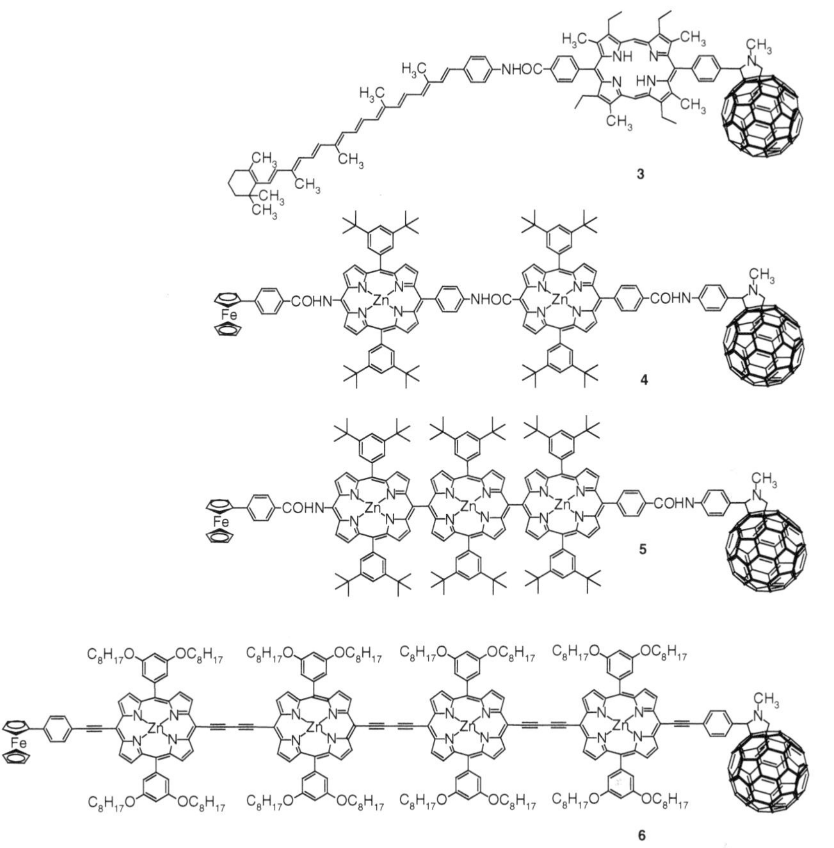

Figure 21-2 Leading examples of donor-fullerene-linked multichromophore systems **3–6**.

prepared in the form of C_{60}-containing tetrads, pentads, and hexads. The final, distant radical ion pair states lived for 1.6 s (**4** at 163 K) and 0.53 s (**5** at 163 K), with total charge-separation efficiencies of 34% and 83%, respectively [61,62]. In stark contrast, **6** gives rise to a very short lifetime of the final radical ion pair state (7.8 ns) probably because of superexchange-mediated charge recombination [65].

21.2.2   Fullerene-Based Electron Donor–Acceptor Hybrids

Among the tools exploitable for the creation of new assemblies are organization principles, such as biomimetic methodologies, that help regulate size, shape, and function down to the molecular scale. In this regard, a variety of non-covalently assembled electron donor–acceptor hybrids have been prepared to probe photoinduced energy and charge transfer processes in solutions toward the realization of efficient solar energy conversion systems [66–75]. Figure 21-3 illustrates leading examples of electron donor–acceptor hybrids (**7–10**). TTF—as

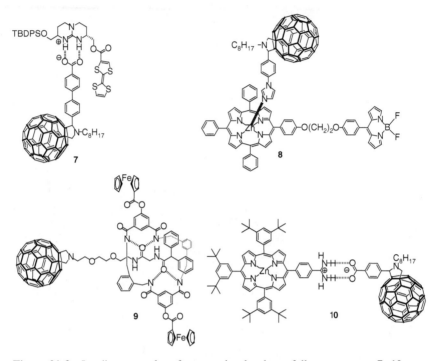

Figure 21-3 Leading examples of supramolecular donor-fullerene systems **7–10**.

electron potent electron donor—was assembled with C_{60} through a complementary guanidinium–carboxylate ion pair to yield **7** [68]. The lifetime measured for the radical ion pair state was 1.0 ms, which is about one order of magnitude less than those reported for covalently linked C_{60}–TTF conjugates [68]. Compound **8**, on the other hand, was assembled by axially coordinating imidazole-containing C_{60} to ZnP of a ZnP–boron dipyrrin system [69]. Selective excitation of boron dipyrrin resulted in efficient energy transfer (9.2×10^9 s^{-1}), with a quantum yield of 83%. In **8**, subsequent charge transfer between the ZnP singlet excited state and C_{60} powered the radical ion pair state formation (4.7×10^9 s^{-1}), for which a quantum yield of 90% was estimated [69]. A molecular rotaxane shuttle (**9**) consisting of two ferrocene-appended macrocycles and a C_{60}-tethered axle was prepared to modulate the kinetics of charge separation and charge recombination [74]. In nonpolar solvents, hydrogen bonding between the macrocycle and the fumaramide template of the axle is strengthened. As a result, the macrocycle shuttles to the opposite end of the thread far from C_{60}, leading to slow charge separation and charge recombination. On the other hand, in polar solvents, the hydrogen-bonding interaction is weakened. Therefore, the macrocycle shuttles to the opposite end of the thread close to the C_{60} moiety, resulting in fast charge separation and charge recombination [74]. A two-point amidinium-carboxylate binding motif guarantees extraordinary stabilization for a set of ZnP–C_{60} hybrids (**10**) [72]. Association constants reach up to 10^7 M^{-1}. Exceptionally strong electronic couplings stem from such binding, which in turn facilitate faster, more efficient, and longer lived formation of radical ion pair states (i.e.,~10 µs in THF (tetrahydrofuran)) when compared to similar covalent ZnP–C_{60} conjugates (i.e., ~1 µs in THF). Most importantly, such remarkable radical ion pair lifetimes outperform previously reported ensembles based on non-amidinium-carboxylatebinding motif by several orders of magnitude. These results point unmistakably to the fundamental advantages of strong and highly directional hydrogen-bonding networks in assisting electron transfer processes.

21.3 FULLERENES IN SOLAR CELLS

Organic solar cells are constituted by semiconducting organic materials formed by contacting electron donor and acceptor compounds (p/n-type). A further improvement in the construction of organic photovoltaic devices (OPVs) consists of the realization of interpenetrated networks of the donor and acceptor materials. In such bulk heterojunction (BHJ) solar cells, the dramatic increase in the contact area between D/A materials leads to a significant increase in the number of generated excitons as well as their dissociation into free charge carriers and hence on the power conversion efficiency (PCE).

Transformation of solar energy into electricity occurs through a series of optical and electronic processes, which basically involve (i) optical absorption of sunlight and formation of the exciton, (ii) exciton migration to the donor–acceptor interface, (iii) exciton dissociation into charges (electron and holes), and (iv) charge transport and collection at the electrodes. All these steps are not totally understood at present, and a number of research groups are currently dedicated to unravel essential aspects in the search for better energy transformation efficiencies [75]. In contrast to inorganic semiconductors, which on light excitation form free electron and hole carriers, organic compounds form excitons whose dissociation into free carriers is not straightforward. Owing to their low dielectric constants, the donor–acceptor interface has to be reached to dissociate into free charges. The driving force for this exciton dissociation is provided by the energy difference between the molecular orbitals of the donor and acceptor.

The transport of the generated free charges toward the electrodes represents another important issue. Organic materials typically show charge carrier mobilities significantly lower (around $10^{-5}-1$ cm²/V/s ranging from amorphous to crystalline materials) than those of inorganic semiconductors (around 10^2-10^3 cm²/V/s). Since charge carrier mobility is strongly dependent on the molecular organization of the material, in order to achieve better efficiencies, a good control on the morphology of the donor–acceptor materials at a nanometer scale is necessary. Furthermore, an efficient charge carrier mobility is essential to prevent charge recombination (geminal or bimolecular) processes, which result in lower energy conversion efficiencies.

In summary, an appropriate choice of the donor and acceptor materials is critical to ensure a good match between them in terms of optical, electronic, and morphological properties, which eventually determine the effective photocurrent and performance of the PV (photovoltaic) device. Therefore, the rational design of new materials able to improve some specific demands within the PV device is critical for the successful development of competitive solar cells. The following sections discuss those materials mainly used in the preparation of "all-organic" solar cells, according to their chemical nature and composition.

The term "plastic solar cells" has been coined for those PV devices using polymers in their constitution. Thus, although a variety of donor and acceptor polymers have previously been used for constructing PV devices, the highest efficiencies have been achieved by mixing electron donor polymers with fullerenes as acceptors. Polymer:fullerene (BHJ)-based solar cells are considered suitable candidates to obtain low-cost renewable energy from a large-area, flexible, plastic material [76].

Fullerenes and their derivatives possess important electronic properties such as small reorganization energy, high electron affinity, ability to transport charge, and stability, which make them one of the best candidates to act as electron acceptor components in BHJ PV devices [77]. On the other hand, MDMO-PPV {poly[2-methoxy-5-(3'7'-dimethyloctyloxy)]-1,4-phenylenevinylene} (**11**) and P3HT [poly(3-hexylthiophene-2,5-diyl)] (**12**) have been by far the most studied π−conjugated polymers as donor components in BHJ devices (Fig. 21-4). Actually, both polymers have been thoroughly studied from a morphological and electronic standpoint to form BHJ devices by mixing with fullerene derivatives, namely, PCBM ([6,6]-phenyl-C_{61} butyric acid methyl ester) (**20**, Fig. 21-5). In fact, the best fullerene/polymer combination has been obtained with PCBM/P3HT(regioregular) mixtures, reaching energy conversion efficiencies in the range of 5% [78].

The following section briefly describes the most relevant achievements both in the chemistry of polymers as well as in fullerene chemical modifications for PV purposes.

21.3.1   Polymers for BHJ Organic Solar Cells

The chemistry of π−conjugated polymers has been focused on tuning their energy levels (HOMO (highest occupied molecular orbital) and LUMO (lowest occupied molecular orbital)) in order to have good control on the band gap of the polymer. This is a key issue to modulate the light harvesting properties of the polymer [low band gap polymers (lower than 1.5 eV) absorb in the visible region] as well as the light energy match with the acceptor moiety, which eventually controls the open-circuit voltage (V_{oc}) values and hence the cell PCE. Since π−conjugated polymers constitute a specific field by their own right, we discuss the trends currently followed in polymer chemistry to improve the performance of those "classical" MDMO-PPV and P3HT. Some of the most relevant polymers synthesized so far are shown in Fig. 21-4. All of them (**13−19**) involve the use of alternating donor and acceptor moieties within the polymer backbone since they allow a fine tune of the band gap of the polymer, as well as an improvement of planarity and hence of the mobility of the charge carriers. Thus, a variety of soluble electroactive carbo- and heterocycle derivatives such as fluorene, thiophene, cyclopentadithiophene, and carbazole have been used as

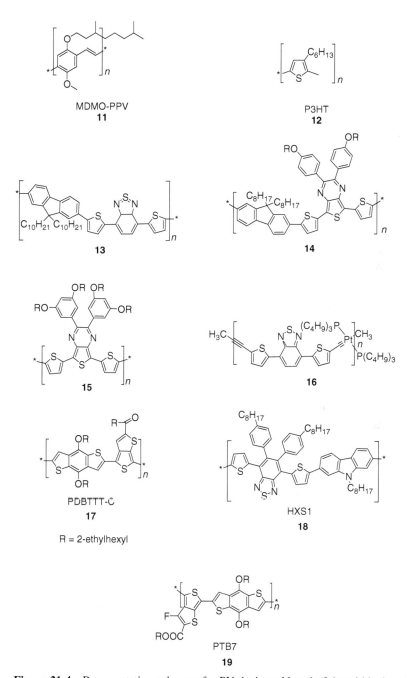

Figure 21-4 Representative polymers for PV devices: **11**, poly(3-hexylthiophene) (P3HT); **12**, poly[2-methoxy-5-(3′7′-dimethyloctyloxy)-1,4-phenylene vinylene] (MDMO-PPV); **13**, poly[*N*-9′-heptadecanyl-2,7-carbazole-*alt*-5,5-(4′7′di-2-thienyl-2′,1′,3′-benzothiadiazole)] [79]; **14**, APFO-green **5** [80]; **15**, poly{5,7-di-2-thienyl-2, 3-bis(3,5-di(2-ethylhexyloxy)phenyl)thieno[3,4-*b*]pyrazine}[81]; **16**, platinum(II) polyyne polymer [82]; **17**, PDBTTT-C [83]; **18**, HXS1 [84]; and **19**, PTB7 [85].

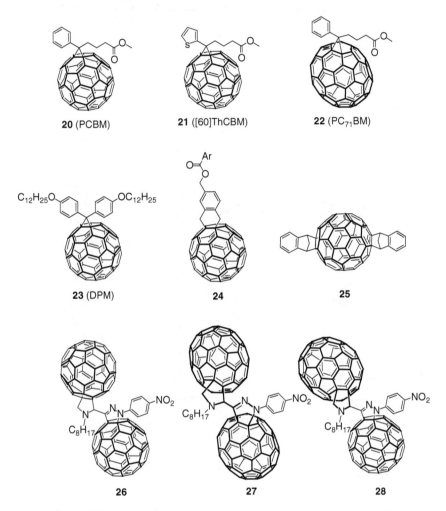

Figure 21-5 Different modified fullerenes used as successful acceptors for PV devices.

electron-rich units, whereas benzothiadiazole and thienopyrazine have recently been used as the electron-poor units.

Among the different electron donor polymers prepared so far, polyfluorene copolymers prepared by Andersson *et al.*, known as *APFO* (*alternating polyfluorene copolymers*) (**14**), have allowed good control on the band gap, covering the solar spectrum up to 1000 nm. Power conversion efficiencies more than 4% have been demonstrated with APFOs in blends with PCBM [86].

Typically, π–conjugated polymers are endowed with flexible lateral chains to improve their solubility. However, in many cases, the side chains prevent the polymer backbones from close packing. Recently, a planar polymer conformation has been achieved in the polymer poly(2-(5-(5,6-bis(octyloxy)-4-(thiophen-2-yl)benzo[*c*][1,2,5]thiadiazol-7-yl)thiophen-2-yl)-9-octyl-9*H* carbazole, HXS-1) bearing two octyloxy chains on the benzothiazole ring and an octyl chain on the carbazole ring. Semiconducting polymers based on alternating thieno[3,4-*b*]thiophene and benzodithiophene units such as **17** and **19** show an excellent PV behavior. The stabilization of the quinoid structure from the thieno[3,4-*b*]thiophene moiety leads to remarkable low-band-gap polymers (around 1.6 eV). A further optimization has been achieved with the new PTB7, in which the presence of the fluorine atom leads to a lower HOMO level, thus enhancing the V_{oc} parameter. With an average molecular weight of about 97.5 kDa and a polydispersity index of 2.1, this soluble polymer exhibits strong absorption from 550 to 750 nm. Interestingly, by blending this polymer with $PC_{71}BM$ (**22**), which complements the absorption in the visible range, it has been possible to prepare a PV device exhibiting an energy conversion efficiency of 7.40%. This value represents the highest efficiency reported so far for a polymeric solar cell [85]. Since the development of new π–conjugated donor polymers for improving the efficiency in BHJ PV devices is currently a broad and active field, for further details the reader is referred to some of the excellent monographs currently available in the literature [87].

21.3.2   Fullerene Derivatives for BHJ Organic Solar Cells

The most widely used configuration of polymer solar cells is based on the use of a fullerene derivative as the acceptor component. Indeed, fullerenes have been demonstrated to be the ideal acceptor because of their singular electronic and geometrical properties and the ability of their chemically functionalized derivatives to form a bicontinous phase network with π–conjugated polymers acting as electron-conducting (n-type) material.

A variety of chemically modified fullerenes were initially synthesized for blending with semiconducting polymers (namely, PPV (poly(*p*-phenylene vinylene)) derivatives) and to prepare PV devices. These fullerene derivatives were covalently linked to different chemical species such as electron acceptors [88], electron donors [89], π–conjugated oligomers [90], and so on (Fig. 21-5). However, in general, the blends obtained resulted in PV devices exhibiting low energy conversion efficiencies [91].

The best known and most widely used fullerene derivative as acceptor for PV devices is PCBM (**20**), first prepared by Hummelen *et al.* in 1995 [92]. Since its first reported application in solar cells [93], it has been by far the most widely used fullerene, being considered as a benchmark material for testing new devices.

This has led to the synthesis of many other PCBM analogs (**21, 22**) [94] in an attempt to increase the efficiencies of the cells by improving the processability, the stability, or PV parameters such as the open-circuit voltage (V_{oc}) by raising the LUMO energies of the fullerene acceptor (Fig. 21-5).

In this regard, only small shifts (<100 meV) of the LUMO level have been obtained by attaching a single substituent on the fullerene sphere, even by using electron-donating groups. In contrast, significantly higher V_{oc} values have been achieved through the polyaddition of organic addends to the fullerene cage ($\sim$100 mV raising the LUMO per saturated double bond). Recently, an externally verified PCE of 4.5% has been reported by Hummelen *et al.* employing a regioisomeric mixture of PCBM bis-adducts as a result of an enhanced open-circuit voltage while maintaining a high short-circuit current (J_{sc}) and fill factor (FF) values [95].

The cyclopropanation of superior fullerenes to form PCBM analogs is more complex than that of C_{60}. Indeed, the low symmetry and the presence of more than one reactive double bond are often responsible for the formation of regioisomeric mixtures. Nevertheless, the loss of symmetry of C_{70} induces a stronger absorption, even in the visible region. As a result, $PC_{71}BM$ [96] is considered a suitable candidate for more efficient polymer solar devices. Moreover, such devices performed with the highest verified efficiency determined so far in a BHJ solar cell, with an internal quantum efficiency approaching 100% [97]. Analogously, $PC_{84}BM$ [98] has been obtained as a mixture of three major isomers. The stronger electron affinity and the diminished solubility gave rise, however, to poor power conversion efficiencies.

Although PCBMs are the acceptors that provide the best performance guarantee at present, it does not mean that they are necessarily the optimal fullerene derivatives. Therefore, a variety of other fullerene derivatives [99] have been synthesized in order to improve device efficiency or to achieve a better understanding of the dependence of the cell parameters on the structure of the acceptor.

Among the different modified fullerenes prepared so far, diphenyl-methanofullerene (**23**) (DPM12) prepared by Martín *et al.* is another type of methanofullerene endowed with two alkyl chains to drastically improve the solubility of the acceptor in the blend, reaching efficiencies of about 3% (Fig. 21-5). Although the LUMO energy level for DPM12 is the same as that for PCBM, an increase in the V_{oc} of 100 mV for DPM12 over PCBM has been observed [100]. This is currently an important issue for improving the design of future fullerene-based acceptors.

Devices based on dihydronaphthylfullerene benzyl alcohol benzoic ester (**24**) synthesized by Fréchet *et al.* reported one of the highest PCEs (up to 4.5%) for a non-PCBM-based polymer–fullerene solar cell [101].

A remarkable bis-adduct fullerene derivative formed by two indene units covalently connected to the fullerene sphere of C_{60} (**25**) has recently been reported by Hou and Li [102]. Interestingly, the presence of two aryl groups improves visible absorption compared to the parent PCBM; it also increases solubility (>90 mg/mL in chloroform) and the LUMO energy level, which is 0.17 eV higher than that of PCBM. Surprisingly, PV devices formed with P3HT as the semiconducting polymer revealed PCE values of 5.44% under illumination of AM1.5, 100 mW/cm^2, thus surpassing PCBM, which afforded an efficiency of 3.88% under the same experimental conditions.

A major drawback in the synthesis of bis-adducts of fullerenes is that the products formed are constituted by a mixture of regioisomers, which are not separated because of experimental difficulties. This fact, however, does not seem to have a strong influence on the PV parameters and power conversion efficiencies. Nevertheless, from a chemical viewpoint, the synthesis of isomerically pure bis-adducts and multiadducts of fullerenes is still an open question that should be properly addressed by the chemical community.

Although fullerene dimers are known to exhibit low solubilities [103], recently a series of soluble homo (**26**: $C_{60}-C_{60}$ and **27**: $C_{70}-C_{70}$) and heterodimers (**28**: $C_{60}-C_{70}$) formed from C_{60} and C_{70} derivatives have been prepared and explored as less known acceptors for PV devices [104]. The synthesis of these pyrrolidino–pyrazolino-fullerene dimers has been carried out in a straightforward manner from the formyl-containing pyrazolino[60] or [70] fullerene [105], which is used as the aldehyde component in the subsequent pyrrolidine formation by 1,3-dipolar cycloaddition of the respective azomethyne ylide with the fullerene. As expected, [70]fullerene dimers showed better absorption in the visible spectrum than the analogs of [60]fullerene. Interestingly, nonoptimized BHJ solar cells formed by blending with P3HT exhibited quantum conversion efficiencies of 37%, with a PCE of 1.0% (Fig. 21-5).

Although some of the fullerene derivatives prepared so far exhibit good performances in PV devices, the synthesis of new fullerene derivatives with stronger visible absorption and higher LUMO energy levels than PCBM is currently a challenge for all those chemists engaged in the chemical modification of fullerenes for PV applications.

21.4 SUMMARY

The most important issue in all the above-mentioned materials for improving the cell performance is the precise control of the energy levels of donor and acceptor components as well as the network morphology to maximize the mobility of the charge carriers. In this regard, the realization of nanostructured materials by means of a bottom-up supramolecular organization is currently in its infancy, and, therefore, more work is need for improving the materials used for PV applications.

In summary, organic solar cells are called to play an important role in satisfying the increasing energy demands of our society. Because of features such as low cost, flexibility, and lightness, these new PV cells—which should complement the commercially available silicon cells—are very appealing for a variety of new practical purposes. The outstanding achievements obtained so far put these solar cells closer to the market.

ACKNOWLEDGMENTS

This work has been supported by the EU (FUNMOLS FP7-212942-2), the MEC of Spain (CT2008-00795/BQU, and Consolider-Ingenio 2010C-07-25200, Nanociencia Molecular), and Comunidad de Madrid (MADRISOLAR-2, S2009/PPQ-1533). J.L.D. thanks the MICINN of Spain for a Ramón y Cajal Fellowship, cofinanced by the EU Social Funds.

REFERENCES

1. Lewis, N. S., Nocera, D. G. (2006) Powering the planet: chemical challenges in solar energy utilization. *Proc. Natl. Acad. Sci. U.S.A.*, *103*, 15729–15735.
2. Nocera, D. G. (2009) Living healthy on a dying planet. *Chem. Soc. Rev.*, *38*, 13–15.
3. Fukuzumi, S. (2008) Bioinspired energy conversion systems for hydrogen production and storage. *Eur. J. Inorg. Chem.*, *9*, 1351–1362.
4. Hambourger, M., Moore, G. F., Kramer, D. M., Gust, D., Moore, A. L., Moore, T. A. (2009) Biology and technology for photochemical fuel production. *Chem. Soc. Rev.*, *38*, 25–35.
5. Fukuzumi, S., Honda, T., Ohkubo, K., Kojima, T. (2009) Charge separation in metallomacrocycle complexes linked with electron acceptors by axial coordination. *Dalton Trans.*, *20*, 3880–3889.
6. Fukuzumi, S., Kojima, T. (2008) Photofunctional nanomaterials composed of multiporphyrins and carbon-based π–electron acceptors. *J. Mater. Chem.*, *18*, 1427–1439.
7. Gust, D., Kramer, D., Moore, A., Moore, T. A., Thomas, A., Vermaas, W. (2008) Engineered and artificial photosynthesis: human ingenuity enters the game. *Mater. Res. Soc. Bull.*, *33*, 383–387.
8. Turner, J. (2008) The other half of the equation. *Nat. Mater.*, *7*, 770–771.
9. Guldi, D. M. (2008) Let there be light—but not too much. *Nat. Nanotechnol.*, *3*, 257–258.
10. Imahori, H., Sakata, Y. (1997) Donor-linked fullerenes: photoinduced electron transfer and its potential application. *Adv. Mater.*, *9*, 537–546.
11. Prato, M. (1997) [60]Fullerene chemistry for materials science applications. *J. Mater. Chem.*, *7*, 1097–1109.
12. Martín, N., Sánchez, L., Illescas, B., Pérez, I. (1998) C_{60}-based electroactive organofullerenes. *Chem. Rev.*, *98*, 2527–2547.
13. Imahori, H., Sakata, Y. (1999) Fullerenes as novel acceptors in photosynthetic electron transfer. *Eur. J. Org. Chem.*, *10*, 445–2457.
14. Diederich, F., Gómez-López, M. (1999) Supramolecular fullerene chemistry. *Chem. Soc. Rev.*, *28*, 263–277.
15. Guldi, D. M. (2000) Fullerene: three dimensional electron acceptor materials. *Chem. Commun.*, 321–327.
16. Guldi, D. M., Prato, M. (2000) Excited state properties of C_{60} fullerene derivatives. *Acc. Chem. Res.*, *33*, 695–703.
17. Gust, D., Moore, T. A., Moore, A. L. (2001) Mimicking photosynthetic solar energy transduction. *Acc. Chem. Res.*, *34*, 40–48.
18. Armaroli, N. (2001) Photoactive mono- and polynuclear Cu(I)-phenanthrolines. A viable alternative to Ru(II)-polypyridines? *Chem. Soc. Rev.*, *30*, 113–124.
19. Guldi, D. M. (2002) Fullerene-porphyrin architectures; photosynthetic antenna and reaction center models. *Chem. Soc. Rev.*, *31*, 22–36.
20. Imahori, H., Mori, Y., Matano, Y. (2003) Nanostructured artificial photosynthesis. *J. Photochem. Photobiol.*, *C*, *4*, 51–83.
21. Fukuzumi, S. (2004) New perspective of electron transfer chemistry. *Org. Biomol. Chem.*, *1*, 609–620.
22. Imahori, H. (2004) Porphyrin-fullerene linked systems as artificial photosynthetic mimics. *Org. Biomol. Chem.*, *2*, 1425–1433.
23. Nierengarten, J.-F. (2004) Chemical modification of C_{60} for materials science applications. *New J. Chem.*, *28*, 1177–1191.
24. El-Khouly, M. E., Ito, O., Smith, P. M., D'Souza, F. (2004) Intermolecular and supramolecular photoinduced electron transfer processes of fullerene-porphyrin/phthalocyanine systems. *J. Photochem. Photobiol.*, *C*, *5*, 79–104.
25. Sánchez, L., Martín, N., Guldi, D. M. (2005) Materials for organic solar cells: the C_{60}/π–conjugated oligomer approach. *Chem. Soc. Rev.*, *34*, 31–47.
26. Imahori, H. (2007) Creation of fullerene-based artificial photosynthetic systems. *Bull. Chem. Soc. Jpn.*, *80*, 621–636.

27. Fukuzumi, S. (2008) Development of bioinspired artificial photosynthetic systems. *Phys. Chem. Chem. Phys.*, *10*, 2283–2297.

28. Araki, Y., Ito, O. (2008) Factors controlling lifetimes of photoinduced charge-separated states of fullerene-donor molecular systems. *J. Photochem. Photobiol., C*, *9*, 93–110.

29. Imahori, H., Hagiwara, K., Akiyama, T., Aoki, M., Taniguchi, S., Okada, T., Shirakawa, M., Sakata, Y. (1996) The small reorganization energy of C_{60} in electron transfer. *Chem. Phys. Lett.*, *263*, 545–550.

30. Guldi, D. M., Asmus, K.-D. (1997) Electron transfer from $C_{76}(C'_{2v})$ and C_{78} (D_2) to radical cations of various arenes: evidence for the Marcus inverted region. *J. Am. Chem. Soc.*, *119*, 5744–6745.

31. Tkachenko, N. V., Guenther, C., Imahori, H., Tamaki, K., Sakata, Y., Fukuzumi, S., Lemmetyinen, H. (2000) Near infra-red emission of charge-transfer complexes of porphyrin-fullerene films. *Chem. Phys. Lett.*, *326*, 344–350.

32. Imahori, H., Tkachenko, N. V., Vehmanen, V., Tamaki, K., Lemmetyinen, H., Sakata, Y., Fukuzumi, S. (2001) An extremely small reorganization energy of electron transfer in porphyrin-fullerene dyad. *J. Phys. Chem. A*, *105*, 1750–1756.

33. Imahori, H., Yamada, H., Guldi, D. M., Endo, Y., Shimomura, A., Kundu, S., Yamada, K., Okada, T., Sakata, Y., Fukuzumi, S. (2002) Comparison of reorganization energies for intra- and inter-molecular electron transfer. *Angew. Chem. Int. Ed.*, *41*, 2344–2347.

34. Fukuzumi, S., Ohkubo, K., Imahori, H., Guldi, D. M. (2003) Driving force dependence of inter-molecular electron-transfer reactions of fullerene. *Chem.—Eur. J.*, *9*, 1585–1593.

35. Kuciauskas, D., Lin, S., Seely, G. R., Moore, A. L., Moore, T. A, Gust, D. (1996) Energy and electron transfer in porphyrin-fullerene dyads. *J. Phys. Chem.*, *100*, 15926–15932.

36. Imahori, H., Hagiwara, K., Aoki, M., Akiyama, T., Taniguchi, S., Okada, T., Shirakawa, M., Sakata, Y. (1996) Linkage and solvent dependence of photoinduced electron transfer in zincporphyrin-C_{60} dyads. *J. Am. Chem. Soc.*, *118*, 11771–11782.

37. Imahori, H., Tamaki, K., Guldi, D. M., Luo, C., Fujitsuka, M., Ito, O., Sakata, Y., Fukuzumi, S. (2001) Modulating charge separation and charge recombination dynamics in porphyrin-fullerene linked dyads and triads: Marcus-normal versus inverted region. *J. Am. Chem. Soc.*, *123*, 2607–2617.

38. Schuster, D. I., Cheng, P., Jarowski, P. D., Guldi, D. M., Echegoyen, L., Pyo, S., Holzwarth, A. R., Braslavsky, S. E., Williams, R. M., Klihm, G. (2004) Design, synthesis, and photophysical studies of a porphyrin-fullerene dyad with parachute topology; charge recombination in the Marcus inverted region. *J. Am. Chem. Soc.*, *126*, 7257–7270.

39. Guldi, D. M., Hirsch, A., Scheloske, M., Dietel, E., Troisi, A., Zerbetto, F., Prato, M. (2003) Modulating charge-transfer interactions in topologically different porphyrin-C_{60} dyads. *Chem.—Eur. J.*, *9*, 4968–4979.

40. Sutton, L. R., Scheloske, M., Pirner, K. S., Hirsch, A., Guldi, D. M., Gisselbrecht, J. P. (2004) Unexpected change in charge transfer behavior in a cobalt(II) porphyrin-fullerene conjugate that stabilize radical ion pair states. *J. Am. Chem. Soc.*, *126*, 10370–10381.

41. De la Torre, G., Giacalone, F., Segura, J. L., Martín, N., Guldi, D. M. (2005) Electronic commu-nication through pi-conjugated wires in covalently linked porphyrin/C_{60} ensembles. *Chem.—Eur. J.*, *11*, 1267–1280.

42. Tkachenko, N. V., Rantala, L., Tauber, A. Y., Helaja, J., Hynninen, P. H., Lemmetyinen, H. (1999) Photoinduced electron transfer in photochlorin-[60]fullerene dyads. *J. Am. Chem. Soc.*, *121*, 9378–9387.

43. Tkachenko, N. V., Lemmetyinen, H., Sonoda, J., Ohkubo, K., Sato, T., aImahori, H., Fukuzumi, S. (2003) Ultrafast photodynamics of exciplex formation and photoinduced electron transfer in porphyrin-fullerene dyads linked at close proximity. *J. Phys. Chem. A*, *107*, 8834–8844.

44. Armaroli, N., Marconi, G., Echegoyen, L., Bourgeois, J. P., Diederlich, F. (2000) Charge-transfer interactions in face-to-face porphyrin-fullerene systems: solvent-dependent luminescence in the infrared spectral region. *Chem.—Eur. J.*, *6*, 1629–1645.

45. Ohkubo, K., Kotani, H., Shao, J., Ou, Z., Kadish, K. M., Li, G., Pandey, R. K., Fujitsuka, M., Ito, O., Imahori, H., Fukuzumi, S. (2004) Production of an ultra-long-lived charge-separated state in a zinc chlorin–C_{60} dyad by one-step photoinduced electron transfer. *Angew. Chem. Int. Ed.*, *43*, 853–856.

46. Gonzalez-Rodriguez, D., Torres, T., Guldi, D. M., Rivera, J., Herranz, M. A., Echegoyen, L. (2004) Subphthalocyanines: tunable molecular scaffolds for intramolecular electron and energy transfer. *J. Am. Chem. Soc.*, *126*, 6301–6313.

47. De la Escosura, A., Martinez-Diaz, M. V., Guldi, D. M., Torres, T. (2006) Stabilization of charge-separated states in phthalocyanine-fullerene ensembles through supramolecular donor-acceptor interactions. *J. Am. Chem. Soc.*, *128*, 4112–4118.

48. Williams, R. M., Koeberg, M., Lawson, J. M., An, Y.-Z., Rubin, Y., Paddon-Row, M. N., Verhoeven, J. W. (1996) Photoinduced electron transfer to C_{60} across extended 3- and 11-bond hydrocarbon bridges: creation of a long-lived charge-separated state. *J. Org. Chem.*, *61*, 5055–5062.

49. Lawson, J. M., Oliver, A. M., Rothenfluh, D. F., An, Y.-Z., Ellis, G. A., Ranasinghe, M. G., Khan, S. I., Franz, A. G., Ganapathi, P. S., Shephard, M. J., Paddon-Row, M. N., Rubin, Y. (1996) Synthesis of a variety of bichromophoric "ball-and-chain" systems based on buckminsterfullerene (C_{60}) for the study of intramolecular electron and energy transfer processes. *J. Org. Chem.*, *61*, 5032–5054.

50. Armaroli, N., Diederlich, F., Dietrich-Buchecker, C. O., Flamigni, L., Marconi, G., Nierengarten, J.-F., Sauvage, J.-P. (1998) A copper(I)-complexed rotaxane with two fullerene stoppers: synthesis, electrochemistry, and photoinduced processes. *Chem.—Eur. J.*, *4*, 406–416.

51. Armspach, D., Constable, E. C., Diederich, F., Housecroft, C. E., Nierengarten, J.-F. (1998) Bucky ligands: synthesis, ruthenium(II) complexes, and electrochemical properties. *Chem.—Eur. J.*, *4*, 723–733.

52. Maggini, M., Guldi, D. M., Mondini, S., Scorrano, G., Paolucci, F., Ceroni, P., Roffia, S. (1998) Photoinduced electron transfer in a tris(2,2′-bipyridine)-C_{60}-ruthenium(II) dyad: evidence of charge recombination to a fullerene excited state. *Chem.—Eur. J.*, 4, 1992–2000.

53. Imahori, H., Cardoso, S., Tatman, D., Lin, S., Noss, L., Seely, G. R., Sereno, L., Silber, J. C., Moore, T. A., Moore, A. L., Gust, D. (1995) Photoinduced electron transfer in a carotenobuckminsterfullerene dyad. *Photochem. Photobiol.*, *62*, 1009–1014.

54. Guldi, D. M., Maggini, M., Scorrano, G., Prato, M. (1997) Intramolecular electron transfer in fullerene/ferrocene based donor-bridge-acceptor dyads. *J. Am. Chem. Soc.*, *119*, 974–980.

55. Marczak, R., Wielopolski, M., Gayathri, S. S., Guldi, D. M., Matsuo, Y., Matsuo, K., Tahara, K., Nakamura, E. (2008) Uniquely shaped double-decker buckyferrocenes-distinct electron donor-acceptor interactions. *J. Am. Chem. Soc.*, *130*, 16207–16215.

56. Martín, N., Sánchez, L., Herranz, M. A., Guldi, D. M. (2000) Evidence for two separate one-electron transfer events in excited fulleropyrrolidine dyads containing tetrathiafulvalene (TTF). *J. Phys. Chem. A*, *104*, 4648–4657.

57. Fujitsuka, M., Ito, O., Yamashiro, T., Aso, Y., Otsubo, T. (2000) Solvent polarity dependence of photoinduced charge separation in a tetrathiophene-C_{60} dyad studied by pico- and nanosecond laser flash photolysis in the near-IR region. *J. Phys. Chem. A*, *104*, 4876–4881.

58. Kuciauskas, D., Liddell, P. A., Lin, S., Stone, S. G., Moore, A. L., Moore, T. A., Gust, D. (2000) Photoinduced electron transfer in carotenoporphyrin-fullerene triads: temperature and solvent effects. *J. Phys. Chem. B*, *104*, 4307–4321.

59. Curiel, D., Ohkubo, K., Reimers, J. R., Fukuzumi, S., Crossley, M. J. (2007) Photoinduced electron transfer in a β,β'-pyrrolic fused ferrocene-(zinc porphyrin)-fullerene. *Phys. Chem. Chem. Phys.*, *9*, 5260–5266.

60. Imahori, H., Guldi, D. M., Tamaki, K., Yoshida, Y., Luo, C., Sakata, Y., Fukuzumi, S. (2001) Charge separation in a novel artificial photosynthetic reaction center lives 380 milliseconds. *J. Am. Chem. Soc.*, *123*, 6617–6628.

61. Guldi, D. M., Imahori, H., Tamaki, K., Kashiwagi, Y., Yamada, H., Sakata, Y., Fukuzumi, S. (2004) A molecular tetrad allowing efficient energy storage for 1.6 s at 163 K. *J. Phys. Chem. A*, *108*, 541–548.

62. Imahori, H., Sekiguchi, Y., Kashiwagi, Y., Sato, T., Araki, Y., Ito, O., Yamada, H., Fukuzumi, S. (2004) Long-lived charge-separated state generated in ferrocene-meso,meso-linked porphyrin trimer-fullerene pentad with a high quantum yield. *Chem.—Eur. J.*, *10*, 3184–3196.

63. Straight, S. D., Kodis, G., Terazono, Y., Hambourger, M., Moore, T. A., Moore, A. L., Gust, D. (2008) Self-regulation of photoinduced electron transfer by a molecular nonlinear transducer. *Nat. Nanotechnol.*, *3*, 280–283.

64. Kuciaukas, D., Liddell, P. A., Lin, S., Johnson, T. E., Weghorn, S. J., Lindsey, J. S., Moore, A. L., Moore, T. A., Gust, D. (1999) An artificial photosynthetic antenna-reaction center complex. *J. Am. Chem. Soc.*, *121*, 8604–8614.

65. Winters, M. U., Dahlstedt, E., Blades, H. E., Wilson, C. J., Frampton, M. J., Anderson, H. L., Albinsson, B. (2007) Probing the efficiency of electron transfer through porphyrin-based molecular wires. *J. Am. Chem. Soc.*, *129*, 4291–4297.

66. Guldi, D. M., Martín, N. (2002) Fullerene architectures made to order; biomimetic motifs—design and features. *J. Mater. Chem.*, *12*, 1978–1992.

67. Sánchez, L., Martín, N., Guldi, D. M. (2005) Hydrogen-bonding motifs in fullerene chemistry. *Angew. Chem. Int. Ed.*, *44*, 5374–5382.

68. Segura, M., Sánchez, L., de Mendoza, J., Martín, N., Guldi, D. M. (2003) Hydrogen bonding interfaces in fullerene-TTF ensembles. *J. Am. Chem. Soc.*, *125*, 15093–15100.

69. D'Souza, F., Smith, P. M., Zandler, M. E., McCarty, A. L., Itou, M., Araki, Y., Ito, O. (2004) Energy transfer followed by electron transfer in a supramolecular triad composed of boron dipyrrin, zinc porphyrin and fullerene: a model for the photosynthetic antenna-reaction center complex. *J. Am. Chem. Soc.*, *126*, 7898–7907.

70. Li, K., Schuster, D. I., Guldi, D. M., Herranz, M. A., Echegoyen, L. (2004) Convergent synthesis and photophysics of [60]fullerene/porphyrin-based rotaxanes. *J. Am. Chem. Soc.*, *126*, 3388–3389.

71. Li, K., Bracher, P. J., Guldi, D. M., Herranz, M. A., Echegoyen, L., Schuster, D. I. (2004) [60]Fullerene-stoppered porphyrinorotaxanes: pronounced elongation of charge-separated-state lifetimes. *J. Am. Chem. Soc.*, *126*, 9156–9157.

72. Sánchez, L., Sierra, M., Martín, N., Myles, A. J., Dale, T. J., Rebek, J., Seitz, W., Guldi, D. M. (2006) Exceptionally strong electronic communication through hydrogen bonds in porphyrin-C_{60} pairs. *Angew. Chem. Int. Ed.*, *45*, 4637–4641.

73. Wessendorf, F., Gnichwitz, J.-F., Sarova, G. H., Hager, K., Hartnagel, U., Guldi, D. M., Hirsch, A. (2007) Implementation of a Hamilton-receptor-based hydrogen-bonding motif toward a new electron donor-acceptor prototype: electron versus energy transfer. *J. Am. Chem. Soc.*, *129*, 16057–16071.

74. Mateo-Alonso, A., Ehli, C., Rahman, G. M. A., Guldi, D. M., Fioravanti, G., Marcaccio, M., Paolucci, F., Prato, M. (2007) Tuning electron transfer through translational motion in molecular shuttles. *Angew. Chem. Int. Ed.*, *46*, 3521–3525.

75. Brabec, C. J., Dyakonov, V., Dcherf, U. *Organic Photovoltaics: Materials, Device Physics and Manufacturing Technologies*. Wiley-VCH Verlag GmbH & co. KGaA, Weinheim, 2008.

76. Po, R., Maggini, M., Camaioni, N. (2010) Polymer solar cells: recent approaches and achievements. *J. Phys. Chem. C*, *114*, 695–706.

77. (a) Hirsch, A. (2005) *The Chemistry of Fullerenes*. Wiley-VCH, Weinheim;   (b) Guldi, D. M., Martín, N. (2002) *Fullerenes: From Synthesis to Optoelectronic Properties*. Kluwer Academic Publishers, Dordrecht, The Netherlands;   (c) Taylor, R. (1999) *Lecture Notes on Fullerene Chemistry: A Handbook for Chemists*. Imperial College Press, London;   (d) Langa, F., Nierengarten, J.-F. (2007) *Fullerenes Principles and Applications*. RSC, Cambridge;   (e) Martín, N. (2006) New challenges in fullerene chemistry. *Chem. Commun.*, 2093–2104.

78. (a) Li, G., Shrotriya, V., Huang, J., Yao, Y., Moriarty, T., Emery, K., Yang, Y. (2005) High-efficiency solution processable polymer photovoltaic cells by self-organization of polymer blends. *Nat. Mater.*, *4*, 864–868;   (b) Ma, W., Yang, C., Gong, X., Lee, K., Heeger, A. J. (2005) Thermally stable, efficient polymer solar cells with nanoscale control of the interpenetrating network morphology. *Adv. Funct. Mater.*, *15*, 1617–1622.

79. Blouin, N., Michaud, A., Leclerc, M. (2007) A low-bandgap poly(2,7-carbazole) derivative for use in high-performance solar cells. *Adv. Mater.*, *19*, 2295–2300.

80. Zhang, F., Mammo, W., Andersson, L. M., Admassie, S., Andersson, M. R., Inganäs, O. (2006) Low-bandgap alternating fluorene copolymer/methanofullerene heterojunctions in efficient near-infrared polymer solar cells. *Adv. Mater.*, *18*, 2169–2173.

81. Wienk, M. M., Turbiez, M. G. R., Struijk, M. P., Fonrodona, M., Janssen, R. A. J. (2006) Low-band gap poly(di-2-thienylthienopyrazine):fullerene solar cells. *Appl. Phys. Lett.*, *88*, 153511–153511-3.

82. Wong, W.-Y., Wang, X.-Z., He, Z., Djurisic, A. B., Yip, C.-T., Cheung, K.-Y., Wang, H., Mak, C. S. K., Chan, W.-K. (2007) Metallated conjugated polymers as a new avenue towards high-efficiency polymer solar cells. *Nat. Mater.*, *6*, 521–527.

83. Hou, J., Chen, H.-Y., Zhang, S., Chen, R. I., Yang, Y., Wu, Y., Li, G. (2009) Synthesis of a low band gap polymer and its application in highly efficient polymer solar cells. *J. Am. Chem. Soc.*, *131*, 15586–15587.

84. Qin, R., Li, W., Li, C., Du, C., Veit, C., Schleiermacher, H.-F., Andersson, M., Bo, Z., Liu, Z., Inganas, O., Wuerfel, U., Zhang, F. (2009) A planar copolymer for high efficiency polymer solar cells. *J. Am. Chem. Soc.*, *131*, 14612–14613.

85. Liang, Y., Xu, Z., Xia, J., Tsai, S.-T., Wu, Y., Li, G., Ray, C., Yu, L. (2010) For the bright future-bulk heterojunction polymer solar cells with power conversion efficiency of 7.4%. *Adv. Mater.*, *20*, E135–E138.

86. (a) Zhang, F., Mammo, W., Andersson, L. M., Admassie, S. Andersson, M. R., Inganäs, O. (2006) Low-bandgap alternating fluorene copolymer/methanofullerene heterojunctions in efficient near-infrared polymer solar cells. *Adv. Mater.*, *18*, 2169–2173; (b) Inganäs, O., Zhang, F., Andersson, M. R. (2009) Alternating polyfluorenes collect solar light in polymer photovoltaics. *Acc. Chem. Res.*, *42*, 1731–1739.

87. (a) Dennler, G., Sariciftci, N. S., Brabec, C. J. (2007) *Semiconducting Polymers*, 2nd edn. Wiley-VCH, Weinheim; (b) Chen, J. Cao, Y. (2009) Development of novel conjugated donor polymers for high-efficiency bulk-heterojunction photovoltaic devices. *Acc. Chem. Res.*, *42*, 1709–1718; (c) Cheng, Y. J., Yang, S. H., Hsu, C. S. (2009) Synthesis of conjugated polymers for organic solar cell applications. *Chem. Rev.*, *109*, 5868–5923.

88. Zerza, G., Scharber, M. C., Brabec, C. J., Saricftci, N. S., Gómez, R., Segura, J. L., Martín, N., Srdanov, V. I. (2000) Photoinduced charge transfer between tetracyano-anthraquino-dimethane derivatives and conjugated polymers for photovoltaics. *J. Phys. Chem. A*, *104*, 8315–8322.

89. (a) Waldauf, C., Graupner, W., Tasch, S., Leising, G., Gügel, A., Scherf, U., Kraus, A., Walter, M., Müllen, K. (1998) Efficient charge carrier transfer from m-LPPP to C_{60} derivatives. *Opt. Mater.*, *9*, 449–453; (b) For a recent review involving exTTF as a donor, see: Martín, N., Sánchez, L., Herranz, M. A., Illescas, B., Guldi, D. M. (2007) Electronic communication in tetrathiafulvalene (TTF)/C_{60} systems: toward molecular solar energy conversion materials? *Acc. Chem. Res.*, *40*, 1015–1024.

90. (a) Nierengarten, J.-F., Eckert, J.-F., Nicoud, J.-F., Ouali, L., Krasnikov, V., Hadziioannou, G. (1999) Synthesis of a C_{60}-oligophenylenevinylene hybrid and its incorporation in a photovoltaic device *Chem. Commun.*, 617–618; (b) Eckert, J.-F., Nicoud, J.-F., Nierengarten, J.-F., Liu, S.-G., Echegoyen, L., Barigelletti, F., Armaroli, N., Ouali, L., Krasnikov, V., Hadziioannou, G. (2000) Fullerene-oligophenylenevinylene hybrids: synthesis, electronic properties, and incorporation in photovoltaic devices. *J. Am. Chem. Soc.*, *122*, 7467–7479; (c) Guldi, D. M., Luo, Ch., Swartz, A., Gómez, R., Segura, J. L., Martín, N., Brabec, C., Saricftci, N. S. (2002) Molecular engineering of C_{60}-based conjugated oligomer ensembles: modulating the competition between photoinduced energy and electron transfer processes. *J. Org. Chem.*, *67*, 1141–1152; (d) Atienza, C., Fernández, G., Sánchez, L., Martín, N., Dantas, I. S., Wienk, M. M., Janssen, R. A. J., Rahman, G. M. A., Guldi, D. M. (2006) Light harvesting tetrafullerene nanoarray for organic solar cells. *Chem. Commun.*, 514–516; (e) Fernández, G., Sánchez, L., Veldman, D., Wienk, M. M., Atienza, C., Guldi, D. M., Janssen, R. A. J., Martín, N. (2008) Tetrafullerene conjugates for all-organic photovoltaics. *J. Org. Chem.*, *73*, 3189–3196.

91. Rispens, M. T., Hummelen, J. C. (2002) *Fullerenes: From Synthesis to Optoelectronic Properties*. Kluwer Academic Publishers, Dordrech, The Netherlands, pp. 387–435.

92. Hummelen, J. C., Knight, B. W., LePeq, F., Wudl, F., Yao, J., Wilkins, C. L. (1995) Preparation and characterization of fulleroid and methanofullerene derivatives. *J. Org. Chem.*, *60*, 532–538.

93. Yu, G., Gao, J., Hummelen, J. C., Wudl, F., Heeger, A. J. (1995) Polymer photovoltaic cells: enhanced efficiencies via a network of internal donor-acceptor heterojunctions. *Science*, *270*, 1789–1791.

94. (a) Zhang, Y., Yip, H. L., Acton, O., Hau, S. K., Huang, F., Jen, A. K.-Y. (2009) A simple and effective way of achieving highly efficient and thermally stable bulk-heterojunction polymer solar cells using amorphous fullerene derivatives as electron acceptor. *Chem. Mater.*, *21*, 2598–2600; (b) Yang, Ch., Kim, J. Y., Cho, Sh., Lee, J. K., Heeger, A. J., Wudl, F. (2008) Functionalized methanofullerenes used as n-type materials in bulk-heterojunction polymer solar cells and in field-effect transistors. *J. Am. Chem. Soc.*, *130*, 6444–6450; (c) Kooistra, F. B., Knol, J., Kastenberg, F., Popescu, L. M., Verhees, W. J. H., Kroon, J. M., Hummelen, J. C. (2007) Increasing the open circuit voltage of bulk-heterojunction solar cells by raising the LUMO level of the acceptor. *Org. Lett.*, *9*, 551–554; (d) Drees, M., Hoppe, H., Winder, C., Neugebauer, H., Sariciftci, N. S., Schwinger, W., Schäffler, F., Topf, C., Scharber, M. C., Zhu, Z., Gaudiana, R. (2005) Stabilization of the nanomorphology of polymer–fullerene bulk heterojunction blends using a novel polymerizable fullerene derivative. *J. Mater. Chem.*, *15*, 5158–5163; (e) Zheng, L., Zhou, Q., Deng, X., Yuan, M., Yu, G., Cao, Y. (2004) Methanofullerenes used as electron acceptors in polymer photovoltaic devices. *J. Phys. Chem. B*, *108*, 11921–11926; (f) Popescu, L. M., Van't Hof, P., Sieval, A. B., Jonkman, H. T., Hummelen, J. C. (2006) Thienyl analog of 1-(3-methoxycarbonyl)propyl-1-phenyl-[6,6]-methanofullerene for bulk heterojunction photovoltaic devices in combination with polythiophenes. *Appl. Phys. Lett.*, *89*, 213507–213507-3.

95. Lenes, M., Wetzelaer, G.-J. A. H., Kooistra, F. B., Veenstra, S. C., Hummelen, J. C., Blom, P. W. M. (2008) Fullerene bisadducts for enhanced open-circuit voltages and efficiencies in polymer solar cells. *Adv. Mater.*, *20*, 2116–2119.

96. Wienk, M. M., Kroon, J. M., Verhees, W. J. H., Knol, J., Hummelen, J. C., van Hal, P. A., Janssen, R. A. J. (2003) Efficient methano[70]fullerene/MDMO-PPV bulk heterojunction photovoltaic cells. *Angew. Chem. Int. Ed.*, *42*, 3371–3375.

97. Park, S. H., Roy, A., Beaupré, S., Cho, S., Coates, N., Moon, J. S., Moses, D., Leclerc, M., Lee, K., Heeger. A. J. (2009) Bulk heterojunction solar cells with internal quantum efficiency approaching 100%. *Nat. Photon.*, *3*, 297–302.

98. Kooistra, B., Mihailetchi, V. D., Popescu, L. M., Kronholm, D., Blom, P. W. M., Hummelen, J. C. (2006) New C_{84} derivative and its application in a bulk heterojunction solar cell. *Chem. Mater.*, *18*, 3068–3073.

99. (a) Wang, X., Perzon, E., Delgado, J. L., de la Cruz, P., Zhang, F., Langa, F., Andersson, M., Inganäs, O. (2004) Infrared photocurrent spectral response from plastic solar cell with low-band-gap polyfluorene and fullerene derivative. *Appl. Phys. Lett.*, 85, 5081–5083; (b) Perzon, E., Wang, X., Zhang, F., Mammo, W., Delgado, J. L., de la Cruz, P., Inganäs, O., Langa, F., Andersson, M. R. (2005) Design, synthesis and properties of low bandgap polyfluorenes for photovoltaic devices. *Synth. Met.*, *154*, 53–56.

100. (a) Riedel, I., von Hauff, E., Parisi, J., Martín, N., Giacalone, F., Diakonov, V. (2005) Diphenyl-methanofullerenes: new and efficient acceptors in bulk-heterojunction solar cells. *Adv. Funct. Mater.*, *15*, 1979–1987; (b) Riedel, I., Martín, N., Giacalone, F., Segura, J. L., Chirvase, D., Parisi, J., Diakonov, V. (2004) Polymer solar cells with novel fullerene-based acceptors. *Thin Solid Films*, *43*, 451–452.

101. Backer, S., Sivula, K., Kavulak, D. F., Fréchet, J. M. J. (2007) High efficiency organic photovoltaics incorporating a new family of soluble fullerene derivatives. *Chem. Mater.*, *19*, 2927–2929.

102. He, Y., Chen, H.-Y., Hou, J., Li, Y. (2010) Indene-C_{60} bisadduct: a new acceptor for high-performance polymer solar cells. *J. Am. Chem. Soc.*, *132*, 1377–1382.

103. (a) For a review on fullerene dimers, see: Segura, J. L., Martín, N. (2000) [60]Fullerene dimers. *Chem. Soc. Rev.*, *29*, 13–25; (b) Segura, J. L., Priego, E. M., Martín, N., Luo, C. P., Guldi, D. M. (2000) A new photoactive and highly soluble C_{60} - TTF-C_{60} dimer: charge separation and recombination. *Org. Lett.*, *2*, 4021–4024; (c) González, J. J., González, S., Priego, E. M., Luo, C. P., Guldi, D. M., de Mendoza, J., Martín, N. (2001) A new approach to supramolecular C_{60}-dimers based in quadrupole hydrogen bonding. *Chem. Commun.*, 163–164.

104. Delgado, J. L., Espíldora, E., Liedtke, M., Sperlich, A., Rauh, D., Baumann, A., Deibel, C., Dyakonov, V., Martín, N. (2009) Fullerene dimers (C_{60}/C_{70}) for energy harvesting. *Chem.—Eur. J.*, *15*, 13474–13482.

105. (a) Delgado, J. L., Cardinali, F., Espíldora, E., Torres, M. R., Langa, F., Martín, N. (2008) Oxidation of 3-alkyl-substituted 2-pyrazolino[60]fullerenes: a new formyl-containing building block for fullerene chemistry. *Org. Lett.*, *10*, 3705–3708; (b) Delgado, J. L., Oswald, F., Cardinali, F., Langa, F., Martín, N. (2008) On the thermal stability of [60]fullerene cycloadducts: retro-cycloaddition reaction of pyrazolino[4,5:1,2][60]fullerenes. *J. Org. Chem.*, *73*, 3184–3188.

CHAPTER 22

RECENT TRENDS IN SUPRAMOLECULAR PHOTOVOLTAIC SYSTEMS

Dario M. Bassani

Institut des Sciences Moléculaires, Université Bordeaux 1, CNRS, Talence, France

22.1 INTRODUCTION

Long-term scenarios for sustainable and efficient sources of energy invariably involve harnessing solar energy on a large scale. This will most likely require the development of alternatives to today's mono- or polycrystalline silicon-based photovoltaic (PV) devices in order to reach the performance/cost objectives necessary for widespread deployment. Organic (and hybrid) solar cells have emerged as a viable alternative to inorganic PV devices, allying desirable mechanical properties with low cost and high-speed throughput reel-to-reel manufacturing processes. Indeed, their low weight and flexibility make them ideal for integration into a variety of real-life environments and architectures. However, the performance and durability of organic photovoltaic (OPV) devices must be greatly improved before they become practical on a large scale. In solar cells, charge separation on light absorption must be followed by transport of the separated hole-electron pair to the corresponding electrodes. This cascade of critical steps must be optimized as a whole rather than individually.

Both the durability and performance of organic solar cells have been improving steadily over the last decade, with recent record performances reaching nearly 8% (Solarmer, Inc.). While the lifespan of OPV devices is still below what is sought for for incorporation in building-integrated photovoltaic (BIPV) devices, other mainstream applications are being developed and commercial production of OPV panels with ca 3% overall conversion efficiency is underway. Kalowekamo and Baker [1] have estimated the module cost of future OPV devices to be between \$50 and \$140/m^2, which, even assuming a 5% module conversion efficiency, would result in prices between \$1.00 and

Supramolecular Soft Matter: Applications in Materials and Organic Electronics, First Edition.
Edited by Takashi Nakanishi.

$2.83/ W-p. Although such prices would be, in principle, competitive with future thin film solar cell (TFSC) technologies, the actual cost of electrical production for OPV devices would be significantly higher unless their lifetime can be extended from 5 to ≥10 years. A common misconception is that the organic material of the active layer is not stable and undergoes light-induced degradation. In fact, major degradation pathways in OPV devices are due to the infiltration of oxygen. Improvements in encapsulation technology (principally driven by the food industry) have already brought the lifespan of OPV devices to 10,000 h while maintaining ca 3% module efficiencies. Besides oxygen, heat is the other enemy of OPV devices. The temperature of a solar panel can reach 80°C during operation, and such temperatures accelerate the macroscopic phase separation of the donor and acceptor materials in the active layer of the device.

22.2 PRINCIPLES OF SUPRAMOLECULAR ORGANIC PHOTOVOLTAIC DEVICES

The basic operating principle of OPV devices is not fundamentally different from their silicon-based inorganic counterparts. Absorption of a photon induces charge separation, and the charges must diffuse to the corresponding electrodes to be collected. The process of converting light radiation to electric current can be divided into four steps (Fig. 22-1a). The first step is exciton formation induced by photon absorption. Owing to high absorption (ca 10^5 au/cm for solid films), π-conjugated polymers used in organic PV cells absorb efficiently at their absorption maximum wavelength. However, the absorption bands of most conjugated polymers are relatively narrow compared to Si-based PV cells. For example, the optical band gap of poly-3-hexylthiophene (P3HT) is $\sim$1.9 eV, which corresponds to the absorption of only about 30% of AM 1.5 solar photon flux. The excitons that are formed are coulombically bound electron-hole pairs with a typical binding energy of several hundreds of millielectron volts in organic materials. Compared to inorganic solar cells, where exciton binding energy is only in the millielectron volt range, this prevents spontaneous charge separation due to thermal energy at room temperature. For example, the exciton binding energies in MEH-PPV and poly(p-phenylphenylenevinylene) (PPPV) were determined to be 200–600 mV [2–4] and 400 mV [5] respectively, resulting in the formation of Frenkel type excitons [6] which decreases the charge separation efficiency. It is estimated that only 10% of the photoinduced excitons in conjugated polymer chains dissociate into electrons and holes [7].

Owing to the large exciton binding energy, excitons diffuse within the material instead of spontaneously dissociating into charges unless the dissociation is assisted by an applied potential or by the presence of electron acceptors. The exciton diffusion lengths are reported to range from 5 to 15 nm from various polymer-based device experiments [8–10]. Because of the low relative dielectric constant (ε_R) of organic materials, ranging from 2 to 4, photogenerated electron-hole pairs remain strongly bonded by coulombic attraction. These electron-hole pairs either dissociate into free electrons and holes or recombine, resulting in

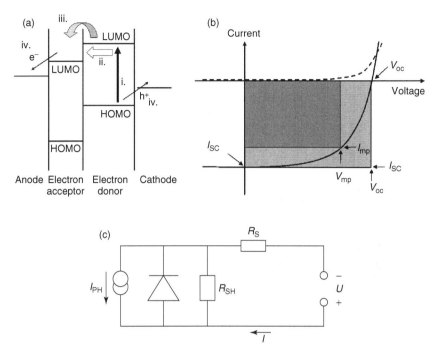

Figure 22-1 (a) Working principle of an organic solar cell. (i) Exciton formation induced by photon absorption, (ii) exciton diffusion to the interface, (iii) charge separation between electron donor molecule and an electron acceptor molecule, and (iv) charge transport to the corresponding electrodes. (b) I–V characteristics of a PV cell. The characteristic intersections with the X and Y axes give the open-circuit voltage (V_{oc}) and short-circuit current (I_{sc}), respectively. The maximum output power is at the point $I_{mp} \times V_{mp}$. The FF is the ratio of the maximum output power delivered by the cell to $I_{sc} \times V_{oc}$ and corresponds to the ratio between light and dark rectangles [14]. (c) Equivalent circuit of a PV cell. I_{PH}, photocurrent; I, total current; R_S, serial resistance; R_{SH}, shunt resistance; U, bias across electrodes.

energy loss. If the interface between the electron D and A materials is located within the diffusion length of the exciton, the electron-hole pair may be separated owing to the assistance of the energetic difference between the lowest unoccupied molecular orbital (LUMOs) of the materials in the interface. Furthermore, theoretical calculations suggest that a dipolar layer is formed at the interface because of partial charge transfer (CT) prior to photon absorption [11], which reduces the back electron transfer and hence the probability of charge recombination [12]. From ultrafast transient absorption experiments, the time scale of charge separation from electron D polymers to electron A fullerene derivatives is estimated to be ∼45 ps [13]. In such cases, CT is faster than other competing relaxation processes, giving efficient charge separation.

Once formed, the separated holes and electrons must be transported through the corresponding n- and p-type materials to the electrodes without being trapped, and a driving force is thus needed to extract the charges from the active layer. In the case of bilayer junction solar cells, the gradient of charge concentration can provide a driving force for the charges to diffuse to the electrodes. However, this is not the case in bulk heterojunction solar cells (*vide infra*), where the asymmetrical contacts (a high-work-function metal for the hole collection electrode and a low-work-function metal for the electron collection electrode) assure directional charge transport by the induction of an internal field.

When a PV cell is irradiated with light, the voltage across the electrodes shifts from 0 V (open circuit) to the open-circuit voltage (V_{oc}), which is the maximum voltage difference attainable between the electrodes. The origin of V_{oc} is still not clear, but it follows an empirical formula that reveals the relationship between the V_{oc}, highest occupied molecular orbital (HOMO) of the donor polymer, and LUMO of PCBM ([6,6]-phenyl-C_{61}-butyric acid methyl ester) based on a statistical analysis of polymer–PCBM OPV cells (Eq. 22.1) [14]. The short-circuit current (I_{sc}) is the maximum current delivered by the cell under short circuit voltage ($V = 0$). The short circuit current gives information about the charge separation and transport efficiency in the cell and is dependent on the illumination intensity.

$$V_{oc} = \frac{1}{e}(|E_{Donor}HOMO| - |E_{PCBM}LUMO|) - 0.3 \text{ V} \qquad (22.1)$$

On illumination, a PV cell generates power that is equal to the product of the current and voltage. At a certain point, this reaches a maximum at $I_{mp} \times V_{mp}$. To determine the efficiency of a PV cell, the power output needs to be compared to the input power (P_{in}), which corresponds to the total radiative energy falling on the device surface. The fill factor (FF) is calculated as FF $= V_{mp} \times I_{mp}/(V_{oc} \times I_{sc})$ to indicate the portion of power that can be extracted from a PV cell. With the value of FF, the power conversion efficiency (η) can be written as

$$\eta_{power} = \frac{P_{out}}{P_{in}} = \frac{I_{mp}V_{mp}}{P_{in}} = \frac{FF \times I_{sc}V_{oc}}{P_{in}} \qquad (22.2)$$

The corresponding equivalent circuit of a PV cell is shown in Fig. 22-1c. An ideal PV cell can be viewed as a current generator, generating photocurrent (I_{PH}) in parallel with a diode. In a real PV cell, however, the power is lost through the resistance of contacts and through leakage current passing the active layer. Power loss can be expressed in serial resistance (R_S) in series and shunt resistance (R_{SH}) in parallel to the equivalent circuit of an ideal PV cell, respectively. The serial resistance is induced from the resistance of the whole device to current flow, while shunt resistance is caused from the leakage of current due to poor insulation between two electrodes. An efficient PV cell requires high R_{SH} in order to block the leakage of current through the active layer and low R_S to generate a sharp rise in current in forward bias.

It is widely accepted that the morphology of the active layer in OPV devices plays a crucial role in determining the overall performance of the material [15–19]. In fact, the biggest single-step improvement in OPVs can be attributed to the introduction of the bulk heterojunction principle by Heeger in the 1990s (Fig. 22-2a) [20]. Compared to the initial bilayer devices pioneered by Tang at

Eastmann Kodak in the mid-1980s [23], the leap in performance is attributed to greater contact between the electron donor and acceptor materials leading to more efficient charge separation. Combined with the latest generation materials, this has allowed current technology to reach overall efficiencies of nearly 8% [24]. Still, the bulk heterojunction is far from the ideal interdigitated structure-optimizing charge separation and collection (Fig. 22-2c). Moreover, its formation is controlled by the underlying chemical and physical properties of the materials, as well as by the technology and the conditions used for the deposition process and postproduction treatment [25]. The biggest drawback is that each and every material must be optimized individually before its full potential can be assessed. In a seminal paper, Shaheen *et al.* showed that the performance of a material could be tripled by adopting conditions conducive to the formation of a fine-grain heterogeneous morphology [26]. Because the exciton diffusion length in organic materials is short, ca 10 nm [27], the finer heterogeneous structure increases the probability that an exciton will encounter a discontinuity and dissociate into free charge carriers. In the case of devices based on P3HT, thermal annealing is required to generate microcrystalline domains exhibiting high hole-transport mobilities. As mentioned above, the domains of electron D and A materials

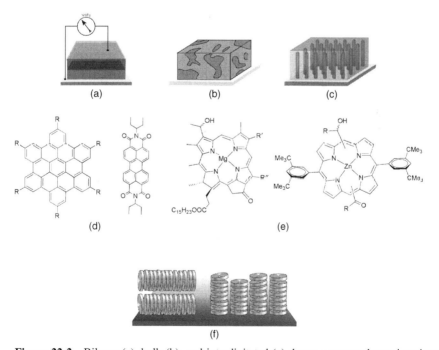

Figure 22-2   Bilayer (a), bulk (b), and interdigitated (c) donor–acceptor heterojunctions. (d) Example of an extended π-conjugated aromatic molecule capable of forming discotic liquid-crystalline phases suitable for the construction of OPV devices when blended with electron acceptor 2 [21]. (e) The natural self-assembling bacteriochlorophyll *c* (left) and a synthetic mimic (right) [22]. (f) Homeotropic versus parallel arrangement of discotic liquid crystals.

in the active layer of a bulk heterojunction cell are randomly oriented, and no internal field with a preferred direction is produced under illumination to direct the charges to the corresponding electrodes. Therefore, the use of asymmetrical contacts is necessary in such cells.

An important advantage of organic semiconductors is their inexhaustible variety. Using chemical synthesis, it is possible to tailor the electronic (energy levels, band gap, absorption cross-section, etc.) and physical properties of a material. However, because many of these properties are interconnected, precise design of a material with predefined characteristics remains a major objective that has yet to be reached. It is interesting that the same forces that make molecules appealing are the cause of their shortcomings. The well-defined electron density map that characterizes a molecule determines its relatively narrow electronic absorption spectra; it also signifies that intermolecular electronic overlap integrals are small except in specific cases, for example, when π-stacking is important. One fortunate outcome of this is that molecule-based electronic materials are much less susceptible to defects and impurities, as their effect is localized to nearby molecules only. However, this lack of intermolecular electronic communication abates the charge carrier mobility, which ultimately clamps device performance. The introduction of long-range crystalline order can restore high charge carrier mobility, but this would be detrimental for the flexibility of the device, which benefits from amorphous materials, and is potentially costly to manufacture.

What then, is the solution to improve the properties of molecular materials for organic electronic applications? How can one introduce long-range order without compromising flexibility and ease of manufacture? The problem involves upscaling the order from the molecular level to—at least—the nano- or microscopic domain. As it turns out, a possible solution to this conundrum has been used by Nature for millennia: supramolecular self-assembly. Using this, Nature builds molecular components that spontaneously self-assemble into ordered architectures with designed functionalities. By controlling the type and number of interactions, it is possible to obtain very precise architectures (e.g., the photosynthetic reaction center in bacteria) or more flexible systems that are tolerant of physical deformations, such as cellular lipid bilayer membranes.

In the field of solar energy conversion, photoactive supramolecular systems are well known for their use in light collection and charge separation molecular devices [28–32]. Such systems have been mainly applied to the design of artificial ensembles mimicking photosynthesis (solar-to-fuel conversion), but the general guidelines can be useful in providing improved materials for OPVs. The principle underlying the use of designed intermolecular interactions in organic electronic devices is to differentiate the fundamental electronic properties of the molecular component (principally determined by its electronic structure) from its self-aggregation properties, which drive the morphology of the molecular active layer of the device. The latter is important in controlling through-space interactions, which in turn determine long-range charge transport. Since device performance is to a large extent determined by the mobility of charge carriers, independent control of the morphology and electronic properties will in principle allow one to separately tune electronic characteristics while optimizing charge transport. Materials capable of self-assembly are of interest here, since they have the potential to form well-defined structures in which molecular ordering facilitates efficient charge transport. The quality of the intermolecular interaction (in terms of the magnitude of the CT integral between adjacent sites) is, of course, dependent on the relative orientation and distance of the molecular components. Depending on the degree of order within the material, different theoretical models can be used to describe the mobility of charge carriers, from band theory for structurally ordered materials to tight-binding models for weakly disordered systems or hopping models for localized charges in strongly disordered materials. An overview of charge-transport models applicable to self-organizing molecular materials is provided by Grozema and Siebbels [33].

Numerous examples attest to the importance of the effect of molecular order on charge carrier mobility. The charge carrier mobility of pentathiophene, for example, increases from 10^{-4} cm^2/V/s in amorphous films to 0.1 cm^2/V/s in the crystalline phase [34]. Liquid-crystalline materials offer the unique opportunity to follow the electronic properties of a device as a function of molecular order by varying the temperature over a narrow range centered on a phase transition in the material. For example, charge-carrier mobilities as high as 1 cm^2/V/s were found (by time-resolved microwave conductivity measurements) for a hexabenzocoronene discotic liquid-crystalline material when in the crystalline phase, which, on heating to a temperature above the $K_2 \rightarrow D_h$ transition (ca 100°C), undergoes a sudden fourfold drop in charge carrier mobility [35]. In another case, the degree of order in the columnar packing of triphenylene liquid-crystalline materials was improved by the introduction of secondary amide groups capable of hydrogen-bonding (H-B) interactions linearly aligned to the direction of the columnar stacks, resulting in a charge carrier mobility that is ca five times higher than for conventional triphenelene-based liquid-crystalline materials [36]. The applicability of discotic liquid crystals for organic electronics has been comprehensively reviewed recently [37–44] and is not further discussed herein except for selected examples involving illustrative cases of supramolecular interactions.

Supramolecular interactions loosely cover a wide range of intermolecular forces, ranging from hydrophobic/dispersion forces to H-B and metal ion coordination. The forces generated by these interactions also span a wide gamut, from <1 to 4 kJ/mol for the relatively weak hydrophobic, aromatic π-stacking interactions and single-point H-B to >250 kJ/mol for very strong coordinative bonds. Besides the free energy, other factors such as kinetic lability and directionality also describe supramolecular interactions. Thus, some (e.g., H-B and coordination bonds) are directional, some are only partly directional (e.g., π-stacking), and others are nondirectional (dispersive forces). Clearly, directionality is desirable in cases in which a bottom-up design is expected to lead to the formation of precise, well-defined architectures. For these applications, hydrogen and coordinative bonds have beenused, with H-B interactions being particularly well-suited toward the construction of photoactive supramolecular assemblies [45,46]. Aromatic π-stacking, a principal driving force behind the formation of discotic liquid crystals, also ensures moderate to good electronic communication between the electron clouds of the molecular components [40]. When the electronic interactions are sufficiently strong or driven by outside forces [39], the formation of J- or H-type aggregate structure and their subsequent effect on the electronic absorption and emission spectra is observed.

Supramolecular chemistry thus offers a promising route to bridge the gap between the single-molecule world and real-life nano- to micron-scale devices, combining the advantages of molecular design (synthesis, purification, and characterization) and solution-based processing (low cost, easily upscalable). Further down the line, behavior specific to dynamic supramolecular components may also be harnessed, such as autonomic regulation and self-healing [47–51]. Supramolecular organization also provides a direct method for assembling large numbers of molecules into structures that can bridge length scales from nanometers to macroscopic dimensions for efficient long-distance charge transport [52,53]. This approach allows the design of extended complex structures built through the ordered assembly of elementary building blocks in solution before—or during—the casting process using secondary interactions such as H-B, electrostatic forces, $\pi-\pi$ interaction, and CH/π interaction. Despite all this, the number of examples involving molecule-based devices in which designed supramolecular self-assembly is used remains relatively limited. The two reasons for this are (i) it is still difficult to relate molecular structure to supramolecular self-assembly in solids, particularly for multifunctional molecules in which more than one interaction is present (e.g., π-stacking by an aromatic core, hydrophobic forces from solubilizing side-chains, and H-B groups for self-assembly) and (ii) design principles relating the morphology of the active layer to the performance of organic devices are poorly understood. Additionally, fine-tuning of the solubility versus aggregation properties is required to obtain materials that spontaneously self-assemble in the solid state while maintaining good processability from solution.

22.3   SELF-ASSEMBLY BASED ON HYDROPHOBIC INTERACTIONS

A basic approach to control the hydrophobic forces in a bulk heterojunction solar cell is by the addition of a third component [54–56]. Heeger and coworkers investigated the effect of adding a small amount (<5% v/v) of substituted alkane derivative on the device properties of typical P3HT and C_{61}-PCBM organic solar cells, which results in improved overall conversion efficiencies. Similar results were also obtained for more advanced devices comprising low-band-gap polymers and C_{71}-PCBM, which attained overall efficiencies of 5.5% [56]. Investigation of the effect of such additives on the device morphology reveals that the action of the additive can be resumed to that of a processing additive, providing enhanced solubility of one of the two phases (the fullerene component in this case), which in turn improves the nanometric phase separation between the hole and electron transporting materials [57]. Further progress in this area might include the use of additives capable of better structuring the heterojunction through their own specific morphology, such as, for example, amphiphillic materials capable of self-assembling into bi-continuous phases. Ideally, vertically aligned lamellar phases would represent the optimal geometry to optimize charge separation and charge transport.

Promising approaches to attain vertically aligned nanostructured active layers include discotic liquid-crystalline materials and self-organized heterojunctions obtained from the sequential assembly of rigid, interdigitated donor–acceptor oligomers. As reported by Müllen and coworkers, discotic liquid-crystalline materials based on extended π-aromatic structures can lead to the formation of 1D columnar superstructures that allow efficient charge transport (Fig. 22-2d) [21,37]. The intermolecular attractive forces resulting from the large π-areas induce a pronounced propensity to self-assemble into highly organized structures. Balaban and coworkers have designed biomimetic chromophores (Fig. 22-2e) in which self-assembly can be controlled by adjusting the transition dipole moment [22,58]. The resulting supramolecular nanostructures are strongly fluorescent and not quenched on anchoring onto nanocrystalline titania with different grain sizes, which makes them promising candidates for the fabrication of dye-sensitized solar cells (DSSCs). Specific helical assemblies of aggregated chromophores displaying strong electronic coupling (for example, as evidenced by the formation of J-aggregates) were obtained using modified amylose polymers as supramolecular templates [59]. Extended columnar architectures are equally interesting for the fabrication of OPV devices, provided that stacking of the material can be controlled to give columns that are formed orthogonal to the surface of the substrate (Fig. 22-2f). Homeotropic alignment can sometimes be favored by tuning the molecular structure [60] or by surface modification [61]. Even when this is not the case, overall power conversion efficiencies of 1.5% were achieved in such devices despite a modest external quantum efficiency of 12% [21]. This suggests that considerable room for improvement is available by optimizing *inter alia* light absorption and device fabrication.

Hydrophobic interactions between aromatic residues need not be confined to a two-dimensional plane. The curved aromatic surface of fullerenes is a topologically interesting target for binding using concave receptors [62,63]. Kennedy *et al.* [64] employed fivefold addition of 4-*tert*-butylphenylmagnesium bromide to fullerene to obtain a fulleroid derivative that possessed Janus-like properties, with one (concave) face being complementary to the other (curved) surface of the molecule. The combination of hydrophobic and π-stacking interactions with shape complementarity ensured the formation of extended C_{60} molecular wires as evidenced by X-ray crystallographic analysis. Although the interfullerene distance thus obtained is significantly longer than that observed in crystals of PCBM (4.0 vs 2.9 Å), the use of this material as a replacement for PCBM in bulk heterojunction solar cells led to devices with efficiencies of ca 1.5% after thermal annealing.

Conjugated rigid-rod molecules provide convenient scaffolds that can be used to enforce the stacked arrangement of aromatic acceptor units while providing a pathway for hole transport. Synthetic accessibility and solubility limit the length of these covalent structures, but it is possible to use a combination of hydrophobic and electronic interactions to promote the formation of interdigitated architectures spanning much greater lengths and incorporation of chromophores covering a greater portion of the solar spectrum. By combining n-semiconducting naphthalenebis(dicarboxyimide) rainbow stacks and p-oligophenylene semiconducting rigid-rod scaffolds, the Matile group has developed a programmed assembly of interdigitating intra- and interlayer recognition motifs on conducting surfaces [65–67]. These elegant self-assembled zipper assemblies offer rapid access to supramolecular cascade n-/p-heterojunctions exhibiting efficient photoconversion in photoelectrochemical setups. Particularly promising is their potential to incorporate redox gradients across the thickness of the active layer, which can serve to guide charges toward the corresponding electrodes for collection in lieu of the semipermeable membranes such as poly(3,4-ethylene- dioxythiophene)/poly(styrensulfonate) used in polymer bulk heterojunction solar cells. The spectral response of the IPCE (incident photon to collected electron) of the devices reveals that both the perylenebis(dicarboximide) (PDI) unit and the p-oligophenylene scaffold contribute to the absorption of light for the purpose of generating photocurrent. Furthermore, the planarization of the p-phenylene backbone and bathochromic shift of the absorption of the PDI unit point to the formation of the proposed interdigitated structure.

The molecular association between fullerene C_{60} and porphyrins can be significantly enhanced by moving on to porphyrin dendrimers [68,69]. Under appropriate conditions (e.g., by employing solvent mixtures such as toluene/acetonitrile), porphyrin dendrimers and C_{60} are readily clusterized [70]. The latter are relatively monodisperse, typically measuring a few hundred nanometers in diameter. Hasobe *et al.* prepared OPV devices based on porphyrin dendrimers/C_{60} clusters using peripherally porphyrin-substituted polypropyleneamine dendrimers of first, second, or third generation [71]. The clusters were deposited onto a transparent tin oxide electrode, and the photocurrent generated was investigated in a photoelectrochemical setup. Interestingly, the IPCE curves showed that the cells were responsive from 400 to 950 nm, and this substantial enhancement of the spectral response in the near-infrared (NIR) was attributed to the use of the porphyrin dendrimers.

22.4 SELF-ASSEMBLY BASED ON H-BONDING

H-B interactions offer a convenient approach toward precisely controlling the geometry of electron donor–acceptor assemblies, and numerous conjugated polymers bearing H-B groups have been reported [72,73]. For example, triaminotriazine-substituted *oligo*-phenylenevinylenes bind perylene bisimides to form H-B trimers, which then self-assemble into stacks conducive for charge separation (Fig. 22-3a) [52,74].

Besides perylene bisimides, other electron acceptors such as fullerene [78] and tetrathiofulvalene (TTF) [79] derivatives incorporating H-B units have been described, although self-doping behavior, such as that observed for TTF-2-carboxylic acid in the presence of ammonia [80], may be a limitation in the latter case. Samorí and coworkers [75] investigated in detail the relationship

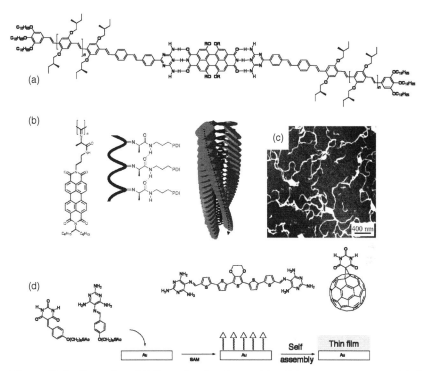

Figure 22-3 Examples of H-B mediated self-assembly of donor–acceptor materials based on (a) OPVs and PDIs [52,74] and (b) polymeric PDIs preorganized through H-B interactions on the polyisocyanide backbone (AFM image of polymeric PDI strands on SiO_2). *Source*: Adapted with permission from reference [75]. Copyright 2008 American Chemical Society. (d) Complementary hydrogen-bonding interactions between oligothiophenes and fullerenes were found to enhance the performance of photoelectrochemical devices [76]. In (d), modification of the substrate with H-B motifs with preference for the donor or acceptor material was found to positively impact the performance of the device [77].

between self-assembled molecular wires and PV activity using Kelvin probe force microscopy (KPFM). Using this technique, which allows the determination of the surface potential with nanoscale resolution, it is possible to map out variations in electrical potential in darkness and under illumination. The molecular wires were composed of PDI chromophores strung onto polyisocyanopeptide polymers, in which intramolecular H-bonding rigidifies a helical conformation that places the chromophores into an infinite π-stacked assembly (Fig. 22-3b). Blended with P3HT, the PDI wires behaved as electron acceptors and were shown to provide better percolation pathways than single PDI chromophores.

Huang *et al.* proposed using non-self-complementary H-B interactions to direct the formation of supramolecular heterojunctions for PV applications [76]. Photoelectrochemical devices incorporating components possessing complementary H-B units were found to give a 2.5-fold enhancement in photocurrent compared to model systems in which self-assembly is suppressed. Subsequent improvement was sought by using H-B-terminated self-assembled monolayers (SAMs) on gold and the combination of a H-B barbituric-acid-appended fullerene and a complementary melamine-terminated thiophene oligomer to promote hierarchical self-assembly (Fig. 22-3c) [77]. The terminal thiol-group interacts with the gold surface, resulting in the formation of SAMs that are proposed to better accommodate the subsequent formation of photo-/electroactive fullerene-containing thin films because of complementary H-B interactions. Intermolecular H-B interactions between perylene tetracarboxylic dianhydride and pentatiophene bearing terminal formyl groups were also shown by Jiang et al. to account for an increase in V_{oc} in flexible OPV devices [81]. However, the absence of H-B interactions, as evidenced by IR absorption, for the analogous tertiophene derivative is puzzling.

The presence of ground-state CT absorption bands in bulk heterojunction (poly[2-methoxy-5-(3,7- dimethyloctyloxy)]-1,4-phenylene-vinylene) or P3HT/PCBM active layers has been observed using Fourier-transfer photocurrent spectroscopy (FTPS) [82,83]. Although very weak, the CT bands correlate with the observed V_{oc} through the determination of an effective band gap that differs from that calculated using electrochemical data obtained in solution. However, the weak nature of the ground-state CT absorption makes direct photophysical studies using conventional spectroscopic techniques difficult. Supramolecular interactions can be used to design architectures in which face-to-face contact between extended π-conjugated polymers or oligomers is favored. In such an assembly, strong ground-state interactions between complementary H-B fullerene and oligo-thiophenevinylenes were observed [84]. This contrasts with the absence of such transitions in an analogous covalent C_{60}-terthiophene derivative prepared by Roncali et al. [85,86] and may provide a means by which it is possible to enhance light absorption in the visible region to enhance photoconversion efficiency. Currently, this is obtained through the use of C_{70} derivatives, which provide higher efficiency solar cells than C_{60} because of increased absorption at the price of much reduced availability. As expected, ultrafast ($k_{ET} = 5 \times 10^{12}$ s^{-1}) electron transfer follows pulsed excitation of the supramolecular architecture.

Precise control of the architecture of the active layer can be achieved by using molecular self-assembly to template the formation of small clusters of defined size and composition, which can then be cast from solution onto transparent conducting electrodes. Imahori et al. applied this technique for the formation of porphyrin$-C_{60}$ clusters using H-B interactions [87] or Au nanoparticles (NPs) [88] as templates. The resulting modified electrodes, when irradiated in a photoelectrochemical setup using I_3^-/I_2 as an electron relay, exhibited maximum IPCE efficiencies of 40–60%. Along similar lines, modified electrodes incorporating semiconducting CdS NPs were prepared by self-assembly using complementary H-B interactions [89]. In this case, melamine- or barbiturate-capped CdS-NP were shown to bind to a gold electrode grafted with a thiol-SAM terminated with the complementary melamine or barbiturate unit. Additionally, a 2.6-fold enhancement in the observed photocurrent was noted when gold NPs were incorporated between the macroscopic Au electrode and the CdS-NP.

22.5   SUPRAMOLECULAR HYBRID SOLAR CELLS

Dye-sensitized electrochemical photovoltaic cell (DSSC) technology has been recognized as a viable alternative to the well-developed solid-state homo- and heterojunction solar cell technologies [90–93]. In this strategy, a network of titania NPs serves as an electron-transporting medium and as a high-surface-area support for dye molecules capable of injecting electrons into the titana on photoexcitation, and an electrolyte solution (typically aqueous I_3^-/I_2) is used to complete the electrical circuit. A fundamental difference between polymer OPV and DSSC devices is that intermolecular charge transport is not necessary in the latter, as charges are directly injected into the semiconductor. Continuous improvement in cell architecture and dye structure have brought the conversion efficiency up to 12% [93], but this could be further increased by augmenting the absorption envelope of the sensitizer and by diminishing unproductive charge recombination processes. Thus, possible strategies to augment device efficiency involve the use of supramolecular interactions to construct composite organic dye layers integrating energy or redox gradients to enhance the absorption envelope in the NIR region and/or direct electron transfer toward the metal oxide while minimizing charge recombination.

To achieve directional energy transfer over long distances, Calzaferri *et al.* designed artificial photonic antenna systems based on the incorporation of chromophores inside metal oxide frameworks [94,95]. These architectures have received considerable attention for their potential in the fabrication of PV devices [96], because their supramolecular architecture leads to the electronic excitation energy being transferred to a well-defined location in the device through sequential resonant energy transfer. In this strategy, light-absorbing molecules are incorporated in the zeolite channels, which corresponds to the first stage of molecular organization. This allows light-harvesting within the volume of the host and radiationless energy transport along the channels. The second stage of organization involves electronic coupling to an external acceptor or donor fluorophore (stopcock) at the channel entrances, which can then act as a sink to trap the electronic excitation energy. The third stage of organization is obtained by interfacing the material to an external device through the stopcock.

Supramolecular chemistry, particularly of coordination compounds, has emerged as a viable route to prepare efficient dye sensitizers of greater complexity [97]. Hardin *et al.* have recently reported that the use of ancillary light-harvesting dyes can lead to a 26% improvement in power conversion efficiency [98], and a similar enhancement was observed by Handa *et al.* when using donor–acceptor dye combinations [99]. Examples of both covalent [100] and noncovalent [87,101] multichromophore dyes, and the effects of H-B on the surface structure of TiO_2 were investigated by Imahori *et al.* [102]. The hierarchical organization of chromophores was illustrated by Kira *et al.* by obtaining bicontinuous vertically aligned donor–acceptor domains through supramolecular organization of Zn-porphyrin wires and fullerenes [103]. The assembly strategy relies on the use of meso-bipyridyl-substituted porphyrins, which can be assembled into wires by coordination to palladium (II) metal ions. Pyridyl-substituted fullerenes may then axially coordinate the chelated zinc metal ion to provide a supramolecular double cable that can be oriented normal to the surface through a step-wise assembly approach. A carboxylic acid group is used to anchor the first pyridyl-porphyrin monolayer onto the SnO_2 surface, which is then used as a template for the subsequent formation of the vertically aligned phases (Fig. 22-4a).

The IPCE value at 440-nm excitation wavelength reaches a maximum of ca 20% for devices that are five porphyrins in thickness and then decreases for thicker samples. It is proposed that this reversal in trend is due to self-aggregation of the porphyrins caused by the simultaneous presence of vacant axial coordination sites and pyridine ligands, which reduces penetration and complexation of the fullerene component. In related work, Imahori et al. [104] took advantage of the formation of porphyrin–C_{60} clusters by employing Au-NP stabilized with porphyrin-appended alkyl thiols. Variations in solvent conditions permitted first the aggregation of C_{60}, then of the Au-NP. Finally, in a last organizational step, the clusterized NPs were deposited onto nanostructured SnO_2/ITO electrodes (Fig. 22-4b). The resulting hierarchically organized assemblies exhibited IPCE values as high as 42% at the absorption maximum (475 nm).

The usefulness of supramolecular self-assembly in the fabrication of hybrid PV devices is of course not limited to the organization of chromophores. In particular, the use of a liquid electrolyte solution is frequently viewed as a potential disadvantage in current DSSC designs. Replacing it with a conducting polymer introduces similar problems related to long-distance charge transport in molecular materials as encountered in OPV devices. One interesting solution involves the use of molecular gels, which rely on supramolecular organization to form a large network capable of imprisoning a large number of solvent molecules and preventing macroscopic flow. Quasi-solid-state DSSC devices were thus prepared by employing an H-B gelator for ionic liquids (Fig. 22-4c) [105]. Just 2% w/w of the gelating compound was sufficient to raise the sol–gel transition temperature to nearly 120°C while maintaining the efficiency of the device (6.3% at the time). Similarly, urea-pyrimidone H-B units were incorporated onto short ethylene-glycol frameworks to prepare supramolecular polymers for ionic electrolytes that retain the rheological advantages of longer polymers while remaining small enough to fit inside the titania pores [106].

22.6 CONCLUSION

Our understanding of the intricate balance of the molecular forces that affect charge separation and transport inside molecular-based PV devices has grown tremendously thanks to experimental results combined with advanced modeling. It is only natural that this understanding should fuel our desire to control these processes, thereby increasing the overall efficiency of future devices by maximizing productive versus unproductive steps in the long chain of events that lead to the conversion of light into electrical energy. Supramolecular interactions

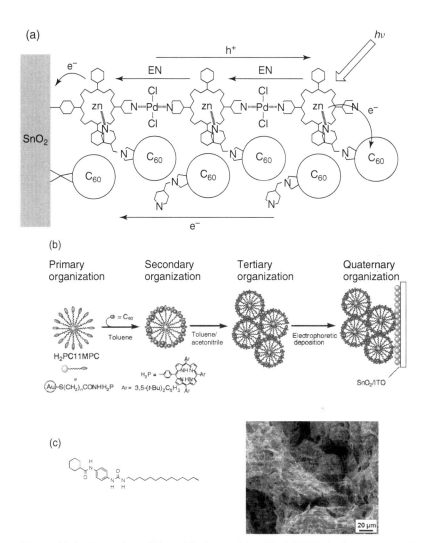

Figure 22-4 Examples of hierarchical assembly of hybrid PV devices. In (a), Pd(II) coordination is used to bind layers of zinc dipyridylporphyrin to a pyridylporphyrin monolayer grafted onto tin oxide. Incubation with pyridyl-C_{60} is proposed to afford supramolecular assemblies reminiscent of double cable architectures. *Source*: Adapted with permission from reference [103]. Copyright 2009 American Chemical Society. (b) Multistep organization of porphyrin-modified Au nanoparticles. Incubation with C_{60} in toluene is followed by precipitation and deposition onto nanostructured SnO2/ITO electrodes. *Source*: Reproduced with permission from Ref. [104]. Copyright Wiley-VCH Verlag GmbH & Co. KGaA. (c) Chemical structure and SEM picture of the xerogel of a hydrogen-bonding organogelator for ionic liquids used for the fabrication of quasi-solid-state DSSCs [105].

allow the dissection of large multicomponent (or multitask) objects into smaller ones, more apprehensible in terms of synthesis and characterization. Their self-assembly into functional architectures opens the way toward future generation of devices that are both efficient and can harness the added benefits of supramolecular materials, such as the ability of autonomic regulation [107] or self-healing [47] present in Nature.

REFERENCES

1. Kalowekamo, J., Baker, E. (2009). Estimating the manufacturing cost of purely organic solar cells. *Sol. Energ.*, *83*, 1224–1231.
2. Yang, Y., Pei, Q., Heeger, A. J. (1996). Efficient blue light-emitting diodes from a soluble poly(para-phenylene): Internal field emission measurement of the energy gap in semiconducting polymers. *Synthetic Met.*, *78*, 263–267.
3. Campbell, I. H., Davids, P. S., Ferraris, J. P., Hagler, T. W., Heller, C. M., Saxena, A., Smith, D. L. (1996). Probing electronic state charging in organic electronic devices using electroabsorption spectroscopy. *Synthetic Met.*, *80*, 105–110.
4. Hagler, T. W., Pakbaz, K., Heeger, A. J. (1994). Polarized-electroabsorption spectroscopy of a soluble derivative of poly(p-phenylenevinylene) oriented by gel processing in polyethylene-polarization anisotropy, the off-axis dipole-moment, and excited-state delocalization. *Phys. Rev. B*, *49*, 10968–10975.
5. Kersting, R., Lemmer, U., Deussen, M., Bakker, H. J., Mahrt, R. F., Kurz, H., Arkhipov, V. I., Bassler, H., Gobel, E. O. (1994). Ultrafast field-induced dissociation of excitons in conjugated polymers. *Phys. Rev. Lett.*, *73*, 1440–1443.
6. Powell, R. C., Soos, Z. G. (1975). Singlet exciton energy-transfer in organic solids. *J. Lumin.*, *11*, 1–45.
7. Miranda, P. B., Moses, D., Heeger, A. J. (2001). Ultrafast photogeneration of charged polarons in conjugated polymers. *Phys. Rev. B*, *64*, 081201.
8. Stubinger, T., Brutting, W. (2001). Exciton diffusion and optical interference in organic donor-acceptor photovoltaic cells. *J. Appl. Phys.*, *90*, 3632–3641.
9. Haugeneder, A., Neges, M., Kallinger, C., Spirkl, W., Lemmer, U., Feldmann, J., Scherf, U., Harth, E., Gugel, A., Mullen, K. (1999). Exciton diffusion and dissociation in conjugated polymer fullerene blends and heterostructures. *Phys. Rev. B*, *59*, 15346–15351.
10. Theander, M., Yartsev, A., Zigmantas, D., Sundstrom, V., Mammo, W., Andersson, M. R., Inganas, O. (2000). Photoluminescence quenching at a polythiophene/c-60 heterojunction. *Phys. Rev. B*, *61*, 12957–12963.
11. Arkhipov, V. I., Heremans, P., Bassler, H. (2003). Why is exciton dissociation so efficient at the interface between a conjugated polymer and an electron acceptor? *Appl. Phys. Lett.*, *82*, 4605–4607.
12. Muller, J. G., Lupton, J. M., Feldmann, J., Lemmer, U., Scharber, M. C., Sariciftci, N. S., Brabec, C. J., Scherf, U. (2005). Ultrafast dynamics of charge carrier photogeneration and geminate recombination in conjugated polymer: Fullerene solar cells. *Phys. Rev. B*, *72*, 195208.
13. Brabec, C. J., Zerza, G., Cerullo, G., De Silvestri, S., Luzzati, S., Hummelen, J. C., Sariciftci, S. (2001). Tracing photoinduced electron transfer process in conjugated polymer/fullerene bulk heterojunctions in real time. *Chem. Phys. Lett.*, *340*, 232–236.
14. Scharber, M. C., Wuhlbacher, D., Koppe, M., Denk, P., Waldauf, C., Heeger, A. J., Brabec, C. L. (2006). Design rules for donors in bulk-heterojunction solar cells—towards 10% energy-conversion efficiency. *Adv. Mater.*, *18*, 789–794.
15. Hoppe, H., Sariciftci, N. S. (2004). Organic solar cells: An overview. *J. Mater. Res.*, *19*, 1924–1945.
16. Roncali, J., Leriche, P., Cravino, A. (2007). From one- to three-dimensional organic semiconductors: in search of the organic silicon? *Adv. Mater.*, *19*, 2045–2060.

17. Gunes, S., Neugebauer, H., Sariciftci, N. S. (2007). Conjugated polymer-based organic solar cells. *Chem. Rev.*, *107*, 1324–1338.

18. Campoy-Quiles, M., Ferenczi, T., Agostinelli, T., Etchegoin, P. G., Kim, Y., Anthopoulos, T. D., Stavrinou, P. N., Bradley, D. D. C., Nelson, J. (2008). Morphology evolution via self-organization and lateral and vertical diffusion in polymer: fullerene solar cell blends. *Nat. Mater.*, *7*, 158–164.

19. Brédas, J.-L., Norton, J. E., Cornil, J., Coropceanu, V. (2009). Molecular understanding of organic solar cells: The challenges. *Accounts Chem. Res.*, *42*, 1691–1699.

20. Yu, G., Gao, J., Hummelen, J. C., Wudl, F., Heeger, A. J. (1995). Polymer photovoltaic cells: Enhanced efficiencies via a network of internal donor-acceptor heterojunctions. *Science*, *270*, 1789–1791.

21. Li, J., Kastler, M., Pisula, W., Robertson, J. W. F., Wasserfallen, D., Grimsdale, A. C., Wu, J., Mullen, K. (2007). Organic bulk-heterojunction photovoltaics based on alkyl substituted discotics. *Adv. Funct. Mater.*, *17*, 2528–2533.

22. Linke-Schaetzel, M., Bhise, A. D., Gliemann, H., Koch, T., Schimmel, T., Balaban, T. S. (2004). Self-assembled chromophores for hybrid solar cells. *Thin Solid Films*, *451*, 16–21.

23. Tang, C. W. (1986). 2-Layer organic photovoltaic cell. *Appl. Phys. Lett.*, *48*, 183–185.

24. Solarmer Energy, Inc. reported an efficiency of 7.9% for OPV devices with a 0.047 cm^2 active area. *Source:* www.Solarmer.com.

25. Peet, J., Heeger, A. J., Bazan, G. C. (2009). "Plastic" solar cells: self-assembly of bulk hetero-junction nanomaterials by spontaneous phase separation. *Accounts Chem. Res.*, *42*, 1700–1708.

26. Shaheen, S. E., Brabec, C. J., Sariciftci, N. S., Padinger, F., Fromherz, T., Hummelen, J. C. (2001). 2.5% efficient organic plastic solar cells. *Appl. Phys. Lett.*, *78*, 841–843.

27. Halls, J. J. M., Pichler, K., Friend, R. H., Moratti, S. C., Holmes, A. B. (1996). Exciton diffusion and dissociation in a poly(p-phenylenevinylene)/c-60 heterojunction photovoltaic cell. *Appl. Phys. Lett.*, *68*, 3120–3122.

28. Wasielewski, M. R. (2009). Self-assembly strategies for integrating light harvesting and charge separation in artificial photosynthetic systems. *Accounts Chem. Res.*, *42*, 1910–1921.

29. Campagna, S., Denti, G., Serroni, S., Juris, A., Venturi, M., Ricevuto, V., Balzani, V. (1995). Den-drimers of nanometer-size based on metal-complexes—luminescent and redox-active polynuclear metal-complexes containing up to 22 metal centers. *Chem. Eur. J.*, *1*, 211–221.

30. Puntoriero, F., Nastasi, F., Cavazzini, M., Quici, S., Campagna, S. (2007). Coupling synthetic antenna and electron donor species: A tetranuclear mixed-metal Os (II)-Ru (II) dendrimer con-taining six phenothiazine donor subunits at the periphery. *Coord. Chem. Rev.*, *251*, 536–545.

31. Balzani, V., Credi, A., Venturi, M. (2008). Processing energy and signals by molecular and supramolecular systems. *Chem. Eur. J.*, *14*, 26–39.

32. Balzani, V., Campagna, S., Denti, G., Juris, A., Serroni, S., Venturi, M. (1995). Harvesting sunlight by artificial supramolecular antennae. *Sol. Energ. Mater. Sol. Cell.*, *38*, 159–173.

33. Grozema, F. C., Siebbeles, L. D. A. (2008). Mechanism of charge transport in self-organizing organic materials. *Int. Rev. Phys. Chem.*, *27*, 87–138.

34. Mallik, A. B., Locklin, J., Mannsfeld, S. C. B., Reese, C., Roberts, M. S., Senatore, M. L., Zi, H., Bao, Z., In *Organic Field-effect Transistors*. (Eds.: Bao, Z., Locklin J.) CRC Press, Boca Raton, Florida, 2007.

35. Craats, A. M. v. d., Warman, J. M., Fechtenkötter, A., Brand, J. D., Harbison, M. A., Mullen, K. (1999). Record charge carrier mobility in a room-temperature discotic liquid-crystalline derivative of hexabenzocoronene. *Adv. Mater.*, *11*, 1469–1472.

36. Paraschiv, I., Giesbers, M., van Lagen, B., Grozema, F. C., Abellon, R. D., Siebbeles, L. D. A., Marcelis, A. T. M., Zuilhof, H., Sudholter, E. J. R. (2006). H-bond-stabilized triphenylene-based columnar discotic liquid crystals. *Chem. Mater.*, *18*, 968–974.

37. Wu, J., Pisula, W., Mullen, K. (2007). Graphenes as potential material for electronics. *Chem. Rev.*, *107*, 718–747.

38. Kumar, S. (2006). Self-organization of disc-like molecules: chemical aspects. *Chem. Soc. Rev.*, *35*, 83–109.

39. Ghosh, S., Li, X. Q., Stepanenko, V., Wurthner, F. (2008). Control of h- and j-type π stacking by peripheral alkyl chains and self-sorting phenomena in perylene bisimide homo- and heteroag-gregates. *Chem. Eur. J.*, *14*, 11343–11357.

40. Chen, Z. J., Lohr, A., Saha-Moller, C. R., Wurthner, F. (2009). Self-assembled pi-stacks of functional dyes in solution: structural and thermodynamic features. *Chem. Soc. Rev.*, *38*, 564–584.
41. Laschat, S., Baro, A., Steinke, N., Giesselmann, F., Hagele, C., Scalia, G., Judele, R., Kapatsina, E., Sauer, S., Schreivogel, A., Tosoni, M. (2007). Discotic liquid crystals: from tailor-made synthesis to plastic electronics. *Angew. Chem. Int. Ed.*, *46*, 4832–4887.
42. Pisula, W., Zorn, M., Chang, J. Y., Mullen, K., Zentel, R. (2009). Liquid crystalline ordering and charge transport in semiconducting materials. *Macromol. Rapid Commun.*, *30*, 1179–1202.
43. Schmaltz, B., Weil, T., Mullen, K. (2009). Polyphenylene-based materials: control of the electronic function by molecular and supramolecular complexity. *Adv. Mater.*, *21*, 1067–1078.
44. Sergeyev, S., Pisula, W., Geerts, Y. H. (2007). Discotic liquid crystals: a new generation of organic semiconductors. *Chem. Soc. Rev.*, *36*, 1902–1929.
45. Chu, C. C., Bassani, D. M. (2008). Challenges and opportunities for photochemists on the verge of solar energy conversion. *Photochem. Photobio. Sci.*, *7*, 521–530.
46. Huang, C. H., Bassani, D. M. (2005). Exciting supramolecular architectures: light-induced processes and synthetic transformations in noncovalent assemblies. *Eur. J. Org. Chem.*, 4041–4050.
47. Cordier, P., Tournilhac, F., Soulie-Ziakovic, C., Leibler, L. (2008). Self-healing and thermoreversible rubber from supramolecular assembly. *Nature*, *451*, 977–980.
48. Burattini, S., Colquhoun, H. M., Greenland, B. W., Hayes, W. (2009). A novel self-healing supramolecular polymer system. *Faraday Discuss.*, *143*, 251–264.
49. Montarnal, D., Tournilhac, F., Hidalgo, M., Couturier, J. L., Leibler, L. (2009). Versatile one-pot synthesis of supramolecular plastics and self-healing rubbers. *J. Am. Chem. Soc.*, *131*, 7966–7967.
50. Reutenauer, P., Buhler, E., Boul, P. J., Candau, S. J., Lehn, J. M. (2009). Room temperature dynamic polymers based on Diels-Alder chemistry. *Chem. Eur. J.*, *15*, 1893–1900.
51. Willner, I., Shlyahovsky, B., Zayats, M., Willner, B. (2008). Dnazymes for sensing, nanobiotechnology and logic gate applications. *Chem. Soc. Rev.*, *37*, 1153–1165.
52. Beckers, E. H. A., Chen, Z. J., Meskers, S. C. J., Jonkheijm, P., Schenning, A., Li, X. Q., Osswald, P., Wurthner, F., Janssen, R. A. J. (2006). The importance of nanoscopic ordering on the kinetics of photoinduced charge transfer in aggregated pi-conjugated hydrogen-bonded donor-acceptor systems. *J. Phys. Chem. B*, *110*, 16967–16978.
53. Bullock, J. E., Carmieli, R., Mickley, S. M., Vura-Weis, J., Wasielewski, M. R. (2009). Photoinitiated charge transport through π-stacked electron conduits in supramolecular ordered assemblies of donor-acceptor triads. *J. Am. Chem. Soc.*, *131*, 11919–11929.
54. Peet, J., Cho, N. S., Lee, S. K., Bazan, G. C. (2008). Transition from solution to the solid state in polymer solar cells cast from mixed solvents. *Macromolecules*, *41*, 8655–8659.
55. Peet, J., Soci, C., Coffin, R. C., Nguyen, T. Q., Mikhailovsky, A., Moses, D., Bazan, G. C. (2006). *Appl. Phys. Lett.*, *89*, 252105.
56. Peet, J., Kim, J. Y., Coates, N. E., Ma, W. L., Moses, D., Heege, A. J., Bazan, G. C. (2007). *Nat. Mater.*, *6*, 497.
57. Lee, J. K., Ma, W. L., Brabec, C. J., Yuen, J., Moon, J. S., Kim, J. Y., Lee, K., Bazan, G. C., Heeger, A. J. (2008). Processing additives for improved efficiency from bulk heterojunction solar cells. *J. Am. Chem. Soc.*, *130*, 3619–3623.
58. Balaban, T. S. (2005). Tailoring porphyrins and chlorins for self-assembly in biomimetic artificial antenna systems. *Accounts Chem. Res.*, *38*, 612–623.
59. Kim, O. K., Melinger, J., Chung, S. J., Pepitonet, M. (2008). Supramolecular device for artificial photosynthetic mimics as helix-mediated antenna/reaction center ensemble. *Org. Lett.*, *10*, 1625–1628.
60. Charlet, E., Grelet, E., Brettes, P., Bock, H., Saadaoui, H., Cisse, L., Destruel, P., Gherardi, N., Seguy, I. (2008). Ultrathin films of homeotropically aligned columnar liquid crystals on indium tin oxide electrodes. *Appl. Phys. Lett.*, *92*, 024107.
61. Archambeau, S., Seguy, I., Jolinat, P., Farenc, J., Destruel, P., Nguyen, T. P., Bock, H., Grelet, E. (2006). Stabilization of discotic liquid organic thin films by ITO surface treatment. *Appl. Surf. Sci.*, *253*, 2078–2086.
62. Perez, E. M., Martin, N. (2008). Curves ahead: molecular receptors for fullerenes based on concave-convex complementarity. *Chem. Soc. Rev.*, *37*, 1512–1519.

63. Perez, E. M., Capodilupo, A. L., Fernandez, G., Sanchez, L., Viruela, P. M., Viruela, R., Orti, E., Bietti, M., Martin, N. (2008). Weighting non-covalent forces in the molecular recognition of C-60. Relevance of concave–convex complementarity. *Chem. Commun.*, 4567–4569.

64. Kennedy, R. D., Ayzner, A. L., Wanger, D. D., Day, C. T., Halim, M., Khan, S. I., Tolbert, S. H., Schwartz, B. J., Rubin, Y., (2008). Self-assembling fullerenes for improved bulk-heterojunction photovoltaic devices. *J. Am. Chem. Soc.*, *130*, 17290–117292.

65. Sisson, A. L., Sakai, N., Banerji, N., Furstenberg, A., Vauthey, E., Matile, S. (2008). Zipper assembly of vectorial rigid-rod π-stack architectures with red and blue naphthalenediimides: toward supramolecular cascade n/p-heterojunctions. *Angew. Chem. Int. Ed.*, *47*, 3727–3729.

66. Sakai, N., Sisson, A. L., Burgi, T., Matile, S. (2007). Zipper assembly of photoactive rigid-rod naphthalenediimide π-stack architectures on gold nanoparticles and gold electrodes. *J. Am. Chem. Soc.*, *129*, 15758–15759.

67. Bhosale, R., Misek, J., Sakai, N., Matile, S. (2010). Supramolecular n/p-heterojunction photosystems with oriented multicolored antiparallel redox gradients (OMARG-SHJs). *Chem. Soc. Rev.*, *39*, 138–149.

68. Sun, D. Y., Tham, F. S., Reed, C. A., Chaker, L., Boyd, P. D. W. (2002). Supramolecular fullerene-porphyrin chemistry. Fullerene complexation by metalated "Jaws porphyrin" hosts. *J. Am. Chem. Soc.*, *124*, 6604–6612.

69. Evans, D. R., Fackler, N. L. P., Xie, Z. W., Rickard, C. E. F., Boyd, P. D. W., Reed, C. A. (1999). π-Arene/cation structure and bonding. Solvation versus ligand binding in iron(III) tetraphenyl-porphyrin complexes of benzene, toluene, p-xylene, and [60]fullerene. *J. Am. Chem. Soc.*, *121*, 8466–8474.

70. Kimura, M., Saito, Y., Ohta, K., Hanabusa, K., Shirai, H., Kobayashi, N. (2002). Self-organization of supramolecular complex composed of rigid dendritic porphyrin and fullerene. *J. Am. Chem. Soc.*, *124*, 5274–5275.

71. Hasobe, T., Kashiwagi, Y., Absalom, M. A., Sly, J., Hosomizu, K., Crossley, M. J., Imahori, H., Kamat, P. V., Fukuzumi, S. (2004). Supramolecular photovoltaic cells using porphyrin dendrimers and fullerenes. *Adv. Mat.*, *16*, 975–979.

72. Hoeben, F. J. M., Jonkheijm, P., Meijer, E. W., Schenning, A. P. H. J. (2005). About supramolecular assemblies of π-conjugated systems. *Chem. Rev.*, *105*, 1491–1546.

73. Mishra, A., Ma, C. Q., Bauerle, P. (2009). Functional oligothiophenes: molecular design for multidimensional nanoarchitectures and their applications. *Chem. Rev.*, *109*, 1141–1276.

74. Würthner, F., Chen, Z., Hoeben, F. J. M., Osswald, P., You, C.-C., Jonkheijm, P., von Herrikhuyzen, J., Schenning, A. P. H. J., van der Schoot, P. P. A. M., Meijer, E. W., Beckers, E. H. A., Meskers, S. C. J., Janssen, R. A. J. (2004). Supramolecular p-n-heterojunctions by co-self-organization of oligo(p-phenylene vinylene) and perylene bisimide dyes. *J. Am. Chem. Soc.*, *126*, 10611–10618.

75. Palermo, V., Otten, M. B. J., Liscio, A., Schwartz, E., de Witte, P. A. J., Castriciano, M. A., Wienk, M. M., Nolde, F., De Luca, G., Cornelissen, J., Janssen, R. A. J., Mullen, K., Rowan, A. E., Nolte, R. J. M., Samori, P. (2008). The relationship between nanoscale architecture and function in photovoltaic multichromophoric arrays as visualized by Kelvin probe force microscopy. *J. Am. Chem. Soc.*, *130*, 14605–14614.

76. Huang, C. H., McClenaghan, N. D., Kuhn, A., Hofstraat, J. W., Bassani, D. M. (2005). Enhanced photovoltaic response in hydrogen-bonded all-organic devices. *Org. Lett.*, *7*, 3409–3412.

77. Huang, C. H., McClenaghan, N. D., Kuhn, A., Bravic, G., Bassani, D. M. (2006). Hierarchical self-assembly of all-organic photovoltaic devices. *Tetrahedron*, *62*, 2050–2059.

78. Sanchez, L., Martin, N., Guldi, D. M. (2005). Hydrogen-bonding motifs in fullerene chemistry. *Angew. Chem. Int. Ed.*, *44*, 5374–5382.

79. Fourmigue, M., Batail, P. (2004). Activation of hydrogen- and halogen-bonding interactions in tetrathiafulvalene-based crystalline molecular conductors. *Chem. Rev.*, *104*, 5379–5418.

80. Kobayashi, Y., Yoshioka, M., Saigo, K., Hashizume, D., Ogura, T. (2009). Hydrogen-bonding-assisted self-doping in tetrathiafulvalene (TTF) conductor. *J. Am. Chem. Soc.*, *131*, 9995–10002.

81. Jiang, C. Y., Liu, P., Deng, W. J. (2009). Synthesis and photovoltaic properties of formyl end-capped oligothiophenes. *Synthetic Commun.*, *39*, 2360–2369.

82. Benson-Smith, J. J., Goris, L., Vandewal, K., Haenen, K., Manca, J. V., Vanderzande, D., Bradley, D. D. C., Nelson, J. (2007). Formation of a ground-state charge-transfer complex in polyfluorene/[6,6]-phenyl-C-61 butyric acid methyl ester (PCBM) blend films and its role in the function of polymer/PCBM solar cells. *Adv. Funct. Mater.*, *17*, 451–457.

83. Vandewal, K., Gadisa, A., Oosterbaan, W. D., Bertho, S., Banishoeib, F., Van Severen, I., Lutsen, L., Cleij, T. J., Vanderzande, D., Manca, J. V. (2008). The relation between open-circuit voltage and the onset of photocurrent generation by charge-transfer absorption in polymer: fullerene bulk heterojunction solar cells. *Adv. Funct. Mater.*, *18*, 2064–2070.

84. McClenaghan, N. D., Grote, Z., Darriet, K., Zimine, M., Williams, R. M., De Cola, L., Bassani, D. M. (2005). Supramolecular control of oligothienylenevinylene-fullerene interactions: evidence for a ground-state EDA complex. *Org. Lett.*, *7*, 807–810.

85. van Hal, P. A., Beckers, E. H. A., Meskers, S. C. J., Janssen, R. A. J., Jousselme, B., Blanchard, P., Roncali, J. (2002). Orientational effect on the photophysical properties of quaterthiophene-C-60 dyads. *Chem. Eur. J.*, *8*, 5415–5429.

86. Martineau, C., Blanchard, P., Rondeau, D., Delaunay, J., Roncali, J. (2002). Synthesis and electronic properties of adducts of oligothienylenevinylenes and fullerene C-60. *Adv. Mater.*, *14*, 283–287.

87. Kang, S. C., Umeyama, T., Ueda, M., Matano, Y., Hotta, H., Yoshida, K., Isoda, S., Shiro, M., Imahori, H. (2006). Ordered supramolecular assembly of porphyrin-fullerene composites on nanostructured SnO2 electrodes. *Adv. Mater.*, *18*, 2549–2552.

88. Imahori, H., Fujimoto, A., Kang, S., Hotta, H., Yoshida, K., Umeyama, T., Matano, Y., Isoda, S., Isosomppi, M., Tkachenko, N. V., Lemmetyinen, H. (2005). Host–guest interactions in the supramolecular incorporation of fullerenes into tailored holes on porphyrin-modified gold nanoparticles in molecular photovoltaics. *Chem. Eur. J.*, *11*, 7265–7275.

89. Baron, R., Huang, C. H., Bassani, D. M., Onopriyenko, A., Zayats, M., Willner, I. (2005). Hydrogen-bonded CdS nanoparticle assemblies on electrodes for photoelectrochemical applications. *Angew. Chem. Int. Ed.*, *44*, 4010–4015.

90. Oregan, B., Gratzel, M. (1991). A low-cost, high-efficiency solar-cell based on dye-sensitized colloidal TiO2 films. *Nature*, *353*, 737–740.

91. Bach, U., Lupo, D., Comte, P., Moser, J. E., Weissortel, F., Salbeck, J., Spreitzer, H., Gratzel, M. (1998). Solid-state dye-sensitized mesoporous TiO2 solar cells with high photon-to-electron conversion efficiencies. *Nature*, *395*, 583–585.

92. Gratzel, M. (2001). Photoelectrochemical cells. *Nature*, *414*, 338–344.

93. Grätzel, M. (2009). Recent advances in sensitized mesoscopic solar cells. *Accounts Chem. Res.*, *42*, 1788–1798.

94. Calzaferri, G., Huber, S., Maas, H., Minkowski, C. (2003). Host–guest antenna materials. *Angew. Chem. Int. Ed.*, *42*, 3732–3758.

95. Calzaferri, G., Li, H. R., Bruhwiler, D. (2008). Dye-modified nanochannel materials for photo-electronic and optical devices. *Chem. Eur. J.*, *14*, 7442–7449.

96. Koeppe, R., Bossart, O., Calzaferri, G., Sariciftci, N. S. (2007). Advanced photon-harvesting concepts for low-energy gap organic solar cells. *Sol. Energ. Mater. Sol. Cell.*, *91*, 986–995.

97. Argazzi, R., Iha, N. Y. M., Zabri, H., Odobel, F., Bignozzi, C. A. (2004). Design of molecular dyes for application in photoelectrochemical and electrochromic devices based on nanocrystalline metal oxide semiconductors. *Coord. Chem. Rev.*, *248*, 1299–1316.

98. Hardin, B. E., Hoke, E. T., Armstrong, P. B., Yum, J. H., Comte, P., Torres, T., Frechet, J. M. J., Nazeeruddin, M. K., Gratzel, M., McGehee, M. D. (2009). Increased light harvesting in dye-sensitized solar cells with energy relay dyes. *Nat. Photon.*, *3*, 406–411.

99. Handa, S., Wietasch, H., Thelakkat, M., Durrant, J. R., Haque, S. A. (2007). Reducing charge recombination losses in solid state dye sensitized solar cells: The use of donor-acceptor sensitizer dyes. *Chem. Commun.*, 1725–1727.

100. Hirata, N., Lagref, J. J., Palomares, E. J., Durrant, J. R., Nazeeruddin, M. K., Gratzel, M., Di Censo, D. (2004). Supramolecular control of charge-transfer dynamics on dye-sensitized nanocrystalline TiO2 films. *Chem. Eur. J.*, *10*, 595–602.

101. Martini, C., Poize, G., Ferry, D., Kanehira, D., Yoshimoto, N., Ackermann, J., Fages, F. (2009). Oligothiophene self-assembly on the surface of ZnO nanorods: toward coaxial p-n hybrid hetero-junctions. *Chemphyschem*, *10*, 2465–2470.

102. Imahori, H., Liu, J. C., Hotta, H., Kira, A., Umeyama, T., Matano, Y., Li, G. F., Ye, S., Isosomppi, M., Tkachenko, N. V., Lemmetyinen, H. (2005). Hydrogen bonding effects on the surface structure and photoelectrochemical properties of nanostructured SnO_2 electrodes modified with porphyrin and fullerene composites. *J. Phys. Chem. B*, *109*, 18465–18474.

103. Kira, A., Umeyama, T., Matano, Y., Yoshida, K., Isoda, S., Park, J. K., Kim, D., Imahori, H. (2009). Supramolecular donor–acceptor heterojunctions by vectorial stepwise assembly of porphyrins and coordination-bonded fullerene arrays for photocurrent generation. *J. Am. Chem. Soc.*, *131*, 3198.

104. Imahori, H., Fujimoto, A., Kang, S., Hotta, H., Yoshida, K., Umeyama, T., Matano, Y., Isoda, S. (2005). Molecular photoelectrochemical devices: supramolecular incorporation of C-60 molecules into tailored holes on porphyrin-modified gold nanoclusters. *Adv. Mater.*, *17*, 1727–1730.

105. Mohmeyer, N., Kuang, D. B., Wang, P., Schmidt, H. W., Zakeeruddin, S. M., Gratzel, M. (2006). An efficient organogelator for ionic liquids to prepare stable quasi-solid-state dye-sensitized solar cells. *J. Mater. Chem.*, *16*, 2978–2983.

106. Kim, Y. J., Kim, J. H., Kang, M. S., Lee, M. J., Won, J., Lee, J. C., Kang, Y. S. (2004). Supramolecular electrolytes for use in highly efficient dye-sensitized solar cells. *Adv. Mater.*, *16*, 1753–1757.

107. Straight, S. D., Kodis, G., Terazono, Y., Hambourger, M., Moore, T. A., Moore, A. L., Gust, D. (2008). Self-regulation of photoinduced electron transfer by a molecular nonlinear transducer. *Nat. Nanotechnol.*, *3*, 280–283.

FUTURE PERSPECTIVE IN SUPRAMOLECULAR SOFT MATERIALS

WHAT WILL BE THE ROSETTA STONE FOR THE NEXT-GENERATION SUPRAMOLECULAR CHEMISTRY?

Takuzo Aida

Department of Chemistry and Biotechnology, The University of Tokyo, Tokyo, Japan

Soft matter is a subfield name of science and technologies on condensed organic materials that comprise a variety of physical states, including colloids, polymers, foams, gels, liquid crystals (LCs), and a number of biological materials, which, in contrast to hard materials such as metals and ceramics, are easily deformed by applied stresses or thermal fluctuations. As an important property of soft materials, they preferably function at an energy scale that is comparable with room-temperature thermal energy. Needless to say, the progress of our society has relied much on the development of hard materials. However, *ab aeterno*, we have recognized the importance of soft materials, as they assemble hierarchically into biological materials that serve as major components of our body. Since Staudinger proved the existence of macromolecules in 1920 and Carothers invented nylon in 1935, polymer chemistry has made tremendous progress both from scientific and technologic points of view. Consequently, a variety of hard polymeric materials have been developed as lightweight alternatives to metallic and ceramic materials. However, the invention of LC displays allowed us to recognize that "being soft" as well as "being hard" is practically important and attractive. In fact, there are currently a number of soft materials that really require "soft" properties for their practical use. Not only tuning such soft properties but also advanced processing technologies have now become important for tailoring hierarchical structures from the molecular to macroscopic level. For example, synthetic mimics of the

Supramolecular Soft Matter: Applications in Materials and Organic Electronics, First Edition. Edited by Takashi Nakanishi.
© 2011 John Wiley & Sons, Inc. Published 2011 by John Wiley & Sons, Inc.

gecko's feet need to carry on their surfaces thin fibrils 100 nm in diameter that are composed of soft polymeric materials with finely tuned mechanical properties.

One of the promising applications of soft condensed materials is to fabricate organic electronics, where electron-donating and/or accepting semiconducting organic molecules are made to align uniformly and bridge over the electrodes macroscopically. For thin-film organic solar cells and organic field-effect transistors, this configuration should be vertical and horizontal with regard to the substrate surface, respectively. How can one tailor such ideal device configurations at the macroscopic level? As described in this book, supramolecular chemistry, so far developed, is mostly based on assembling phenomena in dilute solutions, which are controlled by thermodynamic equilibria. However, assembling phenomena in condensed organic phases are more or less controlled kinetically, and therefore most properties and functions of soft materials are hardly predictable directly from their molecular constituents and even from their nanoscopic structures. Current supramolecular chemistry is well applicable to the fabrication of nanoscopic structures. However, there is no rational design strategy to develop such elaborate structures over a macroscopic length scale. Although molecules in soft materials may respond to external fields such as electric and magnetic fields and align macroscopically, successful examples are extremely limited. Namely, without new breakthroughs that allow for bridging over nano-, meso-, and macroscopic structures hierarchically, supramolecular chemistry might not be able to contribute further to the progress of our society; it would just be a stimulant for the intellectual curiosity of chemists. I would like to emphasize this issue, although it may sound a bit hard to supramolecular chemists including myself, since this scientific challenge in supramolecular chemistry, which is essential for linking molecules to macroscopic properties of materials, has yet been unexplored.

One of the biggest remaining issues may be how to design interfaces. Interfaces always play crucial roles in many aspects of engineering. For example, electrodes sometimes have a certain preference for the orientation of molecules at an interface with organic materials. If such a particular preference leads to improper assembly of conducting molecules, giving rise to an insulating layer at the interface, device performance would be poor, even though bulk electronic properties of the materials are excellent. Equally interesting and important is to design interfaces between synthetic and biological motifs that need to work properly under physiological conditions. Understanding of interfacial phenomena has been a long-term issue in colloid chemistry. However, interfaces of solid substrates with organic soft materials, which are quite important for device applications, have not been well explored. Much bigger challenges for supramolecular chemistry and soft matter may include fabrication of artificial organs that mesoscopically and macroscopically respond to external stimuli in an autonomous manner. Obviously, research activities on regenerative medicines may open a wide window to supramolecular chemists working on soft materials. In these research fields, we, chemists, again have to consider how to translate tiny molecular events into macroscopic phenomena and to transmit the resulting macroscopic motions over a long distance.

In summary, I believe that supramolecular chemistry, which has made a tremendous progress over the past 20 years, is approaching a certain turning point. Recently, we chemists more often handle non-equilibrated processes, affording low-symmetric assembled objects. We are requested more seriously to make a substantial contribution to solving practical or social problems. However, if we accept these situations positively, we might be able to see even more clearly what fundamentals are really lacking. This book not only provides a collection of hot ongoing researches in the field of soft materials but also may surely help readers notice what will be the Rosetta stone for the next-generation supramolecular chemistry.

SUPRAMOLECULAR CHEMISTRY IN MATERIALS SCIENCE

Dirk G. Kurth

Chemische Technologie der Materialsynthese, Universität Würzburg, Würzburg, Germany

The use of ceramics and glasses for food storage marks a significant step in the development of mankind. The discovery of polymers in the 1920s has taken mankind to a new horizon, from landing on the moon to artificial heart valves. Despite the enormous progress, modern materials are generally limited in their functionality. A scratch on a surface may diminish the function of a material; in case of a coating, the underlying metal may corrode and in case of a lens, the visibility may deteriorate. Two different strategies can be adopted in order to resolve this problem: one could develop harder and harder coatings or one could resort to new construction principles. We use an intelligent coating that can sense and heal damage and wear, thus preventing failure of function and structure.

The materials of the future will unify several functions and be able to adapt to external conditions. They will change their mechanical properties, color, or shape as a result of an external stimulus—that is why they are termed as *dynamic materials*. These materials will be able to recognize molecules, generate electrical or optical signals, release a drug, or indicate a load limit. The materials of tomorrow are multifunctional, adaptive, responsive, and dynamic. Their production is as revolutionary as their properties. They form spontaneously according to the principles of molecular self-organization as we know it from Nature.

Molecular self-assembly and self-organization provide an elegant route to materials. The particular advantages of this approach include parallel fabrication, (molecular) dimension control, component alignment, and inclusion of repair mechanisms. The modularity of this approach provides extensive control of structure and function from molecular to macroscopic length scale. In addition, the use of modules provides an unsurpassed degree of synthetic simplicity, diversity,

Supramolecular Soft Matter: Applications in Materials and Organic Electronics, First Edition.
Edited by Takashi Nakanishi.
© 2011 John Wiley & Sons, Inc. Published 2011 by John Wiley & Sons, Inc.

and flexibility. The ability to control the spatial arrangement of functional constituents is of critical importance with respect to the encoding of new (collective) properties and the exploitation of a material's potential. The current challenge in the field of supramolecular materials science is, therefore, the development of strategies to deliberately combine, orient, and order structural and functional modules in predictable ways because material and device performance is critically dependent on the spatial arrangements of the functional modules. The potential lack of long-range order and symmetrical invariance constitute a tremendous challenge in establishing accurate structure–property relationships. The analytical problems of structure determination call for complementary methods and sophisticated molecular modeling. Once the Aufbau principles are understood and structure–property relationships are at hand, it will be possible to assess reliability, lifetime, and possible hazards. It becomes clear that the commercial realization of these materials will rely on interdisciplinary research strategies that involve chemistry, physics, biology, medicine, and engineering.

The concepts of supramolecular chemistry go back to the work of Werner on coordination chemistry in 1893, the lock-and-key model of Fischer in 1894, as well as the magic bullet concept of Ehrlich concerning the selectivity of drugs in the treatment of diseases. Finally, the discovery of crown ethers by Pederson and the eventual contributions by Cram and Lehn stimulated worldwide research in supramolecular chemistry. Today, the principles of molecular recognition on the molecular length scale are well understood. In moving toward systems of more and more complexity, supramolecular chemistry has stimulated much research in materials chemistry.

Liquid crystals constitute the first major commercial success in soft materials. While their discovery dates back to the nineteenth century, their commercial success did not begin until the 1970s with the LCDs (liquid crystal displays). The weak forces that act between the anisotropic liquid crystal molecules not only give rise to short-range ordering but also allow reorganization of the molecules under the influence of an external field. As a result, the optical properties can be altered through an external stimulus and the liquid crystal acts in combination with polarizing sheets as a light valve. The combination of semiconductor lithography and liquid crystals paved the way for large-scale production of ever increasing active matrix TFT-LCDs (thin film transistor liquid crystal displays).

The specificity and selectivity of molecular recognition are widely explored in sensing, for example, metal ions in biomedical applications. The inclusion of lanthanides in suitable hosts can lead to novel optical properties that are useful for time-resolved luminescence immunoassays with monoclonal antibodies [1]. Hosts can be used as odor binders in textiles or house hold cleaners. But hosts can also release a substance, for example, silver ions for antimicrobial textile coatings or a fragrance. The controlled assembly of building blocks can be used to construct porous materials, and the so-called metal organic frameworks are potential candidates for hydrogen or methane storage, for example, for fuel cells in automobiles [2].

Weak forces, such as electrostatic interactions, have been extensively explored to make surface coatings and multilayers [3]. The layer-by-layer method relies on alternating deposition of oppositely charged species, such as polyelectrolytes, nanoparticles, viruses, DNA, enzymes, and many others. The composition, structure, and properties of the resulting films are readily altered by the experimental conditions. The layers can be used in many applications ranging from sensing [4] to biomimetic signal chains [5] to antireflection coatings [6]. In a commercial application, such a multilayer is used to improve the biocompatibility and comfort of contact lenses.

Recently, weak forces have been explored to fabricate a new class of dynamic polymers [7]. The monomers are linked through reversible interactions. If the interactions are of intermediate strength, it is possible to achieve a high degree of polymerization while maintaining the reversibility. Hydrogen bonding [8] and metal ion coordination have been explored to generate supramolecular polymers. The inclusion of metal ions in polymers gives rise to value-adding properties, for example, optical, electrical, reactive, and magnetic properties. Through the choice of metal ions and ligands as well as the external conditions, dynamic properties can be tailored, thus giving rise to interesting materials that can respond to pH, temperature, shear, or external fields.

Justifiably so, we can speak of the next technological revolution. Dynamic materials offer solutions for the technological, economical, and ecological challenges of tomorrow's world and belong, besides nano-, bio-, and gene technology as well as information technology, to the key technology of the twenty-first century. The promotion of research and development of these visionary materials and their corresponding technologies is of central social importance because it provides the necessary force for sustainable innovation and global economic development.

REFERENCES

1. Bunzli J.-C. G. (2009) Lanthanide luminescent bioprobes (LLBs). *Chem. Lett.*, *38*, 104–109.
2. Yaghi, O. M., Li, Q. (2009) Reticular chemistry and metal-organic frameworks for clean energy. *MRS Bulletin*, *34*, 682–690.
3. Decher G. (1997) Fuzzy nanoassemblies: toward layered polymeric multicomposites. *Science*, *277*, 1232–1237.
4. Liu, S., Volkmer, D., Kurth, D. G. (2004) Smart polyoxometalate-based nitrogen monoxide sensors. *Anal. Chem.*, *76*, 4579–4582.
5. Lisdat, F., Dronov, R., Möhwald, H., Scheller, F. W., Kurth D. G. (2009) Self-assembly of electro-active protein architectures on electrodes for the construction of biomimetic signal chains. *Chem. Commun.*, 274–283.
6. Hiller, J., Mendelsohn, J. D., Rubner, M. F. (2002) Reversibly erasable nanoporous anti-reflection coatings from polyelectrolyte multilayers. *Nature Materials*, *1*, 59–63.
7. Kurth, D. G., Higuchi, H. (2006) Metal ions: weak links for strong polymers. *Soft Matter*, *2*, 915–927.
8. De Greef, T. F. A., Smulders, M. M. J., Wolffs, M., Schenning, A. P. H. J., Sijbesma, R. P., Meijer, E. W. (2009) Supramolecular polymerization. *Chem. Rev.*, *109*, 5687–5754.

INDEX

Supramolecular Soft Matter: Applications in Materials and Organic Electronics, First Edition.
Edited by Takashi Nakanishi.
© 2011 John Wiley & Sons, Inc. Published 2011 by John Wiley & Sons, Inc.